Free Radicals,
Lipoproteins, and
Membrane Lipids

NATO ASI Series

Advanced Science Institutes Series

A series presenting the results of activities sponsored by the NATO Science Committee, which aims at the dissemination of advanced scientific and technological knowledge, with a view to strengthening links between scientific communities.

The series is published by an international board of publishers in conjunction with the NATO Scientific Affairs Division

A	**Life Sciences**	Plenum Publishing Corporation
B	**Physics**	New York and London
C	**Mathematical and Physical Sciences**	Kluwer Academic Publishers Dordrecht, Boston, and London
D	**Behavioral and Social Sciences**	
E	**Applied Sciences**	
F	**Computer and Systems Sciences**	Springer-Verlag
G	**Ecological Sciences**	Berlin, Heidelberg, New York, London,
H	**Cell Biology**	Paris, and Tokyo

Recent Volumes in this Series

Series A: Life Sciences

Free Radicals, Lipoproteins, and Membrane Lipids

Edited by

A. Crastes de Paulet

INSERM Unit 58
Montpellier, France

L. Douste-Blazy

INSERM Unit 101
Toulouse, France

and

R. Paoletti

University of Milan
Milan, Italy

Plenum Press
New York and London
Published in cooperation with NATO Scientific Affairs Division

Proceedings of a NATO Advanced Research Workshop on
Action of Free Radicals and Active Forms of Oxygen on
Lipoproteins and Membrane Lipids:
Cellular Interactions and Atherogenesis,
held October 5-8, 1988,
in Ile de Bendor, Bandol, France

Library of Congress Cataloging-in-Publication Data

NATO Advanced Research Workshop on Action of Free Radicals and Active
   Forms of Oxygen on Lipoproteins and Membrane Lipids: Cellular
   Interactions and Atherogenesis (1988 : Bendor, France)
    Free radicals, lipoproteins, and membrane lipids / edited by A.
  Crastes de Paulet, L. Douste-Blazy, and R. Paoletti.
       p.   cm. -- (NATO ASI series. Series A, Life sciences ; vol.
  189)
    "Proceedings of a NATO Advanced Research Workshop on Action of
  Free Radicals and Active Forms of Oxygen on Lipoproteins and
  Membrane Lipids: Cellular Interactions and Atherogenesis, held
  October 5-8, 1988, in Ile de Bendor, Bandol, France"--T.p. verso.
    "Published in coopertion with NATO Scientific Affairs Division."
    Includes bibliographical references.
    ISBN 978-1-4684-7429-9   ISBN 978-1-4684-7427-5 (eBook)
    DOI 10.1007/978-1-4684-7427-5
    1. Free radicals (Chemistry)--Physiological effect--Congresses.
  2. Active oxygen--Physiological effect--Congresses. 3. Membrane
  lipids--Peroxidation--Congresses. 4. Lipoproteins--Metabolism-
  -Congresses. I. Crastes de Paulet, A. II. Douste-Blazy, Louis.
  III. Paoletti, Rodolfo. IV. North Atlantic Treaty Organization.
  Scientific Affairs Division. V. Title. VI. Series: NATO ASI
  series. Series A, Life sciences ; v. 189.
  RB170.N38 1988
  612'.01524--dc20   90-7159
                                                             CIP

© 1990 Plenum Press, New York
Softcover reprint of the hardcover 1st edition 1990

A Division of Plenum Publishing Corporation
233 Spring Street, New York, N.Y. 10013

PREFACE

This book contains the proceedings of the ARW NATO Conference on "Action of Free Radicals and Active Forms of Oxygen on Lipoproteins and Membrane Lipids : Cellular Interactions and Atherogenesis", held in Bendor, France, October 5-8, 1988.

Since the pioneer work of Mc Cord and Fridovitch, growing interest has been focused on the study of the role of oxyradicals role in pathology. This interest is reflected in the exponential increase in the number of papers on free radicals, the success of specialized journals and books on this theme, and the organization of national and international meetings. These meetings have discussed, from a broad point of view, the problems concerning the mechanisms of production of free radicals, their effects on cell constituants (lipids, proteins, nucleic acids) and cell function, the methods of analysis of these phenomena, the pathological states in which free radicals may be involved, natural biological defense systems, and the design of "antiradical" therapies.

But it is now well established that the most common target of oxy free radicals are membrane lipids because of their chemical nature (cholesterol insaturation, malonic linkage of polyunsaturated fatty acids (PUFA)) and of their regular structural arrangement (monolayers in lipoproteins, bilayers in cell membranes). Thus the analysis of the products resulting from the action of oxy free radicals on PUFA is considered the best tool to indirectly evaluate the effects of tissue peroxidations, although the analytical basis for doing so is very questionable.

Similarly, cell disorders caused by free radicals are considered to result essentially and primarily from alterations in membrane lipids : intra- and inter molecular linkages, shortening of PUFA chains giving rise to very reactive aldehydes (malone- dialdehyde, 4-OH nonenal), leading to chemical modifications of the amino group of aminophospholipids, proteins and nucleic acids.

Apart from their action on membrane PUFA, the formation of oxysterols by free radicals in membranes has special importance since some of these oxysterols show cytostatic, angiotoxic, atherogenic and immunomodulating properties.

Finally, a new concept is now emerging, debated with particular interest during the 1987 Congress of the American Heart Association, that of the leading role of peroxidized lipoproteins in the process of atherogenesis.

The workshop held in Bendor on the action of free radicals on lipoproteins and membrane lipids was thus highly pertinent. Its purpose was to bring together, on this very problem of the action of free radicals on living organisms, the most distinguished researchers from various disciplines (physicochemists, chemists, analysts specialized in molecular and cell biology, pathologists, pharmacologists) so that they could pool their knowledge and confront their points of view.

Thirty six lectures were given and have been grouped into the following seven sections :

Section 1 (7 lectures) provides an extensive overview of the various mechanisms of production of free radicals, from physico-chemistry and chemical analysis to physiological or pathological situations.

Section 2 (3 lectures) examines some of the chemical effects of oxy free radicals, with a special emphasis on oxysterols formation.

A critical survey ot the most recent methods of measurement is described in *Section 3* (3 lectures), followed by two lectures dealing with the action of oxy free radicals on cellular structures (plasma membranes, Golgi Apparatus) *(Section 4)*.

An important place is given to the chapter "Oxidized lipoproteins and atherosclerosis *(Section 5,* 5 lectures). The mechanisms of formation of these oxidized lipoproteins, their cytotoxicity and the effect of antioxydants are analysed in detail.

The following section provides a wide overview on the biological effects of oxy free radicals with special emphasis on membrane lipid disorders *(Section 6,* 9 lectures).

The volume ends with four lectures on either intracellular (Vit E, Vit E free radical reductase, DT diaphorase) or extracellular (coeruleoplasmin, Vit C) defense mechanisms.

In conclusion, the physiological and pathological situations in which cellular lipids and lipoprotein oxydations are shown to play a critical role are so varied that biochemists, pharmacologists and pathologists should all find food for thought in these excellent contributions. This was the aim of the symposium.

The editors gratefully acknowledge Mrs Hélène Thaler-Dao for her valuable assistance in the organisation of the meeting and the proof &readings of the manuscripts.

A. Crastes de Paulet
L. Douste-Blazy
R. Paoletti

CONTENTS

FREE RADICAL PRODUCTION MECHANISMS

CHEMISTRY OF FREE RADICAL EFFECTS

METHODS OF MEASUREMENTS

ACTION ON MODEL SYSTEMS AND CELLULAR STRUCTURES

OXIDIZED LIPOPROTEINS AND ATHEROGENESIS

BIOLOGICAL EFFECTS

DEFENCE MECHANISMS

FREE RADICAL PRODUCTION MECHANISMS

RADICAL MECHANISMS IN FATTY ACID AND LIPID PEROXIDATION

Wolf Bors, Michael Erben-Russ[+], Christa Michel and Manfred Saran

Institut für Strahlenbiologie
GSF Forschungszentrum
D-8042 Neuherberg, FRG

INTRODUCTION

Autoxidation of fatty acid moieties of membrane lipids is a process which is generally considered to involve radical chain reactions[1-3]. Since polyunsaturated fatty acids (PUFA) are more sensitive than saturated ones, it is obvious that the activated methylene bridge represents a critical target site. The basic scheme for fatty acid or lipid autoxidation (Scheme I) shows, that in sequence of their appearance, alkyl, peroxyl and alkoxyl radicals are involved. As is typical for a chain process, propagation reactions dominate during the early phase, whereas termination reactions become important only if a sufficient amount of radicals is present - an event which hardly happens under biological conditions.

In the following, these radical species and their properties as well as their reactions in fatty acid peroxidation will be discussed. Since mainly physico-chemical procedures have been successfully employed to gather pertinent data, the methods for such investigations shall first be briefly introduced.

ANALYTICAL PROCEDURES

1. Optical Spectroscopy

Because of the inherent instability of the radical intermediates, it is understandable that rapid kinetic spectroscopy combined with pulse radiolysis or flash photolysis is best suited to determine kinetic data.

Alkyl radicals, as derived from saturated aliphatic compounds, are optically transparent, in contrast to allylic and especially double-allylic PUFA radicals which reveal strong u.v. absorptions. This is due to electron delocalization, which in case of the latter species, the so-called pentadienyl radical, extends over five C-atoms (eq./1/ in Scheme I). This pentadienyl species is the most stable alkyl radical formed from PUFA and, as shown in Fig. 1, can easily be observed after pulse radiolysis or flash photolysis ($\mathcal{E}_{280}$ = 30.000 $M^{-1}cm^{-1}$; ref. 4).

[+] Present address: Sektion Physik, LMU München, Coulombwall, 8046 Garching.

Free Radicals, Lipoproteins, and Membrane Lipids
Edited by A. Crastes de Paulet *et al.*
Plenum Press, New York, 1990

Scheme I – Major chain reactions during PUFA autoxidation.

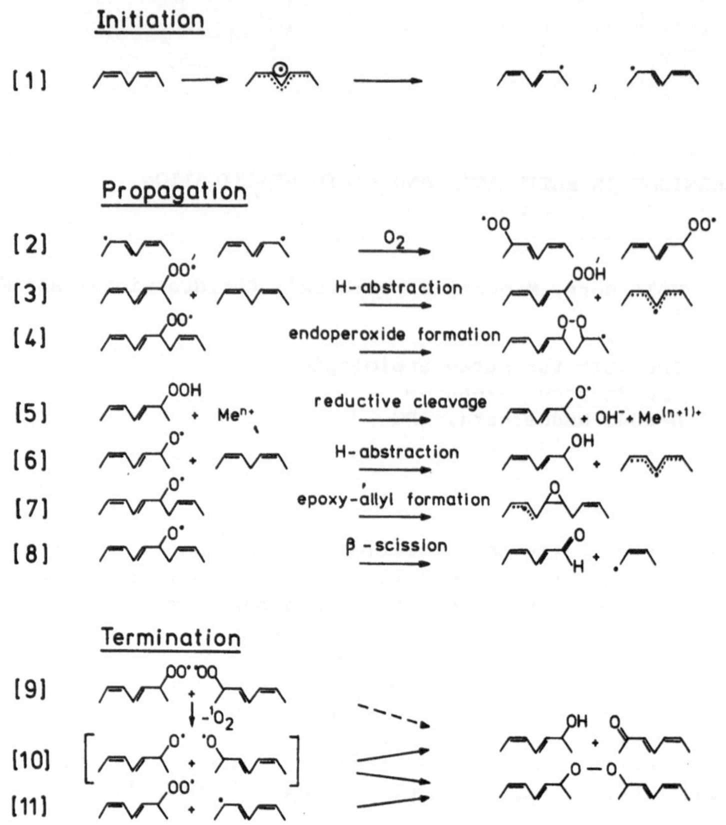

Initiation

[1]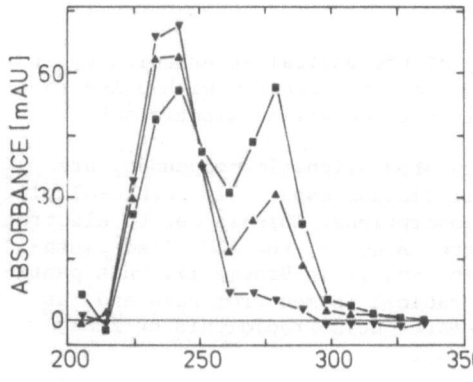

Propagation

[2] O_2

[3]   H–abstraction

[4]   endoperoxide formation

[5] $+ Me^{n+}$ reductive cleavage $+ OH^- + Me^{(n+1)+}$

[6]   H–abstraction

[7]   epoxy-allyl formation

[8] β –scission

Termination

[9]

     $\downarrow -{}^1O_2$

[10]

[11]

Figure 1 – Transient spectrum of pentadienyl radical (280nm) and diene conjugation (230–240nm) after pulse radiolysis of linoleic acid.

Alkaline aqueous solution, pH 10.9, saturated with in N_2O/O_2 (12:1) gas mixture.
Concentration of linoleic acid 1 mM initial OH· radical concentration 29 μM (dose = 49 Gy). (■) 0.017ms; (▲) 0.160 ms; (▼) 8.66 ms after the pulse.

Peroxyl radicals can only be directly observed, if <u>no</u> conjugated double bonds are present in the molecule. For simple aliphatic peroxyl radicals, absorption maxima near 250 nm and molar absorptivities around 1500 $M^{-1}cm^{-1}$ can be expected[5]. When the peroxyl radicals are derived from the pentadienyl radical (eq./2/ in Scheme I), the conjugated diene absorption (ε_{234} = 24.000 $M^{-1}cm^{-1}$, ref. 6 and Fig. 1) completely obscures the weak absorption of the peroxyl radical structure.

Alkoxyl radicals cannot be observed spectroscopically at all, as they show only an unstructured absorption tail in the far u.v.

2. EPR Spectroscopy

The time resolution of this monitoring method being more critical than that of optical spectroscopy combined with pulse radiolysis, only sufficiently stable radicals can be observed. Alkyl, allyl and pentadienyl radicals have been observed directly after <u>in situ</u> photolysis of neat di-<u>tert</u>-butylperoxide, in which the saturated and unsaturated fatty acids and esters were dissolved[7,8]. It is, however, more convenient to study these radicals after spin-trapping to convert them into rather stable nitroxyl radicals[9]. Each of these species has been detected by this method – see the compilation by Buettner[10], which is constantly being updated[11] – offering clear evidence for their intermediary formation during fatty acid peroxidation.

3. Oxygen Consumption Studies

The groups of Ingold[12] and Niki[13] have developed very sensitive monitors to study the consumption of oxygen during fatty acid and lipid peroxidation. In comparison to electrochemical oxygen assays, recording of the signals of pressure-sensitive transducers in a closed vessel allows much more accurate, incremental and time-dependent measurements. The method is an indirect one and depends on continuous generation of radicals, which is achieved by thermolysis of either lipid- or water-soluble azo initiators 13-15.

4. Product Identification Studies

Aside from the EPR/spin-trapping method mentioned before, some other procedures have been developed, which allow <u>chemical</u> trapping of short-lived intermediate radicals. Aliphatic peroxyl radicals combine with phenolic antioxidants in a 2:1 stoichiometric reaction[16], forming a quinone methide product[17]. Unfortunately, it has thus far not been possible to trap <u>fatty acid</u> peroxyl radicals, e.g. by α-tocopherol. With the flavonols kaempferol and quercetin we obtained only kinetic evidence[18] and could not yet isolate and identify reaction products.

Porter and colleagues have introduced the HPLC analysis of the <u>cis</u>, <u>trans</u>- and <u>trans,trans</u>-hydroperoxide isomers[19] to determine the effective ness of hydrogen donors to interfere with the initial reversible oxygen addition (see Scheme II). Yet, the disagreement between the rate constants with α-tocopherol obtained by this method and by the induction time method as used by Niki[20] seems to be stronger than one would expect if it were only caused by the different organic solvents (<u>tert</u>-butanol vs. benzene, see Table III).

1. Alkyl Radicals

The chemical _generation_ of fatty acid alkyl radicals is dependent on the removal of a single hydrogen atom from the aliphatic chain. As shown in Table I, this can be achieved (with varying degrees of efficiency) by a number of oxidizing radicals.

Table I. Absolute rate constants for reactions of various oxidizing radicals and singlet oxygen with fatty acids

Radical	Fatty acid [a]				Dimension $M^{-1}s^{-1}$	References
	18:1	18:2	18:3	20:4		
·OH [b]	–	7.4	7.3	–	$\times 10^9$	21
	–	5.6 [c]	–	–	$\times 10^9$	18
t-BuO· [b]	0.68	1.3	1.6	1.8	$\times 10^8$	23
" [d]	0.38	0.88	1.3	2.05	$\times 10^7$	24
SO_3^- [b]	–	1.8	2.8	3.9	$\times 10^6$	25
$CCl_3OO·$ [e]	1.7	3.9	7.0	7.3	$\times 10^6$	26
$CF_3CHClOO·$ [e]	0.3	0.8	1.3	1.5	$\times 10^6$	26
HO_2 [f]	n.r.	1.2	1.7	3.1	$\times 10^3$	27
$^1O_2$ [g]	0.73	1.3	1.9	2.4	$\times 10^5$	28

a   Abbreviations denote: 18:1 – oleic acid (octadecenoic acid); 18:2 – linoleic acid (octadecadienoic acid); 18:3 – linolenic acid (octadecatritrienoic acid); 20:4 – arachidonic acid (icosatetraenoic acid).

b   Alkaline aqueous solution, pH 11.0 – 11.5

c   The value of 1.1×10^{10} $M^{-1}s^{-1}$ in ref. 22 is probably too high, see discussion in ref. 18.

d   In 1:2 mixture of benzene and di-_tert_-butylperoxide.

e   Air-saturated aqueous _tert_-butanol (50%) at neutral pH; data from competition plots using ABTS (2,2-azino-di(3-ethyl-benzthiazoline-sulfonate) as reference substance; $CF_3CHClOO·$ derived from halothane, $CF_3CHBrCl$.

f   Strongly acidic anaerobic aqueous ethanol (85%), HO_2/O_2^- inserted by stop-flow technique; no reaction was observed with O_2^- in alkaline aqueous ethanol.

g   Photo-sensitized oxidation in pyridine, protoporphyrin IX as sensitizer; data from competition with cholesterol.

–   no value known; n.r. – no reaction (upper limits of 4×10^7 $M^{-1}s^{-1}$ for ·N_3 and 1×10^6 $M^{-1}s^{-1}$ for $(SCN)_2^-$ radicals have also been determined, ref. 18).

Hydroxyl radicals as the most strongly oxidizing species attack the fatty acid molecule rather indiscriminately. Only about 10% of them abstract a hydrogen atom from double-allylic sites, forming the pentadienyl radical directly[4,18,22]. A far higher selectivity is exhibited by t-BuO·

radicals in non-polar solvents, which have been shown independently by laser flash photolysis[24] (FP) and _in situ_ photolysis/EPR[7] to prefer to attack at the activated methylene bridge.

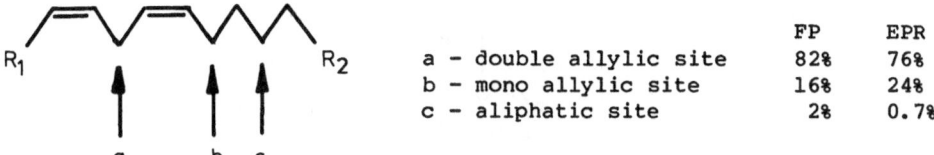

		FP	EPR
a – double allylic site		82%	76%
b – mono allylic site		16%	24%
c – aliphatic site		2%	0.7%

In aqueous solution the attack of _t_-BuO· on PUFA is faster by one order of magnitude and, as might be expected, shows a lower degree of selectivity[23].

Comparing the rate constants for the different PUFA in Table I, a direct correlation of the number of double-allylic sites with the observed rate constants is evident for all slowly reacting radicals. Table I also contains rate constants for the attack of singlet oxygen[28], a non-radical species which directly forms hydroperoxides with fatty acids [29,30]. Not included in the table are data with iron-oxygen complexes, such as ferryl ($[FeO]^{2+}$), perferryl ($[FeO_2]^{2+}$) species or mixed-valence state iron-oxygen complexes which, however, are serious contenders as initiators of lipid peroxidation in biological systems[31]. At least for the ferryl species calculations have indicated that it would be a strongly oxidizing entity[32].

The reactivities of the fatty acid alkyl radicals cannot be easily summarized in a categorical manner. In a biological environment, the fate of the pentadienyl radical will be dominated by the existence of oxygen which will lead in a very rapid reaction to the formation of peroxyl radicals (18,21; eq./2/ in Scheme I). In organized micellar or membrane structures, this process will, according to Ivanov[33], be in competition with a lateral transfer of the radical site by alkyl-alkyl interactions.

2. Peroxyl Radicals

Generation and reactivities of fatty acid peroxyl radicals have been reviewed before[34]. As stated above, alkyl radicals add oxygen in diffusion-controlled reactions to form peroxyl radicals[35]. In the case of PUFA this addition of oxygen has most extensively been investigated by Porter and his colleagues[36]. In aqueous solution pentadienyl radicals add oxygen with a rate constant of $3x10^8$ $M^{-1}s^{-1}$ [18,21].

As shown in Scheme II, this reaction - occasionally termed 'oxygen scrambling'[37,38] - is a reversible process and can be influenced by strong hydrogen donors. The carbon atoms at each end of the pentadienyl radical, e.g. the 9- and 13-position of linoleic acid, are the preferred sites of oxygen addition. Only smaller amounts of peroxyl radicals with non-conjugated double bonds due to oxygen attachment to C-positions 8, 10, 12 and 14 of allylic radicals, are formed during autoxidation[39] (Scheme III).

During photosensitized singlet oxygen reaction, formation of 9- + 13-peroxyl radicals is favored over 10- + 12-peroxyl radicals by a ratio of 2 to 1 (ref. 40, c.f. the discussion on p. 203 of ref. 30).

Scheme II - Reversible oxygen addition to pentadienyl radicals,
leading to stereoisomeric hydroperoxides.

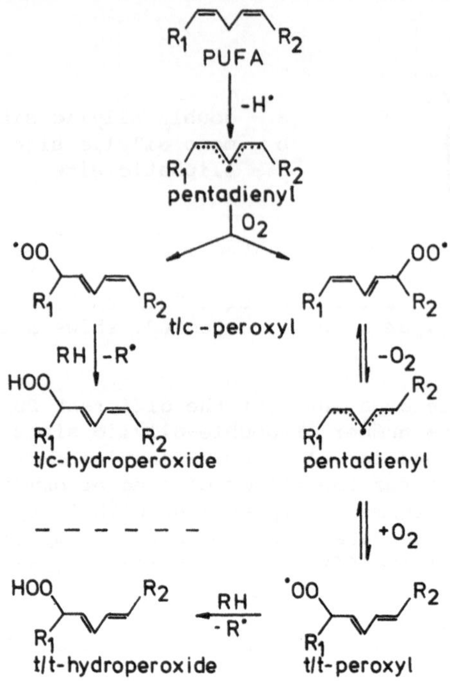

Scheme III - Positional isomers of peroxyl radicals of linoleic acid,
formed from oxygen addition at the pentadienyl radical
(H abstraction at C-11) or the mono-allyl radicals
(H abstraction at C-8 or C-14).

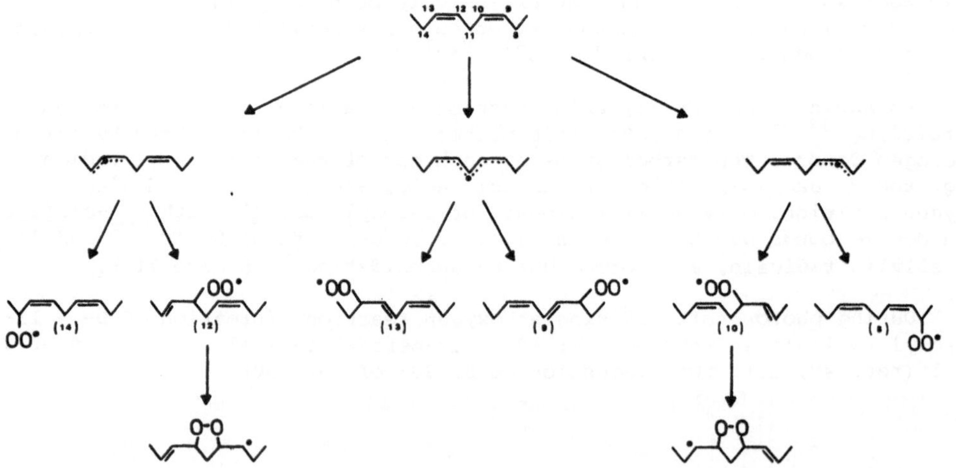

6

Peroxyl radicals can undergo intramolecular rearrangements if certain structural requirements are met. As shown in Scheme III, only the 10- and 12-peroxyl radicals cyclize to endoperoxide alkyl radicals; the corresponding hydroperoxides are consequently found in far smaller amounts than the 8- and 14-hydroperoxides[39,41]. Hydroperoxide epidioxides as important secondary peroxidation products[30] as well as enzymatically produced prostaglandin G_2 (from arachidonic acid, ref. 42) are formed via this pathway.

Bimolecular recombination reactions, which for linoleic acid peroxyl radicals in alkaline aqueous solution have been calculated to proceed with a rate constant of 3×10^8 $M^{-1}s^{-1}$ [18], are almost 100 times faster than in an organic solvent[43]. Such reactions may occur via the so-called 'Russell mechanism'[44] and other decay processes (Scheme IV).

Scheme IV - Bimolecular decay of peroxyl radicals and decay of alkoxyl radicals.

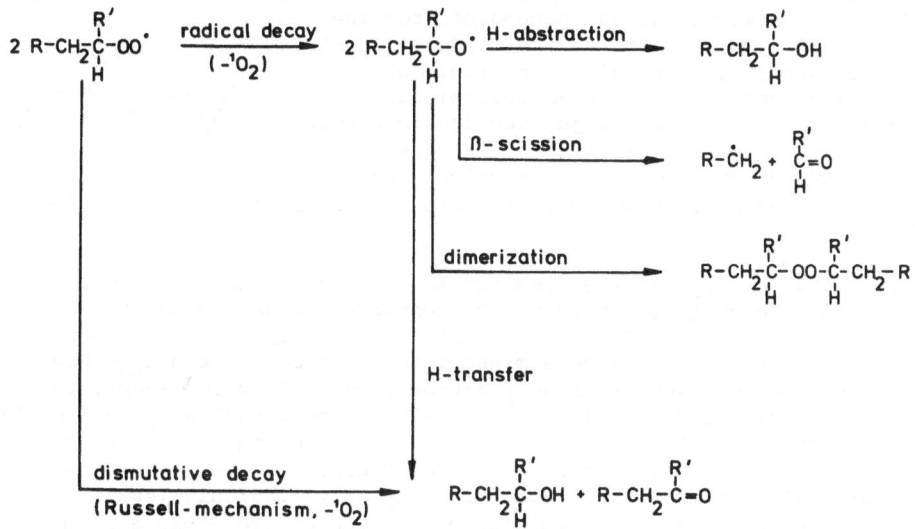

The formation of singlet oxygen in this reaction as one of the tenets of low level chemiluminescence in enzymatic or biological systems[45,46] seems to be questionable as the reaction is energetically unfavorable (see discussion on pp. 14-15 in ref. 47).

Peroxyl radicals are considered to be the major chain carriers during PUFA autoxidation. Hydrogen abstraction from another PUFA molecule, which will certainly entail a selective yet 'stereorandom'[48] reaction at the activated methylene bridge, proceeds very slowly in chlorobenzene (62 M^{-1} s^{-1} ref. 43); no value is known for aqueous solution. Yamamoto et al.[49] proposed that this chain propagation is favored in more polar media due to the generally higher reactivity. They assumed, however, negligible effects of the solvent on the bimolecular decay - which, according to our data, is not the case - and on the ß-scission reaction involving reversal of the initial oxygen addition.

3. Alkoxyl Radicals

Under biological conditions, the generation of alkoxyl radicals is essentially dependent on the existence of hydroperoxide precursors. The latter species, e.g., when generated from peroxyl radicals within a membrane structure, will be recognized by phospholipase A_2 and hydrolytically removed[51]. Whenever such a hydroperoxide – which is capable of diffusing over extended distances in a biological environment – encounters reducing metal ions, it may be cleaved in an 'organic Fenton reaction' (eq./5/ in Scheme I) to the respective alkoxyl radical. Alternative generation by recombination of two peroxyl radicals (see Scheme IV) is of minor importance in biological systems as it occurs only in heavily peroxidized material.

For experimental purposes, fatty acid alkoxyl radicals can be generated in three ways from the respective hydro- or dialkylperoxides:
- metal-induced reductive cleavage[51,52];
- homolytic cleavage by u.v. photolysis[53,54];
- reductive cleavage by hydrated electrons (e_{aq}^-) produced by pulse radiolysis[55].

An excellent application of the first method has recently been reported (52). The alkoxyl radical generated from the photolytically produced 10-hydroperoxide of linoleic acid is evidently less prone to intramolecular rearrangement because no allylic radical can be formed due to the lack of conjugated double bonds. Intermediacy of such an alkoxyl radical after a metal-catalyzed 'organic Fenton reaction' (eq./5/ in Scheme I) leads to 10-hydroxy-linoleic acid which can be identified by HPLC.

Formation by e_{aq}^- has the disadvantage that these strongly reducing radicals also react quite rapidly with conjugated double bonds[56]. Therefore only a fraction of e_{aq}^- may react with the hydroperoxyl group of PUFA and the rate constants determined from the decay of the e_{aq}^--absorption[57] represent the sum of these reactions. In analogy to t-BuO·, the best-investigated alkoxyl radical[58-60], PUFA alkoxyl radicals are probably highly unstable in polar solvents and can be trapped only by very efficient radical scavengers. In the absence of scavengers, the preferred decay processes are (i) intramolecular decay to epoxy allyl radicals (eq./7/ in Scheme I), deduced from the formation of the respective epoxy alcohols [52,61] or the adduct with α-tocopherol[62] and (ii) ß-scission (eq./8/ in Scheme I) which may account for the various off-flavor alkenals[39,63].

Intermolecular reactivities of PUFA alkoxyl radicals have been established using the crocin assay[53] (see also Table IV). Table II lists the relative rate constants for a number of nucleic acid bases and nucleosides.

It is evident, that purine compounds are more effective scavengers of PUFA alkoxyl radicals than pyrimidines. This is in line with the reported selectivity of lipid autoxidation products to form DNA radicals exclusively at the guanine moieties[64].

RADICAL SCAVENGING BY ANTIOXIDANTS

An important aspect of fatty acid autoxidation is the control in a

Table II. Relative rate constants of photolytically generated PUFA alkoxyl radicals with nucleic acid derivatives.[a]

Substrate	9-LO·	13-LO·	9-LnO·	13-LnO·
Nucleic acid bases:				
adenine	0.072	0.057	0.047	0.052
cytosine	0.036	0.032	0.027	0.035
thymine	0.019	0.017	0.020	0.021
uracil	0.042	0.044	0.024	0.036
Nucleosides:				
adenosine	0.078	0.061	0.080	0.055
cytidine	0.033	0.030	0.023	0.035
guanosine	0.086	0.069	0.055	0.066
thymidine	0.034	0.023	0.023	0.015

[a]   the values were determined in the 'crocin assay'[53]; the PUFA hydroperoxide precursors were produced enzymatically - the 9-hydroperoxide with potato lipoxygenase[65] and the 13-hydroperoxide with soybean lipoxygenase[66] - and were a gift of W. Grosch, Dt. Forschungsanstalt für Lebensmittelchemie, Garching, FRG.

biological environment. Hydroperoxides formed from peroxyl radicals in membranes can be reduced to innocuous hydroxy compounds by the glutathione-dependent phospholipid-hydroperoxide peroxidase[67]. Alternatively they are removed by phospholipase A_2 [50], in which case they are either substrates for glutathione peroxidase[68] or a source of alkoxyl radicals (see before). In addition to enzymatic control mechanisms, antioxidants via their radical scavenging properties constitute a second line of defense, and discussion of their important reactions will conclude this review.

1. Kinetics of antioxidant reactions

Only peroxyl and alkoxyl radicals can be scavenged by antioxidants, whereas with alkyl radicals the diffusion-controlled oxygen addition predominates under physiological conditions. Hydrogen donation by antioxidants leads to stable hydroperoxides or hydroxyl compounds and, at the same time, to antioxidant-derived radicals. Since in most cases either the parent antioxidant or its radical have a much stronger absorption than the fatty acid peroxyl and alkoxyl radicals, such scavenging reactions can easily be monitored by kinetic spectroscopy[18,23]. Evaluation may, however, be complicated by competing decay reactions caused by insufficient scavenger concentration due to low solubility - i.e. only a fraction of the fatty acid radicals may be observed to react with the antioxidant. In that case, kinetic modelling by iterative computer programs can help to elucidate the pertinent rate constants[18]. Table III gives a compilation of absolute rate constants for peroxyl radical reactions.

Table III. Absolute rate constants for reactions of fatty acid
peroxyl radicals with antioxidants and hydrogen donors.

Substrate	Rate Constant	Method	References
kaempferol	3.4, 4.2x10^7	a	18
quercetin	1.8, 1.5x10^7	"	"
α-tocopherol	8.0x10^4	b	69
– " –	5.1x10^5	c	20
– " –	8.5x10^3	d	19
9,10-dihydroanthracene	4.0x10^2	"	70
cyclohexadiene	3.6x10^2	"	"
linoleic acid	6.2x10^1	e	43
tetralin	6.7	d	70
cumene	0.6	"	"

a   Pulse radiolysis in alkaline aqueous solution, pH 11.5; first
values for LOO· isomers, second ones for 13-LOO· specifically
b   Pulse radiolysis in aqueous micellar system
c   Induction time in _tert_-butanol
d   HPLC-determination of _cis,trans/trans,trans_-ratio of hydroperoxides formed by autoxidation in benzene at 30°C
e   Rotating disk method in chlorobenzene at 30°C

Depending on the type of antioxidant or hydrogen donor as well as the
solvent, considerable differences in the reactivities are encountered,
again being enhanced in alkaline aqueous solution as compared to organic
solvents.

For fatty acid _alkoxyl_ radicals, only relative rate constants are
known at present. The data in Table IV clearly show, that only the known
phenolic antioxidants nordihydroguaiaretic acid (NDGA), propylgallate,
Trolox c and some isoflavonoids, but to a far lesser extent some other
phenolic compounds are effectively scavenging the PUFA alkoxyl radicals.
It is also evident that not only _t_-BuO·, but PUFA alkoxyl radicals as
well are extremely reactive oxidizing radicals, second only to ·OH
radicals. They are thus very potent chain propagators, albeit with less
selective attack at the methylene bridge.

Thus far, it was not possible to calculate _absolute_ rate constants
from these relative values, because control experiments with radiolytically generated alkoxyl radicals could not be evaluated due to the unknown
fraction of electrons reacting with the conjugated diene group[56]. The
values of Tables II and IV were assessed by the bleaching of the water-
soluble carotenoid crocin caused by the alkoxyl radicals which is inhibited in a competitive manner by radical scavengers[71].

2. Mechanisms of antioxidant action

The antioxidant radicals formed after reaction with peroxyl or alkoxyl radicals may
- decay by a bimolecular dismutation reaction[72];

Table IV. Relative rate constants of photolytically generated PUFA
alkoxyl radicals with phenols and phenolic antioxidants.

Substrate	9-LO·	13-LO·	13-LnO·
Phenols:			
Trolox c	0.43	0.32	0.33
3,4-dihydroxytoluene	0.063 [a]	0.055	0.083 [a]
3,4-dihydroxycinnamic acid (caffeic acid)	0.075	0.072	0.063
nordihydroguaiaretic acid	0.20 [a]	0.33 [a]	0.20 [a]
2',4'-dihydroxyacetophenone	0.012	0.010	0.009
2',5'-dihydroxyacetophenone	0.070	0.063	0.065
3',4'-dihydroxyacetophenone	0.070	0.085	0.085
propylgallate	0.23	0.167	0.20
Flavonoids:			
3,5,7,3',4'-pentahydroxy-flavanol [(+)-catechin]	0.043 [a]	[b]	0.057
3,5,7,3',4'-pentahydroxy-flavanone (dihydroquercetin)	0.040 [a]	[b]	0.026
5,7,3'-trihydroxy-4'-methoxy-flavanone (hesperetin)	0.037	0.032	0.017 [a]
6,7-dihydroxy-4'-methoxy-isoflavanol	0.45	0.51	0.41
6,7-dihydroxy-4'-methoxy-isoflavone (texasin)	0.16	0.15	0.17
5,7-dihydroxy-4'-methoxy-isoflavone (biochanin A)	0.17	0.12	0.12

[a] Competition plot does not intersect ordinate at unity
[b] Non-linear competition plot (see also footnote to Table II)

- regenerate the parent compound if they can be univalently reduced[73] –
the best example being the synergism between vitamin E and vitamin C
20,74;
- if they are stable enough, react with a second radical to exhibit an
overall 1:2 stoichiometry as discussed before (Scheme V, ref. 16).

Products of such a radical-radical recombination reaction have thus
far only been observed for hindered phenols and cumene peroxyl radicals
[17,75] in the case of α-tocopherol the structure of the adduct(s) is still
controversial[76,77]. For the flavonoid antioxidants kaempferol and quercetin in alkaline aqueous solution, this recombination reaction is as rapid
(> 10^8 $M^{-1}s^{-1}$; ref. 18) as for the previously observed analog reaction in
an organic solvent[75].

CONCLUSIONS

During autoxidation reactions of lipid components, three types of
fatty acid radical intermediates occur - alkyl, peroxyl and alkoxyl radi-

Scheme V - Reaction of phenolic antioxidants with peroxyl radicals.

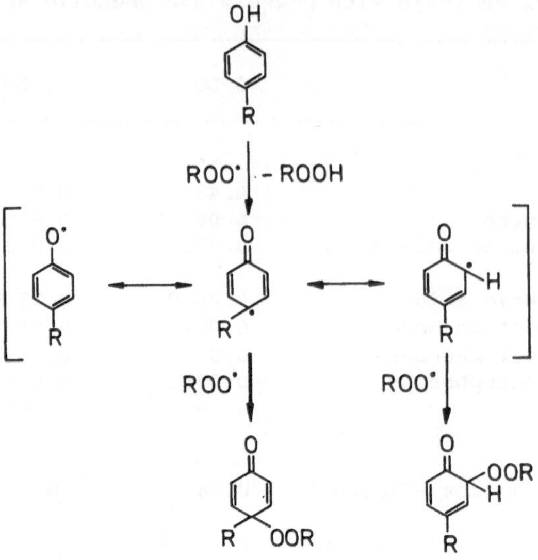

cals - which exhibit quite distinct properties. While interaction between these three types of radicals can practically be excluded because of their short life-times and low concentrations, reaction with radical scavengers and/or antioxidants is the dominant control mechanism. Kinetic parameters pertaining to these types of fatty acid radicals are presently only available for (i) the generation of the initial alkyl radical; (ii) some rate constants of peroxyl radical reactions with antioxidants and (iii) relative rate constants for the scavenging of several fatty acid alkoxyl radicals. Future development is expected to shed more light on the latter reactions and to arrive at a clear identification of scavenging products.

Yet with the present knowledge of the most pressing problems concerning the radical mechanisms during fatty acid autoxidation, interpretation of biological lipid peroxidation processes has been considerably advanced, as will become evident in the subsequent contributions.

REFERENCES

1. Y.A. Vladimirov, V.I. Olenev, T.B. Suslova, and Z.P. Cheremisina, Lipid peroxidation in mitochondrial membrane, Adv. Lipid Res. 17:173 (1980).
2. W. Grosch, Neuere Vorstellungen über die Lipidoxidation, Lebensmittelchem. Gerichtl. Chem. 38:81 (1984).
3. A. Sevanian, and P. Hochstein, Mechanisms and consequences of lipid peroxidation in biological systems, Ann. Rev. Nutr. 5:365 (1985).
4. L.K. Patterson, and K. Hasegawa, Pulse radiolysis studies in model lipid systems. The influence of aggregation on kinetic behavior of OH induced radicals in aqueous sodium linoleate, Ber. Bunsenges. Physik. Chem. 82:951 (1978).
5. W. Bors, C. Michel, and M. Saran, Superoxide anions do not react with hydroperoxides, FEBS-Lett. 107:403 (1979).

6. A.E. Johnston, K.T. Zilch, E. Selke, and H.J. Dutton, Analysis of fat acid oxidation products by countercurrent distribution methods. V. Low temperature decomposition of methyl linoleate hydroperoxide, J. Am. Oil Chem. Soc. 38:367 (1961).

7. E. Bascetta, F.D. Gunstone, and J.C. Walton, An ESR study of fatty acids and esters. I. Hydrogen abstraction from olefinic and acetylenic long chain esters, JCS, Perkin II 603 (1983).

8. E. Bascetta, F.D. Gunstone, and J.C. Walton, An ESR study of fatty acids and esters. II. Hydrogen abstraction from saturated acids and their derivatives, JCS, Perkin II 401 (1984).

9. M.J. Davies, Applications of ESR spectroscopy to the identification of radicals produced during lipid peroxidation, Chem. Phys. Lipids 44:149 (1987).

10. G.R. Buettner, Spin trapping: ESR parameters of spin adducts, Free Radic. Biol. Med. 3:259 (1987).

11. A.S.W. Li, K.B. Cummings, H.P. Roethling, G.R. Buettner and C.F. Chignell, A spin-trapping database implemented on the IBM PC/AT, J. Magn. Reson. in press.

12. G.W. Burton, and K.U. Ingold, Vitamin E: application of the principles of physical organic chemistry to the exploration of its structure and function, Accts. Chem. Res. 19:194 (1986).

13. E. Niki, A. Kawakami, Y. Yamamoto, and Y. Kamiya, Oxidation of lipids VIII. Synergistic inhibition of oxidation of phosphatidylcholine liposomes in aqueous dispersion by Vitamin E and C, Bull. Chem. Soc. Japan 58:1971 (1985).

14. A. Pohlman, and T. Mill, Free radical oxidations in water: decomposition of azoinitiators and oxidation of p-cresol and p-isopropylphenol, J. Org. Chem. 48:2133 (1983).

15. L.R.C. Barclay, S.J. Locke, J.M. MacNeil, and J. Vankessel, Autoxidation of micelles and model membranes. Quantitative kinetic measurements can be made by using either water-soluble or lipid-soluble initiators with water-soluble or lipid-soluble chain-breaking antioxidants, J. Am. Chem. Soc. 106:2479 (1984).

16. C.E. Boozer, G.S. Hamilton, C.E. Hamilton, and J.N. Sen, Air oxidation of hydrocarbons. II. The stoichiometry and fate of inhibitors in benzene and chlorobenzene, J. Am. Chem. Soc. 77:3233 (1955).

17. T.W. Campbell, and G.M. Coppinger, The reaction of tert-butylhydroperoxide with some phenols, J. Am. Chem. Soc. 74:1469 (1952).

18. M. Erben-Russ, W. Bors, and M. Saran, Reactions of linoleic acid peroxyl radicals with phenolic antioxidants: a pulse radiolysis study, Int. J. Radiat. Biol. 52:393 (1987).

19. N.A. Porter, B.A. Weber, H. Weenen, and J.A. Khan, Autoxidation of polyunsaturated lipids. Factors controlling the stereochemistry of product hydroperoxides, J. Am. Chem. Soc. 102:5597 (1980).

20. E. Niki, T. Saito, A. Kawakami, and Y. Kamiya, Inhibition of the oxidation of methyl linoleate in solution by vitamin E and vitamin C, J. Biol. Chem. 259:4177 (1984).

21. K. Hasegawa, and L.K. Patterson, Pulse radiolysis studies in model lipid systems: formation and behavior of peroxy radicals in fatty acids, Photochem. Photobiol. 28:817 (1978).

22. M.G.J. Heijman, A.J.P. Heitzman, H. Nauta, and Y.K. Levine, A pulse radiolysis study of the reactions of OH·/O⁻ with linoleic acid in oxygen-free aqueous solution, Radiat. Phys. Chem. 26:83 (1985).

23. M. Erben-Russ, C. Michel, W. Bors, and M. Saran, Absolute rate constants of alkoxyl radical reactions in aqueous solutions, J. Phys. Chem. 91:2362 (1987).

24. R.D. Small, J.C. Scaiano, and L.K. Patterson, Radical processes in lipids. A laser photolysis study of tert-butoxy radical reactivity towards fatty acids, Photochem. Photobiol. 29:49 (1979).

25. M. Erben-Russ, C. Michel, W. Bors, and M. Saran, Determination of sulfite radical (SO_3^-) reaction rate constants by means of competition kinetics, <u>Radiat</u>. <u>Environ</u>. <u>Biophys</u>. 26:289 (1987).

26. L.G. Forni, J.E. Packer, T.F. Slater, and R.L. Willson, Reaction of the trichloromethyl and halothane-derived peroxy radicals with unsaturated fatty acids: a pulse radiolysis study, <u>Chem.-Biol</u>. <u>Interactions</u> 45:171 (1983).

27. B.H.J. Bielski, R.L. Arudi, and M.W. Sutherland, A study of the reactivity of HO_2/O_2^- with unsaturated fatty acids, <u>J</u>. <u>Biol</u>. <u>Chem</u>. 258:4759 (1983).

28. F.H. Doleiden, S.R. Fahrenholtz, A.A. Lamola, and A.M. Trozzolo, Reactivity of cholesterol and some fatty acids towards singlet oxygen, <u>Photochem</u>. <u>Photobiol</u>. 20:519 (1974).

29. H.R. Rawls, and P.J. van Santen, Singlet oxygen: a possible source of the original hydroperoxides in fatty acids, <u>Ann</u>. <u>New</u> <u>York</u> <u>Acad</u>. <u>Sci</u>. 171:135 (1970).

30. E.N. Frankel, Chemistry of free radical and singlet oxidation of lipids, <u>Prog</u>. <u>Lipid</u> <u>Res</u>. 23:197 (1985).

31. G. Minotti, and S.D. Aust, The role of iron in the initiation of lipid peroxidation, <u>Chem</u>. <u>Phys</u>. <u>Lipids</u> 44:191 (1987).

32. W.H. Koppenol, and J.F. Liebman, The oxidizing nature of the hydroxyl radical. A comparison with the ferryl ion (FeO^{2+}), <u>J</u>. <u>Phys</u>. <u>Chem</u>. 88:99 (1984).

33. I.I. Ivanov, A relay model of lipid peroxidation in biological membranes, <u>J</u>. <u>Free</u> <u>Radic</u>. <u>Biol</u>. <u>Med</u>. 1:247 (1985).

34. W. Bors, M. Erben-Russ, and M. Saran, Fatty acid peroxyl radicals: their generation and reactivities, <u>Bioelectrochem</u>. <u>Bioenerg</u>. 18:37 (1987).

35. G.E. Adams, and R.L. Willson, Pulse radiolysis studies on the oxidation of organic radicals in aqueous solution, <u>Trans</u>. <u>Faraday</u> <u>Soc</u>. 65:2981 (1969).

36. N.A. Porter, Mechanism for the autoxidation of polyunsaturated lipids, <u>Accts</u>. <u>Chem</u>. <u>Res</u>. 19:262 (1986).

37. H.W.-S. Chan, G. Levett, and J.A. Matthew, Thermal isomerisation of methyl linoleate hydroperoxides. Evidence of molecular oxygen as a leaving group in a radical rearrangement, <u>JCS</u>, <u>Chem</u>. <u>Comm</u>. 1978:756 (1978).

38. P. Schieberle, W. Grosch, H. Kexel, and H.L. Schmidt, A study of oxygen isotope scrambling in the enzymic and non-enzymic oxidation of linoleic acid, <u>Biochim</u>. <u>Biophys</u>. <u>Acta</u> 666:322 (1981).

39. P. Schieberle, W. Grosch, Detection of monohydroperoxides with unconjugated diene systems as minor products of the autoxidation of methyl linoleate, <u>Z</u>. <u>Lebensm</u>. <u>Unters.-Forsch</u>. 173:199 (1981).

40. E.N. Frankel, W.E. Neff, E. Selke, and D. Weisleder, Photosensitized oxidation of methyl linoleate: secondary and volatile thermal decomposition products, <u>Lipids</u> 17:11 (1982).

41. F. Haslbeck, W. Grosch, Autoxidation of phenyl linoleate and phenyl oleate: HPLC analysis of the major and minor monohydroperoxides as phenyl hydroxystearates, <u>Lipids</u> 18:706 (1983).

42. P.H. Gale, and R.W. Egan, Prostaglandin endoperoxide synthase-catalyzed oxidation reactions, <u>in</u>:"Free Radicals in Biology," W.A. pryor, ed., Academic Press, New York, Vol. VI, p.1 (1984).

43. J.A. Howard, and K.U. Ingold, Absolute rate constants for hydrocarbon autoxidation. VI. Alkyl aromatic and olefinic hydrocarbons, <u>Can</u>. <u>J</u>. <u>Chem</u>. 45:793 (1967).

44. G.A. Russell, Deuterium-isotope effects in the autoxidation of aralkyl hydrocarbons. Mechanism of the interaction of peroxy radicals, <u>J</u>. <u>Am</u>. <u>Chem</u>. <u>Soc</u>. 79:3871 (1957).

45. E. Cadenas, Oxidative stress and formation of excited species, <u>in</u>: "Oxidative Stress," H. Sies, ed., Academic Press, London, p.311 (1986).

46. J.R. Kanofsky, Red chemiluminescence from ram seminal vesicle microsomes: pitfalls in the use of spectrally resolved red chemiluminescence as a test for singlet oxygen in biological systems, Photochem. Photobiol. 47:605 (1988).

47. D. Schulte-Frohlinde, and C. von Sonntag, Radiolysis of DNA and model systems in the presence of oxygen, in:"Oxidative Stress," H. Sies, ed., Academic Press, London, p.11 (1986).

48. A.R. Brash, A.T. Porter, and R.L. Mass, Investigation of the selectivity of hydrogen abstraction in the non-enzymatic formation of hydroxyeicosatetraenoic acid and leukotrienes by autoxidation, J. Biol. Chem. 260:4210 (1985).

49. Y. Yamamoto, E. Niki, and Y. Kamiya, Oxidation of lipids. III. Oxidation of methyl linoleate in solution, Lipids 17:870 (1982).

50. A. Sevanian, M.L. Wratten, L.L. McLeod, and E. Kim, Lipid peroxidation and phospholipase A_2 activity in liposomes composed of unsaturated phospholipids: a structural basis for enzyme activation, Biochim. Biophys. Acta 961:316 (1988).

51. H.W. Gardner, and R. Kleiman, Degradation of linoleic acid hydroperoxides by a cysteine-$FeCl_3$ catalyst as a model for similar biochemical reactions. II. Specificity in formation of fatty acid epoxides, Biochim. Biophys. Acta 665:113 (1981).

52. R. Labeque, and L.J. Marnett, 10-Hydroperoxy-8,12-octadecadienoic acid: a diagnostic probe of alkoxyl radical generation in metal-hydroperoxide reactions, J. Am. Chem. Soc. 109:2828 (1987).

53. W. Bors, C. Michel, and M. Saran, Inhibition of the bleaching of the carotenoid crocin. A rapid test for quantifying antioxidant activity, Biochim. Biophys. Acta 796:312 (1984).

54. W.E. Neff, and E.N. Frankel, Photosensitized oxidation of methyl linoleate monohydroperoxides: hydroperoxy cyclic peroxides, dihydroperoxides and hydroperoxy-bis-cyclic peroxides, Lipids 19:925 (1984).

55. W. Bors, D. Tait, C. Michel, M. Saran, and M. Erben-Russ, Reactions of alkoxyl radicals in aqueous solutions, Israel J. Chem. 24:17 (1984).

56. V. Madhavan, N.N. Lichtin, and E. Hayon, Electron adducts of acrylic acid and homologues. Spectra, kinetics and protonation reactions. A pulse-radiolytic study, J. Org. Chem. 41:2320 (1976).

57. M. Saran, D. Tait, W. Bors, and C. Michel, Formation and reactivities of alkoxy radicals, in:"Oxy Radicals and Their Scavenger Systems Vol. I. Molecular Aspects," G. Cohen, R.A. Greenwald, eds., Elsevier, New York, p.20 (1983).

58. C. Walling, Some aspects of the chemistry of alkoxy radicals, Pure Appl. Chem. 15:69 (1967).

59. K.U. Ingold, Rate constants for free radical reactions in solutions, in:"Free Radicals," J.K. Kochi, ed., Wiley-Interscience, New York, Vol. I, p.37 (1973).

60. D.G. Hendry, T. Mill, L. Piszkiewicz, J.A. Howard, and H.K. Eigenman, A critical review of H-atom transfer in the liquid phase: chlorine atom, alkyl trichloromethyl, alkoxy and alkylperoxy radicals, J. Phys. Chem. Ref. Data 3:937 (1974).

61. T.A. Dix, and L.J. Marnett, Hematin-catalyzed rearrangement of hydroperoxylinoleic acid to epoxy alcohols via an oxygen rebound, J. Am. Chem. Soc. 105:7001 (1983).

62. H.W. Gardner, K. Eskins, G.W. Grams, and G.E. Inglett, Radical addition of linoleic hydroperoxides to alpha-tocopherol or the analogous hydroxy chroman, Lipids 7:324 (1972).

63. E.N. Frankel, Volatile lipid oxidation products, Prog. Lipid Res. 22:1 (1982).

64. T. Nakayama, M. Kodama, and C. Nagata, Free radical formation in DNA by lipid peroxidation, Agric. biol. Chem. 48:571 (1984).

65. W. Grosch, Abbau von Linol- und Linolensäurehydroperoxyden in Gegenwart von Ascorbinsäure. Analyse der flüchtigen Aldehyde, Z. Lebensm. Unters.-Forsch. 163:4 (1977).

66. P. Schieberle, B. Tsoukalas, and W. Grosch, Decomposition of linoleic acid hydroperoxides by radicals. I. Structures of products of methyl 13-hydroperoxy-cis,trans-9,11-octadecadienoate, Z. Lebensm. Unters.-Forsch. 168:448 (1979).

67. F. Ursini, M. Maiorino, and C. Gregolin, The selenoenzyme phospholipid hydroperoxide glutathione peroxidase, Biochim. Biophys. Acta 839:62 (1985).

68. R. Ladenstein, O. Epp, W. Guenzler, and L. Flohé, Glutathione peroxidase on approval, Life Chem. Rep. 4:37 (1986).

69. L.K. Patterson, Studies of radiation induced peroxidation in fatty acid micelles, in:"Oxygen and Oxy Radicals in Chemistry and Biology," M.A.J. Rodgers, E.L. Powers, eds., Academic Press, New York, p.89 (1981).

70. N.A. Porter, L.S. Lehman, B.A. Weber, and K.J. Smith, Unified mechanism for polyunsaturated fatty acid autoxidation. Competition of peroxy radical hydrogen atom abstraction, ß-scission, and cyclization, J. Am. Chem. Soc. 103:6447 (1981).

71. W. Bors, C. Michel, and M. Saran, Determination of kinetic parameters of oxygen radicals by competition studies, in:"CRC Handbook of Methods for Oxygen Radical Research," R.A. Greenwald, ed., CRC Press, Boca Raton, p.181 (1985).

72. T. Doba, G.W. Burton, K.U. Ingold, and M. Matsuo, Alpha-tocopherol decay: lack of effect of oxygen, JCS, Chem. Comm. 461 (1984).

73. W. Bors, Semiquinone and phenoxyl radicals of phenolic antioxidants and model compounds: generation, spectral and kinetic properties, Life Chem. Rep. 3:16 (1985).

74. E. Niki, Interaction of ascorbate and alpha-tocopherol, Ann. New York Acad. Sci. 498:186 (1987).

75. A.P. Griva, and E.T. Denisov, Kinetics of the reactions of 2,4,6-tritert-butylphenoxyl with cumene hydroperoxide, cumylperoxyl radicals and molecular oxygen, Int. J. Chem. Kinet. 5:869 (1973).

76. J. Tsuchiya, E. Niki, and Y. Kamiya, Oxidation of lipids. IV. Formation and reaction of chromanoxyl radicals as studied by ESR, Bull. Chem. Soc. Japan 56:229 (1983).

77. J. Winterle, D. Dulin, and T. Mill, Products and stoichiometry of reaction of vitamin E with alkylperoxy radicals, J. Org. Chem. 49:491 (1984).

LIPID PEROXIDATION BY PHAGOCYTES

Bernard M. Babior

Department of Molecular and Experimental Medicine
Research Institute of Scripps Clinic
La Jolla, California 92037

Professional phagocytes (neutrophils, eosinophils, mononuclear phagocytes) are uniquely endowed with the capacity to manufacture large quantities of highly reactive oxidizing agents for use in the destruction of invading pathogens, both unicellular and multicellular. This capacity arises because of the presence in these cells of an enzyme known as the respiratory burst oxidase that catalyzes the one-electron reduction of oxygen to O_2^- at the expense of NADPH:

$$O_2 + NADPH \text{ ----------> } O_2^- + NADP^+ + H^+$$

The rapid dismutation of most of this O_2^- gives rise to H_2O_2. From these two starting materials (i.e., O_2^- and H_2O_2), the phagocytes are able to manufacture a remarkably diverse group of reactive oxidants. These include innumerable compounds containing an oxidized halogen atom, all derived from the reaction of amines with the HOCl produced by the myeloperoxidase-catalyzed oxidation of Cl^- by H_2O_2:

$$H_2O_2 + Cl^- \xrightarrow{\text{myeloperoxidase}} OCl^- + H_2O$$

and what is probably an equally extensive battery of oxidizing radicals whose origins are not fully understood but probably include the Haber-Weiss reaction, which yields the hydroxyl radical,

$$O_2^- + H_2O_2 \xrightarrow{\text{Fe or Cu}} OH\cdot + OH^- + O_2$$

and a related reaction in which H_2O_2 is replaced by an alkyl hydroperoxide:

$$O_2^- + ROOH \xrightarrow{\text{Fe or Cu}} RO\cdot + OH^- + O_2$$

With their ability to deploy these unusually potent oxidants, it is not surprising that phagocytes mediate the peroxidation of lipids at sites of inflammation and other regions where their oxidant-generating capacity is brought into play.

Free Radicals, Lipoproteins, and Membrane Lipids
Edited by A. Crastes de Paulet *et al.*
Plenum Press, New York, 1990

Some lipid peroxides of phagocytic origin, however, arise not through non-specific lipid peroxidation occurring in the course of the respiratory burst, but through deliberate biosynthesis by lipoxygenases designed specifically to catalyze their production. Prostaglandin G$_2$, a precursor of the platelet-active lipid mediators thromboxane A$_2$ and prostacyclin, is a cyclic lipid peroxide/hydroperoxide formed by the incorporation of 2 molecules of O$_2$ into arachidonic acid in a reaction catalyzed by the enzyme cyclooxygenase [1]. Other lipoxygenases catalyze the conversion of arachidonic acid into lipid hydroperoxides carrying an -OOH group on the 5-, 12- or 15-carbon. These compounds, known as HPETEs (hydroperoxyeicosatetraenoic acids), are the immediate precursors of the HETE (hydroxyeicosatetraenoic acid) series of inflammatory mediators. In addition, 5-HPETE is the starting material for the biosynthesis of the leukotrienes and lipoxins, a group of very potent mediators with important actions on smooth muscle and phagocytes [2].

RESPIRATORY BURST OXIDASE-ASSOCIATED LIPID PEROXIDATION

In Vitro

It has been known for nearly two decades that oxidants liberated by activated neutrophils are able to mediate lipid peroxidation. One of the earliest demonstrations of lipid peroxidation by neutrophils was that of Mason et al., who reported in 1972 that malondialdehyde was produced by neutrophils when they ingested an emulsion containing linolenic acid [3]. Other reports confirming these observations soon followed. Early work, however, did not deal with mechanisms of neutrophil-mediated lipid peroxidation. Investigation of this topic is a more recent development.

Lipid peroxidation by oxidized halogens. The participation of oxidized halogens in neutrophil-mediated lipid peroxidation has been investigated by several groups. Representative of these investigations are studies from Clark's and Austen's laboratories. Sepe and Clark used the release of label from liposomes containing ^{51}Cr in their aqueous core as a marker of lipid peroxidation [4,5]. Upon exposure to the peroxide-halide-myeloperoxidase system, liposomes composed of either phosphatidylcholine alone or a combination of phosphatidylcholine, dicetylphosphate and cholesterol rapidly released much of their label into the surrounding environment. Release of label did not occur in the absence of any of the 3 components of the oxidizing system, or in the presence of catalase or of myeloperoxidase inhibitors such as CN$^-$ or N$_{3^-}$. Similar results were obtained when the ^{51}Cr-loaded liposomes were incubated with phorbol-activated neutrophils, which produce large amounts of a very complex mixture of oxidants. The role of the peroxide-halide-myeloperoxidase system in this neutrophil-mediated process was shown by the finding that neutrophils from patients with either chronic granulomatous disease (in which phagocytes are unable to manufacture reactive oxidants of any sort) or myeloperoxidase deficiency (in which phagocytes manufacture oxidizing radicals but not oxidized halogens) were unable to release ^{51}Cr from the liposomes. Lipid peroxidation as the mechanism of liposome disruption was suggested by two pieces of indirect evidence: 1) the lipid-soluble antioxidants α-tocopherol and β-carotene protected the liposomes against the effects of the neutrophil oxidants, and 2) liposomes prepared with synthetic dipalmitoyl phosphatidylcholine, which contains no oxidizable unsaturated fatty acids, were completely resistant to the effects of phorbol-activated neutrophils. Despite the lack of direct evidence showing that malonyldialdehyde or other products of lipid peroxidation are formed in this system, these results strongly suggest that oxidized halogens originating from activated neutrophils are able to carry out lipid peroxidation.

More direct evidence for lipid peroxidation by neutrophil-derived oxidized halogens was reported by Corey, Austen and their associates [6]. This group studied the inactivation of the leukotriene thioethers (leukotrienes C_4, D_4 and E_4) by phorbol-activated neutrophils. Exposure of each of these leukotrienes to the activated neutrophils yielded, among other things, major amounts of the lipid peroxidation product 6-trans-leukotriene B_4, which was identified unequivocally by HPLC, UV spectroscopy and mass spectroscopy. This same product was formed when leukotrienes were oxidized with the peroxide-halide-myeloperoxidase system or with purified HOCl. These findings provide direct evidence for the ability of the peroxide-halide-myeloperoxidase system of neutrophils to carry out the peroxidation of lipids containing polyunsaturated fatty acids.

Lipid peroxidation by hydroxyl radicals. Claster et al. found that neutrophils could also carry out lipid peroxidation through the production of hydroxyl radicals (OH·) [7]. Using red cells as targets and malonyldialdehyde production (actually, the production of thiobarbituric acid-reactive material) as the measure of lipid peroxidation, these workers found that oxidants generated by activated neutrophils were able to mediate the peroxidation of lipids. Malonyldialdehyde production was inhibited by both superoxide dismutase and catalase, suggesting the involvement of OH· in the lipid peroxidation reaction (claims by Cohen and Rosen that neutrophils produce no OH· [8] were recently shown to be based on an artifact [9]). It may be that lipid peroxidation by activated neutrophils is involved in the well-known aggravation of hemolysis that occurs when infections arise in patients with chronic hemolytic anemias.

O_2^--dependent lipid peroxidation. Perhaps the most interesting mechanism for neutrophil-mediated lipid peroxidation is the O_2^--dependent mechanism described by Carlin and Arfors [10]. These workers found that phorbol-stimulated neutrophils were able to peroxidize phospholipid liposomes by a process that required O_2^-, but not H_2O_2 or OH· (as demonstrated by the ability of superoxide dismutase but not catalase or dimethylsulfoxide to abolish lipid peroxidation), provided certain iron complexes were present. Oxidation occurred in the presence of iron chelated to pyrophosphate or ADP, as well as ferritin iron, but iron complexed to EDTA, ATP or transferrin were not effective. The mechanism of this reaction is somewhat of a puzzle, because the powerful oxidants ordinarily formed in secondary reactions involving O_2^- were not participants in the lipid peroxidation observed in this system. Both Fe^{2+} and Fe^{3+} are required for the initiation of lipid peroxidation in certain systems [11], and it may be that O_2^- acts to convert some of the Fe^{3+} to Fe^{2+} to fulfill this requirement.

Carlin described a related system in which linolenic acid micelles were peroxidized by neutrophils stimulated to generate oxidants by the micelles themselves [12]. Lipid peroxidation was enhanced by Fe^{3+}, more or less regardless of the ligands in the coordination sphere of the metal: $FeCl_3$, Fe^{3+}-ADP and Fe^{3+}-EDTA were all suitable. In this system, lipid peroxidation was not inhibited by superoxide dismutase, by catalase or by the OH· scavenger dimethylsulfoxide. It was postulated that this lipid peroxidation proceeded by a mechanism analogous to the peroxidation carried out by NADPH-cytochrome P450 reductase.

In vivo

The acute respiratory distress syndrome (ARDS) is thought to be caused in part by systemic activation of the complement system, leading to the sequestration of neutrophils in the pulmonary vasculature with subsequent pulmonary damage thought to be due to proteases and reactive oxidants released by the sequestered neutrophils. In a model of ARDS produced by the intravenous injection of cobra venom factor, pulmonary injury as measured by increased pulmonary vascular permeability was found to correlate with the appearance in the lungs and blood of conjugated dienes presumably arising through lipid peroxidation [13]. Evidence for excess lipid peroxidation was not observed in the liver, kidneys or spleens of the experimental animals, nor did these organs show any evidence of functional damage. Protection against both lipid peroxidation and pulmonary damage was provided by treating the animals with anti-oxidant agents (catalase, OH· scavengers, iron chelators) or by depleting the animals of neutrophils. These findings indicate that in vivo as well as in vitro, activated neutrophils are able to peroxidize tissue lipids, and support the idea that lipid peroxidation is at least partly responsible for the damage inflicted by activated neutrophils on surrounding tissues.

Related to the foregoing are observations made in a group of patients undergoing cardiac surgery [14]. In these patients, the release of the cross-clamp from the aorta was followed almost immediately by the sequestration of neutrophils in the lungs and the simultaneous appearance of products of lipid peroxidation in the plasma. These findings indicate that under appropriate clinical circumstances, neutrophil-mediated lipid peroxidation takes place in man as well as in experimental animals. These findings in turn support the plausibility of a pathogenetic mechanism involving neutrophil-mediated lipid peroxidation as a cause of tissue damage at sites of inflammation, though conclusive evidence for such a mechanism remains to be obtained.

REFERENCES

1. B. Samuelsson, M. Goldyne, E. Granstrom, M. Hamberg, S. Hammarstrom, and C. Malmsten, Prostaglandins and thromboxanes. Ann. Rev. Biochem. 47:997, 1978.
2. B. Samuelsson, S.-E. Dahlen, J.-A. Lindgren, C.A. Rouzer, and C.N. Serhan, Leukotrienes and Lipoxins: Structures, biosynthesis, and biological effects. Science 237:1171, 1987.
3. R.J. Mason, T.P. Stossel, and M. Vaughan, Lipids of alveolar macrophages, polymorphonuclear leukocytes, and their phagocytic vesicles. J. Clin. Invest. 51:2399, 1972.
4. S.M. Sepe, and R.A. Clark, Oxidant membrane injury by the neutrophil myeloperoxidase system. J. Immunol. 134:1888-1895, 1985.
5. S.M. Sepe, and R.A. Clark, Oxidant membrane injury by the neutrophil myeloperoxidase system. II. Injury by stimulated neutrophils and protection by lipid-soluble antioxidants. J. Immunol. 134:1896, 1985.
6. C.W. Lee, R.A. Lewis, A.I. Tauber, M. Mehrotra, E.J. Corey, and K.F. Austen, The myeloperoxidase-dependent metabolism of leukotrienes C4, D4, and E4 to 6-trans-leukotriene B4 diastereoisomers sulfoxides. J. Biol. Chem. 258:15004-10, 1983.
7. S. Claster, D.T. Chiu, A. Quintanilha, B. Lubin, Neutrophils mediate lipid peroxidation in human red cells. Blood 64:1079-84, 1984.

8. M.S. Cohen, B.E. Britigan, D.J. Hassett, and G.M. Rosen, Do humans neutrophils form hydroxyl radical? Evaluation of an unresolved controversy. J. Free Radic. Biol. Med. 5:81-88, 1988.

9. A. Samuni, C.D. Black, C.M. Krishna, H.L. Malech, E.F. Bernstein, and A. Russo, Hydroxyl radical production by stimulated neutrophils reappraised. J. Biol. Chem. 263:13797-13801, 1988.

10. G. Carlin, and K.E. Arfors, Peroxidation of liposomes promoted by human polymorphonuclear leucocytes. J. Free Radic. Biol. Med. 1:437-442, 1985.

11. G. Minotti, and S.D. Aust, The requirement for the iron (III) in the initiation of lipid peroxidation by iron (II) and hydrogen peroxide. J. Biol. Chem. 262:1098-1104, 1987.

12. G. Carlin, Peroxidation of linolenic acid promoted by human polymorphonuclear leucocytes. J. Free Radic. Biol. Med. 1:255-261, 1985.

13. P.A. Ward, K.J. Johnson, and G.O. Till, Animal models of oxidant lung injury. Respiration 50:5-12, 1986.

14. D. Royston, J.S. Fleming, J.B. Desai, S. Westaby, and K.M. Taylor, Increased production of peroxidation products associated with cardiac operations. Evidence for free radical generation. J. Throac. Cardiovasc. Surg. 91:759-766, 1986.

TRIGGERING AND REGULATION OF THE FREE RADICAL PRODUCTION BY PHAGOCYTES

P. Bellavite, M.C. Serra, F. Bazzoni, S. Miron and S. Dusi
Istituto di Patologia Generale, Università di Verona, 37134, Verona, Italy

INTRODUCTION

Phagocytic cells (neutrophils, eosinophils, monocytes and macrophages) are capable of converting oxygen into potentially toxic species such as superoxide anion, hydrogen peroxide and hydroxyl radical. This peculiar metabolic pathway, which is called respiratory burst, is turned on when a membrane-bound enzyme, the NADPH oxidase, is activated. Other reactions, such as those of the glutathione cycle and of the hexose monophosphate pathway, are secondary to the triggering of NADPH oxidase, having the function of continuous supply of reduced NADPH and of intracellular detoxification.[1-3]

The molecular structure of the NADPH oxidase has not completely clarified yet, but there is evidence that it is composed by an electron transport chain, where a flavoprotein, a cytochrome b with low potential (cytochrome b_{558}, or cytochrome b_{-245}) and possibly other proteins of unknown nature are assembled in a functional complex.[4-7] Membrane phospholipids give stability and possibly regulate the function of this complex.[8-9]

The free radical forming system is activated during phagocytosis and the generation of oxygen free radicals significantly contributes to the defensive (bactericidal and tumoricidal) function of neutrophils, eosinophils and macrophages. On the other hand, other agents that are not related to phagocytosis may trigger the respiratory burst. Toxic oxygen derivatives may diffuse into the extracellular space and damage connective tissue macromolecules, cell membranes and even cause DNA mutations.

In this brief review we will consider: I) the agents that are able of triggering the respiratory burst, especially in

Free Radicals, Lipoproteins, and Membrane Lipids
Edited by A. Crastes de Paulet *et al.*
Plenum Press, New York, 1990

relation with lipid metabolism, II) the mechanisms of their action and III) the possibilities of regulation of the oxidative metabolism at cellular level.

I. AGENTS THAT TRIGGER THE RESPIRATORY BURST OF PHAGOCYTES

As it can be seen in Table 1, besides the phagocytosable particles, a large series of substances with different chemical composition are able of interacting with the cell leading to its activation. The action of some of these agents may be related to lipid metabolism and vascular pathology. In fact, arachidonic acid, leucotriene B_4, platelet activating factor are potent stimulants of the burst and at the same time are produced and released by activated leukocytes, therefore acting as messengers and signals for further cell activation and amplification of the inflammatory process. The effect of acetylated LDL is noteworthy, because monocytes and macrophages exert a scavenger function into the vessel intima by taking up excess of modified lipoproteins. However, concomitant activation of oxygen free radical release could be one of the pathogenetic mechanisms of damage to the vessel wall,

Table 1. Some stimulants of phagocyte's metabolism

	REF.
PARTICLES	
Opsonized bacteria, fungi, virus	10,11
Immunoglobulin aggregates	12
LIPIDS AND LIPID DERIVATIVES	
Arachidonic acid and other fatty acids	13,14
Leukotriene B_4	15
Diacylglycerol	16
Platelet activating factor	17
Acetylated LDL	18
Cerebrosides	19
PROTEINS	
Concanavalin A	20
Anti-leukocyte antibodies	21,22
Complement fragments (C5a, C567)	23,24
Tumor necrosis factor	25
Phospholipase C	26
PEPTIDES	
N-formylated peptides	27
Substance P	28
OTHERS	
Calcium ionophores	29
Urate crystals	30
Sodium fluoride	31
Low-sodium solutions	32
Detergents	33,34
Cross-linking reagents	35

inflammation, sclerosis and possibly cell transformation.

Among the proteins, the effect of C5a and of tumor
necrosis factor (also called cachectin) are probably important
in human pathology. By triggering leukocyte metabolism, the
intravascular complement activation and the release of TNF by
activated mononuclear phagocytes may be responsible for wasting
systemic effects and damage to pulmonary microvasculature that
often complicate sepsis, severe burns, shock.

The stimulatory effect of phospholipase C reveals the
importance of membrane phospholipids in the triggering and
regulation of oxidative metabolism. This is confirmed by the
direct stimulatory effect of diacylglycerol, the main product
of phospholipase C activity. Preliminary data from our
laboratory indicate that also phosphatidic acid, that is formed
in the cell both by phosphorylation of diacylglycerol and by
action of phospholipase D, is able of activating H_2O_2
production by human neutrophils.

As far as the peptides are concerned, it has been recently
discovered in our laboratory[28] that substance P (SP) is a
stimulant of H_2O_2 production by human neutrophils. This
undecapeptide is widely distributed in the nervous system and
is particularly concentrated in the peripheral nerve terminals
of small diameter unmyelinated sensory neurons, termed C-
fibers, which terminate in the dorsal horn of spinal cord. SP
containing fibers have been found also into the vessel wall.
Although SP is considered to be a neurotransmitter at the
central terminals of C-fibers, up to 90% of the peptide
synhesized in the cell bodies of these neurons is transported
to the peripheral terminals, from where it can be released by
noxious stimuli. Fig. 1 provides a possible interpretation of

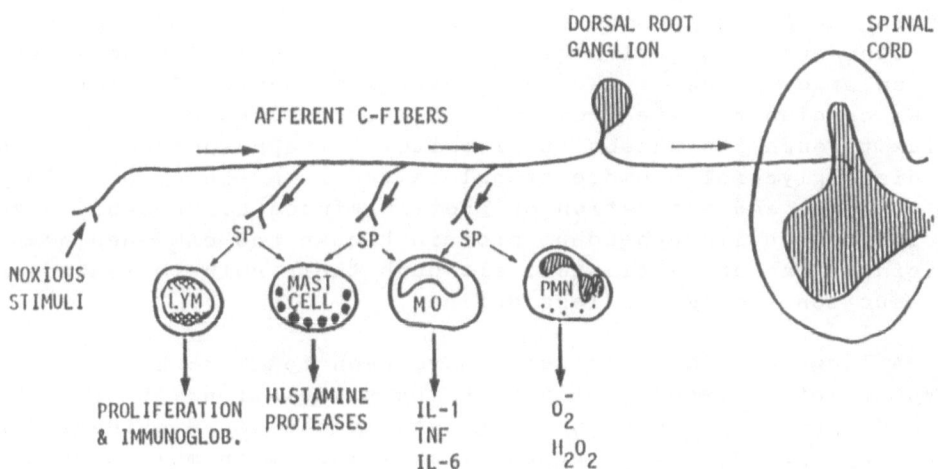

Fig. 1. Effects of substance P on the inflammatory cells

the physiological role of SP. Besides the stimulation of
oxidative metabolism of neutrophils, it is known that SP
produces vasodilatation, it acts as mitogen for lymphocytes, it
degranulates mast cells and activates macrophages. This
neuropeptide therefore meets many of the requirements for a
mediator of the local inflammatory response and represents an
important link between nervous and immunological systems.

II. MECHANISMS OF ACTIVATION OF THE RESPIRATORY BURST

The mechanisms by which agonist-stimulated receptors are
coupled with the terminal effector systems such as phagocytosis,
degranulation, movement, free radical production, gene
expression, etc., are called transduction pathways (or
systems). The matter is very complex because multiple pathways
have been described, that may vary according to the stimulant
used, and also inhibitory mechanisms may be operative in
particular conditions. Clearly, elucidating the transduction
systems is important because the intensity and the duration of
the functional responses, including the respiratory burst, may
be regulated at this level.

In the attempt to simplify the understanding of this
point, three major hypothesis that provide an explanation of
how the NADPH oxidase may be activated are here presented. More
details may be found in other recent reviews.[3,36-39]

II.a. Protein phosphorylation
Phosphorylation and dephosphorylation of specific proteins
regulates a variety of cells responsive to external stimuli.
There is increasing evidence that this mechanism operates also
in neutrophils for the activation of NADPH oxidase and other
functions. As shown in Fig. 2, the ligand-receptor interaction,
through the coupling action of a guanine nucleotide binding
protein, triggers phospholipid hydrolysis in the cell membrane,
with consequent formation of important intracellular messengers
such as diacylglycerol and inositol triphosphate. The latter
causes calcium release from intracellular stores and calcium
influx through its metabolite inositol tetraphosphate. Calcium
and diacylglycerol promote translocation from the cytosol to
the membrane and activation of protein kinase C. Probably also
calcium/calmodulin dependent protein kinase and cAMP dependent
protein kinase are activated, although their role in leukocyte
transduction systems is less defined.

A large series of proteins have been found to be
phosphorylated concomitantly with the stimulation. At least two
of these phosphoproteins are involved in the NADPH oxidase. The
first is a protein, or a group of proteins, with molecular
weight of about 48 kDa. Although their nature is not known, the
participation of these proteins is strongly suggested by the

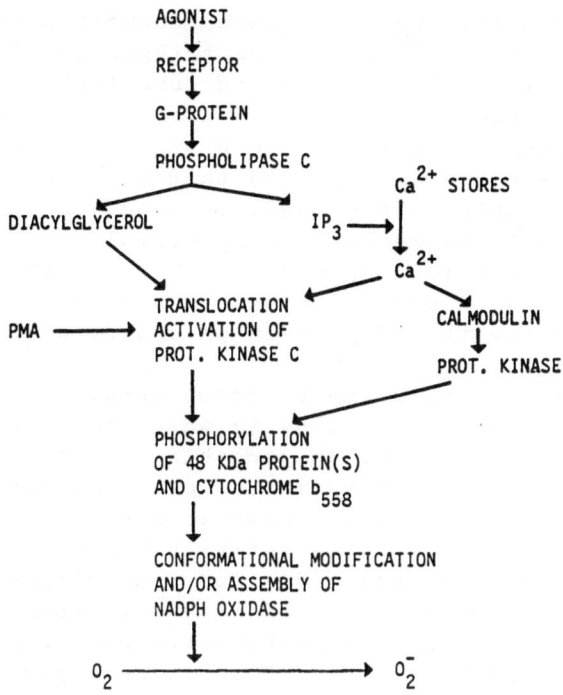

Fig. 2. NADPH oxidase activation by protein phosphorylation

observation that their phosphorylation is lacking in some forms
of a disease - chronic granulomatous disease of childhood (CGD)
- where the respiratory burst fails to be activated.[40-44] The
second relevant protein that is phosphorylated is cytochrome
b_{558}. This cytochrome is actually a component of the enzyme
NADPH oxidase and its phosphorylation suggest a possible
regulatory mechanism at this level. The kinase responsible for
this modification of the protein is probably protein kinase
C.[45]

II.b. Membrane lipid changes

        The relationship between phosphorylation of cytochrome b
and the enzymatic activation is still hypothetical, because
there is no direct demonstration that phosphorylation directly
triggers the enzyme. Studies carried out in our laboratory have
shown that in cells stimulated with phorbol esters or opsonized
zymosam there is marked phosphorylation and a proportional
NADPH oxidase activation, while in cells stimulated with
arachidonic acid a very little phosphorylation is accompanied
by an high activation. We therefore concluded that
phosphorylation is not the only activation mechanism[46] and this
fact was confirmed by others.[44]

        The existence of additional, or alternative, pathways of
oxidase activation is also indicated by studies of activation
mechanism carried out not in intact cells but in cell-free

27

systems. These models that have been recently developed in several laboratories,[47-56] allow the triggering of the enzymatic production of superoxide in subcellular organelles or even in purified fractions by addition of cytosolic components and of fatty acids or detergents such as sodium dodecyl sulphate. We have recently reported that pig neutrophil NADPH oxidase is activatable by phosphatidic acid, an important product of phospholipid metabolism in stimulated cells, even in the absence of cytosolic components.[54] We and others have shown that the activation in cell-free system does not depend on the protein kinase activity and protein phosphorylation.

On the basis of the above reported data, it is possible to construct an hypothesis according to which the terminal modification of the oxidase, responsible for its activation, is caused by changes of the lipid milieu of the membrane where the enzyme complex is embedded. As shown in Fig. 3, the lipid changes that affect the oxidase activity could be either an increase of phosphatidic acid (due to phospholipase D and/or to diacylglycerol kinase) or an increase of arachidonic acid (due to calcium-dependent and perhaps receptor-dependent activation of phospholipase A2). Both these lipid changes have been documented in the membrane of stimulated cells. The alteration of lipid properties (fluidity, electric charges, melting point, etc.) in the enzyme microenvironment may cause conformational modifications and assembly of memebrane and cytosolic components of the oxidase. The electron-transport system can thus start to catalyse superoxide formation.

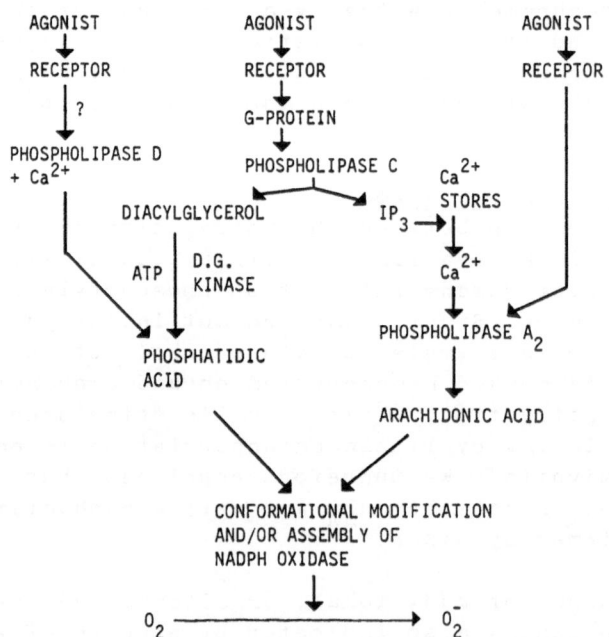

Fig. 3. NADPH oxidase activation by membrane lipid changes

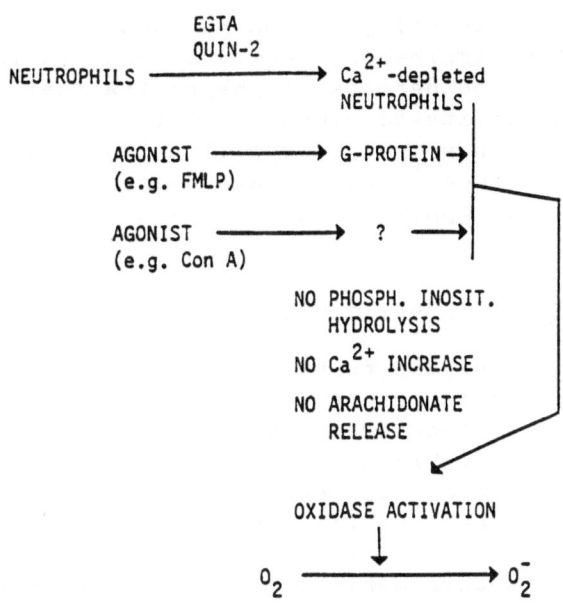

Fig. 4. A third existing pathway of NADPH oxidase activation

II.c. Calcium and phosphoinositide-independent pathway

The existence of a third mechanism that is independent of calcium and lipid changes may be postulated on the basis of recent work from the group of F. Rossi in our laboratory.[57-59] This is schematically represented in Fig. 4. An experimental model has been developed where the neutrophils are completely depleted of intracellular free calcium by the addition of chelators. In these conditions no modifications of free calcium, no phospholipid hydrolysis, no arachidonic acid and phosphatidic acid formation occur. When these cells are challenged with two different agents, either given in sequence or contemporaneously, they undergo to marked metabolic stimulation. Therefore a further and unknown activation mechanism exists and is currently investigated in our laboratory. It remains to be established whether this mechanism, that is operative in calcium depleted cells, is alternative or is additional to the other pathways previously described.

III. REGULATION OF THE RESPIRATORY BURST

On the basis of the knowledge of the structure and the activation mechanism of the NADPH oxidase it is possible to deal with the possible ways of regulating the respiratory burst. This subject is of great interest because it would be useful to decrease, or increase, the intensity and the duration of free radical production when required. An inhibition, or down dregulation of the burst is theoretically desirable during pathologic inflammatory processes in order to decrease free

radical dependent tissue injury. On the contrary, an
enhancement of the respiratory burst is required in the case of
congenital or acquired defects of phagocytes that often cause
increased susceptibility to microbial infections. Here the main
literature data on these subjects will be summarized. it should
be pointed out that most studies have been done on isolated
leukocytes and their application in medical practice is still
hypothetical.

III.a Inhibition of oxidative metabolism of phagocytes

The inhibition of the respiratory burst may be
accomplished both by interference with the activation
mechanism(s) and by blocking the activity of the terminal
oxidase. As shown in Table 2, a large series of inhibitors of
the activation mechanism has been reported, in keeping with the
multiform pathways that are involved.

Some of these agents merit particular discussion. The
homologous pre-stimulation causes de-sensitization of the

Table 2. Inhibitors of the respiratory burst that act on
some step of the activation mechanism

Agent	Possible mechanism	Ref.
Albumin (on arach. acid)	Binding to stimulant	60
H_2O_2 + peroxidase + halide	Inactivation of stimulant	61
-met-mannopyr.(on Con A)	Displacement of ligand	20
Homologous pre-stimulation	Receptor desensitization	62,63
Agonist-coated surfaces	Receptor down-regulation	64,65
Tumor-conditioned medium	?	66
PDGF	Post-recept. deactivation	67
Pertussis toxin	G-protein inactivation	68
Bromophenacylbromide	Phospholipase inhibitor	69
Quinacrine	Membrane perturbation	70
Corticosteroids	Inhibition of Ph.lipase A2	71
Non-ster. antiinfl. agents	Various	72,73
Prostaglandins (E2,D2)	cAMP increase	74,75
Adenosine	cAMP increase	76,77
Nifedipine, Verapamil	Calcium antagonists	78-80
Trifluoperazine	Calmodulin inhibitor	81,82
TPCK, DFP	Protease inhibitors	83,84
Sphinganine, H-7, C-I, etc.	Prot. kinase C inhibitors	85
Nordihydroguaiaretic acid	Lipooxygenase inhibitor	86
Disuccinimidyl suberate	Crosslinking reagent	87
Opioids, benzodiazepines	?	88,89
Anaesthetics (halotane, lidocaine)	?	90,91
Bee venom melittin	?	92
----

receptor. When the cells come in contact with low concentrations of a stimulant they do not activate the burst,
on the contrary they become unresponsive to a second challenge
with the same stimulant. This is an important mechanism that
inhibits the triggering of the burst in phagocytes that are
exposed to a gradient of chemotactic agents, that is during
their movement from vessels to the centre of the inflammatory
site.

Some tumors produce inhibitory factors, whose nature has
to be determined. Recent data suggest that one of these factors
may be transforming-growth factor ß.[93] This mechanism could
protect the tumor cells from the oxidative attack by phagocytes
and therefore could allow them to escape host defence systems.

The effect of platelet-derived growth factor (PDGF) may
have physiological relevance. PDGF inhibits the respiratory
burst at concentrations that are present in serum during the
hemostatic process. PDGF does not inhibit phagocytosis and
chemotaxis. This factor may therefore play an important
regulatory role during hemostasis and wound healing, because it
prevents unsuitable activation of the burst while it does not
affect the scavenger function of these cells.

Phospholipase inhibitors are important tools for
investigating the role of phospholipid hydrolysis in the
activation mechanism, but their specificity is not well
established. Powerful inhibitors such as bromophenacyl bromide
are too toxic for use in vivo. On the other hand,
corticosteroids are poor inhibitors of the respiratory burst,
probably because they do not influence phospholipase C
activity.

A second possibility for down-regulating the respiratory
burst is the inhibition of the NADPH oxidase. The list of
inhibitors is reported in table 3. Most of these agents have
interest for research purposes only. They have been useful for
exploring the participation of individual components in
the catalysis.

A recent advance in the knowledge of the nature of the
oxidase has been provided by the production of antibodies that
inhibit the enzymatic activity. By this way proteins, with
molecular weight of 65 kDa,[105] 70 kDa[106] and a heterodimer of
16/18 and 14 kDa,[107] that participate in the activity of the
oxidase have been identified.

Practical application could have vitamin E, gold salts
(that in rheumatoid arthritis are used as antiinflammatory
agents) and possibly diphenylene iodonium. Imidazole is an

Table 3. Agents tat inhibit the terminal oxidase of the respiratory burst

Agent	Possible mechanism	Ref.
Cibacron blue	NADPH analogue	94
%-carba-deaza FAD	Flavin analogue	95
Diphenylene iodonium	Flavoprotein inhibitor	96
Pyridine, imidazole	Cytochrome b inhibitor	97
Quinones, vitamin E	Interference with electron transport	98-100
EDTA	Ca^{2+} and Mg^{2+} chelation	101
Batophenanthroline sulfonate	Fe^{3+} chelation	5,102
P-chloromercuribenzoate	Sulfhydryl reagent	103,104
Antibodies against proteins of 65-70, 18, 14 kDa	Binding to oxidase	105-107
Strong detergents and salts	Dissociation of complex	108
H_2O_2 + peroxidase + halide	Oxidative inactivation	109
Heat shock	?	110
Gold salts	?	111

inhibitor of NADPH oxidase, but at too high concentration for to be used in vivo.

III.b. Enhancement of the respiratory burst

The response of the phagocyte to a stimulant may be potentiated essentially according to two mechanisms: one is the priming effect, the other is the activation by cytokines. These are physiological phenomena that serve for the enhancement of resistance to infection and there is the hope that in near future they may be utilized also for the pharmacological treatment of immunocompromised host. The priming effect is observed when the cells are exposed either to chemoattractants or to several other compounds (see table V-A) and become more responsive to a second different stimulant. The precise modification that is responsible for the priming is not clear. Theeffect takes place very rapidly in treated cells, but it is not permanent and the increased responsiveness disappears after a few minutes.

These features distinguish priming from the up-regulation of the burst induced by cytokines. This effect was initially thought to be a property of mononuclear phagocytes only but it has been recently described also in neutrophils.[125-127] The activation requires several hours of treatment of the cells and consists in a permanent modification of the responsiveness that probably involves new gene expression.

The cytokines that are able of augment the respiratory burst are interferon-ɣ, granulocyte-macrophage colony stimulating factor, tumor necrosis factor and interleukin 1.

Table 4. Agents that potentiate the respiratory burst

A. Priming effect

Chemotactic factors	(112,113)	Lypopolysaccharide	(114,115)
Phorbol esters	(116)	Leukotriene B4	(117)
Diacylglycerol	(118,119)	Platelet activ. factor	(120)
Concanavalin A	(57)	ATP	(121,122)
Muramyl peptide	(114)		

B. Cytokines

Interleukin-1	(123)	Interferon-γ	(124-127)
GM-Colony stim. factor	(128)	Tumor necrosis factor	(129)

The mechanism of the potentiating effect of cytokines is under active investigation. It has been shown that interferon-γ increases membrane receptors for immunoglobulins,[130] increases gene expression for several proteins including cytochrome b_{558}[131,132] and induces a shift of the oxidase from a form with low affinity for NADPH to a form with high affinity.[133] The availability of human recombinant cytokines has given new support to these studies. Interferon-γ has been already employed in vivo with promising results. This cytokine has been shown to increase H_2O_2 production by phagocytes in patients with tumors[134] and to improve oxidative metabolism of phagocytes in a variant of CGD.[135,136] The study of the effects in vivo and in vitro of cytokines will be one of the most important research fields for leukocytologists in near future.

ACKNOWLEDGEMENTS
The work was supported by grant from C.N.R. n. 87.01513.04

REFERENCES

1. S. J. Klebanoff and R. A. Clark, "The Neutrophil: Function and Clinical Disorders", North Holland Publ. Comp., Amsterdam (1978).
2. B. M. Babior, R. S. Kipnes and J. T. Curnutte, Biological defense mechanisms. The production by leukocytes of superoxide, a potential bactericidal agent, J. Clin. Invest. 52: 7414 (1973).
3. F. Rossi, The O_2^- forming NADPH oxidase of the phagocytes: nature, mechanisms of activation and function, Biochim. Biophys. Acta 853: 65 (1986).
4. P. Bellavite, O. T. G. Jones, A.R. Cross, E. Papini and F. Rossi, Composition of partially purified NADPH oxidase from pig neutrophils, Biochem. J. 223: 639 (1984).
5. G. Berton, E. Papini, M. Cassatella, P. Bellavite and F. Rossi, Partial purification of the superoxide-generating

system of macrophages. Possible association of the NADPH oxidase activity with a low potential cytochrome b, <u>Biochim. Biophys. Acta</u> 810: 164 (1985).

6. J. Doussiere and P. V. Vignais, Purification and properties of O_2^--generating oxidase from bovine polymorphonuclear neutrophils, <u>Biochemistry</u> 24: 7231 (1985).

7. G. A. Glass, D. M. DeLisle, P. DeTogni, T. G. Gabig, B. H. Magee, M. Markert and B. M. Babior, The respiratory burst oxidase of human neutrophils. Further studies of the purified enzyme, <u>J. Biol. Chem.</u> 261: 13247 (1986).

8. T. G. Gabig and B. M. Babior, The O_2^- forming oxidase responsible for the respiratory burst in human neutrophils. Properties of the solubilized enzyme, <u>J. Biol. Chem.</u> 254: 9070 (1979).

9. M. L. Karnovsky, Comparative aspects of the production of oxygen radicals by phagocytic cells, and aspects of other effector substances, <u>Int. J. Tiss. Reac.</u> 8: 91 (1986).

10. R. B.Jr. Johnston, Oxygen metabolism and the microbicidal activity of macrophages, <u>Fed. Proc.</u> 37: 2759 (1978).

11. C. F. Nathan, Regulation of macrophage oxidative metabolism and parasitic activity, <u>In</u>:"Mononuclear Phagocytes. Characteristics, Physiology and Function", R. Van Furth ed., Martinus Niijhoff Publishers, Dordrecht, 411 (1985).

12. P. M. Henson and Z. G. Oades, Stimulation of human neutrophils by soluble and insoluble immunoglobulin aggregates. Secretion of granule constituents and increased oxidation of glucose, <u>J. Clin. Invest.</u> 56: 1053 (1975).

13. K. Kakinuma, Effects of fatty acids on the oxidative metabolism of leukocytes, <u>Biochim. Biophys. Acta</u> 348: 76 (1974).

14. T. Yamaguchi, M. Kaneda and K. Kakinuma, Effect of saturated and unsaturated fatty acids on the oxidative metabolism of human neutrophils. The role of calcium ion in the extracellular medium, <u>Biochim. Biophys. Acta</u> 861: 440 (1986).

15. H. Sumimoto, K. Takeshige and S. Minakami, Superoxide production of human polymorphonuclear leukocytes stimulated by leukotriene B4, <u>Biochim. Biophys. Acta</u> 803: 271 (1984).

16. I. Fujita, K. Irita, K. Takeshige and S. Minakami, Diacylglycerol, 1-oleoyl-2-acetyl-glycerol, stimulates superoxide generation from human neutrophils, <u>Biochem. Biophys. Res. Commun.</u> 120: 318 (1984).

17. H. P.Hartung, M.J. Parnham, J. Winkelman, W. Englberger and U. Hadding, Platelet activating factor (PAF) induces the oxidative burst in macrophages, <u>Int. J. Immunopharmacol.</u> 5: 115 (1983).

18. H. P. Hartung, R. G. Kladetzky, B. Melnik and M. Hennerici, Stimulation of the scavenger receptor on monocytes-macrophages evokes release of arachidonic acid

metabolites and reduced oxygen species, _Lab. Invest._ 55: 209 (1986).

19. T. Chiba, Y. Nagai and K. Kakinuma, Cerebroside sulfuric ester (sulfatide) induces oxygen radical generation in guinea-pig leukocytes, _Biochem. Biophys. Acta_ 930: 10 (1987).

20. D. Romeo, G. Zabucchi and F. Rossi, Reversible metabolic stimulation of polymorphonuclear leukocytes and macrophages by concanavalin A, _Nature_ 243: 111 (1973).

21. F. Rossi, M. Zatti, P. Patriarca and R. Cramer, Stimulation of the respiration of polymorphonuclear leucocytes by antileucocyte antibodies, _Experientia_ 26: 491 (1970).

22. G. Berton, H. Rosen, R. A. B. Ezekowitz, P. Bellavite, M. C. Serra, F. Rossi and S. Gordon, Monoclonal antibodies to a particulate superoxide-forming system stimulate a respiratory burst in intact guinea pig neutrophils, _Proc. Natl. Acad. Sci. USA_ 83: 4002 (1986).

23. F. Tedesco, S. Trani, M.R. Soranzo and P. Patriarca, Stimulation of glucose oxidation in human polymorphonuclear leucocytes by C3-sepharose and soluble C567, _FEBS Lett._ 51: 232 (1975).

24. R. Gennaro, T. Pozzan and D. Romeo, Monitoring of cytosolic free Ca^{2+} in C5a-stimulated neutrophils: Loss of receptor-modulated Ca^{2+} stores and Ca^{2+} uptake in granule-free cytoplasts, _Proc. Natl. Acad. Sci. USA_ 81: 1416 (1984).

25. M. Tsujimoto, S. Yokota, J. Vilcek and G. Weissmann, Tumor necrosis factor provokes superoxide anion generation from neutrophils, _Biochem. Biophys. Res. Commun._ 137: 1094 (1986).

26. P. Patriarca, M. Zatti, R. Cramer and F. Rossi, Stimulation of the respiration of polymorphonuclear leucocytes by phospholipase C, _Life Sci._ 9: 841 (1970).

27. L. C. McPhail and R. Snyderman, Activation of the respiratory burst enzyme in human polymorphonuclear leukocytes by chemoattractants and othe soluble stimuli. Evidence that the same oxidase is activated by different transductional mechanisms, _J. Clin. Invest._ 72: 192 (1983).

28. M. C. Serra, F. Bazzoni, V. DellaBianca, M. Grzeskowiak and F. Rossi, Activation of human neutrophiuls by substance P: Effect on oxidative metabolism, exocytosis, cytosolic Ca^{2+} concentration and inositol phosphates formation, _J. Immunology_ 141: 2118 (1988).

29. E. L. Becker, M. Sigman and J. M. Oliver, Superoxide production induced in rabbit polymorphonuclear leukocytes by synthetic chemotactic peptides and A23187, _Am. J. Pathol._ 95: 81 (1979).

30. C. Salerno, Urate crystal-induced superoxide radical production by human neutrophils, _Adv. Exp. Med. Biol._ 165: 184 (1984).

31. J. T. Curnutte, B. M. Babior and M. L. Karnovsky, Fluoride-mediated activation of the respiratory burst in human neutrophils. A reversible process, J. Clin. Invest. 63: 637 (1979).

32. F. Rossi, V. Della Bianca and A. Davoli, A new way for inducing a respiratory burst in guinea pig neutrophils. Change in the Na^+, K^+ concentration of the medium, FEBS Lett. 132: 273 (1981).

33. R. C. Graham, M. J. Karnovsky, A. W. Shafer, E. A. Glass and M. L. Karnovsky, Metabolic and morphological observations on the effect of surface-active agents on leukocytes, J. Cell. Biol. 32: 629 (1967).

34. F. Rossi and M. Zatti, Mechanism of the respiratory stimulation in saponine-treated leukocytes. The KCN insensitive oxidation of NADPH, Biochim. Biophys. Acta 153: 296 (1968).

35. A. Aviram and I. Aviram, Activation of guinea-pig and bovine neutrophil NADPH oxidase by N,N'-dicyclohexylcarbodiimide, Biochim. Biophys. Acta 844: 224 (1985).

36. L. C. McPhail and R. Snyderman Mechanisms of regulating the respiratory burst in leukocytes, In:" Regulation of Leukocyte Function" R. Snyderman, ed., Plenum Press, New York 247 (1984).

37. A. I. Tauber, Protein kinase C and the activation of the human neutrophil NADPH-oxidase, Blood 69: 711 (1987).

38. P. Bellavite, The superoxide-forming enzymatic system of phagocytes, Free Radical Biol. Med. 4: 225 (1988).

39. R. R. Sandborg and J. E. Smolen, Biology of disease. Early biochemical events in leukocyte activation, Lab. Invest. 59: 300 (1988).

40. A.W. Segal, P.G. Heyworth, S. Cockroft and M.M. Barrowman, Stimulated neutrophils from patients with autosomal recessive chronic granulomatous disease fail to phosphorylate a Mr-44,000 protein, Nature 316: 547 (1985).

41. T. Hayakawa, K. Suzuki, S. Suzuki, P. C. Andrews and B. M. Babior, A possible role for protein phosphorylation in the activation of the respiratory burst in human neutrophils. Evidence for studies with cells from patients with chronic granulomatous disease, J. Biol. Chem. 261: 9109 (1986).

42. I. M. Kramer, A. J. Verhoeven, R. vanderBend, R.S. Weening and D. Roos, Purified protein kinase C phosphorylates a 47-kDa Protein in control neutrophil cytoplasts but not in neutrophil cytoplasts from patients with the autosomal form of chronic granulomatous disease, J. Biol. Chem. 263: 2352 (1988).

43. R.C. Garcia and A.W. Segal, Phosphorylation of the subunits of cytochrome b-245 upon triggering of the respiratory burst of human neutrophils and macrophages, Biochem. J. 252: 901 (1988).

44. N. Okamura, J. T. Curnutte, R. L. Roberts and B. M. Babior,

Relationship of protein phosphorylation to the activation of the respiratory burst in human neutrophils. Defects of the phosphorylation of a group of closely related 48-kDa proteins in two forms of chronic granulomatous disease, J. Biol. Chem. 263: 6777 (1988).

45. E. Papini, M. Grzeskowiak, P. Bellavite and F. Rossi, Protein kinase C phosphorylates a component of NADPH oxidase of neutrophils, FEBS Lett. 190: 204 (1985).

46. P. Bellavite, S. Dusi and M. A. Cassatella, Studies on the nature and activation of O_2^--forming NADPH oxidase of leukocytes II. Relationships between phosphorylation of a component of the enzyme and oxidase activity, Free Rad. Res. Commun. 4: 83 (1987).

47. Y. Bromberg and E. Pick, Unsaturated fatty acids stimulate NADPH-dependent superoxide production by cell-free system derived from macrophages, Cell. Immunol. 88: 213 (1984).

48. R. A. Heyneman and R. E. Vercauteren, Activation of a NADPH oxidase from horse polymorphonuclear leukocytes in a cell-free system, J. Leukocyte Biol. 36: 751 (1984).

49. I. Maridonneau-Parini and A. I. Tauber, Activation of NADPH oxidase by arachidonic acid involves phospholipase A2 in intact human neutrophils but not in the cell-free system, Biochem. Biophys. Res. Commun. 138: 1099 (1986).

50. J. T. Curnutte, Activation of human neutrophil nicotinamide adenine dinucleotide phosphate, reduced (triphosphopyridine nucleotide, reduced) oxidase by arachidonic acid in a cell-free system, J. Clin. Invest. 75: 1740 (1985).

51. L. C. McPhail, P. S. Shirley, C. C. Clayton and R. Snyderman, Activation of the respiratory burst enzyme from human neutrophils ina cell-free system. Evidence for a soluble cofactor, J. Clin. Invest. 75: 1735 (1985).

52. R. A. Clark, K. G. Leidal, D. W. Pearson and W. M. Nauseef, NADPH oxidase of human neutrophils: Subcellular localization and characterization of an arachidonate-activatable superoxide-generating system, J. Biol. Chem. 262: 4065 (1987).

53. R. Seifert, W. Rosenthal and G. Schultz, Guanine nucleotides stimulate NADPH oxidase in membranes of human neutrophils, FEBS Lett. 205: 161 (1986).

54. P. Bellavite, F. Corso, S. Dusi, M. Grzeskowiak, V. Della Bianca and F. Rossi, Activation of a NADPH-dependent superoxide production in plasma-membrane extracts of pig neutrophils by phosphatidic acid, J. Biol. Chem. 263: 8210 (1988).

55. E. Ligeti, J. Doussiere and P. V. Vignais, Activation of the O_2^--generating oxidase in plasma membrane from bovine polymorphonuclear neutrophils by arachidonic acid, a cytosolic factor of protein nature, and nonhydrolyzable

analogues of GTP, Biochemistry 27: 193 (1988).

56. I. Fujita, K. Takeshige and S. Minakami, Characterization of the NADPH-dependent superoxide production activated by sodium dodecyl sulfate in a cell-free system of pig neutrophils, Biochim. Biophys. Acta 931: 41 (1987).

57. F. Rossi, M. Grzeskowiak and V. Della Bianca, Double stimulation with FMLP and Con A restores the activation of the respiratory burst but not of the phosphoinositide turnover in Ca^{2+}-depleted human neutrophils. A further example of dissociation between stimulation of the NADPH oxidase and phosphoinositide turnover, Biochem. Biophys. Res. Commun. 140: 1 (1986).

58. M. Grzeskowiak, V. Della Bianca, M. Cassatella and F. Rossi, Complete dissociation between the activation of phosphoinositide turnover and of NADPH oxidase by Formyl-methionyl-leucyl-phenylalanine in human neutrophils depleted of Ca^{2+} and primed by subtreshold doses of phorbol myristate acetate, Biochem. Biophys. Res. Commun. 135: 785 (1986).

59. F. Rossi, V. Della Bianca, M. Grzeskowiak and F. Bazzoni, Studies on molecular regulation of phagocytosis in neutrophils. Con A-mediated ingestion and associated respiratory burst independent of phosphoinositide turnover, rise in Ca^{2+} i and arachidonic acid release. J. Immunol. (in press)

60. J. A. Badwey, J. T. Curnutte, J. M. Robinson, C. B. Berde, M. J. Karnovsky and M. L. Karnovsky, Effects of free fatty acids on release of superoxide and on change of shape by human neutrophils. Reversibility by albumin, J. Biol. Chem. 259: 7870 (1984).

61. P. De Togni, P. Bellavite, V. Della Bianca, M. Grzeskowiak and F. Rossi, Intensity and kinetics of the respiratory burst of human neutrophils in relation to receptor occupancy and rate of occupation by formyl-methionyl-leucyl-phenylalanine, Biochim. Biophys. Acta 838: 12 (1985).

62. L. Simkowitz, J. P. Atkinson and I. Spilberg, Stimulus-specific deactivation of chemotactic factor-induced cyclic AMP response and superoxide generation by human neutrophils, J. Clin. Invest. 66: 736 (1980).

63. L. A. Sklar, A. J. Jesaitis, R. G. Painter and C. G. Cochrane, The kinetics of neutrophil activation. The response to chemotactic peptides depends upon whether ligand-receptor interaction is rate-limiting, J. Biol. Chem. 256: 9909 (1981).

64. G. Berton and S. Gordon, Desensitization of macrophages to stimuli which induce secretion of superoxide anion. Down-regulation of receptors for phorbol myristate acetate, Eur. J. Immunol. 13: 620 (1983).

65. E. A. Valletta and G. Berton, Desensitization of macrophage oxygen metabolism on immobilized ligand: different effect of immunoglobulin G and complement. J. Immunology 138: 4366 (1987).

66. S. Tsunawaki and C. F. Nathan, Macrophage deactivation. Altered kinetic properties of superoxide-producing enzyme after exposure to tumor cell-conditioned medium, J. Exp. Med. 164: 1319 (1986).

67. E. Wilson, S. M. Laster, L. R. Gooding and J. D. Lambeth, Platelet-derived growth factor stimulates phagocytosis and blocks agonist-induced activation of the neutrophil oxidative burst: a possible cellular mechanism to protect against oxygen radical damage, Proc. Natl. Acad. Sci. U.S.A. 84: 2213 (1987).

68. M. W. Verghese, C. D. Smith, L. A. Charles, L. Jakoi and R. Snyderman, A guanine nucleotide regulatory protein controls polyphosphoinositide metabolism, Ca^{2+} mobilization, and cellular responses to chemoattractants in human monocytes, J. Immunol. 137: 271 (1986).

69. R. E. Duque, Inhibition of neutrophil activation by p-bromophenacyl bromide and its effects on phospholipase A2, Br. J. Pharmacol. 88: 463 (1986).

70. A. I. Tauber and E. R. Simons, Dissociation of human neutrophil membrane depolarization, respiratory burst stimulation and phospholipid metabolism by quinacrine, FEBS Lett. 156: 161 (1983).

71. I. M. Goldstein, D. Roos, G. Weissmann and H. B. Kaplan, Influence of corticosteroids on human polymorphonuclear leukocyte function in vitro: reduction of lysosomal enzyme release and superoxide production, Inflammation 1: 305 (1976).

72. J. C. Gay, J. N. Lukens and D. K. English, Differential inhibition of neutrophil superoxide generation by nonsteroidal antiinflammatory drugs, Inflammation 8: 209 (1984).

73. J. O. Minta and M. D. Williams, Interactions of antirheumatic drugs with the superoxide generation system of activated human polymorphonuclear leukocytes, J. Rheumatol. 13: 498 (1986).

74. R. J. Gryglewsky, A. Szczeklik and M. Wandzilak, The effect of six prostaglandins, prostacyclin and iloprost on generation of superoxide anions by human polymorphonuclear leukocytes stimulated by zymosan or formyl-methionyl-leucyl-phenylalanine, Biochem. Pharmacol. 36: 4209 (1987).

75. P. DeTogni, G. Cabrini and F. Di Virgilio, Cyclic AMP inhibition of fMet-Leu-Phe-dependent metabolic responses in human neutrophils is not due to its effects on cytosolic Ca^{2+}, Biochem. J. 224: 629 (1984).

76. B. N. Cronstein, S. B. Kramer, G. Weissmann and R. Hirshhorn, Adenosine: A physiological modulator of superoxide anion generation by human neutrophils, J. Exp. Med. 158: 1160 (1983).

77. P. A. Ward, T. W. Cunningham, K. K. McCulloch and K. J.

Johnson, Regulatory effects of adenosine and adenine nucleotides on oxygen radical responses of neutrophils, Lab. Invest. 58: 438 (1988).

78. T. Matsumoto, K. Takeshige and S. Minakami, Inhibition of phagocytotic metabolic changes of leukocytes by an intracellular calcium antagonist 8-(N,N-diethylamine)-octyl-3,4,5-trimethoxybenzoate, Biochem. Biophys. Res. Commun, 88: 974 (1979).

79. V. Della Bianca, M. Grzeskowiak, P. DeTogni, M. Cassatella, and F. Rossi, Inhibition by verapamil of neutrophil responses to formyl-methionyl-leucyl- phenylalanine and phorbol myristate acetate. Mechanism involving Ca^{2+} changes, cAMP and protein kinase C, Biochim. Biopys. Acta 845: 223 (1985).

80. Y. Azuma, T. Tokunaga, Y. Takeda, T. Ogawa and N. Takagi, The effect of calcium antagonists on the activation of guinea pig neutrophils, Japan J. Pharmacol. 42: 243 (1986).

81. J. M. Robinson, J. A. Badwey, M. L. Karnovsky and M. J. Karnovsky, Release of superoxide and change in morphology by neutrophils in response to phorbol esters: Antagonism by inhibitors of calcium-binding proteins, J. Cell. Biol. 101: 1052 (1985).

82. K. Takeshige and S. Minakami, Involvement of calmodulin in phagocytotic respiratory burst of leukocytes, Biochem. Biophys. Res. Commun. 99: 484 (1981).

83. B. D. Goldstein, G. Witz, M. Amoruso and W. Troll, Protease inhibitors antagonize the activation of polymorphonuclear leukocyte oxygen consumption, Biochem. Biophys. Res. Commun. 88: 854 (1979).

84. K. M. K. Rao and V. Castranova, Phenylmethyl sulphonyl fluoride (PMSF) inhibits chemotactic peptide-induced actin polymerization and respiratory burst activity in human neutrophils, Fed. Proc. 46: Abs. 3907 (1987).

85. E. Wilson, M. C. Olcott, R. M. Bell, A. H. Merrill and D. J. Lambeth, Inhibition of the oxidative burst in human neutrophils by sphingoid long-chain bases. Role of protein kinase C in activation of the burst, J. Biol. Chem. 261: 12616 (1986).

86. F. Rossi, V. DellaBianca and P. Bellavite Inhibition of the respiratory burst and of phagocytosis by nordihydroguaiaretic acid in neutrophils. FEBS Lett. 127: 183 (1981).

87. I. Aviram and Y. I. Henis, Activation of human neutrophil NADPH oxidase and lateral mobility of membrane proteins, Biochim. Biophys. Acta 805: 227 (1984).

88. G. Goldfarb, J. Belghiti, H. Gautero and P. Boivin, In vitro effect of benzodiazepines on polymorphonuclear leukocyte oxidative activity. Anesthesiol. 60: 57 (1984).

89. P. K. Peterson, B. Sharp, G. Gekker, C. Brummit and W. F.

Keane, Opioid-mediated suppression of cultured peripheral blood mononuclear cell respiratory burst activity, _J. Immunol._ 138: 3907 (1987).

90. J. W. C. White, A. W. Gelb, H. R. Wexler, C. R. Stiller and P. A. Keown, The effects of intravenous anaesthetic agents on human neutrophil chemiluminescence, _Can. Anaesth. Soc. J._ 30: 506 (1983).

91. M. Nakagawara, K. Takeshige, J. Takamatsu, S. Takahasi, J. Yoshitake and S. Minakami, Inhibition of superoxide production and Ca2+ mobilization in human neutrophils by halotane, enflurane, and isoflurane, _Anesthesiol._ 64: 4 (1986).

92. S. D. Somerfield, G. L. Stach, C. Mraz, F. Gervais and E. Skamene, Bee venom melittin blocks neutrophil O2-production, _Inflammation_ 10: 175 (1986).

93. S. Tsunawaki, M. Sporn, A. Ding and C. Nathan, Deactivation of macrophages by transforming growth factor-beta, _Nature_ 334: 260 (1988).

94. T. Yamaguchi and K. Kakinuma, Inhibitory effect of cibacron blue F3GA on the O_2^- generating enzyme of guinea pig polymorphonuclear leukocytes, _Biochem. Biophys. Res. Commun._ 104: 200 (1982).

95. D. R. Light, C. Walsh, A. M. O'Callaghan, E. J. Goetzl and A. I. Tauber, Characteristics of the cofactor requirements for the superoxide-generating NADPH oxidase of human polymorphonuclear leukocytes, _Biochemistry_ 20: 1468 (1981).

96. A. R. Cross and O. T. G. Jones, The effect of the inhibitor diphenylene iodonium on the superoxide-generating system of neutrophils. Specific labelling of a component polypeptide of the oxidase, _Biochem. J._ 237: 111 (1986).

97. T. Iizuka, S. Kanegasaki, R. Makino, T. Tanaka and Y. Ishimura, Pyridine and imidazole reversibly inhibit the respiratory burst in porcine and human neutrophils: evidence for the involvement of cytochrome b558 in the reaction, _Biochem. Biophys. Res. Commun._ 130: 621 (1985).

98. F. Rossi and G. Zoppi, Effect of menadione on the phagocytic activity of guinea pig polymorphonuclear leukocytes, _Experientia_ 22: 433 (1966).

99. C. J. Butterick, R. L. Baehner, L. A. Boxer and R. A. Jersild, Vitamin E-a selective inhibitor of the NADPH oxidoreductase enzyme system in human granulocytes. _Am. J. Pathol._ 112: 287-293 (1983).

100. D. R. Crawford and D. L. Schneider, Evidence that a quinone may be required for the production of superoxide and hydrogen peroxide in neutrophils, _Biochem. Biophys. Res. Commun._ 99: 1277 (1981).

101. T. Yamaguchi, M. Kaneda and K. Kakinuma, Essential requirement of magnesium ion for optimal activity of the NADPH oxidase of guinea pig polymorphonuclear leukocytes, _Biochem. Biophys. Res. Commun._ 115: 261 (1983).

102. P. Bellavite, M. C. Serra, A. Davoli, J. V. Bannister and F. Rossi, The NADPH oxidase of guinea pig polymorphonuclear leucocytes. Properties of the deoxycholate extracted enzyme, Molec. Cell. Biochem. 52: 17 (1983).

103. A. I. Tauber and E. J. Goetzl, Structural and catalytic properties of the solubilized superoxide-generating activity of human polymorphonuclear leukocytes. Solubilization, stabilization in solution and partial characterization, Biochemistry 18: 5576 (1979).

104. G. L. Babior, R. E. Rosin, B. J. McMurrich, W. A. Peters and B.M. Babior, Arrangement of the respiratory burst oxidase in the plasma membrane of the neutrophil, J. Clin. Invest. 67: 1724 (1981).

105. J. Doussiere and P. V. Vignais, Immunological properties of O_2^- generating oxidase from bovine neutrophils, FEBS Lett. 234: 362 (1988).

106. Y. Fukuhara, Y. Ise and K. Kakinuma, Immunological studies on the respiratory burst oxidase of pig neutrophils, FEBS Lett. 229: 150 (1988).

107. G. Berton, S. Dusi, M. C. Serra, P. Bellavite and F. Rossi, Studies on the NADPH oxidase of phagocytes. Production of a monoclonal antibody which blocks the enzymatic activity of pig neutrophils NADPH oxidase, J. Biol. Chem. (submitted for publication).

108. P. Patriarca, R. E. Basford, R. Cramer, P. Dri and F. Rossi, Studies on the NADPH oxidizing activity in polymorphonuclear leukocytes: The mode of association with the granule membrane, the relationship to myelo peroxidase and the interference of hemoglobin with NADPH oxidase determination, Biochim. Biophys. Acta 362: 221 (1974).

109. R. C. Jandl, J. Andre'-Schwartz, L. Borges-Dubois, R. S. Kipnes, B. J. McMurrich and B. M. Babior, Termination of the respiratory burst in human neutrophils . J. Clin. Invest. 61: 1176 (1978).

110. I. Maridonneau-Parini, J. Clerk and B. S. Polla, Heat shock inhibits NADPH oxidase in human neutrophils, Biochem. Biophys. Res. Commun. 154: 179 (1988).

111. M. Miyahara, Watanabe, E. Okimasu and K. Utsumi, Charge-dependent regulation of NADPH oxidase activity in guinea-pig polymorphonuclear leukocytes. Biochem. Biophys. Acta 929: 253 (1987).

112. D. E. VanEpps and M. L. Garcia, Enhancement of neutrophil function as a result of prior exposure to chemotactic factor. J. Clin. Invest. 66: 167 (1980).

113. D. English, J. S. Roloff and J. M. Lukens, Chemotactic factor enhancement of superoxide release from fluoride and phorbol myristate acetate stimulated neutrophils. Blood 58: 129 (1981).

114. M. Kaku, K. Yagawa, S. Nagao and A. Tanaka, Enhanced superoxide anion release from phagocytes by muramyl

dipeptide or lipopolysaccharide, <u>Infect. Immun.</u> 39:
559 (1983).

115. L. A. Guthrie, L. C. McPhail, P. M. Henson and R. B.
     Johnston, Priming of neutrophils for enhanced release
     of oxygen metabolites by bacterial lipopolysaccharide.
     Evidence for increased activity of the superoxide-
     producing enzyme, <u>J. Exp. Med.</u> 160: 1657 (1984).

116. L. C. McPhail, C. C. Clayton and R. Snyderman, The NADPH
     oxidase of human polymorphonuclear leukocytes. Evidence
     for regulation by multiple signals, <u>J. Biol. Chem.</u> 259:
     5768 (1984).

117. J. C. Gay, J. K. Beckman, A. R. Brash, J. A. Oates and J.
     N, Lukens Enhancement of chemotactic factor-stimulated
     neutrophil oxidative metabolism by leukotriene B4,
     <u>Blood</u> 64: 780 (1984).

118. B. Dewald, T.G. Payne and M. Baggiolini, Activation of
     NADPH oxidase of human neutrophils. Potentiation of
     chemotactic peptide by a diacylglycerol, <u>Biochem.</u>
     <u>Biophys. Res. Commun.</u> 125: 367 (1984).

119. R. J. Smith, L. M. Sam and J. M. Justen, Diacylglycerols
     modulate human polymorphonuclear neutrophil
     responsiveness: Effects on intracellular calcium
     mobilization, granule exocytosis, and superoxide anion
     production, <u>J. Leukoc. Biol.</u> 43: 411 (1988).

120. B. Dewald and M. Baggiolini, Activation of NADPH oxidase in
     human neutrophils. Synergism between fMLP and the
     neutrophil products PAF and LTB4, <u>Biochem. Biophys. Res.</u>
     <u>Commun.</u> 128: 297 (1985).

121. P. A. Ward, T. W. Cunningham, B. A. M. Walker and K. J.
     Johnson, Differing calcium requirements for regulatory
     effects of ATP, ATPgammaS and adenosine on O_2^-
     responses of human neutrophils, <u>Biochem. Biophys. Res.</u>
     <u>Commun.</u> 154: 746 (1988).

122. D. B. Kuhns, D. G. Wright, J. Nath, S. Kaplan and R. E.
     Basford, ATP induces transient elevations of $(Ca^{2+})i$ in
     human neutrophils and primes these cells for enhances
     O_2^- generation, <u>Lab. Invest.</u> 58: 448 (1988).

123. Y. Ozaki, T. Ohashi and S. Kume, Potentiation of neutrophil
     function by recombinant DNA-produced interleukin 1a. <u>J.</u>
     <u>Leukoc. Biol.</u> 42: 621 (1987).

124. C. F. Nathan, H. W. Murray, M. E. Wiebe and B. Y. Rubin,
     Identification of interferon gamma as the lymphokine
     that activates human macrophages oxidative metabolism
     and antimicrobial activity, <u>J. Exp. Med.</u> 158: 670 (1983).

125. G. Berton, L. Zeni, M. A. Cassatella and F. Rossi, Gamma
     interferon is able to enhance the oxidative metabolism
     of human neutrophils, <u>Biochem. Biophys. Res. Commun.</u>
     138: 1276 (1986).

126. B. Perussia, M. Kobayashi, M. E. Rossi, I. Anegon and G.
     Trinchieri, Immune interferon enhances functional
     properties of human granulocytes: role of Fc receptors

and effect of lymphotoxin, tumor necrosis factor, and granulocyte-macrophage colony-stimulating factor, J. Immunol. 138: 765 (1987).

127. I. C. Kowanko and A. Ferrante, Stimulation of neutrophil respiratory burst and lysosomal enzyme release by human interferon-gamma, Immunology 62: 149 (1987).

128. R. H. Weisbart, L. Kwan, D. W. Golde and J.C. Gasson, Human GM-CSF Primes neutrophils for enhanced oxidative metabolism in response to the major physiological chemoattractants, Blood 69: 18 (1987).

129. A. Ferrante, M. Nandoskar, E. J. Bates, D. H. B. Goh and L. J. Beard Tumor necrosis factor beta (lymphotoxin) inhibits locomotion and stimulates the respiratory burst and degranulation of neutrophils, Immunology 63: 507 (1988).

130. B. Perussia, E. T. Dayton, R. Lazarus, V. Fanning and G. Trinchieri, Immune interferon induces the receptor for monomeric IgG1 on human monocytic and myeloid cells, J. Exp. Med. 158: 1092 (1983).

131. P. W. Andrew, A. K. Robertson, D. B. Lowrie, A. R. Cross and O. T. G. Jones, Induction of synthesis of components of the hydrogen peroxide generating oxidase during activation of the human monocytic cell line U937 by interferon-gamma, Biochem. J. 248: 281 (1987).

132. P. E. Newburger, R. A. B. Ezekowitz, C. Whitney, J. Wright and S. H. Orkin, Induction of phagocyte cytochrome b heavy chain gene expression by interferon gamma, Proc. Natl. Acad. Sci. U.S.A.: 85, 5215 (1988).

133. G. Berton, M. Cassatella, G. Cabrini and F. Rossi, Activation of mouse macrophages causes no change in expression and function of phorbol diester receptors, but is accompanied by alterations in the activity and kinetic parameters of NADPH oxidase, Immunology 54: 371 (1985).

134. C. F. Nathan, C. R. Horowitz, J. DeLaHarpe, S. Vadhan-Raj, S. A. Sherwin, H. F. Oettgen and S. Krown, Administration of recombinant interferon gamma to cancer patients enhances monocyte secretion of hydrogen peroxide, Proc. Natl. Acad. Sci. U.S.A. 82: 8686 (1985).

135. R. A. B. Ezekowitz, S. H. Orkin and P. E. Newburger, Recombinant interferon gamma augments superoxide production and X-chronic granulomatous disease gene expression in X-linked variant chronic granulomatous disease, J. Clin. Invest. 80: 1009 (1987).

136. J. M. G. Sechler, H. R. Malech, C. J. White and J. I. Gallin, Recombinant human interferon-gamma reconstitutes defective phagocyte function in patients with chronic granulomatous disease of childhood, Proc. Natl. Acad. Sci. U.S.A. 85: 4874 (1988).

TWO STRUCTURALLY UNRELATED PAF ANTAGONISTS, BN 52021 AND BN 52111,

PARTIALLY INHIBIT TNF-INDUCED SUPEROXIDE RELEASE BY HUMAN NEUTROPHILS

Monique Paubert-Braquet, *David Hosford, Philipe Koltz,
Jean Guilbaud and *Pierre Braquet

Burn Centre, H.I.A. Percy, F 92141, Clamart, and *IHB,
17 avenue Descartes, 92350 Le Plessis-Robinson, France

INTRODUCTION

The normal bactericidal response of human neutrophils (PMN) involves cell stimulation, followed by a respiratory burst which results from the activation of NADPH oxidase. This multicomponent electron transport chain transfers electrons from intracellular NADPH to extra-cellular oxygen, reducing molecular oxygen to superoxide. This latter product is rapidly converted to hydrogen peroxide and toxic free radicals (1). Activated PMN also release lysosomal proteases and adhere to the endothelial surface, the cells of which can be severely damaged by the secretion of these toxic products if the local cytotoxic response escalates into a systemic process. Indeed, the microvascular collapse, characterisic of pathologies such as shock, asthma, ischemia and graft rejection, may be partially mediated by dysregulation of PMN functions.

Among the various agonistic mediators released in the inflammatory microenvironment which are capable of modulating PMN activity are platelet-activating factor (PAF) and tumour necrosis factor (TNF). PAF, structurally characterised as 1-0-alkyl-2(R)-acetyl-glycero-3-phospho-choline, is a potent autacoid mediator implicated in a diverse range of human pathologies (2). This alkyl phospholipid is now known to be produced by, and act on, a variety of cell types including eosinophils, monocytes, macrophages, platelets, endothelial cells and PMN themselves. Indeed, PAF is a potent chemotactic agent for neutrophils, inducing superoxide release, aggregation and degranulation in these cells. In vivo the mediator evokes pulmonary sequestration of neutrophils, hypotension, bronchoconstriction, increased vascular permeability and plasma extravasation (2).

TNF is a polypeptide cytokine produced primarily by monocytes and macrophages in response to endotoxin and other immune and inflammatory stimuli (3). Injection of large doses of TNF mimick the effects of endo-toxic shock in a variety of species and can lead to death via circulat-ory collapse, hemorrhage and organ failure (4). Thus, TNF also appears to exert many of its effects through vascular disruption. Similarly to PAF, this cytokine induces superoxide production by PMN and causes adherence of this cell type to the endothelium (5).

Free Radicals, Lipoproteins, and Membrane Lipids
Edited by A. Crastes de Paulet *et al.*
Plenum Press, New York, 1990

45

Although it is well established that endothelial cells produce PAF when stimulated with agonists such as thrombin, very recently, Camussi et al. (6) have shown that _in vitro_ TNF can also induce endothelial cells to synthesise PAF, the majority of which remains associated with the cells. Furthermore, Bonavida et al. (7) have reported that PAF can activate peripheral blood derived monocytes to produce and secrete TNF. In addition to being able to elicit the production of each other from cell types intimately involved in inflammatory reactions, at very low concentrations both PAF and TNF can also "prime" PMN to respond in an ehanced manner to subsequent agonistic stimuli. Following PAF (8) or TNF (9) priming, amplified PMN responses including aggregation, adhesiveness, superoxide production and elastase release have been observed using FMLP as the inducing agonist.

The above data opens the possibility that in inflammatory responses, a positive self-generating feedback process may become operative between TNF and PAF, and that PAF associated to the endothelium may prime or amplify PMN responses elicited by TNF and other agonists. Thus, we were interested to ascertain whether TNF-induced superoxide production by human PMN could be amplified by PAF and if so, whether structually unrelated specific PAF antagonists could modulate superoxide release elicited by these two mediators.

MATERIALS AND METHODS

Chemicals

Platelet-activating factor was purchased from Novabiochem Laboratories and Ficoll Paque from Pharmacia. N-Formyl-L-methionine-L-leucyl-L-phenylalanine (FMLP), cytochrome C, phosphate buffer solution and sodium chloride were obtained from Sigma Chemical Co. Tumour necrosis factor was supplied by Bio-Trans Co. (California). The PAF antagonists, BN 52021 (ginkgolide B), a 20-carbon cage terpene and BN 52111 (2-heptadecyl-2-methyl-4-[5(pyridinium-pentylcarbonyl-oxy-methyl)] 1,3-dioxolane bromide) were gifts from IHB Research Labs, Le Plessis-Robinson, France.

Preparation of PMN

Blood was centrifuged at 1000 x g for 25 min. The platelet-rich plasma was discarded, and the PMN fraction was obtained following 6 % dextran-500 sedimentation, centrifugation over Ficoll Paque and ammonium chloride (160 mM) + Tris (170 mM) treatment. PMN were then resuspended in a standard phosphate-buffered medium (PBS) containing (mM) : Na^+, 146 ; K^+, 4 ; Cl^-, 142 ; $HPO_4{}^{2-}$, 2.5 ; H_2PO_4, 0.5 ; Mg^{2+}, 1 ; glucose, 10 ; (pH 7.4 ; determined osmolarity $\sim$ 284 mOsM).

PMN were counted electronically on a Coulter Counter. The final preparation contained less than 5 % lymphocytes and less than 15 % thrombocytes. Trypan blue staining revealed 95 to 99 % viable cells. Electronic microscopy was carried out on this cell material.

PAF amplification of TNF-induced superoxide generation

TNF was diluted with 1 % bovine serum albumin (BSA) and the solutions used extemporaly. The reaction was performed in phosphate buffer saline (PBS). Cytochrome C (75 μM) was added to 4.10^6 cells at 37° C. TNF was then added to the medium at a concentration of 10 ng/ml and the cells incubated for 1 hour, after which time superoxide production was determined as described below. To examine the potential amplifying effect of PAF, the mediator (10^{-16} to 10^{-8} M) was added 10 minutes before the end of the reaction. The 1 hour incubation was terminated by centrifugation for 7 minutes at 4° C. The supernatant was removed for measurement of cytochrome C reduction. When PAF antagonists were used, they were added at a concentration of 10^{-6} M simultanously with the TNF at the start of the incubation period.

Results are expressed in nanomoles of superoxide anion :

Conc. of $O_2^{\cdot -}$ = $\dfrac{\text{Absorbance}}{21.1}$ x 1000 nM $O_2^{\cdot -}$ = $O_2^{\cdot -}/4.10^6$ cells/hr

where 21.1 μM^{-1} cm^{-1} = the absorption molecular coefficient (Σ)

Statistical Analysis

Results are expressed as mean $\pm$ standard error of the mean (SEM). Statistical analysis was carried out using the Student's test.

RESULTS

TNF-induced superoxide production

Dose-response effect of various concentrations of TNF

PMN (4.10^6 cells/0.5 ml) were incubated for 1 hour with concentrations of TNF ranging between 0.001 ng/ml to 1000 ng/ml. As shown in Figure 1, a dose-response relationship was observed between TNF concentration and superoxide generation. The maximium response was obtained with 10 ng/ml TNF where the $O_2^{\cdot}$ production was 4.45 $\pm$ 0.67 nmoles $O_2^{\cdot -}/4.10^6$ cells/hour. Thus, this concentration of TNF was selected for further studies on PMN with PAF.

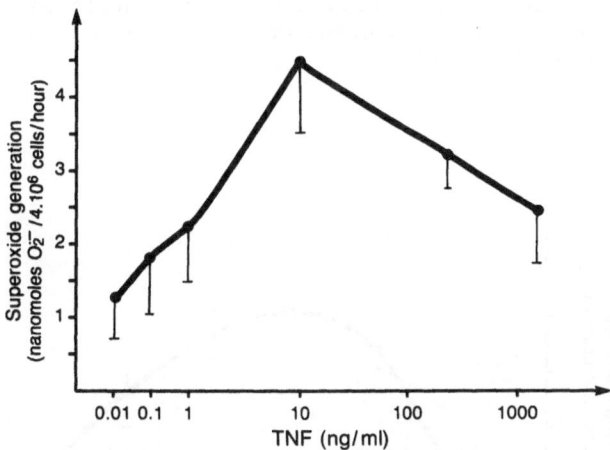

Figure 1   Dose response effect of PMN-superoxide generation induced by TNF (0.01 ng/ml - 1000 ng/ml).

Dose-response effect with various quantities of cells

Superoxide production was also dependent on cell number. With 4.10^6 cells/0.5 ml, the TNF-induced superoxide generation was 4.3 nmole/hour, which was considered adequate to detect any modulatory activity of PAF on superoxide production (Fig. 2). Doubling the cell number only increased the superoxide release by 35 %.

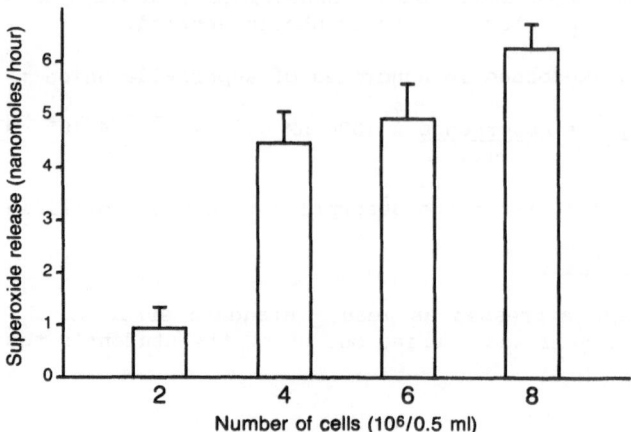

Figure 2   The effect of cell number (2.10^6 - 8.10^6/0.5 ml) on TNF-induced superoxide generation in human PMN.

Time-course of TNF-induced superoxide production

To determine the time-course of TNF-induced superoxide generation we examined $O_2^{\cdot -}$ production by both control PMN (cells not stimulated with TNF) and by PMN incubated with 10 ng/ml TNF for 3 hours at 37 °C. As shown in Figure 3, superoxide generation by TNF-stimulated cells can be described by a parabolic curve ; $O_2^{\cdot -}$ production increased after 15 minutes, reached a maximum value at 90 minutes and then gradually declined. For the studies involving PAF, an incubation period of 1 hour was selected as this marked the onset of a relatively stable and maximal period of superoxide production.

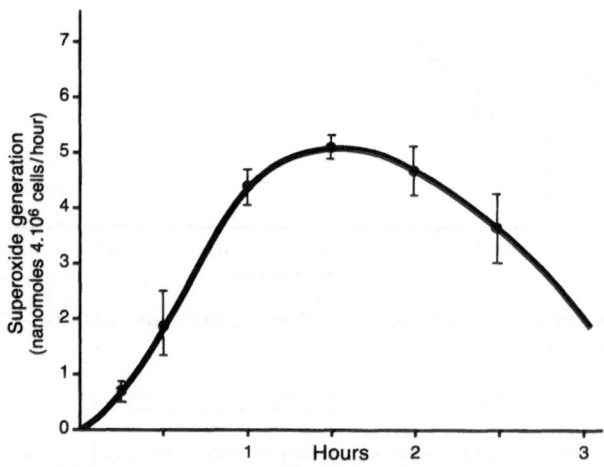

Figure 3   Time-course of TNF-induced superoxide production (10 ng/ml)

PAF-induced superoxide production

PMN were incubated with various concentrations of PAF (10^{-16} M to 10^{-8} M) for periods of between 5 minutes to 3 hours. No significant superoxide production was elicited by PAF at any of the concentrations examined, except 10^{-8} M which elicited the release of a marginally detectable ammount (less than 1 nmole/hr) of superoxide (Fig. 4, bottom curve). This lack of effect of PAF alone on superoxide production occurred regardless of the incubation time or cell number used.

PAF amplification of TNF-induced superoxide production

As shown in Figure 4, when PAF (10^{-16} to 10^{-8} M) was added for 10 minutes to neutrophils which had been previously incubated with 10 ng/ml TNF for 50 minutes, the $O_2^{\cdot -}$ production was significantly amplified (upper curve). Maximum amplification of TNF-induced $O_2^{\cdot -}$ production was elicited by 10^{-12} M PAF, which increased superoxide release by 25 - 30 % (upper curve) relative to that induced by TNF alone after 1 hr (dotted line). In this system, if PAF was added before the TNF at the start of the PMN incubation, no amplification was observed. Furthermore, the amplification effect required an intact PAF molecule since its biologically inactive precursor and metabolite, lyso-PAF, had no effect on TNF-induced $O_2^{\cdot -}$ production.

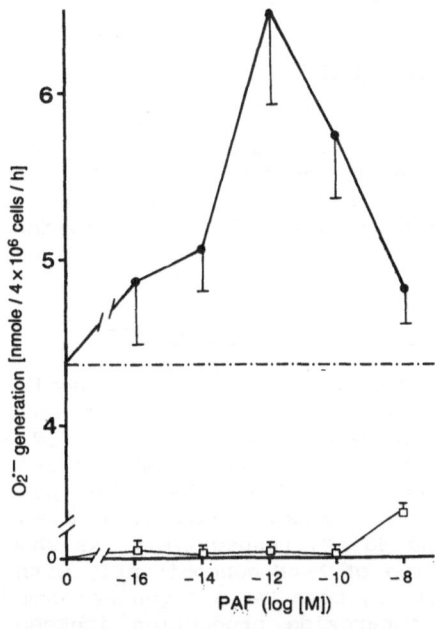

Figure 4   PAF amplification of TNF-induced superoxide generation in human neutrophils. Results show mean ± SEM (n = 10) ; PAF was introduced for 10 minutes before the end of the 1 hour incubation with TNF.
□ PAF alone ; ● TNF (10 ng/ml) + PAF.

Effect of cell washing on the PAF-amplified superoxide production

To determine whether the amplifying effect of PAF on superoxide release was dependent on the continuous presence of exogenous TNF during the incubation with PAF, experiments were performed where the cells were washed twice with buffer to remove the excess of TNF before addition of the PAF. As shown in Figure 5, the amplification by PAF was not reversed by this process. In addition, it can be seen that the PAF effect results from an enhanced rate of superoxide release rather than a prolongation of a relatively short oxidative response induced by PAF.

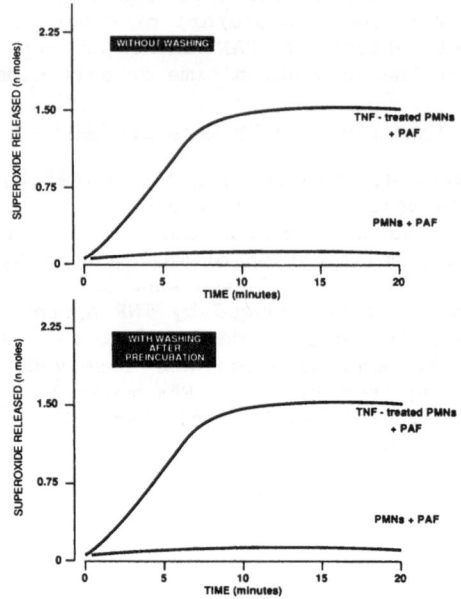

Figure 5   Effect of washing on the amplification by PAF of TNF-induced superoxide production in human neutrophils. PMN (4.10^6 cells) were incubated for 50 minutes with or without (control) TNF (10 ng/ml). Cells were then either directly assayed for PAF (10^{-12} M)-induced superoxide release (upper graph) or washed twice with buffer before the assay (lower graph).

Inhibition of PAF amplification by PAF antagonists

In order to assess the effects of PAF antagonists on the PAF-amplified TNF-induced superoxide production, two structurally unrelated specific antagonists were examined, the ginkgolide, BN 52021 (10), and BN 52111, which is a new constrained PAF-framework related antagonist containing a 1,3 dioxolane ring (11). These two potent antagonists inhibit [^{3}H]-PAF binding to its rabbit platelet membrane receptor with IC_{50} values of 0.25 uM and 35 nM, respectively. As shown in Figure 6, when added at a concentration of 1 uM concomitantly with TNF to the PMN at the start of the incubation, both PAF antagonists completely abolished the amplification of superoxide production induced by subsequent addition of 10^{-10} M PAF (left-hand graph) and 10^{-12} M PAF (centre graph).

Indeed, the inhibition of superoxide release was greater than the amplification effect of PAF, indicating that the antagonists partially inhibited the superoxide production induced by TNF alone. This was confirmed by examining the effect of the two antagonists on TNF-induced superoxide production in PMN which received no exogenously added PAF. Figure 6 (right-hand graph) clearly shows that BN 52021 and BN 52111 inhibit 8 - 20 % of the superoxide release elicited by TNF alone, BN 52111 being the most potent antagonist in this respect.

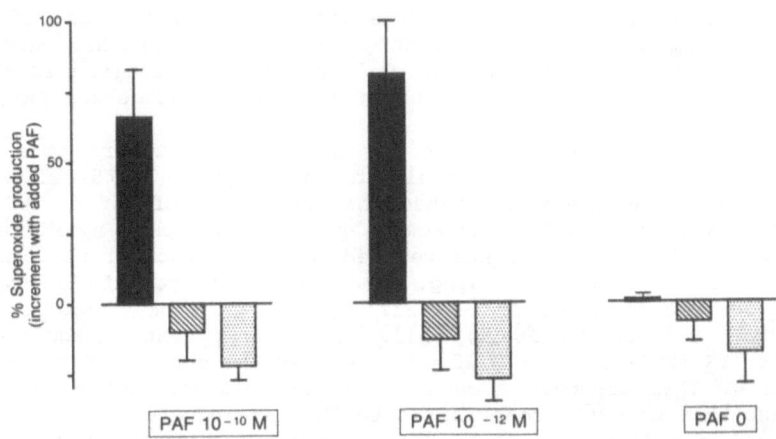

Figure 6 Effect of two chemically unrelated PAF antagonists on PAF-amplified TNF (10 ng/ml)−induced superoxide production in human neutrophils. Results show mean $\pm$ SEM, n = 5 ; both PAF antagonists were used at 10^{-6} M. ■ Control ; ▧ BN 52021 ; ▨ BN 52111

DISCUSSION

Stimulation of PMN with various agonists leads to a respiratory burst with superoxide release, degranulation and cell adhesion. Both TNF and PAF are modulators of the activity of this cell type and play a crucial role in inflammation and host defenses. In our system, TNF induced a time- and dose-dependent production of superoxide from normal human PMN. PAF alone (10^{-16} M to 10^{-8} M) caused no significant release of superoxide from the cells, regardless of cell number or incubation time. However, if PAF was added 10 minutes prior to the end of the 1 hour incubation period with TNF, the superoxide production of the cells was significantly enhanced in comparison with that from cells incubated with TNF alone. Maximum amplification of superoxide production was observed with a concentration of 10^{-12} M PAF.

Several other workers have reported further PAF/TNF interactions <u>in vitro</u>. Valone and Epstein (12) have recently demonstrated that PMN stimulated with TNF are able to produce PAF in a biphasic fashion. The autacoid in the early peak (1-2 hours) is largely retained intra-cellularly, whereas the majority of the PAF in the late peak (6-8 hours) is released into the medium. Studies by Rola-Pleszczynski (13) have also shown that priming of human PMN for 18 hours with various concentrations of PAF (0.1 fM - 1 uM) markedly enhanced their subsequent TNF production in response to interleukin 1 (IL-1). Maximum priming activity was observed at 1 pM - 1 nM PAF, which resulted in a 2 to 3 fold increase in TNF production.

In the present study, PAF had no effect on $O_2^{\cdot-}$ release if it was added to the cells before the TNF at the start of the incubation period, and only a relatively small amplifying effect if it was added before 40 minutes or after 2 hours. The fact that PAF could not influence super-oxide production alone and that its maximum amplifying effect occurred within the period of maximum TNF-induced $O_2^{\cdot-}$ release indicates that PAF must augment some stage of the TNF-dependent $O_2^{\cdot-}$ generation process, which when producing maximum superoxide output can still be amplified to a further degree by a PAF dependent process. Although this mechanism remains undefined, it suggests that $O_2^{\cdot-}$ generation elicited solely by TNF may involve the production and participation of endogenous PAF.

The results obtained with the PAF antagonists support this hypothesis. The two structurally unrelated antagonists, BN 52021 and BN 52111, not only completely abolished the amplifying effect of PAF, but also partially inhibited the superoxide production evoked by TNF alone. The degree of this inhibition was related to the potency of the compound in antagonising $[^3H]$-PAF binding. The extremely potent PAF-related dioxolane-derived antagonist, BN 52111 (IC_{50} = 35 nM), reduced TNF-induced $O_2^{\cdot-}$ release by 20 %, while the less potent ginkgolide PAF antagonist, BN 52021 (IC_{50} = 0.25 uM), reduced the superoxide production by only 8 %. This strongly suggests that TNF elicits PAF synthesis in the PMN and that this PAF contributes to the TNF-induced $O_2^{\cdot-}$ production via a receptor mediated process. Interestingly, a recent study by Sun and Hsueh (14) has also shown that a PAF antagonist can inhibit TNF-induced effects. These latter authors found that in vivo, the PAF antagonist, SRI 63-119, could prevent TNF-induced bowel necrossis in the rat.

The observation that PAF has no amplifying effect if added before the TNF and only a small effect before (or after) the period of maximum TNF-induced $O_2^{\cdot-}$ release may due to the short half-life of the mediator in the cell preparation. Neutrophils rapidly metabolise PAF to lyso-PAF by means of an acetyl-hydrolase (15), although the precise half-life of PAF in our system remains to be determined. Similar findings on the lack of an amplifying effect of PAF if administered before TNF have been obtained in vivo. Heuer and Letts (16) have shown that pretreatment of mice with either S.Typhosa endotoxin or TNF significantly enhanced the mortality induced by PAF. These effects occurred at doses at which PAF, endotoxin or TNF given alone did not significantly affect these parameters, however, the enhancing effect was not recorded when PAF was given prior to TNF.

In conlusion, it appears that an interaction between PAF and TNF, may play an important role in the modulation of neutrophil responses and therefore in inflammatory disorders. Although the biochemical mechanisms involved in the priming and amplification processes remain to be defined, the present study shows that PAF plays an integral role in TNF-induced superoxide production and thus suggests that PAF antagonists may offer some therapeutic value in conditions where this response is excessive and detrimental.

REFERENCES

1. H.L. Malech, and J.I. Gallin, 1987, Neutrophils in human diseases. New Eng J Med., 11:687.
2. P. Braquet, L. Touqui, T.S. Shen, and B.B. Vargaftig, 1987, Perspectives in platelet-activating factor research. Pharmacol Reviews., 39:97.
3. A. Cerami, and B. Beutler, 1988, The role of cachectin/TNF in endotoxic shock and cachexia. Immunology Today., 9:28.
4. K.J. Tracey, B. Beutler, S.F. Lowry, J. Merryweather, S. Wolpe, I.W. Milsark, R.J. Hariri, T.J. Fahey, A. Zentella, J.D. Albert, T. Shires, and A. Cerami, 1986, Shock and tissue injury induced by recombinant human cachectin. Science., 234:4704.

5. Y. Ozaki, T. Ohashi, Y. Niwa, and S. Kume, S., 1988, Effect of recombinant DNA-produced tumour necrosis factor on various parameters of neutrophil function. Inflammation., 12:297.

6. G. Camussi, F. Bussolino, G. Salvidio, and C. Baglioni, 1987, Tumour necrosis factor/cachectin stimulates peritoneal macrophages, polymorphonuclear neutrophils, and vascular endothelial cells to synthesize and release platelet-activating factor. J Exp Med., 166:1390.

7. B. Bonavida, J.M. Mencia-Huerta, and P. Braquet, 1989, Effect of platelet-activating factor (PAF) on monocyte activation and production of tumour necrosis factor (TNF). Int Arch Allergy Appl Immunol., 88:157.

8. G.M. Vercellotti, H.Q. Yin, K.S. Gustafson, P.O. Nelson, H.S. Jacob, 1988, Platelet activating factor primes neutrophil responses to agonists : role in promoting neutrophil-mediated endothelial damage. Blood., 71:1100.

9. R.L. Berkow, D. Wang, J.W. Larrich, R.W. Dodson, and T.H. Howard, 1987, Enhancement of human neutrophil superoxide production by preincubation with recombinant tumour necrosis factor. J. Immunol., 139:3783.

10. P. Braquet, 1987, The Ginkgolides : potent platelet-activating factor antagonists isolated from Ginkgo Biloba L. Drugs of the Future, 12:643.

11. C. Broquet, and P. Braquet, 1988, Compositions thérapeutiques à base de nouveaux aminoacylates d'acétal du glycerol. Fr. Patent. N° 2.616.326.

12. F.H. Valone, and L.B. Epstein, 1988, Biphasic platelet-activating factor synthesis by human monocytes stimulated with interleukin 1 beta (IL 1), tumor necrosis factor (TNF) or gamma interferon (IFN). J. Immunol., 11:3945.

13. M. Rola-Pleszczynski, 1988, Priming of human monocytes with PAF augments their production of tumour necrosis factor in response to interleukin 1. J. Lipid Mediators., in press.

14. X. Sun, and W. Hsueh, 1988, Bowel necrosis induced by tumor necrosis factor in rats is mediated by platelet-activating factor. J Clin Invest., 81:1328.

15. M.L. Blank, T.C. Lee, V. Fitzgerald, and F. Snyder, 1981, Anti-hypertensive activity of an alkyl ether analog of phosphatidyl-choline. Biochem. Biophys. Res. Commun., 29: 2472.

16. H. Heuer, and G. Letts, 1989, Priming of effects of PAF in vivo by tumor necrosis factor and endotoxin. J. Lipid Mediators., in press.

MEMBRANE GLYCOPROTEINS IN SUPEROXIDE RELEASE FROM NEUTROPHILS

Giovanni Ricevuti, Antonino Mazzone, Antonia Notario

Dept. of Internal Medicine and Therapeutics. Section of Medical Pathology, University of Pavia. IRCCS S. Matteo Hospital, 27100 Pavia, Italy

INTRODUCTION

Neutrophils are the cells generally considered to be the first line of defence against invading microorganisms. Stimulation of their surface receptors activates a repertoire of killing functions as phagocytosis, superoxide production, exocytosis of acid hydrolases and proteases, chemotaxis, aggregation and generation of arachidonic acid-derived inflammatory mediators. The major microbicidal mechanisms of the human granulocytes(PMN) require the formation of reactive oxygen metabolites such as superoxide anion (O_2) and hydrogen peroxide (the oxidative burst) and the secretion of granule enzymes such as myeloperoxidase (degranulation). Both of these events can be triggered by various soluble or particulate stimuli which interact with the cell surface and depolarize the cell membrane[1]. The PMN to exposed to the specific murine monoclonal antibodies (MoAbs) exhibit a significant depression in (O_2) generation in response to fMLP to serum-treated zymosan (STZ). Preliminary data suggested that MoAbs depolarizes the PMN cell membrane and that this may be associated with the alterations in oxidative metabolism caused by same MoAbs[2,3]. The purpose of this study was to explore the relationship between surface membrane glycoproteins, membrane depolarization in the human granulocytes and the subsequent oxidative burst Several MoAbs were used for to identifie the membrane glycoprotein fundamental in respiratory burst activation.

MATERIAL AND METHODS

Peripheral blood leukocytes were obtained from 30 ml of heparinized venous blood in a sterile plastic test tube. Twenty ml of 6% dextran were then added and immediately placed on ice.

This procedure was done simultaneously with both patients with pseudo-Pelger granulocytes and control's blood. The polymorphonuclear leukocytes were purified by means of Isopaque-Ficoll system. Contaminating erythrocytes in the PMNs fraction were removed by lysis with 0.75% ammonium chloride solution containing 20 mM Tris HCL buffer (final PH 7.4) and 0.25% autologous plasma. The purity of the cell fractions obtained was 92±2.3% All of these isolation steps were performed at 4°C with cells maintained in buffer containing 150mM NaCl, 3mM Kcl, 8mM $Na_2 HPO_4$, and 1mM KH_2PO_4, pH 7.4 to minimize up-regulation of integrin glycoproteins to the cell surface. For use in studies of up-regulation, the cells were resuspended in 145 mM NaCl containing 10 mM HEPES, 5mM KOH, 1.3 mM $CaCl_2$, 1.2 mM $MgCl_2$ pH, 7.45 (HEPES Buffer). Cell viability was >95% as measured by trypan blue exclusion or release of the cytoplasmic enzyme lactic dehydrogenase. Cells were evaluated within 30 min. from completion of the isolation.

MEASUREMENT OF SUPEROXIDE PRODUCTION

The method used to measure superoxide production was that described by Babior[4], with some modifications. The spectrophotometric measurement is based upon the capacity of superoxide to react with the iron contained in cytochrome type VI (Sigma Chemical USA) both at rest and following stimulation with (fMLP) 10^{-7} M and Phorbol-Myristate-acetate (PMA). This reaction takes place both in the presence and absence of superoxide dismutase, which trasforms O_2 into H_2O_2. Readings were taken with a spectrophotometer (Perkin-Elmer) at wavelegths of 550 nm and 468 nm.The results are expressed in nanomoles of $O_2/1X10^7$ PMNs/15 min. and were obtained by the following formula:A(OD/550-OD/468) - B(OD/550-OD/468) where A=absorbance in the absence of superoxide dismutase and B=absorbance in the presence of superoxide dismutase. The results are multiplied by a correction factor 2.5/0.0245 in relation to the final volume and the coefficient of extintion of iron. Each experiment was done in triplicate.

MONOCLONAL ANTIBODIES

We used for this study the anti-PMN antibodies 60.1, 60.3, 60.5 and 9E8 that were produced by Prof.J.Harlan of Harborview Medical Center,University of Washington,Seattle, USA. The preparation of murine monoclonal antibody (MoAb) 60.1 (IgG_1 kappa), 60.3 (IgG_{2a})and 60.5 (IgG_{2a}) has been described(2). Previous studies indicate that MoAb 60.3 reacts with all human leukocytes. 60.5 reacts with all human leukocytes and was used in these studies as positive control. MoAb 60.1 reacts with all PMN, monocytes, some NK cells, and some Lymphocytes. Monoclonal antibody OKM1 (murine IgG_{2b}; Ortho Diagnostics, Raritan NJ) reacts with human PMN, monocytes, and NK cells.

LFA-1 (murine IgG$_1$) were purchased from Sorin Biomedica, Italy, and reacts with all human leukocytes. Murine MoAb 9E8 recognizes a phage membrane glycoprotein and has no specific reactivity with leukocytes.

RESULTS

The effects of anti PMN integrins MoAbs on superoxide release by PMNs stimulated with FMLP and PMA are shown in Figure 1 and 2.

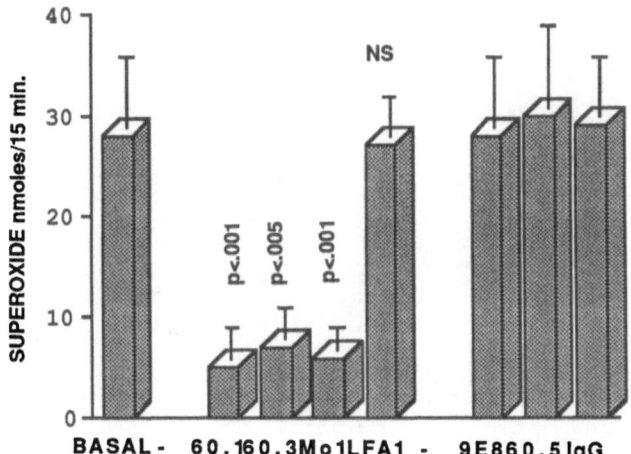

Figure 1 Anion release from PMNs stimulated by FMLP incubated with MoAbs

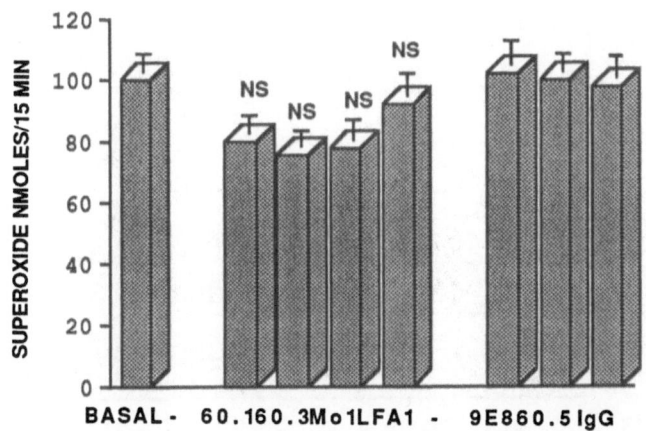

Figure 2 Anion release from PMNs stimulated by PMA incubated with MoAbs

Only stimulation by FMLP is inhibited by MoAbs 60.1 (CD11c), 60.3 (CD11c), Mol (CD11b) involving ß chain CD18 in a dose dependent manner. LFA-1 (CD11a) and other MoAb identifying an inactive phage of PMN membranes are ineffective to block superoxide release.

Oligomers of IgG may bind to the surface of human leukocyte via FcR. The occupation of the FcR by the Fc domain of IgG is not sufficient for activation of the cell and is not obligatory for activation. The cross-linking of cell surface receptors is a general phenomenon involved in the physiologic activation of a variety of cell types.Using PMA as stimulus, any inhibitory effect is shown by MoAbs on superoxide release. This is due probably to a different activation mechanism of FMLP and PMA, as PMA don't act membrane integrins.

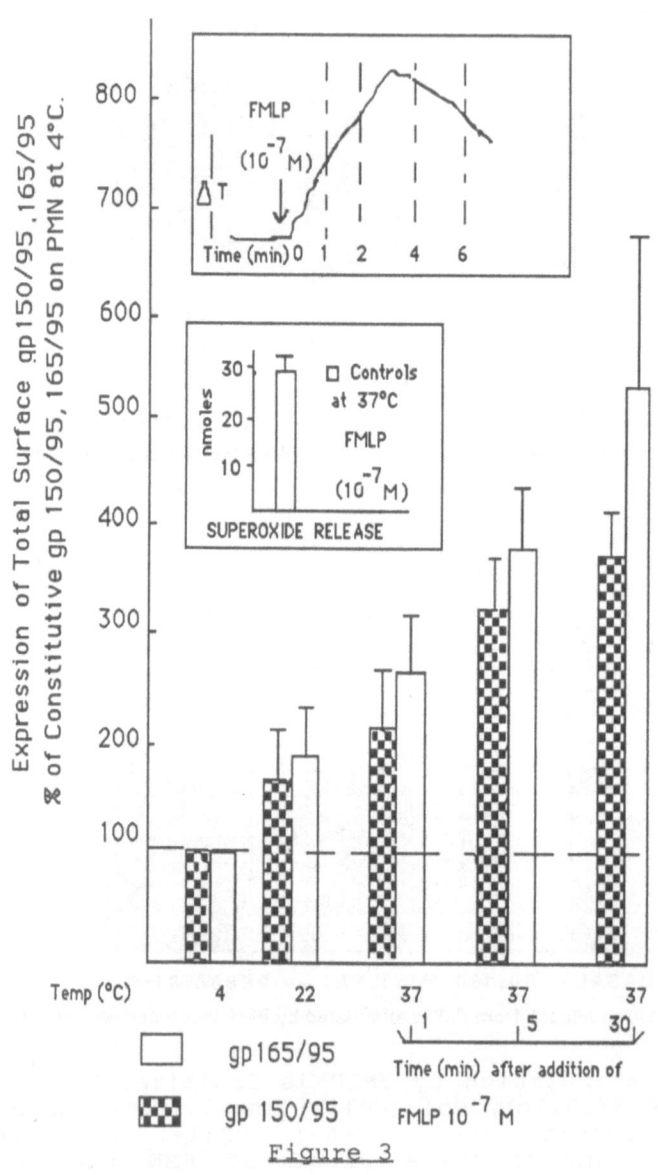

CONTROL GRANULOCYTES

Figure 3

In Figure 3 we can see a standard curve of aggregation and superoxide release from normal PMNs stimulated by FMLP and the effect of temperature and stimulation by FMLP on expression of surface glycoproteins CD11b and CD11c. The expression of surface glycoproteins is strictly related to the temperature of observation and to time of incubation with FMLP. Pseudo Pelger neutrophils don't (Figure 4)

PSEUDO-PELGER GRANULOCYTES

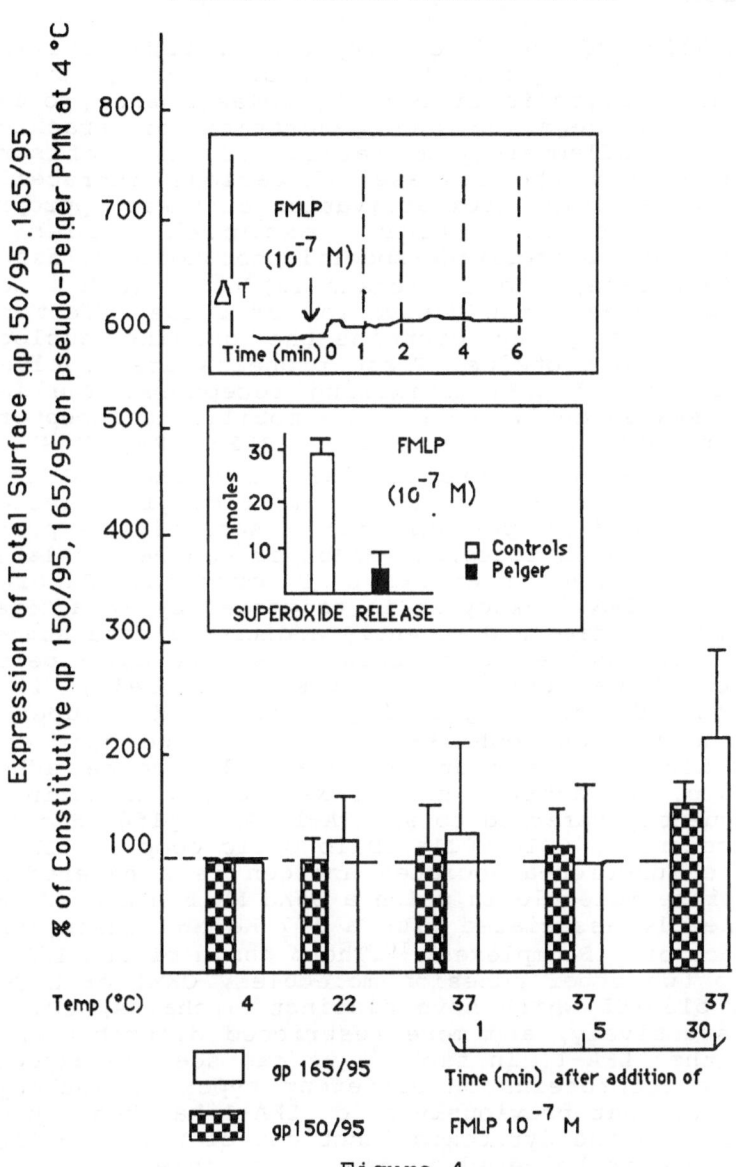

Figure 4

present any response to standard stimulation by FMLP and are not able neither to aggregate nor to release superoxide anion in comparison with normal neutrophils.

Also the expression of integrins CD11b and c is not related neither to temperature nor time of stimulation by FMLP. This is due to deficiency of these glycoproteins on the surface of Pseudo Pelger PMNs.The lack of the stimulatory effect by FMLP, due to the absence of surface receptors is like to the inhibitory effect by MoAbs to CD11b and CD11c on normal PMNs.

DISCUSSION

Stimulation of surface receptors of PMNs activates a repertoire of killing functions as phagocytosis, superoxide production, exocytosis of acid hydrolases and proteases, chemotaxis, aggregation and generation of arachidonic acid-derived inflammatory mediators. Ligation of surface receptors causes the release of various intracellular messengers, resulting from activation of their appropriate effector systems (for example: adenylate cyclase,polyphosphoinositide-specificphosphodiesterase,GMP-phosphodiesterase, and ion channels)[5,6,7]. It has become increasingly clear that activation of these effectors is regulated by coupling proteins-the guanine nucleotide regulatory or G proteins. These proteins are involved in the regulation of Ca++-mobilizing receptors. The second messenger system activated by Ca++-mobilizing receptors is the polyphosphoinositide cycle (PIP_2, IP_3, DAG). IP_3 releases stores of intracellular Ca++ and DAG is the intracellular activator of PKC. Some biological responses can be inhibited by pertuxis toxin sensitive G protein, others by some anti receptors MoAbs or can be regulated by a pertuxis toxin insensitive G protein. Superoxide production in PMN leukocytes is triggered by an activation of the NADPH-oxidase and recently additional data suggest a role for yet another G protein, distinct from pertuxis toxin sensitive and insensitive G protein, in the regulation of the respiratory burst and superoxide generation in neutrophils[8]. In table 1 we can see the presentation of the most important leukocyte adhesion complex g.p. for activation of PMNs, called integrins that also frequently referred to as LFA-1/MAC-1/p150-95 complex, or more recently, CD18,CD11a,CD11b CD11c complex. LFA-1 or lymphocyte function-associated antigen is a heterodimeric cell membrane molecule in which a 95Kd MW ß chain (CD18) is non-covalently associated with a 177 Kd MW chain (CD11a) in the form of /ßcomplexes[9,10].The ß chain of LFA-1/CD18 is common to two other adhesion molecules, OKM1 or MAC-1 or CR3, and p150/95 which have distinct chains, CD11b and CD11c respectively, and more restricted distributions and funtions than LFA-1. In table 1 we can see the expression of these glycoproteins on different types of leukocytes, according to what previously said. LFA-1 has been shown to have a role in the cytotoxic functions of both T cells and NK cells, and is believed to be responssible for mediating firm adherence to target cells prior to cell-mediated killing. LFA-1 may also play a role in the required adherence functions of B and T cells in the immune response. Preliminary studies have suggested that the LFA-1 of neutrophils may be involved in the phagocytosis of

TABLE 1. RELATIONS BETWEEN MEMBRANE ANTIGENS, MONOCLONAL ANTIBODIES, GLYCOPROTEIN SUBUNITS AND CELLULAR FUNCTION

ANTIGEN	MoAb	CHAINS	CELLS	FUNCTIONS
LFA-1 (CD 11a)	LFA-1,6O.3 L1,TS1/22	alpha L	Tcells, NK,PMN, Monocytes	Cytotoxicity T cell proliferation
Mac-1 (CD11 b)	Mo1,Mac1 OKM1	alpha M	PMN,Monocytes	Phagocytosis
Gp 150 (CD 11c)	6O.3,60.1,I B4,MHM23 Ab1-14,Ab1-15	alpha X	PMN, monocytes	Chemotaxis adherence aggregation
CD 18	ß			

bacteria and there is as yet no definitive data on the function of p150/95.Probably, p150/95 may also play a role in the granulocytes adherence and aggregation.CR3 functions as major opsonin and complement receptor on phagocytic cells and has the ability to trigger phagocytosis, a respiratory burst, degranulation, and the release of leukotrienes. In addition, CR3 may play a role in the adherence and aggregation of neutrophils required for chemotaxis. The function of CR3 on T cells and NK cells has not been defined. The formation of reactive oxygen metabolites such as superoxide anion and hydrogen peroxide and other functions of PMN, such as aggregation and degranulation, can be triggered by various soluble or particulate stimuli when they interact with the cell surface and granulocytes are rendered hyper-adhesive. Our studies let us to expolre the relationship between membrane glycoprotein complexes of human PMNs and oxidative burst, using several probes including specific and aspecific MoAbs at various concentrations. Compared with the basal level of FMLP stimulated superoxide production only MoAbs against antigens CD18,CD11c and CD 11b are able to inhibit superoxide production. The gp CD11a is ineffective to mediate superoxide production by FMLP. The monoclonal antibody inhibition studies suggest that LFA-1 is not involved in the oxygen anion release. In human pathology we can found a confirmation to our data by studying aggregation and superoxide release and up regulation of receptors of pseudoPelger neutrophils that is a morphological and functional alteration of PMNs genetically transmitted or acquired in hematological malignancies. FMLP stimulation is not able to increase number expression of surface glycoproteins 150/95 and 165/95, nor to stimulate release of superoxide anion, nor to aggregate cells. When we use PMA stimulus[11], no statistically significative difference was observed between PMA alone and preincubation of cells by MoAbs against CD18,CD11a,CD11b,CD11c and other MoAbs. This different response upon stimulation with FMLP and PMA let us to think that it is possible the presence of two way for superoxide production and this is related to different action mechanism of the two stimulatory agents.PMA bypasse the cascade (IP3) by direct interaction with protein kinase C substituting for 1,2-diacylglycerol[12].In addition FMLP, but not PMA, stimulate the generation of leukotrienes via the 5-lipooxygenase. A membrane depolarization temporally precedes the oxidative burst in association to changes in membrane fluidity after stimulation by FMLP and PMA.Recently Martin[3] explored the relationship between membrane depolarization in the human PMN and the sub sequent oxidative burst. Several probes were used, including anti PMN MoAbs, depolazing high potassium buffers, and glycolytic inhibitors known to inhibit oxygen metabolism. These findings that different classes of probes have similar effects on depolarization of the neutrophil membrane and activation of the respiratory burst provides further evidence for a close association between these two events. Thus, according to Martin[3] the inhibition of PMA and FMLP induced depolarization and respiratory burst activation by PMN MoAbs, high potassium buffers and glycolytic inhibitors cannot likely be explained by an effect on changes in intracellular calcium and may be due to an effect on protein kinase C or on some events distal to protein kinase C activation. In conclusion the production of superoxide stimulated by FMLP from PMNs preincubated with 60.1, 60.3, PMN 7C3 and OKM1 shows a very statistically significant decrease (p<.001)[13]. The incubation with the same MoAbs is able to moderately reduce the production of superoxide after stimulation with PMA (NS). The cells incubated with the other MoAbs and with the aspecific IgG did not present significant differences with respect to the control. In our data the surface membrane antigens very important in modulation superoxide release are:CD18,CD11b(CR3) and CD11c.

ACKNOWLEDGEMENTS

This work was supported by grants from MPI 40% and 60%, Regione Lombardia, IRCCS S. Matteo Hospital Pavia.We aknoweledge Prof. J.M. Harlan from University of Washigton for his kind gift of 60.1,60.3,60.5, 9E8 monoclonal antibodies.

REFERENCES

1.   Goetzl E.J., Goldstein I.M., Cellular components of inflammation: granulocytes.In "TEXTBOOK OF RHEUMATOLOGY" pg.115-144 eds by Kelley W.N.,Harris E.D.,Ruddy S.,Sledge C.B.,pbl by W.B. Saunders Company Philadelphia (1985).
2.   Mauri C.,Rizzo S.C., Ricevuti G. eds. "The Biology of Phagocytes in Health and Disease", Advances in biosciences series, vol.66, Pergamon Press (1987).
3.   Martin M.A., Nauseef W.M., Clark R.A.,Depolarization Blunts the oxidative burst of human neutrophils. J Immunol.140:3928 (1988).

4.   Babior D.M., Kipness R.S.,.Curnutte J.I.Biological
     defence   mechanisms: the   production   by
     leukocytes of   superoxide,a potential bactericidal
     agent. J.Clin. Invest. 52:741 (1973).
5.   Bellavite P.,The superoxide-forming enzymatic system
     of phagocytes Free Radical Biology & Medicine 4:225
     (1988).
6.   Cross C.E., Oxygen Radicals and Human Disease
     Ann. Int. Med. 107:526 (1987).
7.   Klebanoff S. J.. Oxigen metabolism and the toxic
     properties of phagocytes Ann.Int.Med.93:480 (1980).
8.   Koren H.S., Proposed classification of leukocyte-
     associated cytolytic molecules Immunology
     Today 8:139 (1987).
9.   Ricevuti G. Mazzone A. The neutrophils revisited.
     Inflammation 13,475,(1989).
10. Ricevuti G. Mazzone A. Harlan JH. Membrane
     glycoproteins and related disease. Haematologica 73:
     415 (1988).
11. Solomon SS,Palazzolo M. Activation of cyclic AMP
     phosphodiesterase by phorbol and protein kinase C
     pathway.Am.J. Med.Sci. 292: 182(1986).
12. Sedor JR.Free radicals and prostanoid synthesis
     J Lab Clin Med 108:521 (1986).
13. Willis H.E., Browder B., Feister A.J., Mohanakumar
     T. Ruddy S.Monoclonal antibody to human IgG Fc
     receptorsJ Immunol 140:234 (1988)

FREE RADICAL PRODUCTION BY THE MITOCHONDRION

Julio F. Turrens and Joe M. McCord

Department of Biochemistry
College of Medicine
University of South Alabama
Mobile, AL 36688

INTRODUCTION

The electron transport chain located in the mitochondrial inner membrane catalyzes the complete oxidation of NADH to water. This reaction is carried out by a complex system which involves the sequential participation of more than thirty redox-active components including flavoproteins (dehydrogenases), iron-sulfur proteins (dehydrogenases, Rieske protein, etc.), heme-containing proteins (cytochromes), and a lipid (ubiquinone). Some of these components carry only one electron at a time (cytochromes of the b and c type), while others transport two (flavoproteins) and four (cytochrome oxidase) electrons in a single step. As a consequence, there are several points in the chain in which a two-electron carrier must reduce a one-electron carrier, producing free radical intermediates. One of them, ubisemiquinone, is present in relatively large amounts in the respiring respiratory chain, and may be detected by electron spin resonance.

The transport of electrons is favored by a large difference in potential (1.12 V) between the two ends of the respiratory chain, from -0.32 V (NADH) to +0.8 V (oxygen). All the respiratory components located between the negative end of the chain and cytochrome c are thermodynamically capable of partially reducing oxygen to O_2^-, a reaction involving the transfer of a single electron to an oxygen molecule. This reaction is potentially dangerous to the cell. It leads to the formation of O_2^- and H_2O_2, which in the presence of iron chelates may produce hydroxyl radical. The remarkable efficiency of the respiratory chain directs virtually all electrons to the terminal oxidase (reducing oxygen to water), but there are at least two sites where electrons may leak out of the chain to partially reduce oxygen. The first site is located at the NADH dehydrogenase level, and the second one in the ubiquinone-cytochrome b area in the respiratory chain.[1-3]

This paper includes general information about the methods to study these reactions within the respiratory chain, and discusses some of the pathological situations where these species have been implicated.

METHODS FOR DETERMINING MITOCHONDRIAL H_2O_2 PRODUCTION

The partial reduction of oxygen by mitochondrial components generates O_2^- in the mitochondrial matrix, the substrate for the mitochondrial Mn-SOD which catalyzes its dismutation to oxygen and H_2O_2. Hydrogen peroxide permeates freely through biological membranes; thus, when using intact mitochondria, the rate of H_2O_2 release reflects the intramitochondrial generation

Free Radicals, Lipoproteins, and Membrane Lipids
Edited by A. Crastes de Paulet *et al.*
Plenum Press, New York, 1990

of O_2^-. The amount of H_2O_2 production is not stoichiometric with the O_2^- being produced because part of the H_2O_2 is reduced to water by the mitochondrial glutathione peroxidase located in the matrix.

The most sensitive methods for measuring H_2O_2 generation are based on the reaction between a peroxidase and its substrate. Peroxidases catalyze the reduction of H_2O_2 by several hydrogen donors. The specificity for the reductant varies depending on which peroxidase is used. For example, cytochrome c peroxidase is very specific for reduced cytochrome c, catalyzing the reduction of H_2O_2 with a rate constant which is 1000-fold higher than for any other hydrogen donor.[4] Horseradish peroxidase accepts a variety of hydrogen donors including guaiacol, pyrogallol, benzidine, scopoletin, etc. The colorimetric or fluorometric changes that these hydrogen donors undergo during oxidation may be followed to determine rates of H_2O_2 production in the presence of the appropriate enzymes.[5] This constitutes the basis of most methods used for measuring H_2O_2 production. When horseradish peroxidase is used, the concentration of the reductant must be kept relatively high (1 mM or more). Otherwise, other endogenous hydrogen donors would compete with the substrate being monitored, decreasing the sensitivity and accuracy of the method.

The mechanism of peroxidase action usually includes a first step in which the enzyme is oxidized by the substrate (equivalent to the formation of an enzyme-substrate complex, known as Compound I) followed by its reduction by the hydrogen donor. Peroxidases are hemoproteins, and as a consequence their spectrum is modified when undergoing redox changes. The changes occurring with the alpha peaks of the Soret spectrum constitute the basis of other spectrophotometric methods for determining H_2O_2 generation, in which the peroxidases are now used as substrates (undergoing an oxidation to Compound I) rather than as enzymes (catalyzing the decomposition of H_2O_2). Additional sensitivity may be achieved by using dual wavelength spectrophotometry, in which the two wavelengths are close to each other making it possible to use concentrated samples without interference caused by turbidity. This method must be used carefully when horseradish peroxidase is the substrate since endogenous hydrogen donors may decrease the actual concentration of Compound I, underestimating the amount of H_2O_2 being generated. On the other hand, the specificity of cytochrome c peroxidase for cytochrome c makes it ideal for quantitatively scavenging H_2O_2 generated by intact mitochondria, since in this preparation cytochrome c is retained in the inter-membrane space, and is not accessible to the enzyme. The main disadvantage is that cytochrome c peroxidase is not commercially available. It must be isolated in the laboratory from Baker's yeast. The procedure is very simple, consisting of adsorption of the enzyme on DEAE cellulose followed by elution and separation from other contaminants by gel filtration.[6]

MECHANISMS OF H_2O_2 FORMATION BY MITOCHONDRIAL COMPONENTS

The rate of hydrogen peroxide production is a function of the redox steady state of the mitochondrial components. The rate becomes higher as the respiratory chain becomes more reduced. The steady state concentration of a reduced component changes depending on the concentration of oxidizable substrate and on the presence of substrates to phosphorylate. When mitochondria are coupled, electron transport through the mitochondrial respiratory chain is accompanied by a simultaneous movement of hydrogen ions (protons) from the mitochondrial matrix to the inter-membrane space. As the proton gradient increases, the electron flow becomes slower until a steady state is reached in which the rate of proton pumping equilibrates with that of back leakage (which is very low in intact mitochondria). This rate is known as State 4 respiration. Upon addition of ADP, the channel through the enzyme ATPase is opened, allowing the protons to go back in, collapsing the gradient, thus increasing the rate of electron flow (State 3 respiration). The respiratory components are more reduced in State 4 and the rate of H_2O_2 production is maximal. When ADP is added, the rate of H_2O_2 is virtually zero, as the respiratory chain components become more oxidized.[7]

Since the generation of O_2^- by the respiratory chain is a non-enzymatic reaction (i.e., it is a consequence of the autoxidation of reduced respiratory components), the rate of production increases linearly with the concentration of either oxygen or reduced species, according to equation 1:

$$\frac{d[O_2^-]}{dt} = k \, [O_2] \, [AH \cdot] \qquad (1)$$

Experimental data confirmed this hypothesis.[8,9] At 21% oxygen and in the absence of ADP (State 4), the rate of H_2O_2 formation accounts for about 2% of the total oxygen consumption by the respiratory chain. However, under physiological conditions, the proportion of oxygen undergoing partial reduction is expected to be much less, since not only the intracellular concentration of oxygen is lower, but also mitochondria are never under absolute State 4 conditions.

The highest rate of H_2O_2 formation by mitochondria may be achieved by adding inhibitors of the respiratory chain, which block electron flow and produce a full reduction of all the components located on the substrate side of the site of inhibition.[1,8] Thus, addition of rotenone to mitochondria permits an estimation of the highest proportion of O_2^- or H_2O_2 that may be produced at the NADH dehydrogenase site in mitochondria from any given tissue. Similarly, antimycin blocks the respiratory chain between cytochrome b and ubiquinone (Fig. 1), favoring the accumulation of ubisemiquinone and reduced NADH dehydrogenase, thus stimulating the production of O_2^- at both sites of the respiratory chain. The formation of ubisemiquinone from ubiquinol (the fully reduced form of Coenzyme Q) requires the simultaneous transfer of one electron to cytochrome c_1. Thus, the inhibition of the electron flow between cytochrome c_1 and cytochrome oxidase also inhibits O_2^- formation at the ubisemiquinone site.[2,3] This happens when the respiratory chain is supplemented with either cyanide or myxothiazol, or depleted of cytochrome c.

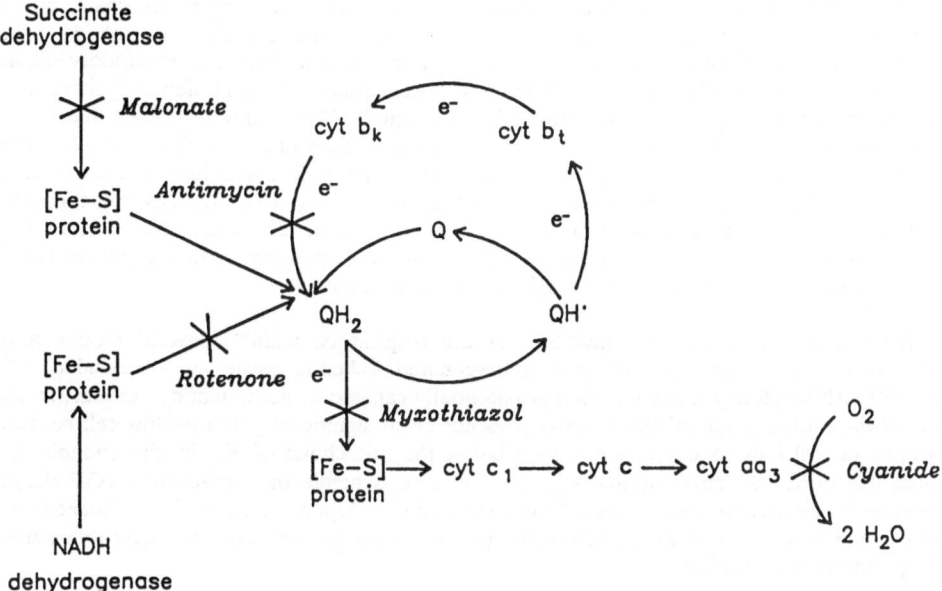

Fig. 1.   Postulated Q-cycle in the mammalian mitochondrial respiratory chain. The figure also indicates the sites where different inhibitors block electron flow.

When the determinations are to be carried out in either broken (or uncoupled) mitochondria or in submitochondrial particles (inside-out vesicles containing fragments of the respiratory chain), the use of inhibitors is required. Otherwise, the respiratory components are mostly oxidized and the amount of O_2^- or H_2O_2 produced is not detectable.

THE ROLE OF MITOCHONDRIAL FREE RADICALS IN CELL INJURY

Mitochondrial damage appears as an early sign of cell injury in various pathological situations including hyperoxic injury and reoxygenation of ischemic tissues. Hyperoxic injury to a tissue occurs when an animal is exposed to a high concentration of oxygen. It affects primarily the lungs, since they are the only organs in direct contact with high oxygen tension. Although the survival time of an animal exposed to 100% oxygen is several days (e.g., 70 hours for the rat), mitochondrial damage can be observed after only 40 to 50 h of exposure. The K_M for oxygen of cytochrome oxidase is very low (5 uM or less). As the intracellular concentration of oxygen increases, the rate of electron leakage to oxygen will increase (see equation 1), as will the proportion of leakage to total oxygen consumption. Indeed, this has been shown in lung submitochondrial particles and in intact mitochondria.[8,9]

In tissues other than lung, pO_2 varies little under hyperoxic conditions because hemoglobin is already saturated with oxygen under normoxia. The additional oxygen dissolved in plasma and carried to the tissues at an oxygen concentration of 1 atm is only around 0.8 mM. This increases the oxygen content of arterial blood by less than 10%. At partial pressures much greater than 1 atm, and as a consequence of the increased generation of oxygen radicals by various intracellular systems including the mitochondria, the oxidative injury may affect tissues other than the lung. If the damage affects the brain, then convulsions are an early symptom and the animal dies soon after.

We are investigating the role of the mitochondrial generation of oxygen radicals during reoxygenation of ischemic tissues. When isolated mitochondria are exposed to cycles of anoxia and reoxygenation, there is no apparent damage to the mitochondrial respiratory chain, nor is there an increased generation of H_2O_2. However, when mitochondria are isolated from reoxygenated ischemic tissues there is severe damage to the mitochondrial respiratory chain. This damage includes inhibition of electron transport between NADH dehydrogenase and ubiquinone and uncoupling of the mitochondrial membrane.[10] This is not an artifact due to the isolation procedure since mitochondrial swelling (a possible cause of uncoupling) has also been seen by electron microscopy in intact post-ischemic heart tissue.[11] The inhibition of electron flow is equivalent to adding rotenone to intact mitochondria, since in both cases the electron transport between NADH and ubiquinone is blocked. The consequence of this is that the generation of H_2O_2 at the NADH dehydrogenase level of the mitochondrial respiratory chain becomes maximal, in a situation in which the cell has plenty of reducing equivalents.

What is the cause for the inhibition of the respiratory chain? Several groups have indicated that mitochondria isolated from reoxygenated ischemic tissue are overloaded with calcium.[11,12] Although any mitochondrion is potentially capable of accumulating calcium (at the expense of the proton gradient) this process does not occur significantly in a healthy cell because the cytoplasmic calcium concentration is kept below the mitochondrial K_M by the endoplasmic reticulum and other calcium-pumping ATPases. During ischemia the intracellular ATP drops, accompanied by an accumulation of cytoplasmic calcium.[12,13] Upon reoxygenation, mitochondrial electron flow is restored, and we postulate that as a consequence, mitochondria load themselves with large amounts of calcium.

We are currently examining the effects of calcium overload (>200 nmol/mg mitochondrial protein) on both mitochondrial function and hydrogen peroxide generation. The results show that as a consequence of calcium overload, the rate of NADH-supported electron flow is inhibited (Table 1). The inhibition of mitochondrial electron flow can also be seen when mitochondrial fragments are isolated from calcium-treated mitochondria, indicating that the inhibition in the rate

Table 1. Effect of Ca^{2+} accumulation on oxygen consumption (State 3) and respiratory control ratio (RCR) in isolated rat heart mitochondria.

Calcium (nmol/mg protein)[a]	RCR	Oxygen consumption (nmol/min/mg protein)	NADH oxidation[b] (nmol/min/mg protein)
0	3.5	165.8	402
43	1.9	121.1	363
128	1.6	106.2	232
256	1.2	55.9	195

[a] Mitochondria (approximately 4 mg/ml) were incubated with the indicated amounts of calcium in the presence of 5 mM malate, 2 mM glutamate and 4 mM phosphate. The suspension was gently shaken at 30° C to allow continuous oxygenation during 15 min.

[b] For determining NADH oxidation mitochondria were frozen and homogenized by passing the suspension through a 27 gauge hypodermic needle three times in order to produce fragments of mitochondrial membranes.

of respiration is the consequence of a blockage in the respiratory chain (Table 1). As a result of calcium overload, those respiratory components located on the substrate side of the site of inhibition are more reduced, stimulating H_2O_2 generation (Table 2). Note that in calcium-loaded mitochondria, oxygen consumption in the presence of ADP (State 3) is decreased from 112 to 24 nmol/min/mg protein. This, in addition to the increase in the *absolute* rate of H_2O_2 production (from less than 0.15 to 0.4 nmol/min/mg protein), results in an overall increase of more than 16-fold in the percentage of oxygen molecules undergoing partial reduction. Furthermore, as a consequence of the decreased oxygen consumption it is likely that the cell experiences a higher pO_2, thus stimulating H_2O_2 formation even more.

Oxygen-derived free radicals generated by the enzyme xanthine oxidase have been implicated as major contributors to reperfusion injury in various tissues of the rat and dog.[14] Many tissues (including heart) of humans and rabbits, however, have been found to contain no detectable xanthine oxidase.[15,16] Despite this fact, oxygen-derived free radicals have been implicated in reperfusion injury to the isolated perfused rabbit heart.[17] We suggest that the mechanism

Table 2. Hydrogen peroxide production by normal and calcium-loaded rat heart mitochondria.

Additions	O_2 consumption (nmol/min/mg)	H_2O_2 production (nmol/min/mg)	% O_2 undergoing partial reduction
Control			
Malate	28.7 ± 2.9	0.33 ± 0.1	1.1
Malate + ADP	112.1 ± 16	< 0.15	< 0.1
Ca^{2+}-loaded (300 nmol/mg protein)			
Malate	22.4 ± 5.6	0.4 ± 0.2	1.5
Malate + ADP	24.3 ± 5.6	0.4 ± 0.2	1.6

described above may provide an alternative source for the production of active oxygen species during reperfusion injury.

CONCLUSIONS

Mitochondria are potential generators of both superoxide anion and hydrogen peroxide. The rate of active oxygen formation depends on both reduction level of respiratory components and oxygen concentration. The production of active oxygen by mitochondria may be significant under pathological conditions involving hyperoxia or calcium overload.

ACKNOWLEDGEMENTS

This work was supported by an Intramural Research Grant Award from the College of Medicine, University of South Alabama, and by a Grant from the American Heart Association, Alabama Affiliate (# AL-G-880007) to J.F.T. The authors thank Mr. Ulises B. Chavez for his technical assistance.

REFERENCES

1. J. F. Turrens and A. Boveris, Generation of superoxide anion by the NADH dehydrogenase of bovine heart mitochondria, *Biochem. J.* 191:421 (1980).

2. A. Boveris and B. Chance, The mitochondrial generation of hydrogen peroxide: general properties and effect of hyperbaric oxygen, *Biochem. J.* 134:707 (1973).

3. J. F. Turrens, A. Alexandre and A. L. Lehninger, Ubisemiquinone is the electron donor for superoxide formation by complex III of heart mitochondria, *Arch. Biochem. Biophys.* 237:408 (1985).

4. T. Yonetani, Studies on cytochrome c peroxidase. II. Stoichiometry between enzyme, H_2O_2, and ferrocytochrome *c* and enzymic determination of extinction coefficients of cytochrome *c*, *J. Biol. Chem.* 240:4509 (1965).

5. B. Chance, H. Sies and A. Boveris, Hydroperoxide metabolism in mammalian organs, *Physiol. Rev.* 59:527 (1979).

6. C. E. Nelson, E. V. Sitzman, C. H. Kang and E. Margoliash, Preparation of cytochrome *c* peroxidase from Baker's yeast, *Anal. Biochem.* 83:622 (1977).

7. A. Boveris, N. Oshino and B. Chance, The cellular production of hydrogen peroxide, *Biochem. J.* 128:617 (1972).

8. J. F. Turrens, B. A. Freeman, J. G. Levitt and J. D. Crapo, The effect of hyperoxia on superoxide production by lung submitochondrial particles, *Arch. Biochem. Biophys.* 217:401 (1982).

9. J. F. Turrens, B. A. Freeman and J. D. Crapo, Hyperoxia increases hydrogen peroxide formation by lung mitochondria and microsomes, *Arch. Biochem. Biophys.* 217:411 (1982).

10. C. D. Malis and J. V. Bonventre, Mechanism of calcium potentiation of oxygen free radical injury to renal mitochondria, *J. Biol. Chem.* 261:14201 (1986).

11. A. D. Romaschin, I. Rebeyka, G. J. Wilson and D. A. G. Mickle, Conjugated dienes in ischemic and reperfused myocardium: an in vivo chemical signature of oxygen free radical mediated injury, *J. Mol. Cell. Cardiol.* 19:289 (1987).

12. P. E. Arnold, V. J. Van Putten, D. Lumlertgul, T. J. Burke and R. W. Schrier, Adenine nucleotide metabolism and mitochondrial Ca^{2+} transport following renal ischemia, *Am. J. Physiol.* 250:F357 (1986).

13. C. Steenbergen, E. Murphy, L. Levy and R. E. London, Elevation in cytosolic free calcium early in myocardial ischemia in perfused rat heart, *Circ. Res.* 60:700 (1987).

14. J. M. McCord, Oxygen-derived free radicals in post-ischemic tissue injury, *N. Engl. J. Med.* 312:159 (1985).

15. L. J. Eddy, J. R. Stewart, H. P. Jones, T. D. Engerson, J. M. McCord, J. M. Downey, Free radical-producing enzyme, xanthine oxidase, is undetectable in human hearts, *Am. J. Physiol.* 253 (Heart 22):H709 (1987).

16. J. M. Downey, T. Miura, L. J. Eddy, D. E. Chambers, T. Mellert, D. J. Hearse, D. M. Yellon, Xanthine oxidase is not a source of free radicals in the ischemic rabbit heart, *J. Mol. Cell. Cardiol.* 19:1053 (1987).

17. J.M. McCord, B.A. Omar and W.J. Russell, Sources of Oxygen-Derived Radicals in Ischemia-Reperfusion, in T. Yoshikawa and E. Niki (eds.), *Chemical, Biochemical and Medical Aspects of Free Radicals*, Elsevier Science, Amsterdam, in press, (1989).

HEMOGLOBIN AS A PROMOTER OF CENTRAL NERVOUS SYSTEM DAMAGE

John W. Eaton, Douglas A. Peterson and S. M. H. Sadrzadeh

University of Minnesota Medical School, and Dight Laboratories

Minneapolis, Minnesota   55455

INTRODUCTION

The full extent of an injury to the central nervous system (CNS) is often not evident until long after the initial insult.   Multiple factors may be involved, but one which appears to be of great importance is hemorrhage into the site of damage.[1]   Indeed, in at least one experimental model of CNS trauma, neutrophils only accumulate in those areas to which blood has extravasated.[2]   We have hypothesized that free hemoglobin (Hb) may promote oxidative reactions and, thereby, certain inflammatory events.   Indeed, in terms of susceptibility to oxidation, the brain would appear to be a combustible tissue, low in oxidant defense enzymes[3] and rich in polyunsaturated fatty acids.   In this paper, we adduce evidence in support of the general proposition that Hb, and especially iron derived therefrom, may be particularly hazardous within the interstitium of the CNS, acting to promote oxidative damage to areas previously subject to hemorrhage.

METHODS

The methods employed in these investigations have been described in detail in earlier publications.[4,12,15,17-19]   Those used in as yet unpublished work are described in the legend of Table 2.

RESULTS AND DISCUSSION

Our interest in the possibility of hemoglobin-potentiated damage to the CNS was sparked by studies of patients with familial idiopathic epilepsy.   We carried out investigations of kindreds in which at least two siblings had a history of seizures not related to prior CNS trauma.   In five of 14 kindreds, most of the affected individuals and about 1/3 of their currently unaffected relatives had anomolously low or absent levels of the serum Hb-binding protein, haptoglobin.[4]   Because hypo- or an-haptoglobinemia is extremely rare in the general population of the United States, this association is highly significant.

Our tentative interpretation of this observation is that the individuals affected by seizures might have pre-existing, congenital hypohaptoglobinemia. Even mild, imperceptible, trauma to the CNS, may be associated with trace hemorrhaging.   The extravasated erythrocytes, incapable of reentering circulation, might lyse and the free Hb would be inefficiently removed because of

Free Radicals, Lipoproteins, and Membrane Lipids
Edited by A. Crastes de Paulet *et al.*
Plenum Press, New York, 1990

the abnormally low serum haptoglobin levels. Free Hb might then be available to potentiate secondary inflammatory reactions, thereby magnifying the extent of damage to the CNS. Indeed, several earlier reports indicate that the intracerebral injection of iron salts or Hb in experimental animals will cause electroencephalographic abnormalities or frank seizure disorders.[5-11]

We began by experimentally examining the idea that hypohaptoglobinemia would, indeed, slow the clearance of free Hb from the CNS. Indeed, in support of this idea, the clearance of radiolabeled (^{51}Cr) Hb intracerebrally injected into mice is significantly retarded in hypohaptoglobinemic animals. Furthermore, the abnormality in clearance is normalized if the Hb is pre-complexed with purified (human) haptoglobin (Figure 1).[4]

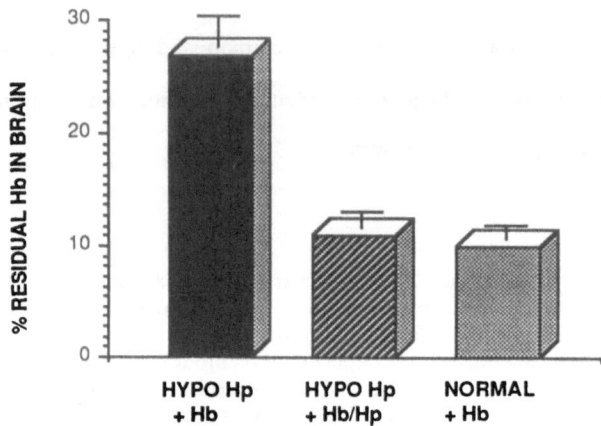

Figure 1. Clearance of ^{51}Cr-labeled Hb from the brains of hypohaptoglobinemic and normal mice. Mice were injected intracerebrally with either ^{51}Cr-labeled Hb alone or Hb complexed to haptoglobin and the amounts of residual Hb were determined by gamma counting of the excised brains after 24 hours as previously described.[4] Note that the clearance of Hb from the brains of hypohaptoglobinemic animals (A) is significantly less than from the brains of normal mice (C) and that, in hypohaptoglobinemic animals, complexation of Hb with haptoglobin corrects this defect (B). Results represent mean ($\pm$ 1 S.D.), n = 6 in all groups. (Redrawn from Panter et al., 1984[4]).

If Hb were allowed to remain within the CNS, rather than being removed, what consequences might accrue? Our earliest experiments on this point were based on the simple (and, in retrospect, naïve) assumption that intact Hb might facilitate the production of dangerous activated oxygen species such as hydroxyl radical (·OH). As a test of this possibility, we incubated purified Hb with a superoxide/hydrogen peroxide (O_2^-/H_2O_2) generating system, xanthine/xanthine oxidase. Dimethylsulfoxide (DMSO) was added to detect any ·OH which might form. DMSO will react with ·OH (and, perhaps, certain complexes of iron and activated oxygen) to yield - among other products - methane and formaldehyde.[12-14] In such an experimental system, Hb produces substantial amounts of formaldehyde from DMSO, in a dose-dependent fashion.[15] Furthermore, when Hb is incubated with a polyunsaturated fatty acid such as arachidonic acid in the presence of a O_2^-/H_2O_2-generating system, lipid peroxidation is stimulated.

In the foregoing experiments, the observed reactions were probably due to Hb itself, rather than Hb-derived iron, because the powerful iron chelator deferoxamine was added in all cases. In binding free iron, deferoxamine occupies all six coordination positions on the iron and effectively neutralizes iron-dependent ·OH generation and lipid peroxidation.[16,17] Therefore, these reactions of Hb with lipid and the apparent generation of ·OH are probably not mediated by free iron released from the Hb. Unfortunately, these observations probably cannot be extended to the situation *in vivo*. First, the amounts of lipid peroxidized and ·OH formed are very small and depend upon the simultaneous presence of artificial O_2^-/H_2O_2-generating systems. Second, although the effects of purported ·OH scavengers indicated that ·OH was being formed, it now appears most likely that we were detecting an activated state of heme iron arising from a single reaction between H_2O_2 and reduced heme, the product of which may not be a biological hazard.

Thus, it remained to be determined how dangerous free Hb might be when in direct contact with cell membranes within the CNS. To investigate this point, we carried out experiments employing crude homogenates of mouse brain. In this case, the simple addition of purified hemoglobin to brain homogenates triggers intense lipid peroxidation (Figure 2).[18] Interestingly, the addition of a superoxide-generating system is not required, and both oxy (Fe^{2+}) and met (Fe^{3+}) Hb are equipotent in causing this peroxidation. However, in contrast with the *in vitro* experiments described above which were carried out with deferoxamine present, the addition of this Fe^{3+} chelator to a Hb-containing murine brain homogenate completely blocks lipid peroxidation (Figure 2).[18] Furthermore, so does the addition of the Fe^{2+} chelator, 3-(2-pyridyl)-5,6-bis(2-[5-furylsulfonic acid]) ('Ferene S').[19] These results imply that free iron, cycling between ferric and ferrous, is directly responsible for the peroxidation touched off by the addition of purified Hb to brain homogenates.

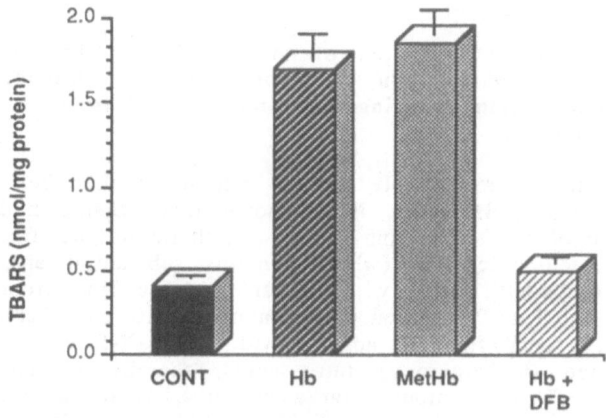

Figure 2. Spontaneous lipid peroxidation (assessed by measurement of thiobarbituric acid reactive substances - TBARS) of lipid in murine brain homogenates incubated with no addition (control) or with the addition of hemoglobin, methemoglobin or hemoglobin and deferoxamine (DFB). Specific conditions of incubation were as previously reported.[18] Note that the extensive peroxidation of CNS polyunsaturated fatty acid caused by the addition of hemoglobin and methemoglobin is completely blocked by the addition of deferoxamine. (Redrawn from Sadrzadeh et al., 1987[18]).

The oxidative damage done to brain homogenates by added Hb is not limited to simple lipid oxidation. We also find a progressive and profound inhibition of the crucial membrane enzyme, Na/K ATPase.[19] It has been known for some time that this enzyme, as well as membrane CaATPase,[20] is inhibited by oxidation of erythrocyte membranes. This marked inhibition of neuronal Na/K ATPase by free Hb has allowed us to directly test the hypothesis that Hb alone may injure the CNS *in vivo*. In this test, small amounts of iron or free Hb were injected stereotaxically into the spinal cords of anesthetized cats. Twenty-four hours after the injection of either iron salts or hemoglobin, there is substantial inhibition of Na/K ATPase (whereas the injection of heme-free globin or of ve-

TABLE 1

Hemoglobin- and Iron-Mediated Inhibition of Feline Spinal Cord
Na/K ATPase*

Injection	Na/K ATPase**	n
Control	116 ± 6.1	3
FeCl$_2$	29 ± 0.9†	10
Hb	73 ± 10.8§	3
Hb + deferoxamine	118, 112	2
Globin	109, 116	2

* Na/K ATPase activity was measured on homogenates of both the injected region and on uninjected regions of the spinal cord of mongrel cats 24 hours
following no injection ("control"), injection of iron FeCl$_2$ (500 nmol), Hb or
heme-free globin (110 nmol as monomer), or Hb with deferoxamine (110
nmol and 18 μmol, respectively).

** Na/K ATPase activity expressed as nmol pyrophosphate min^{-1} mg protein^{-1}.

† Significantly different from control value, $p < 0.001$ (Student's 't' test, two-
tailed).

$\S$ Significantly different from control value, $p < 0.01$ (Student's 't' test, two-
tailed).

hicle alone causes no inhibition of this enzyme). Importantly, simultaneous
injection of deferoxamine completely prevents the Hb-mediated inhibition of
this enzyme once again indicting free iron derived from Hb as the culprit in
the damage (Table 1).

The mechanisms responsible for the release of this free iron from Hb are
not presently clear. However, it is known that organic hydroperoxides stimulate the release of free iron from Hb.[21] Furthermore, we find that commercially available arachidonic acid (which contains substantial amounts of contaminating hydroperoxy fatty acid) will similarly release iron from Hb and cause a
diminution of the heme absorbance maximum at 410 nm (Table 2).[22] Therefore,
it is likely that, once free Hb appears within the CNS interstitium, it may react
with unsaturated (hydroperoxy) fatty acids, thereby releasing iron from the
heme group. These reactions form a partial basis for a self-amplifying pathology: the free iron will intensify lipid peroxidation which, in turn, will cause
the destruction of more Hb and release of more iron.

As noted above, the oxidative processes set off by the co-incubation of Hb
with CNS and CNS homogenates appear to depend upon the repetitive oxidation
and reduction of free iron yet do not require the addition of a O_2^-/H_2O_2-gener-
ating system. This raises the question of the nature of the reducing species
involved in the evident cycling of the iron. In fact, the reduction of iron by
murine brain homogenates can be experimentally demonstrated. During incubation of ferric iron or free Hb with crude murine brain homogenate, ferrous
iron appears as reflected by the accumulation of a ferrous iron:ferenc chelate
which has a readily detectable absorbance.[19] Therefore, brain contains one or
more substances which reduce iron.

These substances are water soluble; peroxidation of murine brain lipids is
greatly lessened by washing of the membranes from these homogenates. However, if the supernatant from the original homogenate is added back to the
washed membranes, brisk iron-driven peroxidation once again occurs. The
water soluble principle involved has been identified as ascorbic acid (probably

TABLE 2

Destruction of the Heme Group of Partially Purified Hemoglobin by Partially Peroxidized Arachidonic Acid*

Treatment	Heme (µM)	n
Hb	18.4 ± 1.7	22
Hb + 0.8 mM arachidonic acid	14.3 ± 1.6**	16
Hb + 1.6 mM arachidonic acid	10.8 ± 3.4**	18

---

* Partially purified human hemoglobin A, approximately 18 µM monomer, was incubated in 50 mM Tris buffer, pH 7.4 at 25°C, with the indicated final concentrations of commercially obtained (Sigma Chemical Co.) arachidonic acid. Note that this latter material had substantial peroxide content as estimated by direct spectrophotometric measurement of absorbance at 234 nm. Arachidonic acid obtained from another supplies (NuChek Prep, Elysian, MN) had no such contamination and did not cause destruction of heme. Concentration of heme was estimated by measurement of absorbance at 412 nm after 5 minutes of incubation. No further reaction was determined beyond this time.

** Differs significantly from Hb alone, $p < 0.01$ (Student's 't' test, two-tailed).

the most prevalent reducing substance in mammalian CNS).[19] This is supported by two observations. First, crude homogenates - but not washed membranes - contain millimolar amounts of ascorbate. If similar amounts of ascorbate are added back to the washed membranes, Hb- or iron-driven peroxidation is restored. Second, if the brain homogenate is first incubated with the enzyme, ascorbate oxidase, this completely abrogates all subsequent peroxidation, whether caused by the addition of iron or Hb.[19]

In summary, it appears the Hb released by red cells extravasated into the CNS following a hemorrhagic event may promote a certain fraction of the ensuing inflammation and damage. Interactions between Hb and CNS unsaturated fatty acids lead to breakdown of the heme group. The iron released upon destruction of the heme engages in cyclic oxidation/reduction reactions involving both ascorbate and fatty acids. This is indicated by the fact that the addition of either ferrous or ferric iron chelators will completely stop the process. The most important effector of the cyclic reduction of iron is endogenous brain ascorbate. Enzymatic or physical removal of ascorbate will block this peroxidation and the addition of exogenous vitamin C to normal levels completely restores the usual brisk rate of Hb or iron-driven lipid peroxidation.

These results raise a paradox and suggest a possible therapeutic approach to the management of patients with hemorrhagic injury to the CNS. The paradox is that the brain normally employs ascorbic acid as an effective reducing agent especially to protect important neurotransmitters from spontaneous oxidation. Furthermore, the very survival of the metabolically fragile cells of the CNS depends on continued and efficient oxygen delivery by Hb. However, when free Hb is present within the brain, the important reducing substance and the critical oxygen delivery device may combine to form an efficient engine of oxidative destruction of the injured tissue. The therapeutic approach which these results suggest is obvious but yet to be tested: Administration of chelating agents such as deferoxamine may help moderate the sequellae of hemorrhagic injury to the CNS.

REFERENCES

1. M. Oehmichen, Inflammatory cells in the central nervous system: An integrating concept based on recent research in pathology, immunology, and forensic medicine, in: "Progress in Neurology," Vol. 5, H. M. Zimmerman, ed., Raven Press, New York (1983).

2. E. D. Means and D. K. Anderson, Neurophagia by leukocytes in experimental spinal cord injury, J. Neuropath. Exp. Neurol. 42:707 (1983).

3. G. Cohen, Oxidative stress in the nervous system, in: "Oxidative Stress," H. Sies, ed., Academic Press, London (1985).

4. S. S. Panter, S. M. H. Sadrzadeh, P. E. Hallaway, J. L. Haines, V. E. Anderson and J. W. Eaton, Hypohaptoglobinemia associated with familial epilepsy, J. Exp. Med. 161:748 (1985).

5. L. M. Kopeloff, S. E. Barrera and N. Kopeloff, Recurrent convulsive seizures in animals produced by immunologic and chemical means, Am. J. Psychol. 98:881 (1942).

6. L. J. Willmore, G. W. Sypert, J. B. Munson and R. W. Hurd, Chronic focal epileptiform discharges induced by injection of iron into rat and cat cortex, Science 200:1501 (1978).

7. L. J. Willmore, R. W. Hurd and G. W. Sypert, Epileptiform activity initiated by oral iontophoresis of ferrous and ferric chloride on rat cerebral cortex, Brain Res., 52:406 (1978).

8. A. D. Rosen and N. V. Frumin, Focal epileptogenesis after intracortical hemoglobin injection, Exp. Neurol. 66:277 (1979).

9. L. J. Willmore and J. J. Rubin, Antiperoxidant pretreatment and iron-induced epileptiform discharges in the rat: EEG and histopathologic studies, Neurology 31:63 (1981).

10. P. Levitt, W. Wilson and R. Wilkins, The effects of subarachnoid blood on the electrocorticogram of the cat, J. Neurosurg. 35:185 (1971).

11. E. J. Hammond, R. E. Ramsay, J. H. Villarreal and G. J. Wilder, Effects of intracortical injection of blood and blood components on the electrocorticogram, Epilepsia 21:3 (1980).

12. J. E. Repine, J. W. Eaton, J. W. Anders, J. R. Hoidal and R. B. Fox, Generation of hydroxyl radical by enzymes, chemicals and human phagocytes *in vitro*: Detection using the anti-inflammatory agent - dimethyl sulfoxide (DMSO), J. Clin. Invest. 64:1642 (1979).

13. S. M. Klein, G. Cohen and A. I. Cederbaum, The interaction of hydroxyl radicals with dimethylsulfoxide produces formaldehyde, FEBS Lett 116:220 (1980)

14. S. M. Klein, G. Cohen and A. I. Cederbaum, Production of formaldehyde during metabolism of dimethyl sulfoxide by hydroxyl radical-generating systems, Biochemistry 20:6006 (1981).

15. S. M. H. Sadrazdeh, E. Graf, S. S. Panter, P. E. Hallaway and J. W. Eaton, Hemoglobin: A biologic fenton reagent, J. Biol. Chem. 259:14354 (1984).

16. J. M. C. Gutteridge, R. Richmond and B. Halliwell, Inhibition of the iron-catalyzed formation of hydroxyl radicals from superoxide and of lipid peroxidation by desferrioxamine, Biochem. J. 184:469 (1979).

17. E. Graf, J. R. Mahoney, R. G. Bryant and J. W. Eaton, Iron-catalyzed hydroxyl radical formation: Stringent requirement for free iron coordination site, J. Biol. Chem. 259:3620 (1984).

18. S. M. H. Sadrzadeh, D. K. Anderson, S. S. Panter, P. E. Hallaway and J. W. Eaton Hemoglobin potentiates central nervous system damage, J. Clin. Invest. 79: 662 (1987).

19. S. M. H. Sadrzadeh and J. W. Eaton, Hemoglobin-mediated oxidant damage to the central nervous system requires endogenous ascorbate, J. Clin. Invest. 82:1510 (1988).

20. O. Shalev, M. Leida, R. P. Hebbel, H. S. Jacob and J. W. Eaton, Abnormal erythrocyte calcium homeostasis in oxidant-induced hemolytic disease, Blood 58:1232 (1981).

21. J. M. C. Gutteridge, Iron promoters of the Fenton reaction and lipid peroxidation can be released from haemoglobin by peroxides, FEBS Letts. 201:291 (1986).

22. D. A. Peterson and J. W. Eaton, A possible mechanism for the tumor cell cytotoxicity of arachidonic acid, manuscript in preparation (1989).

Role of Xanthine Oxidase and Granulocytes in

Ischemia/Reperfusion Injury

Matthew B. Grisham, Barbara J. Zimmerman and D. Neil Granger

Department of Physiology and Biophysics LSU Medical Center
P.O. Box 33932, Shreveport, Louisiana 71130-3932

INTRODUCTION

The concept that xanthine oxidase-derived oxidants mediate the microvascular injury associated with reperfusion of the ischemic intestine was first proposed in 1981 (1). Since its inception most of the assumptions inherent in this concept have been tested and it has been extended to a number of other organ systems (2). The information derived from several studies performed in our laboratory and by others have led us to revise the biochemical scheme originally proposed to explain oxygen-dependent reperfusion injury. The revised scheme (Figure 1) assumes that xanthine oxidase-derived oxidants produced following reoxygenation of ischemic intestine play an important role in recruiting and activating granulocytes, which ultimately mediate reperfusion-induced microvascular injury. The foregoing discussion will summarize the supporting evidence presented in Figure 1.

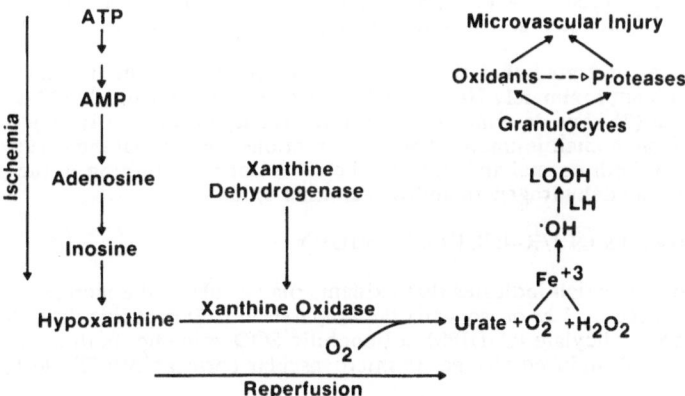

Figure 1. Proposed relationship among xanthine oxidase-derived oxidants, neutrophils and microvascular injury in post-ischemic tissue.

Free Radicals, Lipoproteins, and Membrane Lipids
Edited by A. Crastes de Paulet *et al.*
Plenum Press, New York, 1990

RELATIVE CONTRIBUTIONS OF ISCHEMIA AND REPERFUSION TO MICROVASCULAR INJURY

An important assumption in the oxygen radical hypothesis of ischemia/reperfusion (I/R) injury is that the tissue injury observed following reperfusion is due to the reintroduction of oxygen rather than a delayed manifestation of injury incurred during the ischemic period. Although the validity of this assumption has not been definitively resolved there are several lines of evidence that tend to support it. Parks et al. (3) has demonstrated that the mucosal injury produced by 3 hours of ischemia and 1 hour of reperfusion is significantly greater than that produced by 4 hours of ischemia without reperfusion. He also found that reperfusion of intestine with deoxygenated perfusate after 3 hr of ischemia produced significantly less injury than that observed following reperfusion with oxygenated whole blood.

Microvascular permeability to plasma proteins has proven to be a useful and sensitive index for evaluating the influence of ischemia/reperfusion on microvascular integrity. For example, if the cat small bowel is subjected to one hour of ischemia (blood flow reduced to about 20% of control) without reperfusion, a doubling (0.08 ± 0.01 vs 0.15 ± 0.03) of microvascular permeability is noted. However, the same period of ischemia followed by reperfusion causes a 5-fold (to 0.40 ± 0.03) increase in microvascular permeability. The assumption that reoxygenation accounts for the greater rise in permeability after reperfusion is supported by the consistent observation that antioxidants and inhibitors of oxy-radical formation (e.g., allopurinol) attenuate only that component of the increased permeability that is manifested following reperfusion. Another argument that is used to defend the position that reperfusion *per se* is largely responsible for the injury observed in intestinal models of ischemia-reperfusion is that inhibitors of oxy-radical production (e.g., allopurinol, oxypurinol) offer protection when administered at the time of reperfusion.

PURINE METABOLISM DURING ISCHEMIA/REPERFUSION (I/R)

The concept that cellular ATP levels fall while hypoxanthine accumulates in ischemic tissues is well documented for many organs. However, only recently has such documentation been provided for the intestine (4,5). Schoenberg et al (5) have demonstrated that 2 hours of ischemia (intestinal arterial pressure decreased to 25-30 mmHg) reduces ATP concentration to approximately 40% of the control (preischemic) value. The depletion of ATP is associated with increases in tissue levels of AMP (7.6-fold), hypoxanthine (10-fold) and uric acid (4-fold). Ten minutes following reperfusion, the levels of AMP and hypoxanthine remain significantly elevated.

The relationship between duration of ischemia and the extent of ATP depletion has not been defined for the cat intestine. However, ^{31}P nuclear magnetic resonance spectroscopy has been used to follow the ATP depletion that occurs in the ischemic rat intestine (4). The results of this analysis reveal that depletion of mucosal ATP is rapid and complete within 20 minutes of total ischemia, i.e., longer periods of ischemia do not produce a further decline in ATP levels. This observation is consistent with reports that only 30 minutes of ischemia are needed to produce prolonged functional and structural changes in the rat intestine (6).

It has been demonstrated that the tissue hypoxanthine level in the normally perfused intestinal mucosa is approximately 20 μM and it increases to more than 200 μM during ischemia (5). Mousson et al (7) have determined that the K_m for hypoxanthine is 11 μM for xanthine oxidase isolated from human jejunum. These observations suggest that the tissue concentration of hypoxanthine in both normal and ischemic bowel is not rate-limiting in the production of uric acid by xanthine dehydrogenase and/or xanthine oxidase.

ROLE OF OXIDANTS IN I/R-INDUCED INJURY

There is evidence which indicates that oxidants play a role in the increased microvascular permeability produced by 1 hr of ischemia then reperfusion. Superoxide dismutase (SOD) and copper di-isopropyl salicylate (CuDIPS), a lipophilic SOD-mimetic, both provide significant protection against I/R-induced changes in microvascular permeability (Table 1).

Although the beneficial effects of SOD and CuDIPS implicate the superoxide anion in I/R-induced microvascular injury, there is evidence suggesting that oxidants derived from superoxide play a more important role in the injury process. Detoxification of hydrogen peroxide with catalase significantly attenuates the I/R-induced increase in microvascular permeability. Similar results are obtained with dimethylsulfoxide (DMSO), a hydroxyl radical

Table 1. Modulation of ischemia/reperfusion-induced increase in microvascular permeability.

Condition	Microvascular Permeability $(1 - \sigma)$	Reference
Control	0.08 ± 0.005	2
Ischemia (1 hr) + reperfusion (I/R)	0.41 ± 0.02	2
I/R + treatment with:		
Allopurinol	0.18 ± 0.01	19
Folic acid	0.16 ± 0.04	19
Pterin aldehyde	0.15 ± 0.02	19
Tungsten-supplemented diet	0.20 ± 0.02	20
Superoxide dismutase	0.14 ± 0.01	2
Catalase	0.19 ± 0.01	2
Dimethyl sulfoxide	0.19 ± 0.02	2
Desferoxamine	0.15 ± 0.01	9
Iron-loaded desferoxamine	0.44 ± 0.03	9
Apotransferrin	0.17 ± 0.01	9
Antineutrophil serum	0.13 ± 0.01	28
Monoclonal antibody 60.3	0.12 ± 0.01	28
Adenosine	0.13 ± 0.02	35

scavenger (see Table 1). The protective effects of SOD, catalase, and DMSO are consistent with the view that a secondarily-derived radical such as the hydroxy radical is formed during reperfusion by the iron-catalyzed interaction between xanthine oxidase-generated superoxide and hydrogen peroxide. However it should be noted that DMSO has other biological effects including inhibiting neutrophil adherence (Sekizuka, Benoit, Grisham and Granger, Am. J. Physiol. in press).

The fact that O_2^-, H_2O_2 and possibly ·OH may be involved in I/R-induced microvascular injury suggested a role for iron as well. It is known that O_2^- and H_2O_2 interact in the presence of iron to yield ·OH and the intestinal mucosa is a rich source of iron. Although iron is normally stored in enterocytes in the form of ferritin micelles, superoxide can reduce Fe^{3+} to liberate Fe^{2+} (8). Thus, the reducing power of superoxide could release ferrous ions (from ferritin) for reaction with hydrogen peroxide to form the hydroxyl radical during reperfusion of ischemic intestine. We examined the role of iron in I/R-induced hydroxyl radical production by determining whether deferoxamine (an iron chelator) or apotransferrin (an iron binding protein) provide protection against the increased intestinal microvascular permeability produced by I/R (9). Vascular permeability in deferoxamine- and apotransferrin-treated animals was lower than in untreated animals. The observation that iron-loaded deferoxamine or transferrin did not offer protection against I/R injury argues against a nonspecific protective influence of these iron binding substances. Although our observations are consistent with hydroxyl radical formation by the iron-catalyzed Haber-Weiss reaction, a role for iron-centered radicals such as the ferryl or perferryl radicals cannot be excluded.

A characteristic feature of the hydroxyl radical is its ability to abstract methylene hydrogen atoms from membrane-associated polyunsaturated fatty acids. This reaction initiates the process of lipid peroxidation and leads to the formation of lipid-derived free radicals such as conjugated dienes, lipid hydroperoxide radicals and hydroperoxides (LOOH). Measurements of conjugated dienes are commonly used as an index of lipid peroxide formation. Schoenberg and coworkers (5,10) have reported increased levels of conjugated dienes in the reperfused cat intestine. The reperfusion-induced increase in mucosal conjugated dienes was not observed in animals treated with either superoxide dismutase or allopurinol. These observations suggest that powerful oxidants (e.g., ·OH) mediate the lipid peroxidation associated with reperfusion of the small intestine.

ROLE OF XANTHINE OXIDASE IN I/R - INDUCED INJURY

In early attempts to determine the source of the reactive oxygen species produced during reperfusion, xanthine oxidase was considered a primary candidate. In comparison to other tissues the intestinal mucosa has a tremendous capacity to oxidize hypoxanthine via the enzyme xanthine oxidase. Xanthine oxidase (dehydrogenase plus oxidase) activity in the intestinal mucosa approaches or exceeds 100 mU per gram wet weight of tissue in most species (11). The cytotoxic potential of this high enzyme activity is exemplified by the observation that isolated cells are injured when exposed to xanthine oxidase levels as low as 2 mU/ml (12).

Intestinal xanthine oxidase activity is found primarily in the mucosal layer, with an increasing gradient of activity from villus base to tip (13). This is consistent with the observation that the villus tip is more sensitive to ischemic injury than the base. Immunolocalization studies suggest that mucosal xanthine oxidase is found exclusively in endothelial cells lining the microvasculature in the villus core (14). However, histochemical studies demonstrate preferential localization of the enzyme in epithelial cells (13). We find that fresh isolated villus epithelium accounts for the majority of mucosal xanthine oxidase in cat ileal mucosa (15). Nonetheless, the significant enzyme activity remaining in the lamina propria may be concentrated in microvascular endothelium.

Xanthine oxidase (XO) exists in normal healthy cells predominantly as a NAD^+-reducing xanthine dehydrogenase (XD). Conversion of XD to XO (D-to-O) can be initiated either by limited proteolysis, oxidation of sulfhydryl groups or both (11). The XO formed by limited proteolysis is irreversible, while sulfhydryl-reducing agents (e.g., dithiothreitol) can reverse the XO formed by sulfhydryl oxidation to XD. Relatively little is known about the mechanisms and kinetics of XD-to-XO conversion in the ischemic small intestine. Even after extensive efforts to limit artifactual conversion during tissue preparation (16), 10-20% of the enzyme exists in the oxidase form in normal bowel. Using extensive controls and careful maintenance of temperature during ischemia, Parks et al (16) found that only 50% of XD is converted to XO in rat bowel after 2 hours of ischemia. This represents a 2- to 3-fold increase in activity of the oxidant producing form of the enzyme. A slower rate of conversion was noted in cat ileal mucosa.

The assumption that xanthine oxidase plays a critical role in I/R injury is supported primarily by studies using the inhibitor allopurinol. Parks (17) and others (5,18) have reported a marked reduction in the severity of mucosal lesions induced by I/R in allopurinol-treated animals. Similar protection is noted following treatment with oxypurinol, the long-lived metabolite of allopurinol. Allopurinol has also been shown to attenuate the increased intestinal microvascular permeability to plasma proteins induced by I/R (Table 1). Enteral administration of either pterin aldehyde or folic acid also attenuates the I/R-induced increase in microvascular permeability. Both compounds are effective inhibitors of xanthine oxidase, with the potency of pterin aldehyde comparable to that of oxypurinol (19).

Another approach that has been used to assess the role of xanthine oxidase in I/R injury is to place animals on a molybdenum-deficient, tungsten (W) supplemented diet. This regimen leads to incorporation of tungsten, rather than molybdenum, into XO and thereby inactivates it. Using the W-supplemented diet, Parks et al (20) observed a 75% reduction in mucosal xanthine oxidase (D+O) activity and a corresponding attenuation of the I/R-induced increase in intestinal microvascular permeability. A limitation of this approach is that other molybdenum containing enzymes (e.g., aldehyde oxidase) are also inactivated by tungsten supplementation.

Although the protection afforded by allopurinol has generally been attributed to decreased production of xanthine oxidase-derived oxidants, preservation of the nucleotide pool and free radical scavenging are frequently invoked as alternate explanations. Allopurinol does appear to attenuate the reduction in cellular ATP produced by intestinal ischemia, however, maintenance of ATP levels by administration of exogenous purines (e.g., inosine) does not afford protection against I/R injury in cat intestine (5). The observation that administration of either allopurinol or oxypurinol at the onset of reperfusion is equally effective as pretreatment (administration before ischemia) in attenuating reperfusion injury also argues against a major role for purine salvage (18). However, the latter observation does not negate the possibility that allopurinol affords protection by acting as a free radical scavenger.

There are several reports which suggest that the beneficial effects of allopurinol in models of I/R may be due to direct free radical scavenging properties of the drug rather than its ability to inhibit xanthine oxidase. Moorhouse et al (21) have reported that allopurinol and oxypurinol are powerful scavengers of hydroxly radicals in vitro. They also reported that oxypurinol is a scavenger of the neutrophil-derived oxidant hypochlorous acid, which may be involved in the microvascular injury associated with I/R. We recently examined the possibility that the regimen of allopurinol administration used in most I/R studies leads to a significant modification of the antioxidant properties of extracellular fluid, i.e., plasma and lymph (22). Our results indicate that this regimen of allopurinol administration leads to an extracellular allopurinol/oxypurinol concentration of 10-20 μM, which effectively inhibits XO activity but does not significantly enhance the scavenging properties of extracellular fluid. In vitro studies suggest that the allopurinol/oxypurinol concentration must exceed 500 μM to significantly enhance the radical scavenging potential of extracellular fluid (21).

ROLE OF GRANULOCYTES IN I/R-INDUCED INJURY

Another potential source of oxidants in the small intestine is the polymorphonuclear leukocyte. Measurements of mucosal myeloperoxidase activity in cat intestine suggest that there are approximately 10 million granulocytes per gram tissue (23). When maximally activated, these cells can generate a superoxide flux of approximately 35 nmoles/min/g tissue, a rate which is highly cytotoxic. Activated granulocytes also secrete a variety of enzymes (myeloperoxidase, collagenase, elastase) that can injury parenchymal cells and the microvasculature (8).

We have examined the influence of ischemia and reperfusion on granulocyte fluxes in the cat intestinal mucosa using tissue associated myeloperoxidase (MPO) activity (23). We observed a significant increase in mucosal MPO during the ischemic period while reperfusion produced an even more dramatic enhancement of MPO activity. In an attempt to determine whether I/R-induced tissue granulocyte infiltration is related to the formation of xanthine oxidase-derived oxidants we examined the influence of treatment with either SOD or allopurinol on the I/R-induced increase in mucosal MPO activity (23). The results of these studies indicate that both SOD and allopurinol significantly attenuate the increased mucosal MPO activity observed after reperfusion. This attenuation of MPO activity reflected a decrease in the amount of enzyme rather than inhibition of MPO catalytic activity by SOD and allopurinol. We have recently extended these observations by demonstrating that the reperfusion-induced granulocyte infiltration in intestine is largely prevented by pretreatment with either catalase, deferoxamine, or dimethylthiourea, a hydroxyl radical scavenger (24).

In an attempt to directly observe the response of granulocytes to ischemia and reperfusion, we applied intravital microscopic techniques to study adherence, rolling velocity and extravasation of leukocytes in cat mesenteric venules during low flow ischemia and reperfusion (25,26). The cat mesentery contains 50-80 mU/g tissue xanthine oxidase activity. Leukocyte adherence increases 3.5-, 6-and 5-times control during ischemia and 10 min and 50 min after reperfusion, respectively. The number of extravasated leukocytes increases 3.4-, 5- and 8-times control during the same periods. The responses of venular blood flow, wall shear rate, and leukocyte rolling velocity to ischemia and reperfusion do not differ between control (untreated) animals and animals treated with either allopurinol or superoxide dismutase. However, the numbers of adherent and extravasated leukocytes following reperfusion are significantly lower in both allopurinol and SOD treated animals. These preliminary observations support the view that xanthine oxidase-derived oxidants plays a critical role in the attraction and activation of granulocytes following reperfusion of ischemic tissues.

The ability of xanthine oxidase inhibitors, oxy-radical scavengers and an iron chelator to interfere with I/R-induced granulcoyte infiltration suggests that xanthine oxidase-derived oxidants play a role in the recruitment of granulocytes in the post-ischemic intestine. Our working hypothesis is that xanthine oxidase-derived oxidants, produced in epithelial and endothelial cells, initiate the production and release of proinflammatory agents which subsequently attract and activate granulocytes. This scheme would explain why agents such as SOD, catalase, deferroxamine and allopurinol attenuate both the granulocyte infiltration and microvascular injury induced by I/R.

There are several naturally occurring substances that could mediate the granulocyte infiltration associated with reperfusion of the ischemic intestine. These include bacterial products that gain access to the mucosal interstitium via the gut lumen (N-formylated peptides, endotoxin), leukotrienes, activated complement components (C5a), lipid hydroperoxides, and superoxide-dependent chemoattractants. We have recently examined the chemotactic potential of feline plasma exposed to superoxide using in vitro and in vivo methods (27). The results obtained from these studies indicate that the chemotactic potential of feline extracellular fluid is not enhanced by exposure to superoxide. Extracellular fluid levels of leukotrienes (LTB_4) and activated complement components have not been determined in intestines subjected to I/R. We are currently attempting to determine which proinflammatory agent links xanthine oxidase-derived oxidants to reperfusion-induced granulocyte infiltration.

An important question that arises from the data relating granulocyte infiltration to xanthine oxidase-derived oxidants is whether neutrophils are a cause or an effect of ischemia/reperfusion injury. Hernandez et al (28) have recently assessed the importance of granulocytes in mediating the microvascular injury associated with ischemia/reperfusion in the small intestine using two approaches, i.e., neutrophil depletion and prevention of neutrophil adherence with a monoclonal antibody (Mo 60.3) directed against a specific membrane-associated glycoprotein that modulates granulocyte adherence to endothelium. Our results indicate that both granulocyte depletion and prevention of granulocyte adherence attenuate the I/R-induced increase in microvascular permeability (Table 1). The observation that granulocyte depletion and prevention of granulocyte adherence are equally effective in attenuating the injury suggests that adherence of granulocytes to endothelium is a limiting factor in I/R-induced microvascular injury. In some tissues, the protective effect of granulocyte depletion could be attributed to an increased perfusion during the ischemic phase (29), however, this potential influence was eliminated in our studies by precise control of total intestinal blood flow during the ischemic period in all experimental groups.

The relative importance of xanthine oxidase-derived oxidants and granulocytes in mediating the increased intestinal microvascular permeability produced by ischemia per se remains undefined. Although there is a 5-7-fold increase in mucosal granulocytes during the ischemic period (23), the altered chemical composition of extracellular fluid induced by ischemia may lead to a suppression of granulocyte function. Neutrophilic superoxide production can be inhibited by hypoxia, acidosis and adenosine (30,31). Neutrophilic superoxide production is half maximal at oxygen tensions of 3-10 mmHg (31,32). Inasmuch as resting tissue pO_2 at the apex of rat intestinal villi (the region that is most vulnerable to ischemia-reperfusion injury) is as low as 5 mmHg (mean value = 13 mmHg), it is likely that tissue pO_2 falls well below the k_m for neutrophilic superoxide production when intestinal blood flow is reduced to 20-30% of normal (33). Estimates of intestinal mucosa pH during ischemia generally fall between 6.0-6.7 (34). In vitro studies indicate that acidosis dramatically suppresses superoxide production by human neutrophils, such that O_2^- production is reduced by 60% and 80% at a pH of 6.5 and 6.0, respectively (31). Finally, adenosine, which accumulates in the extracellular fluid of ischemic tissues, has been shown to suppress superoxide production and inhibit the adherence of neutrophils to microvascular endothelium (30). We have observed that intra-arterial infusion of adenosine (to achieve a blood level of 2 nmole/ml) significantly attenuates the intestinal microvascular injury induced by I/R (35). Whether adenosine affords this protection by suppressing neutrophilic superoxide production or prevention of adherence remains unclear.

Although the results of our studies support a role for granulocytes in I/R-induced microvascular injury, the chemical mediators of this injury remain undefined. It is tempting to attribute the injury process entirely to oxidants since oxy-radical scavengers also protect against ischemia/reperfusion injury. However, this observation alone does not constitute strong support for oxidants as final mediators of injury since superoxide dismutase and catalase also prevent neutrophil infiltration, indicating that oxidants may function primarily to recruit granulocytes into postischemic tissue (Figure 1). In addition to oxidants, activated granulocytes release a variety of proteins that are capable of damaging the microvasculature. These include lactoferrin, elastase, collagenase, and cationic proteins. There is also evidence indicating that oxidant production is required in order for some of neutrophilic proteases to produce tissue injury (e.g., collagenase activation requires hypochlorous acid) (36).

REFERENCES

1. D.N. Granger, G. Rutili, and J.M. McCord. Superoxide radicals in feline intestinal ischemia. Gastroenterology 81:22 (1981).
2. R.J. Korthuis and D.N. Granger. Ischemia-reperfusion injury: Role of oxygen-derived free radicals. In Physiology of Oxygen Radicals. Bethesda, MD: Am. Physiol. Soc. chap. 17, p. 217-249 (1986).
3. D.A. Parks and D.N. Granger. Contributions of ischemia and reperfusion to mucosal lesion formation. Am. J. Physiol. 250 (Gastrointest. Liver Physiol. 13): G749 (1986).
4. H. Blum, J.J. Summers, M.D. Schnall, C. Barlow, J.S. Leigh, B. Chance and G.P. Buzby. Acute intestinal ischemia studies by phosphorous nuclear magnetic resonance spectroscopy. Ann. Surg. 204:83 (1986).
5. M.H. Schoenberg, B.B. Fredholm, U. Haglund, H. Jung, D. Sellin, M. Younes, and F.W. Schildberg. Studies on the oxygen radical mechanism involved in small intestinal reperfusion damage. Acta Physiol. Scand. 124:581 (1985).
6. J.W.L. Robinson, V. Mirkovitch, B. Winistorfer, and F. Saegesser. Response of the intestinal mucosa to ischemia. Gut 22:512 (1981).
7. B. Mousson, P. Desjacques, and P. Baltasatt. Measurement of xanthine oxidase activity in some human tissues. Enzyme 29:32 (1983).
8. S.J. Weiss. Oxygen, ischemia and inflammation. Acta Physiol. Scand. Suppl. 548:9 (1986).
9. L.A. Hernandez, M.B. Grisham, and D.N. Granger. A role for iron in oxidant-mediated ischemic injury to intestinal microvasculature. Am. J. Physiol. 253:G49 (1987).
10. M. Younes, A. Mohr, M.H. Schoenberg, and F.W. Schildberg. Inhibition of lipid peroxidation by superoxide dismutase following regional intestinal ischemia and reperfusion. Res. Exp. Med. 187:9 (1987).
11. D.A. Parks and D.N. Granger. Xanthine oxidase: Biochemistry, distribution and physiology. Acta Physiol. Scand. Supp. 548:97 (1986).
12. R.H. Simon, C.H. Scoggin, and D. Patterson. Hydrogen peroxide causes the fatal injury to human fibroblasts exposed to oxygen radicals. J. Biol. Chem. 256:7181 (1981).
13. D.N. Granger, M.E. Hollwarth, and D.A. Parks. Ischemia-reperfusion injury: role of oxygen-derived free radicals. Acta Physiol. Scand. Suppl. 548:47 (1986).
14. E.D. Jarasch, G. Bruder, and H.W. Heid. Significance of xanthine oxidase in capillary endothelial cells. Acta Physiol. Scand. Suppl. 548:39 (1986).
15. L.A. Hernandez, M.B. Grisham, C. von Ritter, and D.N. Granger. Biochemical localization of xanthine oxidase in the cat small intestine. Gastroenterology 92:1433 (1987).
16. D.A. Parks, T.K. Williams and J.S. Beckman. Conversion of xanthine dehydrogenase to oxidase in ischemic rat intestine: A re-evaluation. Am. J. Physiol. 254:G768 (1988).

17. D.A. Parks, G.B. Bulkley, D.N. Granger, S.R. Hamilton, and J.M. McCord. Ischemic injury in the cat small intestine: Role of superoxide radicals. Gastroenterology 82:9 (1982).

18. J.B. Morris, U. Haglund, and G.B. Bulkley. The protection from postischemic injury by xanthine oxidase inhibition: Blockade of free radical generation or purine salvage. Gastroenterology 92:1542 (1987).

19. D.N. Granger, J.M. McCord, D.A. Parks, and M.E. Hollwarth. Xanthine oxidase inhibitors attenuate ischemia-induced vascular permeability changes in the cat intestine. Gastroenterology 90:80 (1986).

20. D.A. Parks and D.N. Granger. Role of oxygen radicals in gastrointestinal ischemia. In Superoxide and Superoxide Dismutase in Chemistry, Biology and Medicine, edited by G. Rotilio. Amsterdam:Elsevier, 1986, p. 614-617.

21. P.C. Moorhouse, M. Grootveld, B. Halliwell, J.G. Quinlan, and J.M.C. Gutteridge. Allopurinol and oxypurinol are hydroxyl radical scavengers. FEBS Letters 213(1): 23 (1987).

22. B.J. Zimmerman, D.A. Parks, M.B. Grisham, and D.N. Granger. Allopurinol does not enhance the antioxidant properties of extracellular fluid. Am. J. Physiol. 255:H202 (1988).

23. M.B. Grisham, L.A. Hernandez, and D.N. Granger. Xanthine oxidase and neutrophil infiltration in intestinal ischemia. Am. J. Physiol. 251 (Gastrointest. Liver Physiol. 14): G567 (1986).

24. B.J. Zimmerman and D.N. Granger. Role of hydrogen peroxide, iron, and hydroxyl radicals in ischemia/reperfusion-induced neutrophil infiltration. The Physiologist 31:000, 1988.

25. J. Russell, J.N. Benoit, M.B. Grisham and D.N. Granger. The Physiologist 31:A31 (1988).

26. D.N. Granger, J. Russell, M.B. Grisham and J.N. Benoit. The Physiologist 31:A31 (1988).

27. B.J. Zimmerman, M.B. Grisham, and D.N. Granger. Role of superoxide-dependent chemoattractants in ischemia-reperfusion induced neutrophil infiltration. Fed. Proc. 46:1124 (1987).

28. L.A. Hernandez, M.B. Grisham, B. Twohig, K.E. Arfors, J.M. Harlan and D.N. Granger. Role of neutrophils in ischemia-reperfusion induced microvascular injury. Am. J. Physiol. 253:H699 (1987).

29. G.W. Schmid-Schonbein and R.L. Engler. Granulocytes as active participants in acute myocardial ischemia and infarction. Am. J. Cardiovasc. Path. 1:15 (1987).

30. B.N. Cronstein, R.I. Levin, J. Belanoff, G. Weissmann, and R. Hirschhorn. Adenosine: an endogenous inhibitor of neutrophil mediated injury to endothelial cells. J. Clin. Invest. 78:760 (1986).

31. T.G. Gabig, S.I. Bearman, and B.M. Babior. Effects of oxygen tension and pH on the respiratory burst of human neutrophils. Blood 53:1133 (1979).

32. D.P. Jones. The role of oxygen concentration in oxidative stress: Hypoxic and hyperoxic models. In: Oxidative Stress (H. Sies, ed), Academic Press pp. 152-195 (1985).

33. H.G. Bohlen. Intestinal tissue pO_2 and microvascular responses during glucose exposure. Am. J. Physiol. 238:H164 (1980).

34. H. Blum, B. Chance, and G.P. Buzby. In vivo noninvasive observation of acute mesenteric ischemia in rats. Surg. Gyn. Obstet. 164:409 (1987).

35. D.N. Granger, M.B. Grisham, and L.A. Hernandez. Ischemia-reperfusion injury in the small intestine: Effect of adenosine. Fed. Proc. 46:1124 (1986).

36. S.J. Weiss, G. Peppin, X. Oritz, C. Ragsdale, and S.T. Test. Oxidative autoactivation of latent collagenase by human neutrophils. Science 227:747 (1985).

EVIDENCE FOR FREE RADICAL GENERATION IN VIVO DURING CARDIAC ISCHEMIA AND REPERFUSION

Roberto Bolli[*], Bharat S. Patel[*], Mohamed O. Jeroudi[*],
Edward K. Lai, and Paul B. McCay

Oklahoma Medical Research Foundation, Oklahoma City, OK,
U.S.A. and [*]Baylor College of Medicine, Houston, TX, U.S.A.

INTRODUCTION

Myocardial stunning results from reperfusion of a zone of heart muscle that has been subjected to ischemia for a period of less than 20 min (10). Although stunning usually results in temporary disruption of heart function, it does not appear to result in death of myocytes which is characteristic of longer periods of ischemia followed by reperfusion (21). Recovery of viable tissue in the reperfused zone after more extensive periods of ischemia is a function of the time of coronary occlusion. Based on the observation of a number of investigators, it became apparent that reoxygenation of anoxic tissues was detrimental to those tissues, but not as detrimental as would be the case if there were no reoxygenation at all. The oxygen paradox concept was set forth by Hearse and colleagues (8) who determined that there were ultrastructural changes in myocytes which occurred as a result of reoxygenation of ischemic heart tissue (9). Other investigations suggested that oxygen free radicals might be involved in reperfusion injury to tissues in general due to the conversion of xanthine dehydrogenase to xanthine oxidase during ischemia (7). In addition, other possible sources of radical production have been implicated, including activated neutrophils (22,23) and cellular redox functions such as the mitochondrial transport system (17) and eicosanoid metabolism (5,24). Factors which inhibit or modulate the in vitro activities of these various sources of radical production and which have been reported to provide some protection against reperfusion tissue injury when administered in vivo constitute the basis for the implications (prostaglandin synthesis inhibitors, superoxide dismutase (15), oxypurinol (19), free radical scavengers (16), iron chelators (3), etc.).

Various efforts have been made to demonstrate the production of free radicals in the myocardium during ischemia and reperfusion. Some positive findings have been reported with isolated heart preparations and spin trapping agents (6,11,25). These heart preparations lacked neurohumoral influences, were globally ischemic, were non-working, and were perfused with buffer so that the relevance of the findings with these models to the stunning phenomenon in the intact animal is uncertain. A report of direct observation of radicals using freeze-clamping and electron spin resonance (26) has been rendered doubtful because the method employed produces artifacts (1). In addition the contribution of the ascorbate radical in the various direct observations on freeze-clamped tissues does not seem to have been dealt with.

In no case have free radicals been detected in the regionally ischemic-reperfused, working heart in an intact animal.

The procedures employed in the investigation combined the advantages of a well-established in vivo model of myocardial reperfusion injury which has been shown to significantly protected by treatment with various anti-oxidants (4,15,16), with the development of an in vivo technique for trapping and detecting highly reactive free radicals in tissues of intact animals (12,13,18). The observation of free radicals in blood emerging from an ischemic zone of a dog heart at the time of reperfusion is described in this report. The radicals appear to be a mixture of carbon-centered spin adducts and at least one other adduct species. The spin trapping technique employed enabled us to observe production of these radicals at various intervals over an extended time period. Initially, there was an abrupt increase in radical production within 4 min. of initiating reperfusion, followed by a gradual decrease during an ensuing 3 hr. period.

METHODS

An open-chested, pentobarbital-anesthetized dog model which had previously been shown to benefit from antioxidant treatment during coronary occlusion and reperfusion (4,15,16), was employed in these experiments. The technical details of instrumentation and measurements of physiological indicators of myocardial function of this model including the extent of collateral circulation in the ischemic zone have been described by Bolli et al (2). In brief, regional ischemia was produced by snaring a coronary artery (either the posterior descending or the obtuse marginal) into which a catheter with a 27-gauge needle was inserted so that the end of the needle was just distal to the snare. The spin trapping agent, PBN (or vehicle) was infused through this catheter to provide an intra-arterial concentration of 285 ug/ml (1.6 mM). A catheter was also inserted through the coronary sinus into the corresponding coronary vein accompanying the artery to be occluded for periodic collection of blood samples. The blood samples were immediately centrifuged in the cold to obtain the plasma fraction which was then frozen (-70°) while being transferred to the laboratory where analysis of spin-trapped free radicals were done. The plasma was thawed and 3.0 ml was subjected to total lipid extraction with chloroform-methanol (2:1). Phase separation was achieved by addition of one-fifth volume of 0.5% saline, and the chloroform layer, which contains the PBN and adducts thereof, was collected and concentrated to recover the extracted materials. Samples of the lipid extract were placed in Pasteur pipettes sealed at the tip end and deposited at that end centrifugal force. The pipettes were then scanned in an electron paramagnetic resonance spectrometer for evidence of free radical adducts of PBN (14). The spectrometer settings were: microwave power, 19.7 mW; modulation amplitude, 1 G; time constant, 1.35 s; scan range, 100; scan time, 8 min; temperature, 25°. Free radical production was determined from the observed adduct spectra and expressed as intensity as measured by the height of first peak of the second set of doublets of the PBN adduct spectral, adjusting all values to the same gain. Radical generation intensity values are expressed as mean ± SEM.

Three types of experiments were done: 1) the spin trapping agent was given 5 min before beginning ischemia and continuing until 10 min after reperfusion (n = 5); 2) spin trap administration was begun 20 sec before reperfusion and continued for 10 min (n = 5); and 3) PBN was administered for 10 min at 30 min (n = 3), 1 hr (n = 3), or 2 hrs (n = 3) after reperfusion. Control studies were performed by administering PBN at the same rate to animals which were not subjected to ischemia and reperfusion (n = 11) to determine whether or not free radical adducts were observed under these conditions. An additional control was to perform ischemia and reperfusion in animals which were not administered PBN.

The physiological parameters of regional myocardial function which were measured during the experiments included systolic wall thickening. Postmortem examination of the occluded-reperfused zone was performed to establish that no irreversible tissue injury resulted from the experimental treatment.

RESULTS

Group I animals were the first to be studied (those give PBN before occlusion of the coronary artery period began, with the blood sample being drawn 3 min after reperfusion). EPR analysis of a lipid extract of the plasma fraction of blood drawn from the corresponding coronary vein exhibited the signal shown in Figure 1A. Computer assisted analysis of this spectrum (which was typical of those in Group I animals) was consistent with it being a mixture of carbon-centered radical adducts of PBN and at least one other unknown adduct, with the latter predominating by a ratio of approximately 1:3. All of the adducts observed in these and later experiments were lipid-soluble, suggesting that they may be PBN-trapped lipid radicals produced during lipid peroxidation. Positive identification will probably depend on isolation of the adducts and determination of their structure.

In order to establish that the observed radicals were produced as a consequence of the ischemia-reperfusion episode, similar experiments were carried out with control animals (treated in the same manner but not subjected to ischemia/reperfusion) in which blood samples were taken at a corresponding time interval after administration as in Figure 1A. Assay of the lipid extract of plasma from this sample showed no evidence of an EPR signal in this control animal (Figure 1B).

Figure 2C shows EPR spectra of coronary venous plasma samples in animals from Group II (those given PBN 20 sec before reperfusion began, with the blood sample being taken 5 min after reperfusion). The paramagnetic species responsible for these EPR spectra are tentatively assigned as being PBN adducts of a mixture of carbon-centered radicals and one or more additional adducts in a 1:1 ratio. Spin adducts were not detected in the plasma of corresponding control animals that not been subjected to ischemia and reperfusion (Figure 2D).

Figure 3E shows the EPR signal observed in the coronary venous plasma of Group III animals (those given PBN 30 minutes after reperfusion, with the blood sample being taken 5 min after PBN administration). This signal, although having only about one-fourth the strength of that observed in Group I and II dogs, is still reasonably strong, and indicated that free radicals were being produced for at least 35 min after reperfusion was initiated. The control animal which was prepared in the same manner but not subjected to ischemia and reperfusion showed no evidence of an EPR signal (Figure 3F). In light of these initial results in which the EPR signal appeared to be most intense at 5 min after the initiation of reperfusion, a determination was made to investigate radical production at more time points over a much longer period of time. In this set of studies, PBN was given either 5 minutes before reperfusion (Group I) or 20 sec before reperfusion (Group II). Figure 4 shows that in either case, the intensity and duration of radical production was essentially the same, indicating that most of the radicals are generated during the first few minutes of reperfusion, and continue to be detected for as long as 3 hours after reperfusion. These results, together with the data shown in Figure

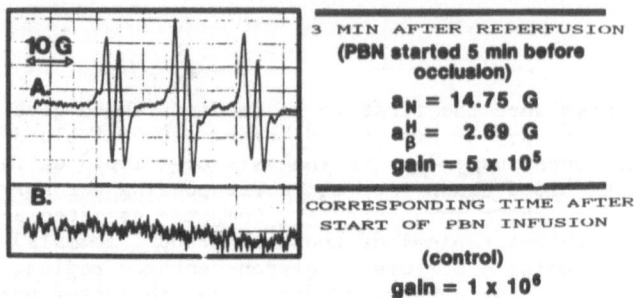

Figure 1

A Electron spin resonance spectrum of PBN adducts observed in a lipid extract of plasma obtained from coronary venous blood emerging from and ischemic/reperfused zone of the heart. PBN was administered in the selected coronary artery 5 min. before occlusion and then the blood sample was taken from the corresponding coronary vein 3 min after reperfusion (Group I). B Control in which PBN was administered and blood sample taken at the same time intervals but the heart was not subjected to occlusion/reperfusion treatment.

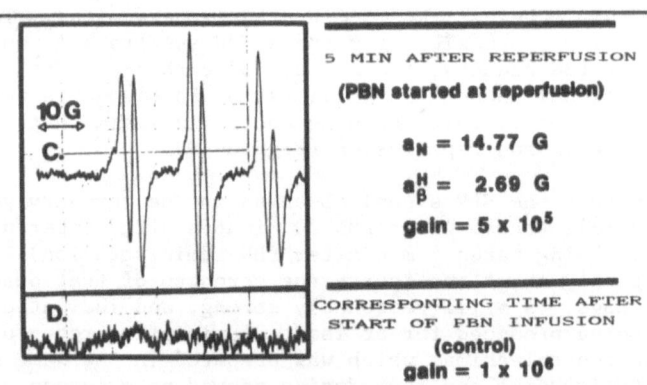

Figure 2

C As in Figure 1A excepting that PBN was not administered until 20 sec prior to reperfusion. The blood sample was taken from the corresponding coronary vein 5 min after reperfusion (Group II). D Control in which PBN was administered and blood sample taken at the corresponding times but the heart was not subjected to ischemia/reperfusion treatment.

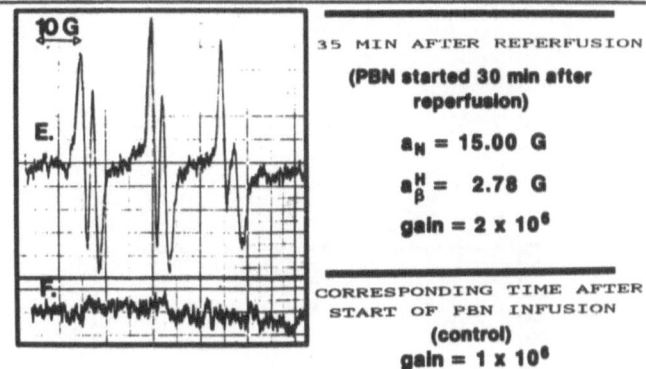

Figure 3

E As in Figure 1A excepting that PBN was not administered until 30 min after the beginning of reperfusion. The blood sample was taken from the coronary vein 5 min after reperfusion began (Group III). F Control in which PBN was administered and blood sample taken at the corresponding times but the heart was not subjected to the ischemia/reperfusion treatment.

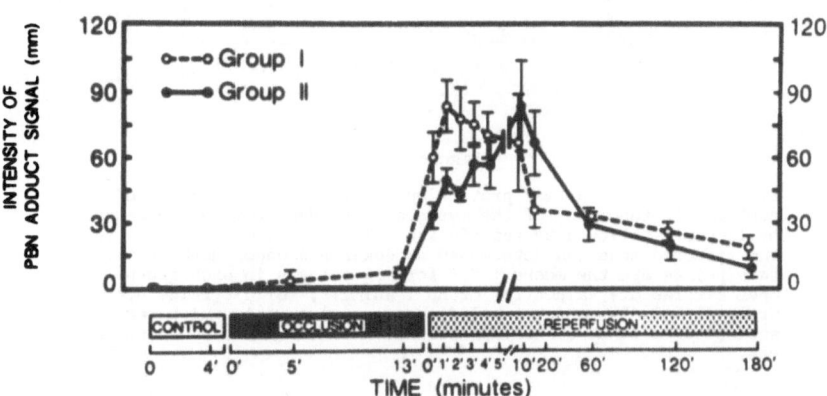

Figure 4

Time course of free radical production in the affected myocardial zone following ischemia and reperfusion. The intensity of radical production at each interval was determined by measuring the peak height (in mm) of the first line of the second doublet in each spectrum. PBN was administered in Group I 5 min before occlusion and the venous plasma sample was taken 3 min after reperfusion. PBN was administered to Group II animals 20 sec before reperfusion and the plasma sample taken 5 min after reperfusion.

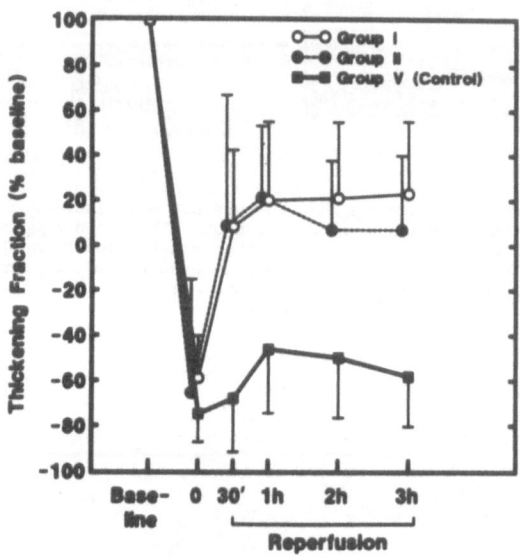

Figure 5

Evidence for the partial protection (manifested as recovery of
contractile function) of the myocardium by PBN from injury caused
by ischemia followed by reperfusion. The systolic thickening in
the affected zone was determined as described under Methods. The
data points are the means ± SEM for all animals in each group.
Open circles are values for Group I animals, solid circles are
Group II animals, and the solid squares are animals which were
subjected to zonal ischemia/reperfusion but were not administered
PBN.

1, clearly indicate that free radical generation in the ischemic-reperfused heart does not occur as a brief burst of activity, but appears to be prolonged, diminishing relatively slowly after a sudden early burst of radical production after reperfusion. Very little radical production was detected during the ischemic period (Fig. 4). Figure 5 shows the systolic wall thickening values as percent of the baseline measurement. These analyses were made as an assessment of myocardial function in the affected region. Systolic thickening fraction was computed from the ratio of net systolic thickening (defined as the maximal systolic increase in wall thickness from the the end diastolic value (20) to end diastolic wall thickness multiplied by 100 (20). In instances where paradoxical wall thinning occurred for 50% or more of systole, the maximum wall thinning value was subtracted from wall thickening to give net systolic thickening (4,15,16). It is apparent from the data shown that PBN itself had a measureable effect on recovery of contractile function. In the control group of animals which were subjected to ischemia and reperfusion but were not administered PBN, (Group V) there was very little recovery of contractile function during the three hour period of observation. However, in groups I and II, which exhibited the same degree of paradoxical wall thinning (dsykinesis) during the ischemic episode, both groups showed significant recovery of contractile function during reperfusion. Group II, in spite of the fact that infusion of PBN into the coronary artery in these animals did not begin until just before reperfusion was initiate, showed essentially the same degree of recovery as Group I.

DISCUSSION

The coupling constants (a^H and a_N) given in Figures 1, 2, and 3 show that the type of radicals being produced at the 3 and 5 min reperfusion time points appear to be quite similar, but in reality the signal shows a shift in adduct composition during the first 35 minutes. The changing values for the coupling constants indicate that either the ratio of the carbon-centered to the unidentified adduct species is changing, or that different types of radicals are being produced, or both. This cannot be determined without positive identification of the radical adducts. The latter will require isolation of the adducts and determination of structure.

Two different types of control studies were performed to verify that the observed EPR signals required both myocardial ischemia and reperfusion and also required the infusion of the spin trapping agent. No signals were observed in the plasma from the ischemic-reperfused zone of the hearts of animals not infused with PBN. Likewise, in PBN-infused animals, no signals were observed in the plasma of leaving a myocardial zone which had not been subjected to ischemia and reperfusion. The first type of control (ischemia-reperfusion but no infusion of the spin trapping agent) indicates that whatever radicals are generated by the affected zone of tissue do not persist long enough to be detected by the EPR analytical process. This observation together with the fact that the radicals can form adducts with PBN indicate that they are extremely reactive chemical species.

The site of production of the observed radicals has not yet been determined. Although they were detected by withdrawal of blood leaving the affected zone of the heart, it is possible that they could have been generated and trapped in myocardial tissue with subsequent diffusion of the trapped radicals into the blood stream. On the other hand, they may have been produced intravascularly. Conceivably, both compartments could be involved. It has also not been established whether or not the presence of oxygen in the reperfusing blood flow is required for the production of the radicals. Experiments designed to acquire information about site of radical production and the possible requirement for oxygen are currently being developed.

PBN, being a radical trapping agent, might be expected to exert some protective effect on myocardial function if the free radicals which are generated during ischemia-reperfusion are essential for some aspect of the stunning phenomenon. The data illustrated in Figure 5 indicate that significant protection of myocardial function is exerted by the infusion of PBN, lending further credence to the idea that free radicals are active participants in the stunning phenomenon.

This is the first unequivocal demonstration of free radical generation in myocardial tissue in vivo. The experimental conditions employed in these investigations retained the interactions between neural, hormonal, and hematological elements that are missing in studies using isolated heart preparations for the investigation of reperfusion injury. These interactions may be critical to the events that occur when heart tissue is subjected to a brief ischemic episode.

REFERENCES

1. Baker, J.E., C.C. Felix, G.N. Olinger, and B. Kalyanaraman. Myocardial ischemia and reperfusion: Direct evidence for free radical generation by electron spin resonance spectroscopy. Proc. Natl. Acad. Sci. USA 85:2786-2789, 1988.
2. Bolli, R., B.S. Patel, M.O. Jeroudi, E.K. Lai, and P.B. McCay. Demonstration of free radical generation in "stunned" myocardium of intact dogs with the use of the spin trap alpha-phenyl-N-tert-butylnitrone. J. Clin. Invest. 82:476-485, 1988.
3. Bolli, R., B.S. Patel, W.X. Zhu, P.G. O'Neill, C.J. Hartley, M.L. Charlat, and R. Roberts. The iron chelator desferrioxamine attenuates postischemic ventricular dysfunction. Am. J. Physiol. 253:H1372-H1380, 1987.
4. Bolli, R., W.X. Zhu, C.J. Harley, L.H. Michael, J. Repine, M.L. Hess, R.C. Kukreja, and R. Robergs. Dimethylthiourea attenuates dysfunction in the postischemic "stunned" myocardium. Circulation 76:458-468, 1987.
5. Emancipator, S.N., and M.E. Lamm. Pathways of tissue injury initiated by humoral immune mechanisms. Lab. Invest. 54:475-478, 1986.
6. Garlick, P.B., M.J. Davies, D.J. Hearse, and T.F. Slater. Direct detection of free radicals in the reperfused rat heart using electron spin resonance spectroscopy. Circ. Res. 61:757-760, 1987.
7. Granger, D.N., G. Rutilli and J.M. McCord. Superoxide radicals in feline intestinal ischemia. Gastroenterology. 81:22-29, 1981.
8. Hearse, D.J., S.M. Humphrey, and E.B. Chain. Abrupt reoxygenation of the anoxic potassium-arrested perfused heart: a study of myocardial enzyme release. J. Mol. Cell. Cardiol. 81:22-29, 1981.
9. Hearse, D.J., S.M. Humphrey, W.G. Nayler, A. Slade, and D. Border. Ultrastructural damage associated with reoxygenation of the anoxic myocardium. J. Mol. Cell. Cardiol. 7:315-324, 1975.
10. Heyndrickx, G.R., H. Baig, P. Nellens, I. Leusen, M.C. Fishbein, and S.F. Vatner. Depression of regional blood flow and wall thickening after brief coronary occlusions. Am. J. Physiol. 234:H653-H659, 1978.
11. Kramer, J.H., C.M. Arroyo, B.F. Dickens, and W.B. Weglicki. Spin-trapping evidence that graded myocardial ischemia alters post-ischemic superoxide production. Free. Radic. Biol. Med. 3:153-159, 1987.
12. Lai, E.K., C. Crossley, R. Sridhar, H.P. Misra, E.G. Janzen, and P.B. McCay. In vivo spin trapping of free radicals generated in brain,

spleen, and liver during gamma radiation of mice. Arch. Biochem. Biophys. 244:156-160, 1986.

13. Lai, E.K., P.B. McCay, T. Noguchi, and K.-.L. Fong. In vivo spin trapping of trichloromethyl radicals formed from CC14. Biochem. Pharmacol. 28:2231-2235, 1979.

14. McCay, P.B., E.K. Lai, J.L. Poyer, C.M. DuBose, and E.G. Janzen. Oxygen- and Carbon-centered Free Radical Formation During Carbon Tetrachloride Metabolism. Observation of Lipid Radicals In Vivo and In Vitro. J. Biol. Chem. 259:2135-2143, 1984.

15. Myers, M.L., R. Bolli, R.F. Lekich, C.J. Hartley, and R. Roberts. Enhancement of recovery of myocardial function by oxygen free radical scavengers after reversible regional ischemia. Circulation 72:915-921, 1985.

16. Myers, M.L., R. Bolli, R.F. Lekich, C.J. Hartley, and R. Roberts. N-2-mercaptopropionylglycine improves recovery of myocardial function after reversible regional ischemia. J. Am. Coll. Cardiol. 8:1161-1168, 1986.

17. Nohl, H., W. Jorday, and D. Hegner. Identification of free hydroxyl radicals in respiring rat heart mitochondria by spin trapping with the nitrone DMPO. FEBS. Lett. 123:241-244, 1981.

18. Poyer, J.L., P.B. McCay, C.C. Weddle, and P.E. Downs. In vivo spin-trapping of radicals formed during halothane metabolism. Biochem. Pharmacol. 30:1517-1519, 1981.

19. Puett, D.W., M.B. Forman, C.U. Cates, B.H. Wilson, K.R. Hande, G.C. Friesinger, and R. Virmani. Oxypurinol limits myocardial stunning but does not reduce infarct size after reperfusion. Circulation 76:678-686, 1987.

20. Raon, P., F. Scales, S. Saffer, L.M. Buja, and J.R. Willerson. Functional characterization of left ventricular segmental responses during the initial 24 hours and first week after experimental canine myocardial infarction. J. Clin. Invest. 64:1074-1088, 1979.

21. Reimer, K.A., J.E. Lowe, M.M. Rasmussen, and R.B. Jennings. The wavefron phenomenon of ischemic cell death. I. Myocardial infarct size vs duration of coronary occlusion in dogs. Circulation 56:786-794, 1977.

22. Simpson, P.J., and B.R. Lucchesi. Free radicals and myocardial ischemia and reperfusion injury. J. Lab. Clin. Med. 110:13-30, 1987.

23. Thompson, J.A., and M.L. Hess. The oxygen free radical system: a fundamental mechanism in the production of myocardial necrosis. Prog. Cardiovasc. Dis. 28:449-462, 1986.

24. Vapaatalo, H. Free radicals and anti-inflammatory drugs. Med. Biol. 64:1-7, 1986.

25. Zweier, J.L. Measurement of superoxide-derived free radicals in the reperfused heart. Evidence for a free radical mechanism of reperfusion injury. J. Biol. Chem. 263:1353-1357, 1988.

26. Zweier., J.T. FLAHERTY, and M.L. WEISFELDT. Direct Measurement of free radical generation following reperfusion of ischemic myocardium. Proc. Natl. Acad. Sci. 84:1404-1407, 1987.

SUPEROXIDE GENERATION BY CLASTOGENIC FACTORS

Ingrid Emerit, Arlette Levy and Shahid Khan

Institut biomédical des Cordeliers, Université Paris VI
15, rue de l'Ecole de Médecine,
75006 Paris, France.

Chromosome breakage factors or clastogenic factors (CF) were first described by radiobiologists, who reported a chromosome damaging effect of plasma from irradiated persons (for review see 1). These authors insisted on the long lasting nature of this phenomenon and suggested that CF may be risk factors in the development of late neoplasias resulting from therapeutic or accidental irradiaton . Renewed interest in diffusable CF paralleled the recognition of clastogenic activity in the plasma of patients with the socalled congenital breakage syndromes, ataxia telangiectasia (2) and Bloom's syndrome (3), and in acquired diseases with increased chromosome damage, such as chronic inflammatory diseases of the connective tissue, the intestinal tract and the nervous system (for review see 4). CF may be isolated not only from plasma, but also from synovial fluid and cerebrospinal liquor.

Previous word of our laboratory has shown that in all these above mentioned disorders, the clastogenic activity is found in ultrafiltrates through Diaflo filters with a cut off at 1 000 daltons. Another common feature of CF is the possibility to prevent their formation and their clastogenic effect by superoxide dismutase (5). The protective effect of SOD was a first indication for implication of superoxide radicals in this phenomenon. Another argument herefore came from the observation that CF can be generated in vitro by exposure of cells to a source of superoxide. CF was extracted from culture media of lymphocytes treated with TPA, a tumor promoter know to stimulate a respiratory burst and to initiate the arachidonic acid cascade (6). SOD prevented CF formation. Similarly the action of xanthine oxidase on hypoxanthine, a widely used method to generate O_2^- in vitro, led to CF formation. Again SOD was protective (7)

The xanthine oxidase reaction is supposed to be a source of oxiradical generation in eschemia-reperfusion injury, and recent work of our laboratory confirmed the formation of CF in patients submitted to coronary bypass surgery (8). This CF was detectable in the coronary sinus blood 10 to 20 minutes after reperfusion, but not during reperfusion or immediately after unclamping of the aorta. No CF was found in the majority of patients receiving the xanthine oxidase inhibitor allopurinol.

These examples show that CF formation is related to oxidative stress, for which increased superoxide production by inflammatory cells, by the xanthine oxidase reaction or from other sources may be responsible. However not only CF formation, but also CF action seems to be related to oxiradical generation, as indicated by the protective effect of

Free Radicals, Lipoproteins, and Membrane Lipids
Edited by A. Crastes de Paulet et al.
Plenum Press, New York, 1990

SOD in cell cultures derived from healthy subjects. It seemed therefore of interest to examine if CF contain substances capable of stimulating superoxide production by phagocytosing cells. In the following, we report the results of experiments, in which polymorphonuclear neutrophils (PMN) were exposed to CF from three different sources. Superoxide production was measured with the cytochrome C assay.

CLASTOGENIC FACTORS

The CF used for this study were derived from cell cultures exposed to a xanthine oxidase reaction, from patients suffering from progressive systemic sclerosis (scleroderma) or from patients submitted to ischemia-reperfusion during coronary bypass surgery. They will be called in the following : X-X0 factor, scleroderma factor and reperfusion factor. CF preparation followed previously described procedures (4). Briefly this consists in ultrafiltration of culture supernatants (collected after 48 h cultivation) or patients' plasma through Diaflo filter YM 10 (cut off 10000 dalton). Similarly handled ultrafiltrates from untreated cultures or plasma from age and sex matched healthy persons served as controls. For the reperfusion factor, blood was drawn in the coronary sinus before and after clamping of the aorta. In this case, each patient was his own control. Patients operated either without or with the xanthine oxidase inhibitor allopurinol were compared ; each group consisted of 5 or 6 patients.

The clastogenic activity of the CF preparations was tested on blood cultures of healthy persons according to conventional cytogenetic procedures. Fifty mitoses were examined for the presence of breaks, fragments and structurally rearranged chromosomes. The results were expressed as the percentages of mitoses with aberrations. The absence of clastogenic activity in the control samples was ascertained by the same cytogenetic test.
Aliquots were frozen until use.

MEASUREMENT OF SUPEROXIDE PRODUCTION

PMN were obtained from the cell pellet after Ficoll-Hypaque density gradient centrifugation by lysis of the erythrocytes with 0.83 % NH_4CL.

The O_2^- production was measured with the SOD-inhibitable cytochrome C reduction assay. The cells were incubated in 1 ml of 80 µM ferricytochrome C solution in PBS during various exposure times (15 to 90 minutes). In general, 20 µl of CF or control samples were added to the wells. All experiments were done in duplicate. Additional mixtures were prepared containing 10 µg/ml of SOD (Boehringer Mannheim) in order to determine the SOD-inhibitable fraction of cytochrome C reduction. The absorbance was measured in a spectrophotometer at 550 nm using reaction mixtures from wells without cells as blanks. The concentration of reduced cytochrome C was determined using an extinction coefficient of 21000 M^{-1} cm^{-1}. The number of nmoles O_2^- produced was calculated assuming that the reduction reaction is stoichiometric (9).

RESULTS OBTAINED WITH X-X0 FACTOR

The O_2^- production of PMN exposed to X-X0 factor was consistently higher than that observed with ultrafiltrates from untreated cultures. The results of 35 independent experiments are shown in table 1.

Table 1 . Variation observed in 35 experiments for superoxide production by PMN in comparison to CF activity.

n	nmoles $O_2^{\bullet}$	activity
3	0 - 0.9 *	±
10	1 - 2.9	+
11	3 - 4.9	++
11	5 - 10 (or >10)	+++

* The figures represent the increase observed between wells exposed to X-X0 factor or control samples (20 µl each, 2 x 10^6 PMN, measurement after 90 min).

The mean increase observed was 5.1 nmoles $O_2^{\bullet}$, the extremes being 0.7 and 17.9 nmoles. Higher values than 10 nmoles were indeed observed for 3 of the 11 experiments. SOD (10 µg/ml) inhibited completely the cytochrome C reduction indicating that $O_2^{\bullet}$ was the reducing agent. There was a positive correlation between $O_2^{\bullet}$ production and the clastogenic activity of the CF preparation.

The optimal quantity of CF for stimulation of $O_2^{\bullet}$ production had been determined in preliminary experiments. They showed an increase in $O_2^{\bullet}$ production up to 20 µl CF for a total of 1 ml reaction mixture. Higher quantities (30-50 µl) yielded lower values and were probably cytotoxic.

Parallel experiments with increasing cell numbers showed a corresponding increase in O_2^- production (mean 3.3 nmoles, 3.8 nmoles and 4.5 nmoles respectively for 2.5,5 and 7.5 millions of PMN). No further increase was observed for 10 millions of cells. For practical reasons the smallest cell number (2.5 x 10^6) was adopted, since this seemed to be sufficient to obtain statistically significant increases in parallel experiments.

Experiments done with various exposure times showed a linear increase in cytochrome C reduction up to 90 minutes. This seemed to be the optimal exposure time in agreement with data from the literature (9), prolongation up to 120 minutes resulted in lower values.

No cytochrome C reduction was observed with CF added to the reaction mixture in absence of PMN. Filtration of CF through sterilizing Millipore filters did not influence the results. In addition, there were no differences between fresh and frozen samples.

Lymphocytes free of other contaminating blood cells did not produce $O_2^{\bullet}$ after exposure to CF. Lymphocytes separated by density gradient centrifugation are usually contaminated with monocytes, which may represent up to 10 % of the mononuclear cell population. The results obtained with these "regular" lymphocytes varied between 0.5 to 4.2 nmoles $O_2^{\bullet}$, for a total of 7.5 x 10^6 cells. This variation was correlated with the proportion of monocytes in the cell suspension. In experiments in which the lymphocytes and monocytes were separated by attachment of the latter to plastic dishes, the monocyte monolayers produced $O_2^{\bullet}$, while the remaining lymphocytes gave again negative results.

RESULTS OBTAINED WITH SCLERODERMA FACTOR

CF preparations from scleroderma patients were added to PMN from healthy subjects. The experimental conditions were adopted from the study with X-X0 factor i.e. 20 μl of CF were used for 2.5×10^6 PMN. The time, when measurements were performed, varied between 15 to 60 minutes. As may be seen on figure 1, the $O_2^{\bullet}$ production in presence of scleroderma factor was regularly higher than that in the parallel dishes, to which ultrafiltrates from control plasma had been added. The differences were similar for the three exposure times and were statistically significant (p inf 0.01 to 0.05). For each exposure time, 7-9 experiments were done in duplicate with CF from different patients with high chromosome breakage rates.

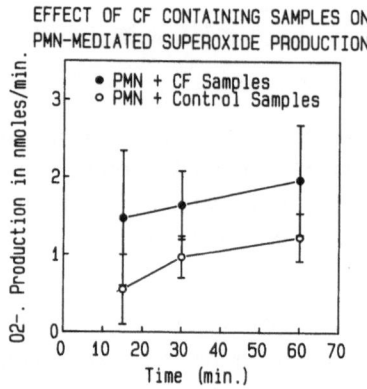

Figure 1. Exposure of normal PMN to scleroderma factor. Differences observed with respect to plasma ultrafiltrates from healthy persons.

RESULTS OBTAINED WITH REPERFUSION FACTOR

As expected from our previous study (8), CF was found in the coronary sinus blood, collected 10 min after reperfusion, in all 5 patients operated without allopurinol. On the other hand, the controls obtained from the blood samples taken before clamping the aorta were not clastogenic. Parallel measurements of $O_2^{\bullet}$ production by PMN from healthy subjects exposes to pre- and postischemic samples showed regularly higher values for the latter samples. The differences are shown in table 2. They vary between 0.9 and 7.7 nmoles $O_2^{\bullet}$ according to patients.

Among the 6 patients receiving allopurinol, only one patient had CF. This sample had moderate $O_2^{\bullet}$ stimulating properties (2.1 nmoles). The samples from the 5 other patients were not clastogenic and except for one (2.5 nmoles) did not stimulate $O_2^{\bullet}$ production.

Since blood collected during ischemia or immediately after reperfusion did not contain detectable levels of CF in our previous study (8), such samples were no taken for the present investigations.

Table 2. Increase in $O_2^{\cdot}$ production observed for
postischemic samples compared to
preischemic samples.

I.		II.	
	7.7		2.1*
	7.3		0.3
	0.9		2.5
	2.1		0
	5.9		0
			0

* The figures represent nmoles $O_2^{\cdot}$ differences
measured after 90 min. Number of PNM : 2.5 x
10^6

Quantities of samples added : 20 μl.
I = 5 patients operated without allopurinol
II = 6 patients operated with 100 mg/l allopurinol
in the cardioplegic solution.

DISCUSSION AND CONCLUSIONS

The results of this study demonstrate that CF stimulate $O_2^{\cdot}$ production by resting
PMN and also by monocytes, as indicated by the still limited data obtained with X-XO
factor. The $O_2^{\cdot}$ production is not important, if compared to that observed with agonists such
as zymosan and TPA. However it was consistently higher than in the corresponding
controls and observed with CF of different origin. Results of experiments in which PMN
were incubated with CF during 30 min prior to stimulation with zymosan or TPA did not
indicate an amplified response to these agonists and are therefore not in favor of a priming
effect.

On the basis of the present observations, we hypothesize that $O_2^{\cdot}$ generated upon
reperfusion (possibly by a xanthine oxidase reaction) initiates the formation of more
longlived, transferable chemical species called CF. They may be responsible for
intravascular activation of phagocytosing cells and play a role in tissue injury. This may be
relevant also for scleroderma patients, where Raynaud's phenomenon and endothelial cell
damage are early findings. Complete arrest of capillary blood flow followed by suffusion
and erythema may create a situation comparable to that in ischemia-reperfusion.

The fact that CF formed via the intermediacy of $O_2^{\cdot}$ stimulate $O_2^{\cdot}$ release and
therefore further CF formation results in an autosustained process. This may explain the
longevity of CF, already observed after irradiation. The clastogenic activity detectable in the
plasma immediately after irradiation persisted in the circulation over years (1).

The exact composition of CF is not know and may be variable according to their
origin. A correlation between CF formation and lipid peroxidation of cell membranes is
documented by the presence of TBA-reactive material and substances with conjugated diene
structure in the clastogenic supernatants of TPA and X-XO treated cultures (6, 7). Also the
possibility to extract CF with ethyl acetate suggested their lipid nature. The aldehyde 4-
hydroxynonenal was detected in 10 of 20 X-XO factor preparations (Emerit, Khan and

Esterbauer in prep.). It was not found in 20 scleroderma factors, which in turn, contained hydrophilic clastogenic material detected by HPLC fractionation.

At present it is not possible to say which component of CF is responsible for the $O_2^{\cdot-}$ stimulating effect.

ACKNOWLEDGEMENT

This study was supported by the GEFLUC (Groupement des entreprises françaises dans la lutte contre le cancer).

REFERENCES

1 . G.B. Faguet, S.M. Reichard and D.A. Welter. radiation-induced clastogenic plasma factors. Cancer Genetics and Cytogenetic, 12:73-83 (1984).
2 . M.Shaham, Y. Becker and M.M. Cohen. A diffusable clastogenic factor in ataxia telangiectasia. Cytogenetic, cell genetic, 27:155-161 (1980).
3 . I. Emerit and P. Cerutti. Clastogenic activity from Bloom's fibroblast cultures. Proc. Natl. Acad. Sci. USA, 78:1868-1872 (1981).
4 . I. Emerit. Chromosome breakage factors : origin and possible significance. Progr. Mutat. Res., 4:61-74 (1982).
5 . I. Emerit. Oxygen-derived free radicals and DNA damage in autoimmune diseases, in "Free Radicals, Aging and Degenerative Diseases.", J.E. Johnson ed. Alan R.Liss, New York pages 307-324 (1986).
6 . S.H. Khan and I. Emerit. Lipid peroxidation products and clastogenic material in culture media of human leukocytes exposed to the tumor promoter phorbol-myristate acetate. J.Free Rad. Biol. Med. 1:443-449 (1985).
7 . I. Emerit, S.H. Khan, and P. Cerutti. Treatment of lymphocyte cultures with a hypoxanthine - xanthine oxidase system induces the formation of transferable clastogenic material. J.Free Rad. Biol. Med. 1:51-57 (1985).
8 . I. Emerit, J.N. Fabiani, O. Ponzio, A. Murday, F. Lunel, A. Carpentier. Clastogenic factor in ischemia-reperfusion. The Annals of Thoracic Surgery, in press.
9 . B.Johnston. Measurment of $O_2^{\cdot-}$ secreted by monocytes and macrophages. In "Methods in Enzymology", L.Packer ed. Academic Press Inc. New York 105 : 215-220 (1984).

CHEMISTRY OF FREE RADICAL EFFECTS

DEGRADATION OF MEMBRANE PHOSPHOLIPIDS BY A DIRECT NUCLEOPHILIC

ACTION OF SUPEROXIDE ANION

DEBY C., BOES M., PINCEMAIL J., BOURDON-NEURAY J. and
G. DEBY-DUPONT

Laboratoire de Biochimie et de Radiobiologie, Université de
Liège, Institut de Chimie, B6, Sart Tilman, 4000 Liège I,
Belgium

At the beginning of the eighties, two contradictory currents of opinions appeared in the literature, about superoxide anion reactivity. For many physicochemists, $O_2^{\cdot-}$ could be considered as a chemical curiosity, presenting only weak reactivity and being thus a "relatively innocuous" species (1). But for biologists and biochemists, $O_2^{\cdot-}$ continued to be regarded, following McCord and Fridovich (2), as an important intermediate, which could be dangerous for aerobic organisms. Recent works, establishing the protective role of superoxide dismutase in grafts and transplantations, as well as in ischemia-reperfusion syndrome, afforded a strong support to that point of view (3-6). However, it was well established that $O_2^{\cdot-}$ cannot act through its oxidant properties (7,8).

It is now commonly admitted that $O_2^{\cdot-}$ does not act by itself, but by the way of the Haber-Weiss cycle, where it reduces iron implicated in a Fenton reaction, together with H_2O_2 (itself issued from $O_2^{\cdot-}$ dismutation), producing $\cdot OH$, which is considered as the real responsible agent of oxygen toxicity (9). The efficiency of SOD could consist in the inhibition of Fe^{3+} reduction, limiting the Fenton reaction, by $O_2^{\cdot-}$ suppression.

Nevertheless, in a recent review, Fridovich sustained the opinion that $O_2^{\cdot-}$ can exert a direct effect without involvement of the Haber-Weiss cycle (10). His assumptions were founded on biological experiments in which SOD was an efficient protector, while catalase failed to modify the phenomenon. But, apart the possibility (very limited in biological conditions) of formation of $HO_2\cdot$ (a powerful oxidant) by protonation of $O_2^{\cdot-}$ in acidic mediums, Fridovich did not give another chemical mechanism to explain a direct effect of $O_2^{\cdot-}$.

However, since 1970 (11), an important property of $O_2^{\cdot-}$ was emphasized : it is its strong nucleophilicity in aprotic mediums. The general nucleophilic reactions can be represented as follows :

Free Radicals, Lipoproteins, and Membrane Lipids
Edited by A. Crastes de Paulet *et al.*
Plenum Press, New York, 1990

$$\left[\text{Nu:}\right]^{-} + \text{b-}\overset{\overset{a}{|}}{\underset{\underset{c}{|}}{\text{C}}}\text{-A} \longrightarrow \text{Nu-}\overset{\overset{a}{|}}{\underset{\underset{c}{|}}{\text{C}}}\text{-b} + \left[\text{A:}\right]^{-}$$

Where $\left[\text{Nu:}\right]^{-}$ is an electrophilic agent provided with an electron doublet. Superoxide anion behaves as a potent nucleophile, particularly able to attack the carbonyl carbon of esters (1,12).

$$\left[\cdot\ddot{\text{O}}:\ddot{\text{O}}:\right]^{-} + \text{R-C}\overset{\nearrow^{\text{O}}}{\searrow_{\text{O-R'}}} \longrightarrow \text{R-C}\overset{\nearrow^{\text{O}}}{\searrow_{\text{O-O}^{\cdot}}} + \text{R'O}^{-}$$

A very transient radicalar intermediate is thus formed, which reacts with a new $\text{O}_2^{\bar{}}$ molecule to form a peroxyacid (12,13).

$$\text{R-C}\overset{\nearrow^{\text{O}}}{\searrow_{\text{OO}^{\cdot}}} + \text{O}_2^{\bar{}} \longrightarrow \text{R-C}\overset{\nearrow^{\text{O}}}{\searrow_{\text{OO}^{-}}} + \text{O}_2$$

which, by a cascade of reactions, involving the peroxy acid, ester molecules, and a new $\text{O}_2^{\bar{}}$ molecule, give the classical products of deesterification : a carboxylic acid and an alcohol (1,13).

In 1978, in a paper which remained largely unknown, Niehaus (14) described the hydrolysis of pure phospholipids, such as dipalmitoyllecithin, by potassium superoxide. The demonstration of a release of non esterified fatty acids (NEFA) was made by gas-liquid chromatography. Niehaus hypothesized that appreciable amounts of superoxide can exist in the hydrophobic region of biological membranes, and that such nucleophilic reactions can thus occur *in vivo*. Effectively, aprotic conditions are found in the intramembranar spaces, limited by the two phospholipidic leaflets; $\text{O}_2^{\bar{}}$ can "survive" several hours in these aprotic conditions (15).

The aim of the present work is to verify if NEFA are released by erythrocyte ghosts submitted to the action of $\text{O}_2^{\bar{}}$, in aprotic conditions, in the same manner as they are released by synthetic phospholipids, as observed by Niehaus. In the same experiments, we monitorized the lipid peroxidation.

METHODS

Human erythrocytes ghosts were prepared according to the technique of Steck and Kant (16), modified following Lynch and Fridovich (17). After hemoglobin elimination, the ghosts were carefully lyophilized and maintained in strict conditions of dryness at -80°C until utilization.

Superoxide anion was electrochemically generated in dimethylformamide (DMF), in the presence of tetrabutylaminobromide Aldrich (TBAB), an electron carrier at the concentration of 0.1M in DMF following a procedure described by Maricle and Hogdson (18). The used device is represented on fig.1; a DC current of 20 V, 50 mA was applied on the Pt electrodes, and a flux of oxygen (dessicated on concentrated H_2SO_4) bubbled at the cathode. After 25 minutes, the $\text{O}_2^{\bar{}}$ concentration in the negative compartment reached up to 10^{-4} M. The decay of superoxide at 37°C in our experimental conditions was studied by electron spin resonance, on a Varian EPR spectrometer, according to Knowles et al (19) (figure 2)

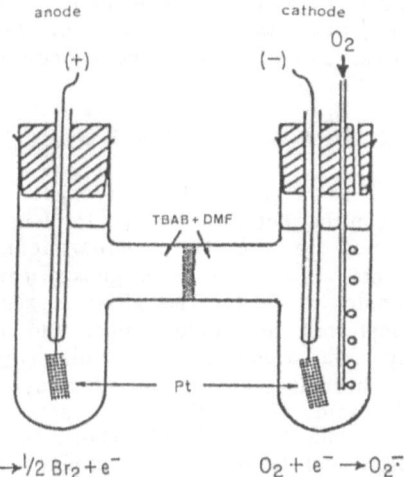

anode   cathode

(+)   (−) O_2

TBAB + DMF

Pt

$Br^- \rightarrow 1/2\, Br_2 + e^-$ $O_2 + e^- \rightarrow O_2^{\cdot -}$

Figure 1   Generator of $O_2^{\cdot-}$ TBAB : tetrabutylaminobromide; DMF : dimethylformamide. O_2 is dessicated by bubbling in concentrated H_2SO_4

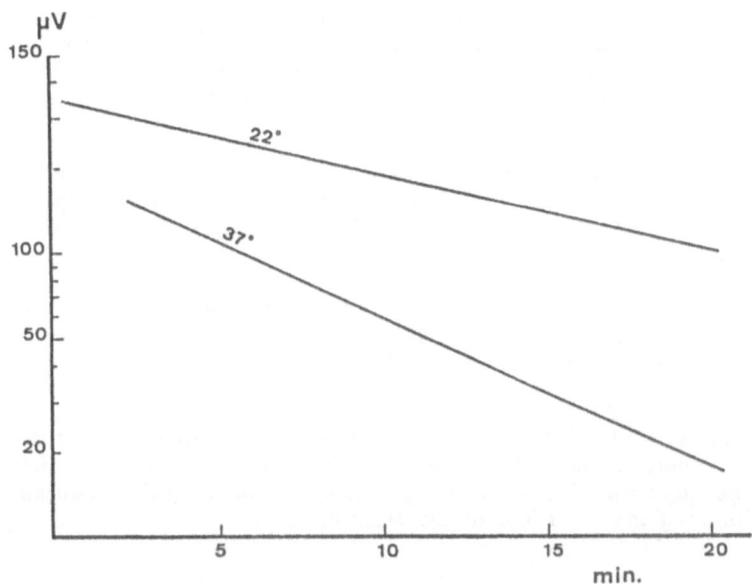

Figure 2   Decay of $O_2^{\cdot-}$ in DMF at 22°C and 37°C studied by EPR. In ordinate, amplitude of the EPR signal. Physical conditions are described in (19).

Immediately after the end of electrolysis, the superoxide solution was used for experiments. An aliquot of this solution was used for superoxide anion concentration determination, which was effected by spectrophotometry with tetranitromethane (20). The conditions and the time of electrolysis were carefully determined, in order to avoid the contamination of the $O_2^{\bar{\cdot}}$ solution by bromine, which results from the electrolysis of TBAB in the positive compartment. In some cases, Br_2 diffused in the $O_2^{\bar{\cdot}}$ compartment and reacted with it :

$$Br_2 \ + \ 2 \ O_2^{\bar{\cdot}} \longrightarrow 2 \ Br^- \ + \ 2 \ O_2$$

Bromine can be considered as a potent $O_2^{\bar{\cdot}}$ scavenger. Its concentration in the negative compartment can be detected by spectrophotometry, at 271 nm. When bromine was detected in the $O_2^{\bar{\cdot}}$ compartment, the solution was discarded. In our experiments, 10 mg of dried ghosts were submitted to 1 ml of $O_2^{\bar{\cdot}}$ solution, for 20 minutes at 37°C. The ghost suspension was centrifuged; the supernatant was transferred in another tube, and DMF was evaporated under dry N_2. The residue was freeze-dried and dissolved in 200 µl $CHCl_3$. Aliquots of 20 µl were chromatographied on thin-layer silicagel plates in hexane/diethyl ether/acetic acid (70:30:2). The plates were then sprayed with a solution of copper acetate (3 gr) in phosphoric acid (14 ml), diluted ad 100 ml with distilled water, and heated for 20 minutes at 160°C. The NEFA were quantified by densitometry with an external standard of 3,4 µg linoleic acid, on a Vitatron Densitometer. Another 40 µl aliquot of chloroform was added with 10 µg of enneadecanoic acid ($C_{19:0}$) as internal standard and chromatographied on silicagel plates in the same conditons. After visualization of the spots with iodine vapours, the NEFA spots were automatically eluted with chloroform, using an Eluchrom CAMAG device. After methylation by diazomethane (21), the samples were analyzed on a GIRA gas-chromatograph, equipped with FID, on a stationnary phase of chromosorb W (100/120 mesh) impregnated with SUPELCO 2330. The elution temperature was programmed between 180°C and 210°C.

We also determined the total NEFA composition of a ghost sample extracted according to Folch et al (23) and transesterified according to Luddy et al (22), which consists in a rupture of the ester bonds, followed by a direct methylation.

The presence of lipoperoxide in ghosts treated with superoxide was searched by iodometry following Buege and Aust (24), and by our technique, using N,N' dimethylparaphenylenediamine (25).

The release of NEFA by superoxide was compared to the release mediated by phospholipase A_2, which hydrolyzes exclusively the fatty ester bonds at the glycerol C-2 position generally occupied by unsaturated fatty acids. We followed the procedure of De Haas et al (26).

RESULTS

a) Decay of $O_2^{\bar{\cdot}}$ in our experimental conditions :
   At 37°C (fig.2), the half-life time of superoxide anion is approximately 10 minutes, as observed by EPR and confirmed by spectrophotometric data with tetranitromethane.

b) <u>Total fatty acids composition of the ghosts</u>, determined by transesterification (expressed in µg/10 mg of freeze-dried ghosts).
$C_{14:0} = 36.3$; $C_{16:0} = 654.7$; $C_{16:1} = 7.9$; $C_{18:0} = 402$; $C_{18:1} = 263$; $C_{18:2} = 384$; $C_{20:3} = 60$; $C_{20:4} = 337$; $C_{22:0} = 81$; $C_{22:4} = 54$.

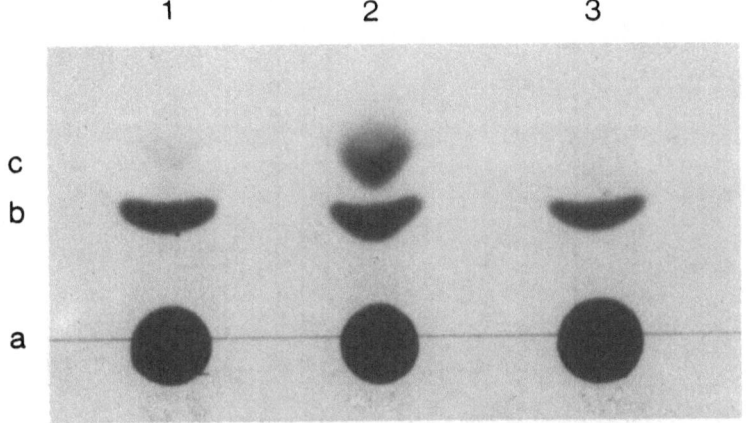

Figure 3   NEFA release from ghosts after treatment with $O_2^{\cdot}$. (TLC on silicagel plates) 1. Control B : DMF + TBAB electrolyzed under argon. 2.$O_2^{\cdot}$ $4.10^{-4}M$. 3. $O_2^{\cdot}$ $4.10^{-4}M + 10^{-4}M$ Br_2. A : phospholipids. B : cholesterol   C : NEFA

c) <u>Effect of superoxide anion solution on ghosts</u> :
The release of NEFA induced by $O_2^{\cdot}$ is proven in a particularly simple and conclusive manner by TLC (fig.3). The total quantity of released NEFA calculated by densitometry on the TLC plates are shown in fig.4. In control A (DMF + TBAB without electrolysis) and control B (DMF + TBAB, submitted to electrolysis with a bubbling of argon), 7 ± 1.8 µg (n=8) and 6.1 ± 1.4 µg (n=8) of total NEFA were respectively released from 10 mg of freeze-dried ghosts, while from the same ghost amounts submitted to $O_2^{\cdot}$ ($4.10^{-4}M$ and $2.10^{-4}M$) the total released NEFA were respectively 56 ± 8 µg (n=8)   and 18 ± 3 µg (n=8). The results calculated from GLC data were similar : the total released NEFA from 10 mg of ghosts were in control A : 5.8 ± 2.1 µg (n=6), in control B : 4.3 ± 1.8 µg (n=5), in $O_2^{\cdot}$ ($4.10^{-4}M$) treated ghosts : 54 ± 10.5 µg (n=6) and for $O_2^{\cdot}$ $2.10^{-4}M$: 14.2 ± 4.5 µg (n=6) (fig.4).

d) <u>Effect of an $O_2^{\cdot}$ scavenger</u> :
Electrophilic agents, such as bromine,   when added before the assay on ghosts, dramatically reduced the NEFA release induced by solutions of $O_2^{\cdot}$. This effect clearly appears on fig. 3 (spot 3 : $O_2^{\cdot}$ $10^{-4}M$ in DMF + TBAB + $10^{-4}M$ Br_2) and is quantified on fig.5. On this figure, the NEFA release obtained from 10 mg of dried-ghosts with $4.10^{-4}M$ $O_2^{\cdot}$ (n=8) is 56 ± 8 µg (n=8) but it is reduced to 21 µg (n=5) and 16 µg (n=5) by 2.10^{-5} and $4.10^{-5}M$ of bromine.

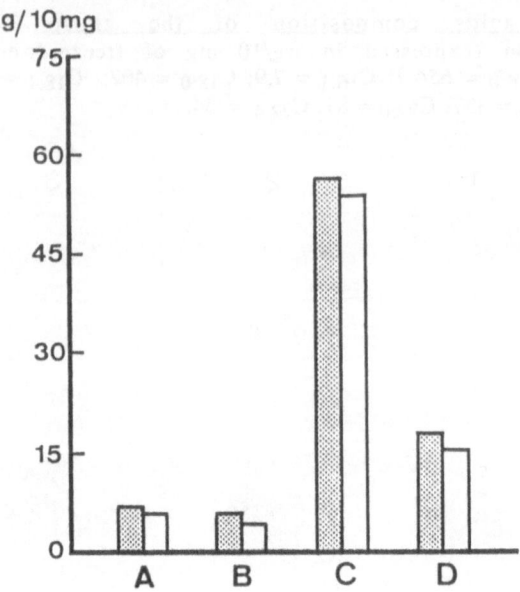

Figure 4 O_2^- induced NEFA release, measured by TLC-densitometry (gray columns) and by GLC (white columns). A. Control A : TBAB + DMF. B : Control B : TBAB + DMF, electrolyzed under argon. C : O_2^- solution 4.10^{-4}M. D : O_2^- solution 2.10^{-4}M

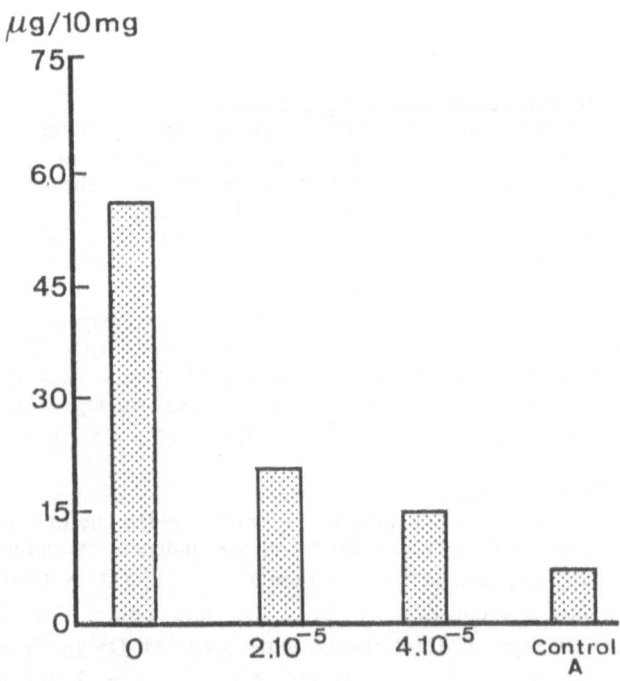

Figure 5   Inhibition of O_2^- -induced NEFA release by Br_2 , an electrophilic reagent.

e) <u>Comparaison between $O_2^{\cdot-}$ and phospholipase A_2 on the NEFA release</u>

On table I, the relative compositions in the different NEFA are presented, for the different experiments. It appears clearly that $O_2^{\cdot-}$ releases the same kinds of NEFA than phospholipase A_2, which releases essentially polyunsaturated fatty acids.

TABLE I

	Total FA	Control B	$O_2^{\cdot-}$	Phl A_2
16:0	29	26	34	11
16:1	0.3	2	0.8	0.3
18:0	17.9	6.2	11	6
18:1	11.7	5.6	9.2	21.1
18:2	17	27	21	28
20:3	2.7	3.7	3.7	3.9
20:4	15.6	22	15.4	22
22:0	3.4	7.5	4.9	4.8
22.4	2.4			2.9

Individual values (expressed in percents of each total NEFA value), of the NEFA released in control B and in $O_2^{\cdot-}$ and in phospholipase A_2 (Phl A_2) -treated ghosts, and compared to the total fatty acids (FA) of lipid ghosts.

f) <u>Presence of lipoperoxide in $O_2^{\cdot-}$ -treated ghosts</u> :

Using the two techniques before-mentioned, we were unable to evidence any lipoperoxide formation.

DISCUSSION

The procedure used to generate $O_2^{\cdot-}$ electrochemically is efficient, provided that some precautions are taken, to avoid diffusion of bromine in the $O_2^{\cdot-}$ compartment. Fee and Hildebrand (27) noticed that generation of superoxide anion in DMF was irregular, a fact which was not interpreted. We think that these irregularities arise from the diffusion of bromine, when the electrolysis is followed up for a long time. A suitable concentration of $O_2^{\cdot-}$ can be rapidly reached with a current of 20 V during 20 minutes, and the Br_2 diffusion can be limited at temperature below 20°C.

The use of SOD as a specific inhibitor to unequivocally proven that the esterolysis was an effect of $O_2^{\cdot-}$,was not possible because the experiments were performed in aprotic conditions, without possibility of dismutation. However, the proposed mechanism of a nucleophilic deesterification of phospholipids is supported by the identification of $O_2^{\cdot-}$ by EPR, and its inhibition by an electrophilic agent.

The action of superoxide anion seemed not to be exerted on a specific position, as it is the case for phospholipase A_2, which releases unsaturated fatty acids linked on the C-2 of glycerol. The composition of the NEFA released by $O_2^{\cdot-}$ approximately reflects that of the ghost lipid fatty acids.

Can these findings be extended to biological conditions ? Can $O_2^{\overline{\cdot}}$ diffuse into the aprotic regions of membranes, as proposed by Niehaus (14) ? More recent papers claimed that $O_2^{\overline{\cdot}}$ can rapidly cross through the cell membranes (28,29). The existence of anion channels, allowing the crossing of $O_2^{\overline{\cdot}}$ through the membrane, is well documented (30).

Our results afford new proofs to support the hypothesis of a direct reactivity of $O_2^{\overline{\cdot}}$ in biological material, and are in agreement with recent observations, reporting such reactions at the level of DNA (31). Bagley et al (32) reported toxic effects of superoxide anion on cell cultures, which can be inhibited by SOD, but not by catalase.

These experiments underline the usefulness of electrophilic reagents, as superoxide anion scavengers.

CONCLUSIONS

Superoxide anion deesterifies membrane phospholipids in erythrocyte ghosts maintained in aprotic conditions. There are many reasons to estimate that such nucleophilic actions can be produced *in vivo*. The hypothesis of a direct toxic effect of superoxide anion, SOD-inhibitable, appears now more likely.

REFERENCES

1. D. T. Sawyer and J.S. Valentine, How super is superoxide ? Acc. Chem. Res. 14:393 (1981)
2. J. M. McCord, and I. Fridovich. Superoxide dismutase. J. Biol. Chem. 244:6049 (1969).
3. J. Bergsland, L. Lo Balsamo, P. Lajos, and B. Moorkerjee, Post-anoxic hemodynamic performances. The effect of allopurinol and superoxide dismutase/catalase. Transpl. Proc. 19:4165 (1987).
4. M. L. Myers, R. Bolli, R.F. Lekich, C.J. Hartley, and R. Roberts,Enhancement of recovery of myocardial function by oxygen free-radical scavengers after reversible regional ischemia. Circulation 72:915 (1985).
5. S. L. Atalla, L. H. Toledo-Pereyra, G. H. McKenzie and J. P. Cederna., Influence of oxygen-derived free radical scavengers on ischemic livers. Transplantation 40:584 (1985).
6. R. J. Korthuis and D. N. Granger. Ischemia-reperfusion injury : role of oxygen-derived free radicals, in :"Physiology of oxygen radicals" A. Taylor, S. Matalon, P. Ward. ed., Amer. Physiol. Soc., p 217 (1986).
7. J.A. Fee., Is superoxide toxic ? in:"Biological and clinical aspects of superoxide and superoxide dismutase', J. V. Bannister, ed., Elsevier-North Holland, pp 41 (1980).
8. D. T. Sawyer, M. J. Gibian, M. M. Morrison, and E.T. Seo., On the chemical reactivity of superoxide anion, J Amer. Chem. Soc. 100:627 (1978).
9. B. Halliwell, and Gutteridge J. M. C., Oxygen toxicity, oxygen radicals, transition metals and disease. Biochem. J. 219:1 (1984).
10. I. Fridovich, Biological effects of the superoxide radical. Arch. Biochem. Biophys.S 247:1 (1984).
11. R. Dietz, A. E. S. Forno, P. E. Larcombe, and M. E. Peover. Nucleophilic reactions of electrogenerated superoxide ion. J. Chem. Soc. B:816 (1970).
12. J. San Filippo, L. J. Romano, C. I. Chern, and J. S. Valentine, Cleavage of esters by superoxide. J. Org. Chem 41:586 (1976).

13. M. J. Gibian, D. T. Sawyer, T; Ungermann, R. Tangpoonpholvivat, and M. M. Morrisen. Reactivity of superoxide with carbonyl compounds in aprotic solvents. J. Amer. Chem. Soc. 101:640 (1979).

14 W. G. Niehaus, A proposed role of superoxide anion as a biological nucleophile in the deesterification of phospholipids. Bioorganic Chem 7:77 (1978).

15. J. S. Valentine, and A. B. Curtis, A convenient preparation of solutions of superoxide anion and the reaction of superoxide anion with a copper (II) complex. J. Am. Chem. Soc. 97:224 (1975).

16. T. L. Steck, and J.A. Kant, Preparation of impermeable ghosts and inside out vesicles from human erythrocyte membranes. Meth. Enzymol 31:172 (1974).

17. R. E. Lynch, and I. Fridovich. Effects of superoxide on the erythrocyte membrane. J. Biol. Chem. 253:1838 (1978).

18. D. L. Maricle and W. G. Hogdson, Reduction of oxygen to superoxide anion in aprotic solvents. Anals. Chem 37:1562 (1965).

19. P. F. Knowles, J. F. Gibson, F. M. Pick, and R. C. Bray, Electron-spin resonance evidence for enzymic reduction of oxygen to a free radical, the superoxide ion. Biochem. J. 111:53 (1969).

20. M. J. Green, and H. A. O. Hill, Chemistry of dioxygen. Methods in Enzymol 105:3 (1984).

21. H. Schenk and J. L. Gellermann, Esterification of fatty acids with diazomethane on small scale. Anal. Chem. 32:1312 (1960).

22. F. E. Luddy, R. A. Bradford, S. F. Herb and P. Magidman, A rapid and quantitative procedure for the preparation of methyl esters of butter oil and other fats. J. Amer. Oil. Chem. Soc. 45:549 (1968).

23. J. Folch, M. Lees and G. H. S. Stanley, A simple method for the isolation and purification of total lipid in animal tissue. J. Biol. Chem. 266:497 (1957).

24. J. A. Buege and S. D. Aust, Microsomal lipid peroxidation. Methods Enzymol. 52:302 (1978).

25. C. Deby, G. Deby-Dupont, P. Hans, J. Pincemail, J. Neuray and R. Goutier, Complementary procedures for pro- and antilipoperoxidant activities measurements. Experientia 39:1113 (1983).

26. G. H. De Haas, N. M. Postema, W. Nieuwenhuyzen and L.L.M. Van Deenen, Purification and properties of phospholipase A_2 from porcine pancreas. Bioch. Biophys. Acta. 159:103 (1968).

27. J. A. Fee and P. G. Hildebrand, On the development of a well-defined source of superoxide ion for studies with biological systems. FEBS Lett. 39:79 (1974).

28. G. M. Rosen and B. A. Freeman, Detection of superoxide generated by endothelial cells. Proc. Natl. Acad. Sci. USA 81:7269 (1984).

29. G. L. Babior, R. E. Rosin, B. J. McMurrich, W. A. Peters and B. M. Babior, Arrangement of the respiratory burst oxidase in the plasma membrane of the neutrophil. J. Clin. Invest 67:1724 (1981).

30. D. Roos, C. M. Eckmann, M. Yazdanbakhsh, M. N. Hamers and M. De Boer, Excretion of superoxide by phagocytes measured with cytochrome C entrapped in resealed erythrocyte ghosts. J. Biol. Chem. 259:1770 (1984).

31. H. C. Birnboim and M. Kanabus-Kaminska, The production of DNA strand breaks in human leukocytes by superoxide anion may involve a metabolic process. Proc. Natl.Acad. Sci. USA 82:6820 (1985).

32. A. C. Bagley, J. Krall and R. E. Lynch, Superoxide mediates the toxicity of paraquat for chinese hamster ovary cells. Proc. Natl Acad. Sci. USA 83:3189 (1986).

MECHANISMS OF FORMATION OF OXYSTEROLS: A GENERAL SURVEY

Leland L. Smith

Department of Human Biological Chemistry & Genetics
University of Texas Medical Branch
Galveston,TX 77550

INTRODUCTION

Oxysterols, the simple oxidation products of the common sterols of biological membranes and tissues, are of current interest for their crucial role in the biosynthesis of other steroids and for their divers biological activities with implications of relevance to human health.[1-4] On the one hand oxysterols include derivatives of cholesterol (cholest-5-en-3ß-ol, **1**)

1 Cholesterol

formed in the initial regulated enzymic steps directed to biosyntheses of bile acids and steroid hormones required for metabolism in mammals and are thus necessarily present in mammalian tissues. By contrast, oxysterols exhibit toxic manifestations in intact animals and in tissue and cultured cell bioassay systems and have been suggested as being agents that cause or exacerbate human chronic health disorders such as atherosclerosis and cancer. Also, oxysterols have been spectulatively implicated in the regulation of de novo sterol biosynthesis and metabolism. Each of these topics is of great interest, and there is a regular increase in efforts to examine these matters in depth. Together these points make a strong case for thorough understanding of what oxysterols are and how they are derived.

In order to manage disparate and confusing aspects of the topic it is helpful to provide a specific definition of the term oxysterol and to examine systematically the recognized means of oxysterol derivation, thus their origins. It is necessary to consider sterol biochemistry in general as oxysterols have metabolic origins, and it is also crucially important to review sterol oxidation chemistry, as nonenzymic processes are also involved. Oxysterols have been known by other terms (cholesterol autoxidation products, cholesterol oxidation products, oxidized cholesterol derivatives, cholesterol oxides, etc.) heretofore, but it is now important to expand the

Free Radicals, Lipoproteins, and Membrane Lipids
Edited by A. Crastes de Paulet *et al.*
Plenum Press, New York, 1990

definition to include oxidized sterols derived from other sterols as well as from cholesterol.

Although most of what we know about the chemistry and biochemistry as well as biological activities of the class deals with oxysterols derived from cholesterol, the present definition of oxysterols encompasses simple oxidized derivatives of all tissue sterols recognized as crucial to membrane stability and function and known to be precursors of more highly oxidized steroids implicated in other physiological functions of living systems. The common sterols include cholesterol, lanosterol (5α-lanosta-8,24-dien-3ß-ol), dihydrolanosterol (5α-lanost-8-en-3ß-ol), lathosterol (5α-cholest-7-en-3ß-ol), desmosterol (cholesta-5,24-dien-3ß-ol), etc. of mammalian metabolism, but sitosterol (24-ethyl-(24R)-cholest-5-en-3ß-ol), stigmasterol, ergosterol (ergosta-5,7,22E-trien-3ß-ol), and a host of other sterols of animal, plant, and microbial life are also implicated. Saturated sterols such as 5α- and 5ß-cholestan-3ß-ol do not appear to be oxidized in animal metabolism and are not precursors of oxysterols.

It is important to limit the definition of oxysterols and thereby the scope of interest to simple sterol oxidation products derived by relatively few processes from the parent sterol. Thus the more highly oxidized bile alcohols, calciferols, sapogenins, ecdysterols, antheridiols, brassinolides, withanolides, etc., the plethora of polyhydroxylated sterols from marine creatures, and yet more highly oxidized degraded steroids (bile acids, steroid hormones) are not oxysterols by the present definition.

It is also important to consider physiological relevance or purposefulness. The more highly oxidized steroids excluded as oxysterols by present definition either serve recognized physiological functions or are biosynthesis precursors of such functional steroids. By contrast, despite the array of biological activities demonstrated for oxysterols,[1,3,4] no uniform concept of purpose or physiological relevance for the occurrence of oxysterols in human tissues, for example, has emerged. It is possible that the oxysterols of current interest will ultimately be shown to have purpose as well, but for the nonce, there is uncertainty, and this uncertainty is a significant aspect of the definition of tissue oxysterols.

As the chemistry of oxysterols is now fairly well understood,[1,2] interest in the class is centered on the issues of how these oxysterols come to be present in human tissues, what physiological function they may serve, and what influence they may have on metabolic processes and human health matters. The present account reviews pertinent chemistry and biochemistry and addresses the question of origins of oxysterols in human tissues. That micromolar amounts of a variety of oxysterols regularly occur in human blood is evinced in Table 1. The matter of the biological activities of oxysterols and potential health effects are beyond the scope of the present treatment.

TYPES OF OXYSTEROLS

Two separate approaches are taken here to examine human tissue oxysterols. The different chemical structural features of oxysterols are initially considered, followed by examination of the recognized processes implicated in their derivation. Although oxysterols are oxidized derivatives of sterols, it seems helpful to recall first the other recognized modes of sterol metabolism. Sterols are subject to four major kinds of metabolic processes: (1) 3ß-hydroxyl esterification, glycosylation, conjugation, etc., (2) side-chain alkylation and dealkylation (in plants and invertebrates), (3) double bond reduction, and (4) oxidation. Other transformations of sterols such as carbon-carbon bond scission (cholecalciferol

Table 1. Oxysterols of Human Serum[a]

Oxysterol	Level, ng/mL	Comments	Reference
7α-OH **19**	20-165	Elevated in CTX	5
	30 ± 4	Gallstone patients	6
7α-OH **19** esters	19	Human plasma	7,8
7β-OH **20**	12-265	Elevated in CTX	5
7β-OH **20** esters	11	Human plasma	7,8
7-Ketone **21**	6 VLDL	Type II hyper-	9
	336 LDL	cholesterolemia	
	32 HDL		
	24-60	BHT added	10
	50-300	Frozen 1-6 months	
7-Ketone **21** esters	6	Human plasma	7,8
5α,6α-Epoxide **8**	22 VLDL	Type II hyper-	9
	493 LDL	cholesterolemia	
	132 HDL		
5β,6β-Epoxide **9**	24 VLDL	Type II hyper-	9
	470 LDL	cholesterolemia	
	101 HDL		
5,6-Epoxides **8,9**	67-293	Healthy individuals	11
24-OH esters	7	Human plasma	7,8
25-OH **25** esters	5	Human plasma	7,8
26-OH	43-130	Normals	12,13
	0-6	CTX patients	
	120 VLDL	Total 310 ng/mL	14
	65 LDL		
	100 HDL		
	140-260	Young adults	15
	335-950	Older patients with CHD	
	30-129	Abolished in CTX	5
26-OH esters	106	Human plasma	7,8
	60-170	Normals	12

[a] Abbreviations: 7α-OH, cholest-5-ene-3β,7α-diol (**19**); 7β-OH, cholest-5-ene-3β,7β-diol (**20**); 7-Ketone, 3β-hydroxycholest-5-en-7-one (**21**); 24-OH, cholest-5-ene-3β,24-diol; 25-OH, cholest-5-ene-3β,25-diol (**25**); 26-OH, cholest-5-ene-3β,26-diol; CTX, cerebrotendinous xanthomatosis; CHD, cardiovascular heart disease; BHT, butylated hydroxytoluene; VLDL, very low density lipoprotein; LDL, low density lipoprotein ; HDL, high density lipoprotein.

formation) and double bond and skeletal rearrangement (ring contraction, expansion) also occur _in vivo_, and food processing and diagenesis processes in the biosphere may also generate other altered sterol derivatives, which if ingested in the diet could pose an additional source of odd components.

A close examination of the structural features of oxysterols and the relevant oxidation processes implicated in their derivation provides an improved definition of oxysterols and a good grasp of their origins. Both oxygen-dependent dehydrogenations and the overt introductions of oxygen into the sterol are involved.

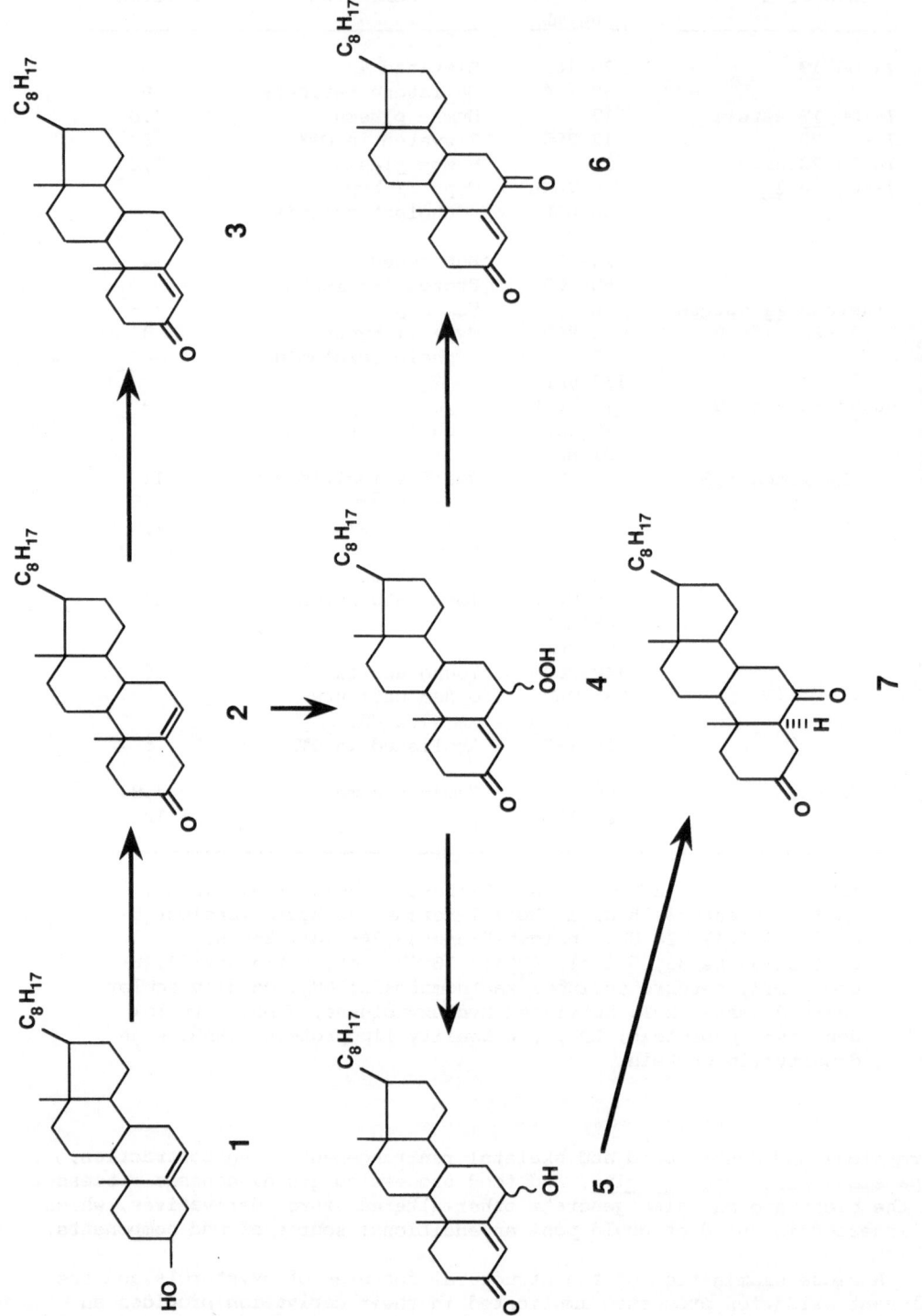

Fig. 1. Oxysterols derived by 3β-hydroxyl dehydrogenation.

Oxygen-Dependent Dehydrogenations

The oxygen-dependent dehydrogantion (whether enzymic and nonenzymic) of sterols occurs by two different processes: by introduction of an additional olefin double bond or by transformation of the 3ß-hydroxyl group to the corresponding 3-ketone. Both types of products may in turn be subject to yet further oxidative alterations. As example, dehydrogenation of the cholesterol 3ß-hydroxyl group yields cholest-5-en-3-one (<u>2</u>) (Fig.1).

Introduction of Oxygen into Sterols

In addition to dehydrogenation there are at least six separate means (both enzymic and nonenzymic) by which one, two, or three atoms of oxygen may be introduced into sterols to produce oxysterols of interest. The oxysterols thereby derived have added hydroxyl, ketone, epoxide, hydroperoxide, epidioxide, and ozonide features, generally at or near the original olefin double bond of the parent sterol.

One Oxygen Atom. One atom of oxygen may be introduced into sterols by four different processes: by hydroxylation, by epoxidation of the olefin bond, by hydration of olefin or epoxide features, and by addition of hypohalous acid to the olefin bond. The most prominent means of enzymic introduction of one oxygen atom is by cytochrome P-450 mixed function oxidase hydroxylation of methyl, methylene, and methine carbons. Epoxidation of double bonds by mixed function oxidases also occurs, and both processes contribute to the oxysterols present in human blood (Table 1). Furthermore, the isomeric cholesterol 5,6-epoxides 5,6α-epoxy-5α-cholestan-3ß-ol (<u>8</u>) and 5,6ß-epoxy-5ß-cholestan-3ß-ol (<u>9</u>) are found in human breast[16] and prostatic [17] fluids. The subsequent metabolism of these oxysterols to fatty acyl and sulfate[18] esters may also occur. Other analyses also support the presence of oxysterols in mammalian tissues,[1,2] but some of these reports are compromised by the methods used (or not used).

These processes afford sterol alcohol and epoxide derivatives only; other oxygen-functionalized derivatives (ketones, aldehydes) are formed by subsequent transformations of primary products, from alcohols by dehydrogenation and from peroxides via peroxide bond homolysis and linked ß-scission reactions. Sterol alcohols and epoxides may be formed by other means as well, as enzymic epoxide reduction to alcohol has been observed in vitro,[19] and water may also serve as source of one atom of oxygen generating oxysterol alcohols. The hydration of the olefin double bond of cholesterol yielding 5α-cholestane-3ß,6ß-diol in aqueous hydrogen peroxide (H_2O_2) has been described,[20] and the hydration of the isomeric 5,6-epoxides <u>8</u> and <u>9</u> occurs by enzymic and non-enzymic processes, yielding as common hydration product 5α-cholestane-3ß,5,6ß-triol (<u>10</u>) (Fig.2).

Oxysterol epoxides may also form from sterol halohydrins in the alkaline medium of the small intestine.[21] Halohydrin formation from oxidation of cholesterol by hypohalous acid must also be included in processes introducing one atom of oxygen, as cholesterol chlorohydrins are formed by human blood polymorphonuclear leukocyte myeloperoxidase acting in vitro on cholesterol.[22] Oxysterol halohydrins have not been found in tissues.

Two Oxygen Atoms. Both atoms of molecular oxygen ($^3O_2$) are introduced into monoolefinic sterols by peroxidation yielding allylic hydroperoxides (Fig.3) and into conjugated sterol 5,7-dienes by 1,4-addition yielding cyclic 5α,8α-epidioxides. Both processes are observed enzymically in vitro and in autoxidations. Additionally, hydroperoxide formation at nonallylic side-chain carbons occurs in autoxidation (Fig.4). Moreover, both formal peroxidation processes may co-occur, as the recently described biosynthesis

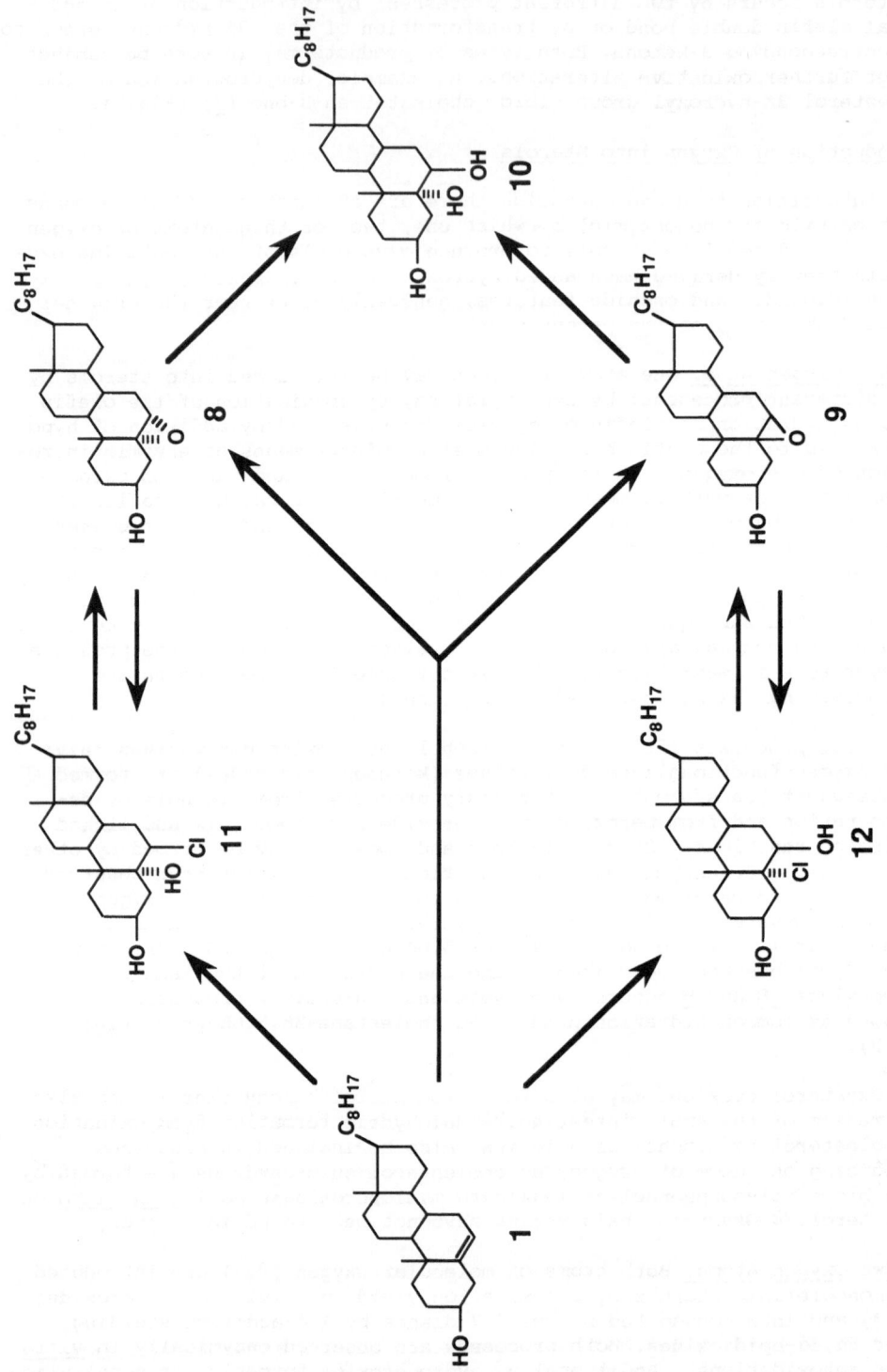

Fig. 2. Oxysterols derived by oxidations of the Δ^5-double bond.

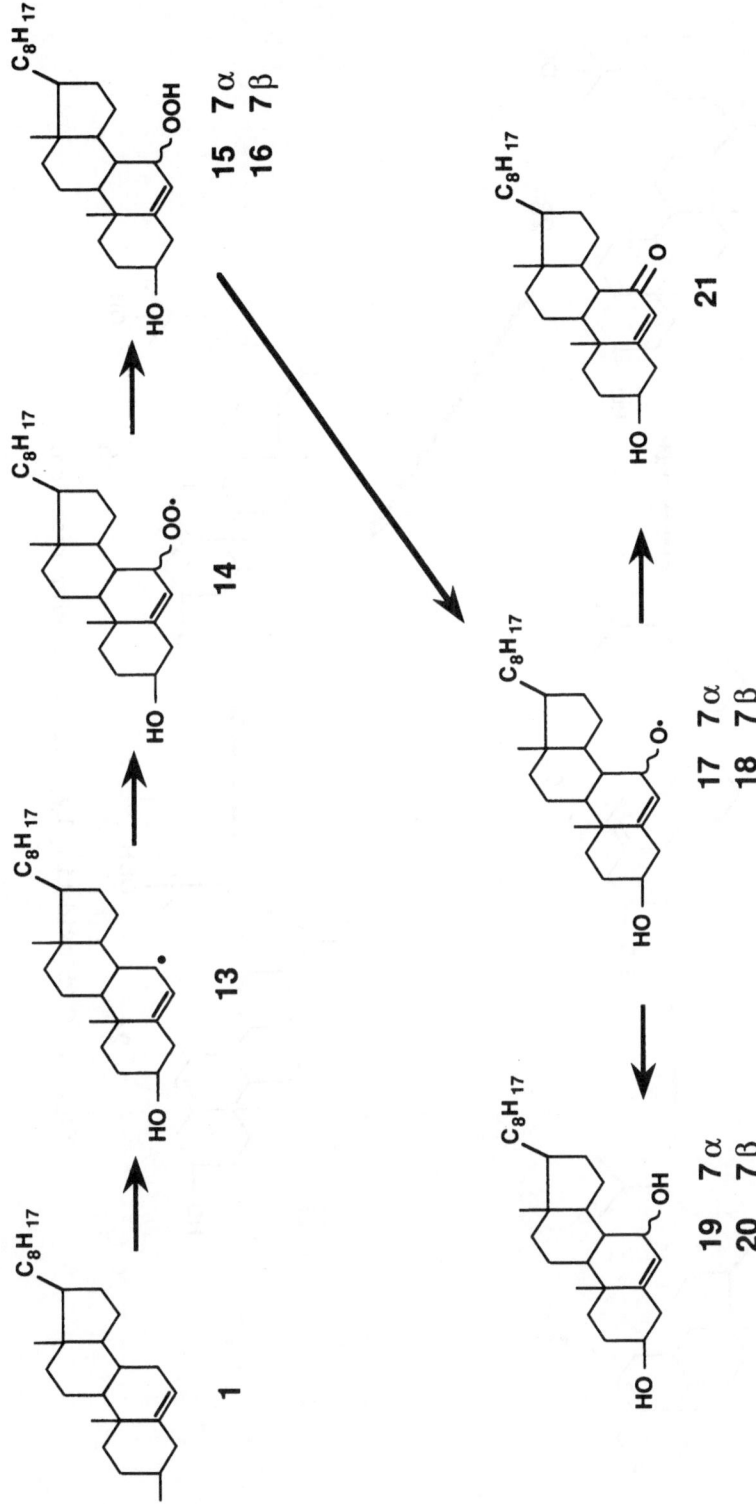

Fig. 3. Major oxysterols derived initially from cholesterol by autoxidation.

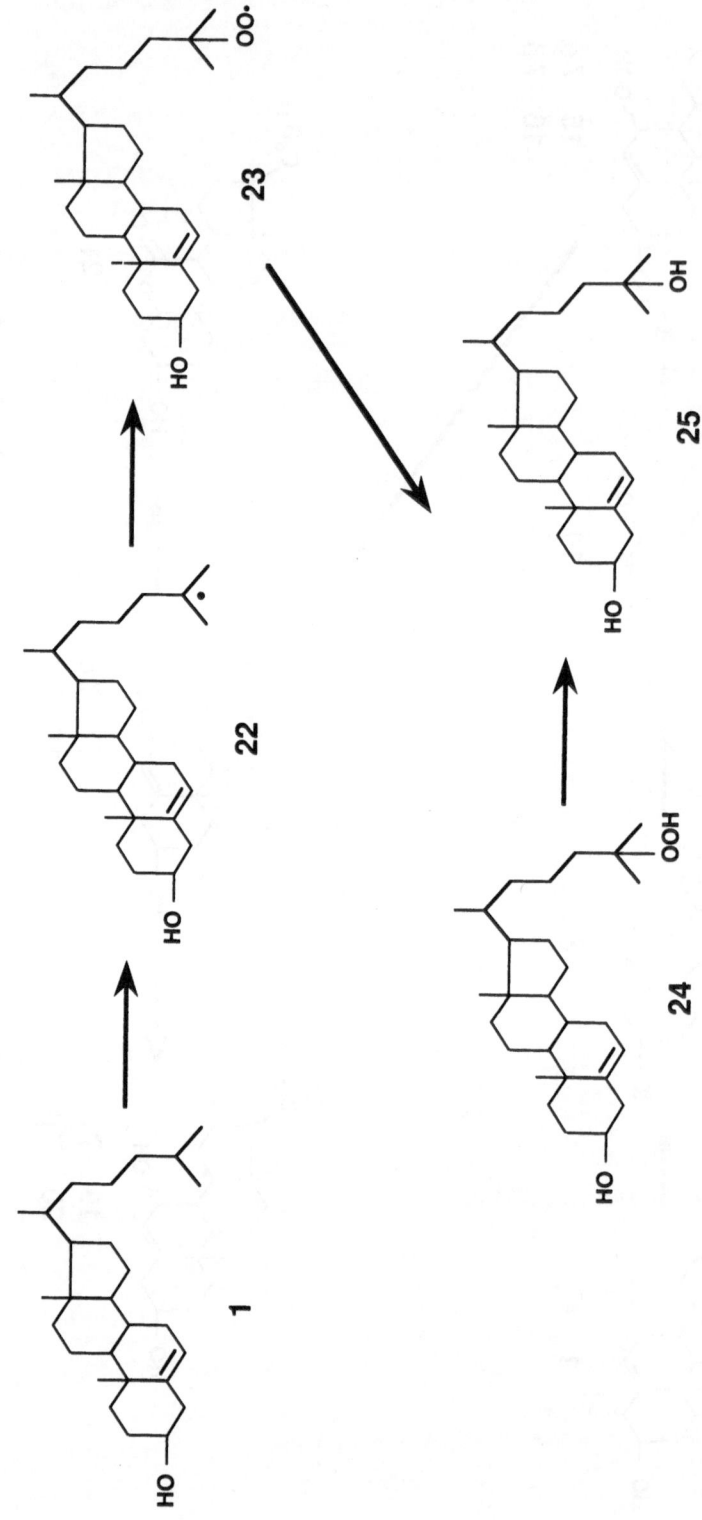

Fig. 4. C-25 oxidation by radical transfer reaction.

of the nonsterol epidioxide ascaridole involving participation of an iodoperoxidase is thought to proceed along the lines of iodination, hydroperoxidation, and ultimate hydroperoxide cyclization with iodide elimination yielding the epidioxide product.[23]

Sterol hydroperoxides and epidioxides give rise to many other kinds of oxysterols, some of which may also derive from oxysterol intermediates formed intiially by introduction of but one atom of oxygen. Thus, a ketone derived by alcohol dehydrogenation may also derive by β-scission of a secondary oxyl radical formed from a sterol secondary hydroperoxide.

Three Oxygen Atoms. Although metabolic processes have yet to be described that introduce three oxygen atoms into sterols, ozone (O_3) is effective in that matter, yielding the classic 1,2,4-trioxolane ozonides of cholesterol in aprotic media, substituted secosterol epidioxides in the presence of water or alcohols.[24,25] These kinds of oxysterols have not been discovered in animal systems, but two related triterpenoid 1,2,4-trioxolane ozonides have been recovered from plants![26,27] The possible ozonization of tissue unsaturated lipids by O_3 in inspired air should not be overlooked.

ORIGINS OF TISSUE OXYSTEROLS

There are at least six potential origins or mechanisms of formation of oxysterols found in human tissues that require consideration: enzymic oxidation, autoxidation, lipid peroxidation, photochemical oxidation, oxidation by other active oxygen species, and formation in foods. Emphasis is placed on cholesterol as substrate, but the sources to be described apply to many other tissue unsaturated sterols as well.

Enzymic Oxidation

The most obvious and certain source is the enzymic metabolism of cholesterol whereby cholesterol is hydroxylated by cytochrome P-450 mixed function oxidases. Examples are: (1) cholesterol 7α-hydroxylation yielding cholest-5-ene-3β,7α-diol (7α-hydroxycholesterol, **19**) as the rate-limiting step in the hepatic biosynthesis of bile acids, (2) hydroxylation at the (22R)-or 22βF-position, yielding (22R)-cholest-5-ene-3β,22-diol and (20R,22R)-cholest-5-ene-3β,20,22-triol involved in side-chain scission in endocrine tissues, (3) 25- and 26-hydroxylations in liver, yielding cholest-5-ene-3β,25-diol (25-hydroxycholesterol, **25**) and (25R)- and (25S)-cholest-5-ene-3β,26-diols implicated in bile acid biosynthesis.

These processes may be regarded as established and purposeful. There is also the established (24S)- or 24βF-hydroxylation of cholesterol in brain yielding (24S)-cholest-5-ene-3β,24-diol (cerebrosterol) for which there is no recognized physiological function. Moreover, there is a suggestion that cholest-5-ene-3β,4β-diol detected in human serum and in butter may be a heretofore unrecognized cholesterol metabolite.[2] There are also enzymic oxidations of other sorts, including the oxygen-dependent dehydrogenation of the sterol 3β-hydroxyl group to the corresponding 3-ketone already mentioned and 5α,6α-epoxidation by liver enzymes demonstrated in vitro with no recognized purpose.

It is also necessary to include among the enzymically derived oxysterols those oxidized derivatives of lanosterol and dihydrolanosterol that are implicated in cholesterol biosynthesis from lanosterol, such items as the 32-hydroxy- and 32-oxo derivatives having biological activities that justify their inclusion as oxysterols. Moreover, yet a third enzymic process must also be included, that of an alternative de novo sterol biosynthesis from (2S,22S)-squalene 2,3;22,23-bisepoxide yielding a lanosterol 24,25-

epoxide (24,25-epoxy-5α-(24S)-lanost-8-en-3β-ol). This alternative de novo biosynthesis does not involve (2S)-squalene 2,3-epoxide cyclization to lanosterol. The lanosterol 24,25-epoxide is transformed in vitro to desmosterol 24,25-epoxide (24,25-epoxy-(24S)-cholest-5-en-3β-ol), and the enzymic reduction in vitro of these sterol 24,25-epoxides to alcohols may potentially give rise to another class of oxysterols. The physiological significance of the alternative biosynthesis, if it operate in vivo, or of the oxysterols thereby formed remains obscure. Another alternative de novo sterol biosynthesis process implicating sesterterpenoids could also contribute oxysterols if such postulated process indeed occur in vivo.[28]

Autoxidation

In addition to these enzymic processes that appear to involve classic enzyme-substrate interactions producing metabolites destined for other fates, there are processes that occur in living systems that do not involve these restraints but that generate oxysterols under other circumstances, oxysterols presently without recognized purpose. Chief among nonenzymic processes is autoxidation, the apparently uncatalyzed oxidation by ground-state dioxygen ($^3O_2$).

Autoxidation occurs at the same sites as enzyme oxidations; indeed, the resemblence of the two processes is uncanny. Exactly the same oxysterols result in many cases. We may pose that enzymes evolved that merely adapted the autoxidative steps yielding oxysterols useful in some manner to primitive life.

Cholesterol autoxidation proceeds via initial formation of the C-7 allylic carbon-centered radical **13** (Fig.3) generated by radiation or by other radical species, with fast interception of $^3O_2$ generating the cholesterol 7-peroxyl radical **14** stabilized by hydrogen abstraction yielding the epimeric cholesterol 7-hydroperoxides **15** and **16**. Radical transfer reactions then can lead to formation of other sterol peroxyl radicals and the corresponding hydroperoxides, for instance at the side-chain sites chief among which is the terminal tertiary C-25 site (Fig.4). The same process occurs for the tertiary C-20, secondary C-24, and primary C-26/C-27 sites, as the requisite cholesterol hydroperoxides have been recovered from air-aged cholesterol. Degraded oxysterols isolated from air-aged cholesterol support the additional radical oxidation of cholesterol in the tertiary C-17 and secondary C-22 and C-23 positions. The cleavage of nuclear carbon-carbon bonds has not been observed. It is uncertain whether side-chain radical oxidation occur in the absence of the more susceptible allylic C-7 reaction, and one may project that it be the B-ring chemistry that provokes the side-chain oxidations by radical inter-changes among sterol molecules.

A myriad of secondary and higher order oxysterols is generated from the thermal decomposition of initially formed hydroperoxides **15** and **16**, the first step of which being peroxide bond homolysis yielding the corresponding cholesterol 7-oxyl radicals (**17**, **18**) then subject to hydrogen abstraction yielding the alcohols 3β,7α-diol**19** and cholest-5-ene-3β,7β-diol (7β-hydroxycholesterol, **20**) and to β-scission yielding the ketone 3β-hydroxy-cholest-5-en-7-one (7-ketocholesterol, **21**) (Fig.3). Dehydration of the 7-ketone **21** yields cholesta-3,5-dien-7-one. The same two degradation processes act on side-chain hydroperoxides, but in this case β-scission of carbon-carbon bonds also occurs, yielding degraded C_{19}-C_{25} derivatives in complex fashion.

As previously outlined, there is one other autoxidation reaction, that of the oxygen-dependent dehydrogenation of the 3β-hydroxyl group, with H_2O_2 also as a product (Fig.1). The product Δ^5-3-ketone **2** is readily isomerized to cholest-4-en-3-one (**3**) and is subject to peroxidation, yielding the 6-hydroperoxycholest-4-en-3-one epimers (**4**) from which the corresponding epimeric 6-hydroxycholest-4-en-3-ones (**5**) and cholest-4-ene-3,6-dione (**6**) derive by the two processes already mentioned. The 6-alcohols **5** may be further isomerized to 5α-cholestane-3,6-dione (**7**) (Fig.1).

The H_2O_2 and sterol hydroperoxides formed in these initial processes in turn serve as oxidants of cholesterol, yielding isomeric 5,6-epoxides **8** and **9**, the 5β,6β-epoxide **9** predominating. Hydration of either epoxide then yields the 3β,5α,6β-triol **10**, their common hydration product (Fig.2). The 3β,5α,6β-triol **10** is further oxidized to 3β,5-dihydroxy-5α-cholestan-6-one.

Exactly the same events occur for other unsaturated sterols, with the initial hydroperoxides being decomposed to alcohol and ketone derivatives by these two fundamental processes. Subsequent processes also may transform initially formed oxysterols to more highly oxidized derivatives; for example, the epoxidation of an oxysterol Δ^5-7-alcohol and subsequent epoxide hydration yields an oxysterol 3β,5α,6β,7-tetraol. Several of the competing and subsequent reactions have been followed, affording understanding of the origins of many of the six dozen or so cholesterol autoxidation products now recognized.

Sitosterol and similar Δ^5-sterols are attacked at the allylic position in the same manner as is cholesterol. Lathosterol yields the epimeric Δ^7-6-hydroperoxides. Dihydrolanosterol yields 7β- and 11β-hydroperoxides; in this case the epimeric hydroperoxides appear not to have formed.

For unconjugated dienes desmosterol and fucosterol (24-ethylcholesta-5,24(28)E-dien-3β-ol) it is the side-chain double bond that is the more sensitive to air oxidations. Desmosterol yields allylic hydroperoxides with double bond rearrangement, but the 3β,25-diol **25** and desmosterol 24,25-epoxide are also recorded as products. For conjugated dienes 7-dehydrocholesterol (cholesta-5,7-dien-3β-ol) and ergosterol the 1,4-addition of dioxygen occurs, yielding 5α,8α-epidioxides in a different reaction mode. Subsequent decompositions lead to very complex oxysterol mixtures that have never been properly examined. Cholecalciferol (9,10-secocholesta-5E,7E,10(19)-trien-3β-ol) likewise has a complex oxidation chemistry that remains undescribed.

Lipid Peroxidation

Lipid peroxidation is a very similar means of forming the same B-ring oxysterols from Δ^5-sterols cholesterol and sitosterol. Three enzyme systems have been shown to peroxidize cholesterol in vitro, the rat liver NADPH-dependent microsomal lipid peroxidation system, soybean lipoxygenase, and horseradish peroxidase. Only the B-ring oxysterols 7-hydroperoxides **15**, **16**, 7-alcohols **19** and **20**, 7-ketone **21**, and 5,6-epoxides **8** and **9** are formed.

Distinction is made here between autoxidation and lipid peroxidation. Lipid peroxidation is not merely the autoxidation of lipids but rather is associated with enzyme-driven processes that initiate the peroxidation. One may recognize several possible processes, one involving dioxygenases and suitable lipid substrates forming enzyme-substrate complexes from which are formed hydroperoxide or cyclic peroxide metabolites capable of initiating lipid peroxidation of other unsaturated lipids. The classic liver microsomal system and soybean lipoxygenase are examples, both systems peroxidizing cholesterol and sitosterol, yielding the same products as found in autoxidation.

Other enzyme systems involving active oxygen species, such as horseradish peroxidase action on H_2O_2, may initiate the same process. Yet another mode of enzyme dependence is the reduction of iron or other transition metal ions (or some other agent), thereby leading to dioxygen activation and lipid peroxidation.

This item brings up two more important points to be made in the understanding of oxysterols in Nature: (1) whether the oxysterols were present in the living system and (2) whether the oxysterols were generated in sample handling and analysis. Unfortunately, both possibilities exist and the usual report of trace oxysterol composition in a new biological setting must be very carefully evaluated. Oxysterols are so frequently discovered in _post mortem_ tissues that have not been protected from light and air, autolysis, etc. Indeed, many plant materials and marine creatures are sun- or air-dried (!) prior to recovery of sterols. We may dismiss these reports are vitrually worthless with respect to evincing underlying causation.

The second item of oxysterol generation during analysis is more subtle but quite important, as oxysterol levels may increases many fold during analysis. Oxysterol levels in human blood are of the same order as those of the circulating steroid hormones and pose the same problems of reliable analysis. Even the most careful analysis must take these points into account in judging the true oxysterol content of biological samples. For example, human serum held without antioxidant at room temperature for one hour has more than double the 7-ketone **21** content in comparison with the same sample protected by antioxidant analyzed directly. Several-fold increases in 7-ketone **21** levels have been observed in serum frozen one to six months without antioxidant in comparison with fresh serum protected by antioxidant.[10] We have also observed even greater differences in analyses of fatty acyl esters of the 7-oxygenated oxysterols **19-21** from stored (frozen) human plasma in comparison with fresh human plasma.[7,8]

In addition to the established oxysterol metabolites (Table 1) human serum contains a trio of oxysterols for which no proper metabolism source has been identified. Whereas the 3ß,7α-diol **19** and 5α,6α-epoxide **8** have demonstrated enzymic origins in liver, the companion 3ß,7ß-diol **20**, 7-ketone **21**, and 5ß,6ß-epoxide **9** do not. Indeed, other than autoxidation and lipid peroxidation origins, these oxysterols have no origin. Yet they occur in human serum at the same levels of the established oxysterol metabolites of cholesterol. We are faced with the inescapable conclusion that these other oxysterols are (1) metabolites of cholesterol by undiscovered enzyme processes, (2) formed from enzymic metabolites by _in vivo_ processes yet to be demonstrated, or (3) products of _in vivo_ lipid peroxidation. By addressing each point in turn, we may be able to make further conclusions.

Despite the ready demonstration of a membrane-bound cholesterol 7α-hydroxylase in liver and oddly in human blood leukocytes (monocytes and lymphocytes),[29] no such cholesterol 7ß-hydroxylase has been suggested. The 3ß,7ß-diol **20** may accompany metabolite 3ß,7α-diol **19** _in vitro_, but accompanying _in vitro_ lipid peroxidation also occurs. Likewise, no cholesterol 5ß,6ß-epoxidase has been described, and as in the case for the 7-hydroxylated oxysterols, the 5ß,6ß-epoxide **9** may accompany the 5α,6α-epoxide **8** _in vitro_, but again lipid peroxidation appears to be the cause. Direct formation of the 7-ketone **21** enzymically from cholesterol would be an unprecedented metabolic process.

As for the second matter, the epimerization of the 3ß,7α-diol **19** might account for the presence of the epimeric 3ß,7ß-diol **20**, and an hepatic 7ß-hydroxysteroid dehydrogenase interconverting the 3ß,7ß-diol **20** and 7-ketone **21** _in vitro_ has been described. Thus, these processes could account for the presence of the 3ß,7ß-diol **20** and 7-ketone **21** _in vitro_, but no such processes have been demonstrated _in vivo_. Nonetheless, there is no compelling reason to reject or accept these processes. Formation of the 5ß,6ß-epoxide **9** is another matter, as no ready isomerization of the 5α,6α-epoxide **8** may be envisaged chemically, let alone metabolically. It has been shown that stomach hydrochloric acid may add to the sterol 5,6-epoxides forming chlorohydrins. The 5α,6α-epoxide **8** thereby yields 6ß-chloro-5α-cholestane-3ß,5-diol (**11**); the 5ß,6ß-epoxide **9** yields the isomeric 5-chloro-5α-cholestane-3ß,6ß-diol (**12**).[21] The reactions are reversed in alkaline medium (Fig.2), but epoxide isomerization does not appear to be prominent in such processes.

These arguments support more acceptably the existence of _in vivo_ lipid peroxidation of cholesterol. The lipid peroxidation of cholesterol _in vitro_ by rat liver microsomal preparations, soybean lipoxygenase, and horseradish peroxidase yielding the 7-oxygenated oxysterols **19-21** and the isomeric 5,6-epoxides **8** and **9** as secondary products following initial formation of the epimeric cholesterol 7-hydroperoxides **15** and **16** is well established. Also, much evidence of other sorts supports _in vivo_ lipid peroxidation of poly-unsaturated fatty acid derivatives.

Photochemical Oxidation

Skin and hair cholesterol exposed to light and air is photochemically oxidized to the common B-ring oxysterols but may also be oxidized by electronically excited (singlet) oxygen ($^1O_2$) created photochemically. Singlet oxygen oxidizes cholesterol via the "ene" reaction mechanism to isomeric allylic hydroperoxides, the major one being 3ß-hydroxy-5α-cholest-6-ene-5-hydroperoxide (**26**). Minor products include the epimeric 3ß-hydroxycholest-

26

4-ene-6-hydroperoxides and thermal decomposition products from 5α-hydroperoxide **26** (cholesta-4,6-dien-3-one, 5α-cholest-6-ene-3ß,5-diol) and 6-hydroperoxides (cholest-4-ene-3ß,6-diols, 3ß-hydroxycholest-4-en-6-one). The 5α-hydroperoxide **26** also rearranges readily to the 7α-hydroperoxide **15** and decomposes to the 3ß,7α-diol **17** and 7-ketone **21**, with thermal elimination products such as cholesta-2,4,6-triene and cholesta-3,5,7-triene also formed.

The photochemical oxidation of cholesterol in erythrocytes and in synthetic liposomes is observed to yield the 5α-hydroperoxide **26**, and in oxidations involving both radical and photochemical processes the 7-hydroperoxides **15** and **16** and the 5α-hydroperoxide **26** may form.[30] Although the $^1O_2$ oxidation of cholesterol may occur in photochemical systems and indications of such oxidation is had in foods and tissues exposed to air and light, the 5α-hydroperoxide **26** has not been detected as an oxysterol of living tissues. However, as it now appears that $^1O_2$ may indeed be formed _in vitro_ by human blood polymorphonuclear leukocytes,[31] and $^1O_2$ may form under unprecedented and heretofore unrecognized biochemical ways.[32] it may be that there be a potential _in vivo_ origin for the 5α-hydroperoxide **26**. The ready isomerization of the 5α-hydroperoxide **26** to the 7α-hydroperoxide **15** could prevent detection of the 5α-hydroperoxide and thus obscure evidence of past action of $^1O_2$.

Oxidation by Other Active Oxygen Species

The oxidation of cholesterol by $^3O_2$ is by far the most obvious, the most well studied, and quantitatively the predominant process. However, other active oxygen species also oxidize cholesterol, and to the extent that active oxygen species be encountered in biological systems, their product oxysterols may also be present. These active oxygen species may be generated in the normal course of in vivo metabolism but may be generated by xenobiotic interference in normal metabolism.

Of the six other active oxygen species currently recognized, only the one-electron reduced dioxygen product superoxide (O_2^-) fails to oxidize cholesterol. The dioxygen cation (O_2^+) oxidizes cholesterol to the common B-ring oxysterols, but O_2^+ is not a likely biological species. Singlet oxygen already mentioned also attacks cholesterol, forming uniquely the 5α-hydroperoxide **26** that is considered diagnostic of the participation of $^1O_2$ in the process. The two-electron reduced species H_2O_2 epoxidizes cholesterol but may also be source of the oxidants $^3O_2$ and $^1O_2$ via disproportionation as well as of hydroxyl radical (HO) by peroxide bond homolysis. Other reactions of H_2O_2 with transition metal ions may lead to yet other complex oxidizing systems. Hydroxyl radical generated radiolytically or by Fenton chemistry attacks cholesterol in the B-ring forming the common 7-oxygenated oxysterols **17-21**, 5,6-epoxides **8** and **9**, and the 3β,5α,6β-triol **10**. Ozone as previously mentioned attackes the Δ^5-double bond in two ways, forming the secosterol ozonides but also the 5,6-epoxides **8** and **9**.[33]

There may be yet other active oxygen species recognized in metabolism, particularly those in which transition metal ions participate in catalytic actions. Furthermore, we must now include among active oxygen species the hypohalous acids and N-halogenated aminoacids potentially generated by blood leukocytes. Hypochlorous acid (HOCl) in organic solvent solutions transforms cholesterol to chlorohydrins **11** and **12** inter alia but autoxidizes cholesterol in water.[34,35] However, an isolated human blood polymorphonuclear leukocyte myeloperoxidase oxidizes cholesterol to the common B-ring oxysterols and to chlorohydrins in vitro.[22]

In concept cholesterol halohydrins may form by other means in vivo. As previously mentioned, stomach acid acting on dietary 5,6-epoxides may yield cholesterol chlorohydrins,[21] for example. Furthermore, recent demonstration of bromide ion dependent oxidations by blood eosinophiles suggests that hypobromous acid (HOBr) may also need consideration as an oxidant of endogenous sterols.[36] It is uncertain whether such oxidations of cholesterol occur in vivo, but a directed search for sterol halohydrins has not been mounted.

Other Xenobiotic Oxidants Nitrogen dioxide (NO_2) in inspired air is an oxidant causing increased levels of isomeric cholesterol 5,6-epoxides **8** and **9** in rat lung in vivo,[37] and cholesterol 3β-nitrite, a potential nitrosating agent, is formed in rat skin exposed to atmospheric NO_2.[38] Cholesterol in organic solvent solution exposed to NO_2 also forms the 3β-nitrite and 6α-nitro-5α-cholestane-3β,5-diol,[39] whereas monomolecular films of cholesterol on water is autoxidized and forms cholesterol 3β-nitrate.[40]

Other xenobiotic influences may oxidize cholesterol. Rats treated with CCl_4 appear to form a 7-trichloromethyl adduct of cholesterol in liver.[41] Cholesterol in CCl_4 solution subjected to ionizing radiation in the presence of $^3O_2$ (putatively generating trichloromethylperoxyl radical Cl_3COO.) yields the isomeric 5,6-epoxides **8** and **9** and chlorohydrins **11** and **12**.[42] Additionally, nitrosobenzene reacts with cholesterol in organic solvent solutions to yield a stable aminoxyl (nitroxyl) radical.[43] Even the influence of dietary cholesterol as xenobiotic appears to cause an increase in liver cholesterol epoxidation[44] and possibly the generation of other polar oxysterols.[45] This matter of xenobiotic influence on cholesterol metabolism and oxysterol formation obviously demands a lot more study.

Dietary Origins of Oxysterols

Many recent reliable analyses of cholesterol-rich processed foods have been reported, and we may accept with confidence that such foods contain traces of oxysterols, including cholesterol metabolites and autoxidation products.[1,2] The presence of oxysterols in foods poses an obvious source of such oxysterols in animals eating such foods, as intestinal absorption of oxysterols appears to occur. However, studies of the presence of oxysterols in fresh foods, of the survival of oxysterols in cooked foods, and of intestinal absorption of oxysterols from real foods ingested in the diet remain to be recorded.

Although dietary sources may contribute to oxysterols of human blood, for example, the presence of oxysterols in foods must ultimately arise via the several enzymic and nonenzymic means already discussed.

CONCLUSIONS

Many studies support the concept of oxysterols being present in human blood at a basal level that may be subject to fluctuations with diet, health state, or other factors. The presence of oxysterols in other tissues is also demonstrated. Some tissue oxysterols have recognized physiological roles; others do not. It is the class of oxysterols with no recognized purpose that is under present scrutiny, as these oxysterols appear to serve no purpose and exhibit toxicities in bioassay systems. Current speculations suggest that these oxysterols be implicated in the etiology of such human disorders as atherosclerosis and cancer.

However, as it now appears prudent to expand the class of oxysterols to include derivatives formed by reactions with xenobiotic agents in vivo, it may be that the sterol-xenobiotic conjugates pose not a toxic hazard in vivo but rather a protective metabolism that detoxifies the reactive, toxic agent. By analogy, a parallel concept of protection by cholesterol against reactive oxidants must also occupy our attentions. Oxysterols of tissues may reflect no more important a process than mere interception of oxidants as a detoxification process, the product oxysterols then being subject to esterification, transport, and ultimate excretion. In this view, tissue oxysterols may not be aberrant toxic agents in their own right in vivo.

The concept of cholesterol as an antioxidant protecting tissues against oxidations has been previously advanced,[46-48] but the more general concept that cholesterol serve as protection agent against xenobiotic insult by interception of toxic agent as oxysterol or sterol conjugate to be excreted from the system should now also receive attention. In this regard cholesterol esters of xenobiotic substances form under some circumstances in vivo.[49]

The fundamental chemistry of cholesterol oxidation by divers oxygen species is now fairly well understood, and the enzymic oxidation of cholesterol in vitro has likewise received sound examination. One must conclude that the oxidation of cholesterol occurs inexorably in Nature, in biological material dead or alive, and that given the considerable evidence that oxysterols possess biological activities per se as well as serve as precursors of the more established biologically active steroids, that the oxysterol story may yet unfold into a more significant matter.

The role of oxysterols in human health matters, if any, remains obscure despite the efforts at adumbration. Although oxysterols are predominantly toxic in bioassays and acute in vivo toxic effects have been demonstrated, physiological relevance remains to be established. Furthermore, a speculative role of selected oxysterols as regulators of de novo sterol biosynthesis continues to attract adherents. Finally, it is possible that sterols act as protective agents to sequester toxic oxygen species and xenobiotic agents. This concept has not been emphasized heretofore, but as it appears that oxysterols be metabolized in vivo and presumably eliminated, their potential role in detoxification should not be overlooked.

REFERENCES

1. L. L. Smith, "Cholesterol Autoxidation", Plenum Press, New York, 1981.
2. L. L. Smith, Cholesterol Autoxidation 1981-1986, Chem. Phys. Lipids 44:87 (1987).
3. L. L. Smith and B. H. Johnson, Biological Activities of Oxysterols, Free Radical Biol. Med. in press.
4. J. P. Beck and A. Crastes de Paulet, eds., "Activités Biologiques des Oxystérols", Editions INSERM, Paris, 1988.
5. B. J. Koopman, J. C. van der Molen, and B. G. Wolthers, Determination of Some Hydroxycholesterols in Human Serum Samples, J. Chromatog. 416:1 (1987).
6. I. Björkhem, E. Reihnér, B. Angelin, S. Ewerth, J.-E. Akerlund, and K. Einarsson, On the Possible Use of the Serum Level of 7α-Hydroxycholesterol as A Marker for Increased Activity of the Cholesterol 7α-Hydroxylase in Humans, J. Lipid Res. 28:889 (1987).
7. L. L. Smith, J. I. Teng, Y. Y. Lin, P. K. Seitz, and M. F. McGehee, Lipid Peroxidations of Cholesterol, in: Lipid Peroxides in Biology and Medicine, K. Yagi, ed., Academic Press, New York, 1982, pp.89-105.
8. L. L. Smith, J. I. Teng, Y. Y. Lin, P. K. Seitz, and M. F. McGehee, Sterol Metabolism. XLVII. Oxidized Cholesterol Esters in Human Tissues, J. Steroid Biochem. 14:889 (1981).
9. C. J. W. Brooks, R. M. McKenna, W. J. Cole, J. MacLachlan, and T. D. V. Lawrie, 'Profile' Analysis of Oxygenated Sterols in Plasma and Serum, Biochem. Soc. Trans. 11:700 (1983).
10. I. Björkhem, Assay of Unesterified 7-Oxocholesterol in Human Serum by Isotope Dilution-Mass Spectrometry, Anal. Biochem. 154:497 (1986).
11. I. Björkhem, O. Breuer, B. Angelin, and S.-A. Wikström, Assay of Unesterified Cholesterol-5,6-epoxide in Human Serum by isotope Dilution Mass Spectrometry. Levels in the Healthy State and in Hyperlipoproteinemia, J. Lipid Res. 29:1031 (1988).
12. N. B. Javitt, E. Kok, S. Burstein, B. Cohen, and J. Kutscher, 26-Hydroxycholesterol. Identification and Quantitation in Human Serum, J. Biol. Chem. 256:12644 (1981).
13. N. B. Javitt, E. Kok, B. Cohen, and S. Burstein, Cerebrotendinous Xanthomatosis: Reduced Serum 26-Hydroxycholesterol, J. Lipid Res. 23:627 (1982).
14. C. J. W. Brooks, W. J. Cole, J. MacLachlan, and T. D. V. Lawrie, Some Aspects of the Analysis of Minor Oxygenated Sterols in Serum and in Serum Lipoprotein Fractions, J. Am. Oil Chem. Soc. 62:622 (1985).
15. I.-L. Kou and R. P. Holmes, The Analysis of 26-Hydroxycholesterol in Plasma by High Performance Liquid Chromatography, Federation Proc. 45:311 (1986).
16. L. D. Gruenke, J. C. Craig, N. L. Petrakis, and M. B. Lyon, Analysis of Cholesterol, Cholesterol-5,6-Epoxides and Cholestane-3β,5α,6β-Triol in Nipple Aspirates of Human Breast Fluid by Gas Chromatography/Mass Spectrometry, Biomed. Environm. Mass Spectrom. 14:335 (1987).
17. D. R. Brill, "Presence of Cholesterol Epoxide in the Human Prostate Gland", Ph.D. Dissertation, Rutgers University, New Brunswick,NJ, 1981.

18. W. R. Eberlein and A. A. Patti, Steroids and Sterols in Umbilical Cord Blood, <u>J. Clin. Endocrinol.</u> 25:1101 (1965).

19. S. R. Steckbeck, J. A. Nelson, and T. A. Spencer, Enzymic Reduction of an Epoxide to an Alcohol, <u>J. Am. Chem. Soc.</u> 104:893 (1982).

20. L. L. Smith, M. J. Kulig, D. Miiller, and G. A. S. Ansari, Oxidation of Cholesterol by Dioxygen Species, <u>J. Am. Chem. Soc.</u> 100:6206 (1978).

21. G. Maerker, E. H. Nungesser, and F. J. Bunick, Reaction of Cholesterol 5,6-Epoxides with Simulated Gastric Juice, <u>Lipids</u> 23:761 (1988).

22. C. S. Foote, R. B. Abakerli, R. L. Clough, and R. I. Lehrer, On the Question of Singlet Oxygen Production in Polymorphonuclear Leucocytes, In: Bioluminescence and Chemiluminescence, Basic Chemistry and Analytical Applications, M. A. DeLuca and W. D. McElroy, eds., Academic Press, New York, 1981, pp.81-88.

23. M. A. Johnson and R. Croteau, Biosynthesis of Ascaridole: Iodide Peroxidase-Catalyzed Synthesis of a Monoterpene Endoperoxide in Soluble Extracts of <u>Chenopodium ambrosioides</u> Fruit, <u>Arch. Biochem. Biophys.</u> 235:254 (1984).

24. J. Gumulka and L. L. Smith, Ozonization of Cholesterol, <u>J. Am. Chem. Soc.</u> 105:1972 (1983).

25. K. Jaworski and L. L. Smith, Ozonization of Cholesterol in Nonparticipating Solvents, <u>J. Org. Chem.</u> 53:545 (1988).

26. H. Itokawa, Y. Tachi, Y. Kamano, and Y. Iitaka, Structure of Gilvanol, A New Triterpene Isolated from <u>Quercus gilva</u> Blume, <u>Chem. Pharm. Bull.</u> 26:331 (1978).

27. H. Ageta, K. Shiojima, R. Kamaya, and K. Masuda, Fern Constituent: Naturally Occurring Adian-5-ene Ozonide in the Leaves of <u>Adiantum monochlamys</u> and <u>Oleandra wallachii</u>, <u>Tetrahedron Lett.</u> 899 (1978).

28. E. J. M. van Haren and A. D. Tait, Inhibition of Cholesterol Side-Chain Cleavage by Intermediates of an Alternative Steroid Biosynthesis Pathway, <u>FEBS Lett.</u> 232:377 (1988).

29. N. K. Dodd, C. E. Sizer, and J. Dupont, Cholanoic Acids and Cholesterol 7-α-Hydroxylase Activity in Human Leucocytes, <u>Biochem. Biophys. Res. Commun.</u> 106:385 (1982).

30. A. W. Girotti, G. J. Bachowski, and J. E. Jordan, Lipid Peroxidation in Erythrocyte Membranes: Cholesterol Product Analysis in Photosentitized and Xanthine Oxidase-Catalyzed Reactions, <u>Lipids</u> 22:401 (1987).

31. J. R. Kanofsky, H. Hoogland, R. Wever, and S. J. Weiss, Singlet Oxygen Production by Human Eosinophils, <u>J. Biol. Chem.</u> 263:9692 (1988).

32. D. Galaris, D. Mira, A. Sevanian, E. Cadenas, and P. Hochstein, Co-oxidation of Salicylate and Cholesterol during the Oxidation of Metmyoglobin by H_2O_2, <u>Arch. Biochem. Biophys.</u> 262:221 (1988).

33. J. Gumulka, J. S. Pyrek, and L. L. Smith, Interception of Discrete Oxygen Species in Aqueous Media by Cholesterol: Formation of Cholesterol Epoxides and Secosterols, <u>Lipids</u> 17:197 (1982).

34. B. O. Lindgren, Reactions of Sterols with Bleaching Agents: Reactions of Cholesterol and its Acetate with Aqueous Chlorine Solutions, <u>Svensk Papperstidn.</u> 70:532 (1967).

35. C. S. Foote, R. B. Abakerli, R. L. Clough, and F. C. Shook, On the Question of Singlet Oxygen Production in Leucocytes, Macrophages and the Dismutation of Superoxide Anion, In: Biological and Clinical Aspects of Superoxide and Superoxide Dismutase, W. H. Bannister and J. V. Bannister, eds., Elsevier/North-Holland, New York/Amsterdam/Oxford, 1980, pp.222-230.

36. S. J. Weiss, S. T. Test, C. M. Eckmann, D. Roos, and S. Regiani, Brominating Oxidants Generated by Human Eosinophils, <u>Science</u> 234:200 (1986).

37. A. Sevanian, J. F. Mead, and R. A. Stein, Epoxides as Products of Lipid Autoxidation in Rat Lungs, <u>Lipids</u> 14:634 (1979).

38. S. S. Mirvish, D. M. Babcock, A. D. Deshpande, and D. L. Nagel, Identification of Cholesterol as a Mouse Skin Lipid that Reacts with Nitrogen Dioxide to Yield A Nitrosating Agent, and of Cholesterol

Nitrite as the Nitrosating Agent Produced in a Chemical System from Cholesterol, Cancer Lett. 31:97 (1986).

39. T. Kobayashi and K. Kubota, The Reaction of Nitrogen Dioxide with Lung Surface Components: The Reaction with Cholesterol, Chemosphere 9:777 (1980).

40. A. M. Kamel. N. D. Weiner, and A. Felmeister, Identification of Cholesterol Nitrate as a Product of the Reaction between NO_2 and Cholesterol Monomolecular Films, Chem. Phys. Lipids 6:225 (1971).

41. G. A. S. Ansari, M. T. Moslen, and E. A. Reynolds, Evidence for in vivo Covalent Binding of CCl_3 Derived from CCl_4 to Cholesterol of Rat Liver, Biochem. Pharmacol. 31:3509 (1982).

42. J. E. van Lier and R. Langlois, Chlorinated Hydrocarbon Mediated Cholesterol Degradation, J. Am. Oil Chem. Soc. 62:622 (1985)

43. R. Sridhar and R. A. Floyd, An Electron Paramagnetic Resonance Study of the Reaction of Nitrosobenzene with Cholesterol, Can. J. Chem. 60:1574 (1982).

44. T. Watabe, M. Isobe, and M. Kanai, Cholesterol Diet Increases Plasma and Liver Concentrations of Cholesterol Epoxides and Cholestanetriol, J. Pharm. Dyn. 3:553 (1980).

45. C. Marco de la Calle, and G. F. Gibbons, Hepatic and Intestinal Formation of Polar Sterols in vivo in Animals Fed on a Cholesterol-supplemented Diet, Biochem. J. 252:395 (1988).

46. S. H. Kon, Biological Autoxidation. II. Cholesterol Esters as Inert Barrier Antioxidants. Self-Assembly of Porous Membrane Sacs. An Hypothesis, Medical Hypotheses 4:5569 (1978).

47. S. K. Jain and S. B. Shohet, Apparent Role of Cholesterol as an Erythrocyte Membrane Anti-Oxidant, Clin. Res. 29:336A (1981).

48. U. L. Bereza, G. J. Brewer, and G. M. Hill, Effect of Dietary Cholesterol on Erythrocyte Peroxidant Stress in vitro and in vivo, Biochim. Biophys. Acta 835:434 (1985).

49. G. B. Quistad and D. H. Hutson, Lipophilic Xenobiotic Conjugates, In: Xenobiotic Conjugation Chemistry, G. D. Paulson, J. Caldwell, D. H. Hutson, and J. J. Menn, eds., American Chemical Society, Washington,DC, 1986, pp.204-213.

ROLE OF OXYGEN AND OF SUPEROXIDE RADICAL

IN THE MECHANISMS OF MONOELECTRONIC

ACTIVATION OF VARIOUS XENOBIOTICS :

A RADIOLYSIS STUDY

Monique Gardès-Albert,
Chantal Houée-Levin,
Abdelhafid Sekaki,
Christiane Ferradini

Laboratoire de Chimie Physique
UA 400, Université Paris V
45, rue des Saints Pères,
75270 Paris Cedex 06 France.

INTRODUCTION

Many xenobiotics need a free-radical mediated metabolism activation
to exhibit their biological properties. Among them, antitumor drugs such
as anthracycline or ellipticine derivatives are known to be enzymically
activated by one-electron transfer. We present here two anticancer agents:
an ellipticine substituted derivative, EH_2 (scheme 1) which has a para-
aminophenol structure and daunorubicin, DOS (scheme 2) which is an anthracycline antibiotic with a quinone group. Both these anticancer agents possess
a resonant plane structure which allows their intercalation in DNA and each
drug has a side group which enhances this intercalation ability (an amino
side chain for EH_2 and a sugar for DOS). During the metabolism of these
antitumor agents, oxygen and/or reactive oxygen species are required for
the development of their cytotoxic properties. Hence we have investigated
the interaction of O_2 and/or $O_2^{\cdot -}$ with the radical transient of each drug.
In both cases pulse radiolysis combined with kinetic spectroscopy is well
appropriated for such studies of model in vitro systems. Indeed this technique gives free radicals in homogeneous solution in a determined amount
and allows the direct observation of the free radical reactions with substrates.

We have reported here the detailed study of the ellipticine derivative and
a summary of the results obtained with daunorubicin.

Free Radicals, Lipoproteins, and Membrane Lipids
Edited by A. Crastes de Paulet *et al.*
Plenum Press, New York, 1990

9. OH ELLIPTICINE SUBSTITUTED DERIVATIVE, EH_2

EH_2 (scheme 1), ((γ – diethylaminopropyl-amino) 1 dimethyl – 5, 11 hydroxy – 9 – 6H – pyrido (4, 3 – b) carbazole) belongs to the family of 9 – hydroxyellipticine (9 OH E, scheme 1).

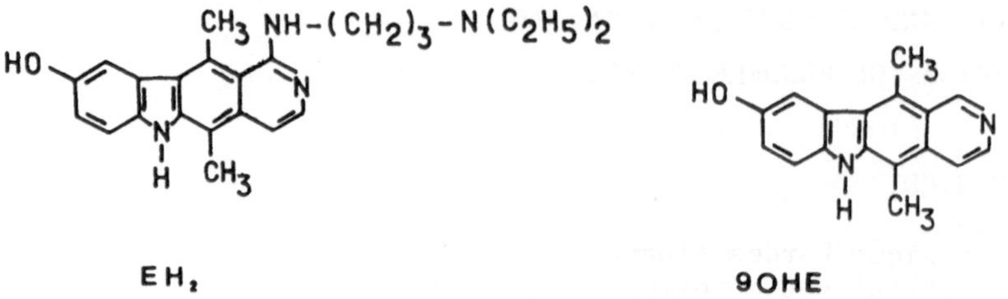

Scheme 1   Ellipticine derivatives – EH_2 :
            9 – hydroxyellipticine substituted derivative,
            9 OHE : 9 – hydroxyellipticine.

It has been shown in vivo that celiptium (another 9 – hydroxyellipticine) might be activated in the human urinary[1], in biliary metabolism[2] and in human red blood cells[3] through an oxidative process.

It has been suggested that the cytotoxic and antitumor properties of this drug could be related to its bioactivation by peroxidase – H_2O_2 systems via the production of strong electrophile species like the semi-quinone-imine free radical and the quinone-imine[4]. Furthermore it is known that oxygen is implicated in the oxidative process of 9 OHE[5].

Consequently it appeared of interest to study the one-electron oxidation of EH_2 by $OH^\cdot$ radicals in order to characterize the reactive oxidized transient(s) and to precise the role of $O_2/O_2^{\bar{\cdot}}$ in this system.

MATERIALS AND METHODS

EH_2 was synthesized according to Bisagni et al.[6] and its purity was estimated to be higher than 99%. Other chemicals were either Merck PA grade ($NaOH, HCO_2Na$) or Prolabo normapur (NaH_2PO_4, HCl) and used as received. The gases O_2 and N_2O were from CFPO (France) and their purity was higher than 99.99%. Water was triply distilled and its purity controlled by conductimetry (resistivity $> 10^6 \Omega$ .cm). pH7 was adjusted with phosphate buffer. EH_2 was first dissolved in concentrated HCl solution. The resulting dichloride salt solution was diluted in phosphate buffer $5 \times 10^{-2} mol.dm^{-3}$ at pH7. EH_2 solutions did not undergo autoxidation in these conditions.

Pulse radiolysis experiments were performed using a febetron 707 (CEN

Saclay, France) which delivers electron pulses (1.8 MeV, 8 ns at half-height). The dose was equal to 30–80 Gy according to experiments. The content of the irradiation cell (1.6 cm^3, optical path 2.5 cm) was irradiated by a single pulse. The variation of dose from pulse to pulse were monitored by measuring the total pulse charge with a charge integrating current and the doses were calibrated with KSCN (ε_{472}(SCN)$_2^{\overline{\cdot}}$ = 7.580 $dm^3 mol^{-1} cm^{-1}$; G = 5.8 x 10^{-7} $mol.J^{-1}$ = 5.6 molec (100eV^{-1})).

It is known that water radiolysis leads to :

$$H_2O \longrightarrow e^-_{aq}, OH^{\cdot}, H^{\cdot}, H_2O_2, H_2, H^+$$

The free radicals e^-_{aq}, $OH^{\cdot}$ and $H^{\cdot}$ can be either used directly as reactants or transformed by suitable scavengers into other radical species.

In N_2O – saturated solutions, hydrated electrons are transformed into hydroxyl radicals :   (1) e^-_{aq} + N_2O $\longrightarrow$ $OH^{\cdot}$ + N_2 + OH^-

Thus $OH^{\cdot}$ free radicals are selectively produced with a yield equal to 5.8 x 10^{-7} $mol.J^{-1}$ (5.6 molec (100eV)$^{-1}$).

In oxygen-saturated solutions, hydrated electrons and $H^{\cdot}$ free radicals react with oxygen :   (2) e^-_{aq} + O_2 $\longrightarrow$ $O_2^{\overline{\cdot}}$

(3) $H^{\cdot}$ + O_2 $\longrightarrow$ $HO_2^{\cdot}$

(4) $HO_2^{\cdot}$ $\rightleftharpoons$ $O_2^{\overline{\cdot}}$ + H^+ pK_a = 4.7

Thus $OH^{\cdot}$ and $O_2^{\overline{\cdot}}$ free radicals are produced with similar yields respectively equal to 2.9 x 10^{-7} $mol.J^{-1}$ and 3.4 x 10^{-7} $mol.J^{-1}$ (2.8 and 3.25 molec.(100 eV)$^{-1}$).

To select superoxide anions, oxygen-saturated solutions contain formate ions (0.1 $mol.dm^{-3}$) : (5) $H^{\cdot}$ + $HCOO^-$ $\longrightarrow$ H_2 + $COO^{\overline{\cdot}}$

(6) $OH^{\cdot}$ + $HCOO^-$ $\longrightarrow$ H_2O + $COO^{\overline{\cdot}}$

(7) $CO_2^{\overline{\cdot}}$ + O_2 $\longrightarrow$ $O_2^{\overline{\cdot}}$ + CO_2

According to reactions (5)(6)(7) and (2), superoxide anions are selectively produced with a yield equal to 6.2 x 10^{-7} $mol.J^{-1}$ (6 molec (100 eV)$^{-1}$).

RESULTS AND DISCUSSION

We have already shown by pulse radiolysis of N_2O saturated solutions that EH_2 is oxidized by $OH^{\cdot}$ radicals giving the radical transient $EH^{\cdot}$ characterized by its difference absorption spectrum[7].

We have then investigated the reaction of $O_2^{\overline{\cdot}}$ anions with EH_2 in oxygenated solutions containing formate ions in the absence or in the presence of SOD. We have found that superoxide radicals do not react neither with EH_2 nor with $EH^{\cdot}$ radical transient.

In order to study the role of O_2 during the one-electron oxidation of the drug by $OH^{\cdot}$ radicals, aqueous solutions of EH_2 were saturated with oxygen. When such EH_2 solutions are irradiated by a single electron pulse

$((EH_2)_0 = 2.5 \times 10^{-4} mol.dm^{-3}$, (phosphate) $= 5 \times 10^{-2} mol.dm^{-3}$, pH7, dose : 72 Gy, $(OH)_0 = 2.1 \times 10^{-5} mol.dm^{-3}$), we can distinguish three steps in the kinetic evolution of absorbances (fig.1). The first step ended in 1.5 μs (inset (a) fig.1) corresponds to the formation of the radical EH$^{\cdot}$ by OH$^{\cdot}$ attack on the substrate. The difference absorption spectrum I belongs to EH$^{\cdot}$ transient. The second step was not seen in the absence of oxygen. It is only observed in the wavelength range from 530 to 750 nm (inset (b) fig.1) and the spectrum II taken 400 μs after the pulse characterizes a second transient. The third step can be seen in the wavelength region from 400 to 500 nm and it reaches a plateau value in 9 ms (inset (c) fig.1). Spectrum III taken at this time (fig.1) is unchanged until 100 ms.

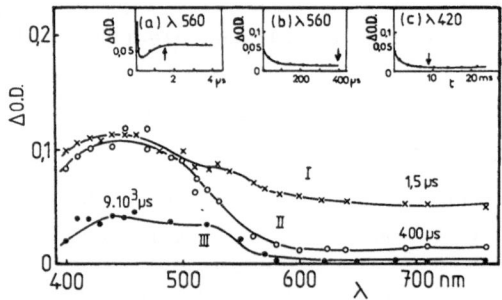

Fig. 1 . Difference absorption spectra of EH_2 aqueous solutions irradiated by electron pulses in an atmosphere of oxygen. $(EH_2)_0 = 2.5 \times 10^{-4}$ mol.dm^{-3} phosphate buffer 5×10^{-2} mol.dm^{-3}, pH7, dose = 72 Gy per pulse, l = 2.5 cm , I 1.5 μs after the pulse, II 400 μs after the pulse, III 9×10^3 μs after the pulse.
Inset (a) and (b) : absorbances at 560 nm versus time for two time scales ; inset (c) : absorbance at 420 nm versus time.

The kinetic analysis of EH$^{\cdot}$ decay from spectrum I to II has shown a first order process in the wavelength range 400–500 nm. The apparent rate constant is equal to $1.2 \times 10^4 s^{-1}$. We attributed this first order reaction to the oxygen action on EH$^{\cdot}$ radical leading perhaps to a peroxy radical whose absorption spectrum would be spectrum II. The related second order rate constant is deduced from $k_{app} = k(O_2)$, ca $k = (1.4 \pm 0.2) \times 10^7 mol^{-1} dm^3 s^{-1}$. The last kinetic studied at several wavelengths between 530 and 750 nm is a second order process whose rate constant is $2k = (1.2 \pm 0.2)$

$\times 10^8 \text{mol}^{-1}\text{dm}^3\text{s}^{-1}$. It would correspond to the self recombination of the peroxyradical $\text{EHO}_2\cdot$.

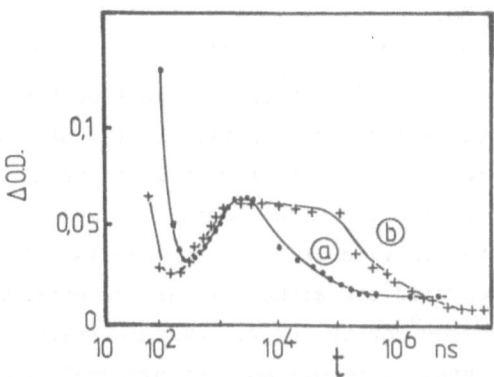

Fig. 2 . Difference absorbances versus time of EH_2 aqueous solutions irradiated under an atmosphere of oxygen.
$(\text{EH}_2)_\circ = 2.5 \times 10^{-4} \text{mol.dm}^{-3}$, phosphate buffer $5 \times 10^{-2} \text{mol.dm}^{-3}$, pH7 , (a) 560 nm, $(\text{OH})_\circ = 2.2 \times 10^{-5} \text{mol.dm}^{-3}$ (b) 420 nm, $(\text{OH})_\circ = 1.2 \times 10^{-5} \text{mol.dm}^{-3}$. In solid line, calculated curves.

Thus the one-electron oxidation mechanism in the presence of oxygen is the following :

$$\text{EH}_2 + \text{OH}\cdot \xrightarrow{(8)} \text{EH}\cdot + \text{H}_2\text{O}$$

$$\text{EH}\cdot \diagup \diagdown \text{O}_2$$

$$(9)(10)$$

$$\text{Product(s)} \qquad \text{EHO}_2\cdot$$

$$(11) \downarrow \text{EHO}_2\cdot$$

$$\text{Product(s)}$$

$\text{EH}\cdot$ radical generated by $\text{OH}\cdot$ attack on the substrate (reaction (8)) can react with O_2 (reaction (10)) in competition with its self-recombination (reaction (9)). Then the peroxy radical recombinates in the absence of other(s) substrate(s) susceptible to be oxidized. All the rate constants have been experimentally determined and the kinetic curves shown on fig.2 have been calculated according to the above kinetic scheme with the following rate constants :

$$k_8 = (4.0 \pm 0.8) \times 10^9 \text{ mol}^{-1}\text{dm}^3\text{s}^{-1}$$
$$2\,k_9 = (3.5 \pm 0.2) \times 10^8 \quad \text{"}$$
$$k_{10} = (1.4 \pm 0.2) \times 10^7 \quad \text{"}$$
$$2\,k_{11} = (1.2 \pm 0.2) \times 10^8 \quad \text{"}$$

and taking into account the well known reactions of e^-_{aq}, $H^\cdot$, $OH^\cdot$ and $O_2^{\cdot-}$:

(2) $e^-_{aq} + O_2 \longrightarrow O_2^{\cdot-}$ $\qquad k_2 = 1.9 \times 10^{10} \text{mol}^{-1}\text{dm}^3\text{s}^{-1}$

(3) $H^\cdot + O_2 \longrightarrow HO_2$ $\qquad k_3 = 1.2 \times 10^{10} \quad \text{"}$

(12) $OH^\cdot + O_2^{\cdot-} \longrightarrow OH^- + O_2$ $\qquad k_{12} = 10^{10} \quad \text{"}$

We can notice that 5 orders of magnitude in time have been explored. The calculated curves (in solid line) on fig. 2 ((a) at 560 nm and (b) at 420 nm) are in good agreement with experimental absorbances. Hence the above kinetic scheme explains the experimental results.

It is known that the in vivo oxidative metabolism of 9OHE (and its derivatives) might be carried out either by various extrahepatic enzymic systems such as peroxidases[4] or by several enzymic systems endowed with oxidant properties (oxygenase, peroxidase, or oxidase) as, for example, hemoglobin contained in red blood cells[3]. We have shown in our one-electron oxidation model that $EH^\cdot$ radical can rapidly react with oxygen to give another transient which could be a peroxy radical. It is known that similar peroxy radicals can abstract H-atom in the presence of other(s) substrate(s) (which is the case of biological material where self-recombination radical reactions are disfavoured). In addition the reactions of $RO_2^\cdot$ are more specific and consequently more damaging to the cell than the ones of hydroxyl radicals. Hence the cytotoxic and antitumor effect of EH_2 could be related to the reactions of such peroxy radicals. Auclair et al.[5] have shown that 9OHE can be oxidized by molecular oxygen via a free radical mechanism giving superoxide anion and the free radical of the drug. This autoxidation reaction is an equilibrium[5] which means that the reverse reaction, ie the reduction of the drug free radical by superoxide anion may occur. EH_2 does not undergo an autoxidation reaction at pH7 and we have no kinetic evidence of the opportunity of a reaction between the free radical transient $EH^\cdot$ and superoxide anion.

DAUNORUBICIN

A considerable amount of research has been devoted to the molecular mode of action of DOS. Although several mechanisms have been proposed, there seems to be a general agreement about the fact that this drug can undergo bioreductive activation leading to semiquinone and hydroquinone forms (scheme 2). These transients can have two fates : in the presence of oxygen, they can generate reactive oxygen species such as superoxide anion, hydrogen

peroxide and consequently OH⋅ free radicals by a Fenton-like mechanism.
On the other hand they can generate alkylating forms such as quinone methi-
des. These possibilities do not exclude each other : their occurence is
linked to tissue oxygenation conditions and to the relative rates of both
processes. Therefore it is important to know the rate constants of the different reactions involved in these phenomena. We have first investigated
by γ and pulse radiolysis daunorubicin reduction by e^-_{aq} and COO^- free radicals in anaerobic conditions[8,9]. This process ends up by sugar loss giving
7-deoxy daunomycinone. We have determined the kinetic scheme of the reduction of the drug and the absorption spectra of all the transients : the semiquinone, hydroquinone and aglycone intermediates including the so-called
quinone methide. The structures of these forms are given in scheme 2.

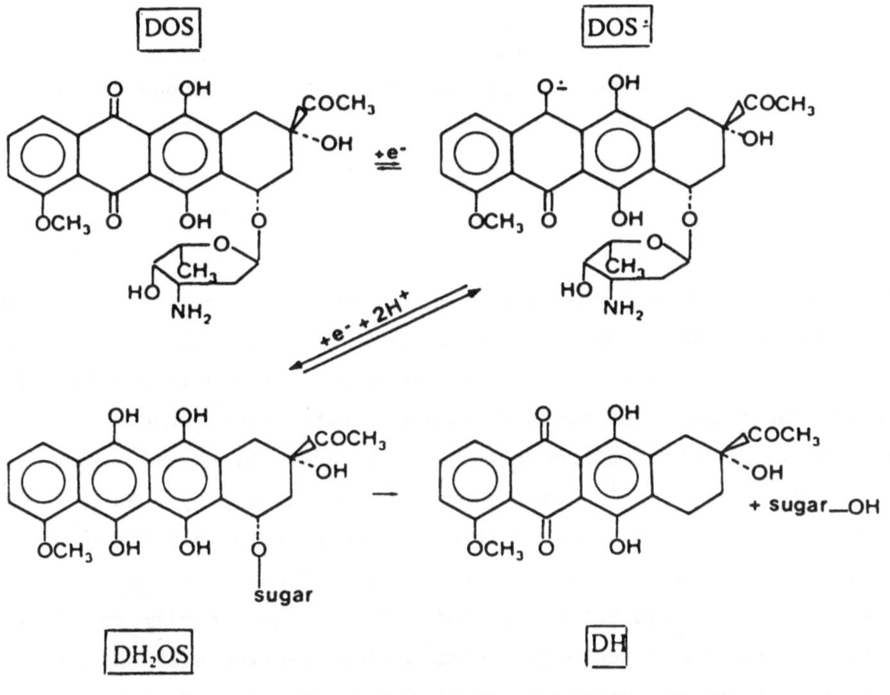

Scheme 2   Danunorubicin one electron reduction in anaerobic conditions.

The study of the reactions of these transients with O_2^- and O_2 has
shown that the semiquinone and the hydroquinone forms were oxidized by O_2
giving O_2^- radicals[10]. However the back reactions which are the reduction
by O_2^- of DOS and DOS^- also happen[10]. In addition O_2^- can also oxidize the
same transients[10]. These reactions are given in scheme 3. A high steady
state of oxygen reactive species is thus established wich can be computed
quantitatively knowing the initial cellular conditions (drug and reductant
concentrations, oxygen amount)[10].

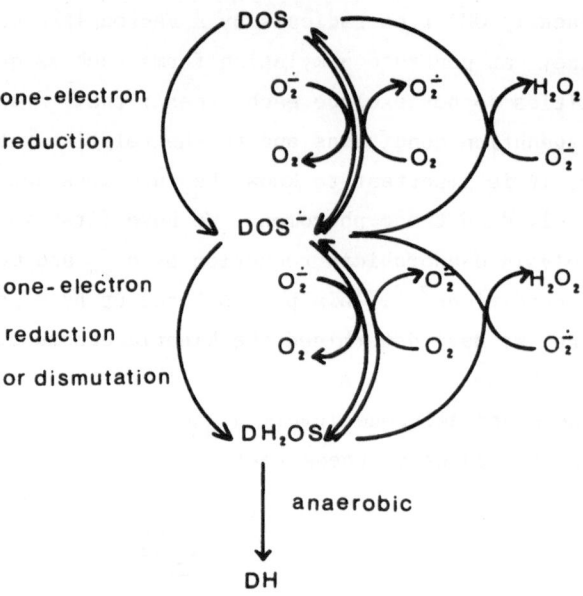

Scheme 3   Redox cycles of daunorubicine in aerobic conditions.

It is known that quinones such as daunorubicin can be reduced enzymically by numerous NAD(P)H and flavine enzymes[11,12]. Reductases can be found everywhere in the cell including in the nucleus[13]. Indeed it has been demonstrated that semiquinone and hydroquinone states from daunorubicin and from adriamycin, another anthracycline, are produced by enzymic reduction in the nucleus[14]. Our work[10] has shown that daunorubicin hydroquinone can either undergo glycosidic cleavage leading perhaps to an alkylating form, or be oxidized by oxygen, producing superoxide radicals and H_2O_2 and provides a way to estimate quantitatively their amount. Our results provide an explanation to the increased oxygen consumption observed with anthracyclines [15] and to the increased superoxide anion formation observed in heart submitochondrial particles[16].

CONCLUSION

The method of pulse radiolysis allowed us to investigate the role of O_2/O_2^- on different drugs known for their radical bioactivation. We have shown that two antitumor drugs, daunorubicin and an ellipticine derivative exhibit specific ways of interaction with O_2 and O_2^-. Daunorubicin redox cycles involve an enhancement of the O_2^- steady state, ellipticine radical transient reacts preferentialy with O_2 producing peroxyradicals and hence increases oxygen consumption. Such interactions with oxygen and reactive oxygen species might play a role in the biological antitumor and cytotoxic effects of these drugs.

Acknowledgments

This work was supported by the "Ligue Nationale Française contre le Cancer" and by "The Fondation de France". We are indebted to C. Rivalle and E. Bisagni for the synthesis and purification of EH_2. We are greatful to B. Hickel for the use of the Febetron 707 (CEN Saclay France). We acknowledge J. Chevrel and J. Potier for technical assistance in pulse radiolysis experiments.

REFERENCES

1. B.Monsarrat, M.Maftouh, G.Meunier, B.Dugué, J.Bernadou, J.P.Armand, C.Picard-Fraire, B.Meunier and C.Paoletti, Human and rat urinary metabolites of the antitumor drug celiptium (N2-methyl -9- hydroxyellipticinium acetate, NSC 264 137). Identification of cysteine conjugates supporting the "bioxidative alkylation" hypothesis, Biochem. Pharmacol. 32 : 3887 (1983).

2. J.Bernadou, B.Monsarrat, H.Roche, J.P.Armand, C.Paoletti and B.Meunier, Evidence for electrophilic properties of N^2 - methyl - 9 - hydroxyellipticinium acetate (celiptium) from human biliary metabolites, Cancer Chemother. Pharmacol. 15 : 63 (1985).

3. T.Ha, J.Bernadou, E.Voisin, C.Auclair and B.Meunier, Hemoglobin - catalyzed transformation of ellipticinium acetate into electrophilic species. Evidences for oxidative activation of the drug in human red blood cells, Chem. Biol. Interactions 65 : 73 (1988).

4. C.Auclair and C.Paoletti, Bioactivation of the antitumor drugs 9 - hydroxyellipticine and derivatives by a peroxidase - hydrogen peroxide system, J. Med. Chem. 24 : 289 (1981).

5. C.Auclair, K.Hyland and C.Paoletti, Autoxidation of the antitumor drug 9 - hydroxyellipticine and its derivatives J. Med. Chem. 26 : 1438 (1983).

6. E.Bisagni, C.Ducrocq, J.M.Lhoste, C.Rivalle and A.Civier, Synthesis of 1 - substituted ellipticines by a new route to pyrido (4,3-b)- carbazoles, J. Chem. Soc. Perkin Trans. I, 1706 (1979).

7. A.Sekaki, M.Gardès-Albert, C.Houée-Levin, C.Ferradini, C.Rivalle, E.Bisagni and B.Hickel, Influence of oxygen and superoxide free radicals on the oxidative activation of several antitumor drugs. In "Medical, Biochemical and chemical aspects of free radicals" Elsevier Science Publishes. In Press (1988).

8. C.Houée-Levin, M.Gardès-Albert, and C.Ferradini. Reduction of daunorubicin aqueous solutions by COO^- free radicals. Reactions of reduced transients with H_2O_2. FEBS Lett. 173 : 27 (1984).

9. C.Houée-Levin, M.Gardès-Albert, C.Ferradini, M.Faraggi, and M.Klapper. Pulse-radiolysis study of daunorubicin redox cycles. Reduction by e_{aq} and COO^- free radicals. FEBS Lett. 179 : 46 (1985).

10. C.Houée-Levin, M.Gardès-Albert, and C.Ferradini. Daunorubicin redox cycles and glycosidic cleavage. J. Free Rad. Biol. Med. 2 : 89 (1986).

11. A.Arcamone, Doxorubicin anticancer antibiotics, Academic Press (1981).

12. J.Fischer, B.R.J.Abdella, and K.E.McLane. Anthracycline antibiotics reduction by spinach ferredoxin, $NADP^+$ reductase and ferredoxin. Biochemistry 24 : 3562 (1985).

13. N.R.Bachur, S.L.Gordon, and M.V.Gee. A general mechanism for microsomal activation of quinone anticancer agents to free radicals. Cancer Res. 38 : 1745 (1978).

14. N.R.Bachur, M.V.Gee and R.D.Friedman, Nuclear catalyzed antibiotic free radical formation. Cancer Res. 42 : 1078 (1982).

15. J.H.Doroshow. Effect of anthracycline antibiotics on oxygen radical formation in rat heart. Cancer Res. 43 : 460 (1983).

16. J.W.Lown, H.H.Chen, J.A.Plambeck, and E.M.Acton. Diminished superoxide anion generation by reduced 5-iminodaunorubicin relative to daunorubicin and the relationship to cardiotoxicity of the anthracycline antitumour agents. Biochem. Pharmacol. 28 : 2563-2568 (1979).

METHODS OF MEASUREMENT

METHODS OF MEASURING LIPID PEROXIDATION IN BIOLOGICAL SYSTEMS:

AN OVERVIEW

Kevin H Cheeseman

Department of Biology and Biochemistry
Brunel University
Uxbridge, England UB8 3PH

INTRODUCTION

Lipid peroxidation is an extremely complicated process, especially in biological systems where one is dealing with mixtures of different lipids in matrices also containing proteins, carbohydrates, nucleic acids and catalytically-active enzymes and trace metals. With regard to measuring lipid peroxidation, the complexity of the process may be viewed by an optimist as presenting the investigator with a large number of opportunities: it is a multi-stage process and each stage can be investigated by at least one technique. The view of the pessimist, or perhaps realist, would be that the complexity of lipid peroxidation is a problem, in that no single assay method can fully describe what is happening or has happened in a system undergoing lipid peroxidation. In considering the large range of assay methods concerned with measuring lipid peroxidation, it is readily obvious that some are straightforward to perform, others tedious; some are more sensitive than others, some more prone to interference; some measure end-products, some measure unstable intermediates, and some measure specific identified compounds whilst others measure poorly characterised entities.

In choosing which assay method to use, the investigator must first decide if the method is applicable to the system under investigation; some methods, for example, are applicable in vitro but not in vivo. Secondly, does the assay tell the investigator what he wants to know in the most straightforward way? For example, if the investigator is actually interested in the generation of biologically-active aldehydes then these should be specifically measured. If one only requires a quantitative index of the extent of lipid peroxidation then a much simpler test may suffice. Other factors to be considered are the sensitivity of the assay the effort and facilities required to perform the assay.

The process of lipid peroxidation begins with the abstraction of a hydrogen atom from a polyunsaturated fatty acid (PUFA) to form a lipid radical and proceeds with the addition of oxygen to form a lipid hydroperoxide. Lipid peroxides are readily degraded to a range of scission products, some of which are sufficiently reactive to form tertiary products. I shall use this sequence of events as a framework for discussion of the available assay methods.

Free Radicals, Lipoproteins, and Membrane Lipids
Edited by A. Crastes de Paulet *et al.*
Plenum Press, New York, 1990

Disappearance of PUFAs

The measurement of PUFA utilisation as an index of lipid peroxidation has been used widely in biological systems in vitro. In peroxidizing rat liver microsomes, the disappearance of arachidonic acid and docosahexaenoic acid was first described by May and McCay some twenty years ago[1]. That these PUFAs are the most susceptible to oxidation has also been demonstrated in rat liver plasma membranes peroxidized in vitro[2]. The advantage of this is that, if other modes of PUFA disappearance can be disregarded, it is fairly definitive as an index of peroxidation and can provide useful qualitative information as to which fatty acids are being used. The disadvantages of this method are that it is cumbersome, as it involves the extraction, saponification, esterification and GC separation of the PUFAs and also it is relatively insensitive. It is inappropriate where other metabolic routes of PUFA utilisation are likely to occur and is therefore not recommended as an index of lipid peroxidation in vivo.

Oxygen utilisation

Apart from PUFAs, the other substrate of lipid peroxidation is oxygen. Clearly, the measurement of oxygen utilisation as an index of lipid peroxidation is not applicable in vivo or in cells or organelles undergoing respiration. This method can, however, be most usefully employed in measuring lipid peroxidation in non-respiring sub-cellular fractions. The first description of such an application was by Hochstein and Ernster using liver microsomes some 25 years ago[3,4]. This is a simple measurement to perform using an oxygen-electrode and has the added advantage of allowing continuous monitoring of the lipid peroxidation process. Monitoring the time-course is especially important in determining the inhibitory effects of chain-breaking antioxidants whose action in such systems is characterised by the presence of an induction period. In carefully controlled systems the length of the induction period is directly proportional to the concentration of antioxidant[5,6]. In biological systems the rate of initiation can seldom be so carefully controlled and this proportionality is less exact. Nevertheless, the oxygen electrode allows a clear picture to be obtained as to the involvement of antioxidants[7].

A considerable refinement of the basic oxygen electrode system is the 'oxystat' system of De Groot and Noll[8,9] in which the oxygen tension is maintained at a constant level by the computer-controlled injection of small volumes of oxygen saturated buffer. Lipid peroxidation in systems of various constant oxygen tensions can thus be investigated.

Lipid free radicals

Lipid peroxidation is a chain-reaction carried by free radicals including carbon-centred acyl radicals and oxygen-centred alkoxyl and peroxyl radicals. The only technique available to measure directly these radicals is electron spin resonance spectroscopy (ESR). At present, this is a technique of limited usefulness in the context of measuring lipid peroxidation. It is not a very sensitive technique, requiring radical concentrations of around $10^{-6}M$. Spin-trapping can, however, usefully increase the effective radical concentration. Identification of the products responsible for the ESR signals is very difficult and it must be remembered that the signal only represents the most stable radical and not necessarily the most 'important'. However, this technique can yield useful information as explained in a recent review by Davies[10].

Conjugated dienes

In the formation of a lipid hydroperoxide the original system of bis-allylic methylenes rearranges to a conjugated diene system (Fig 1).

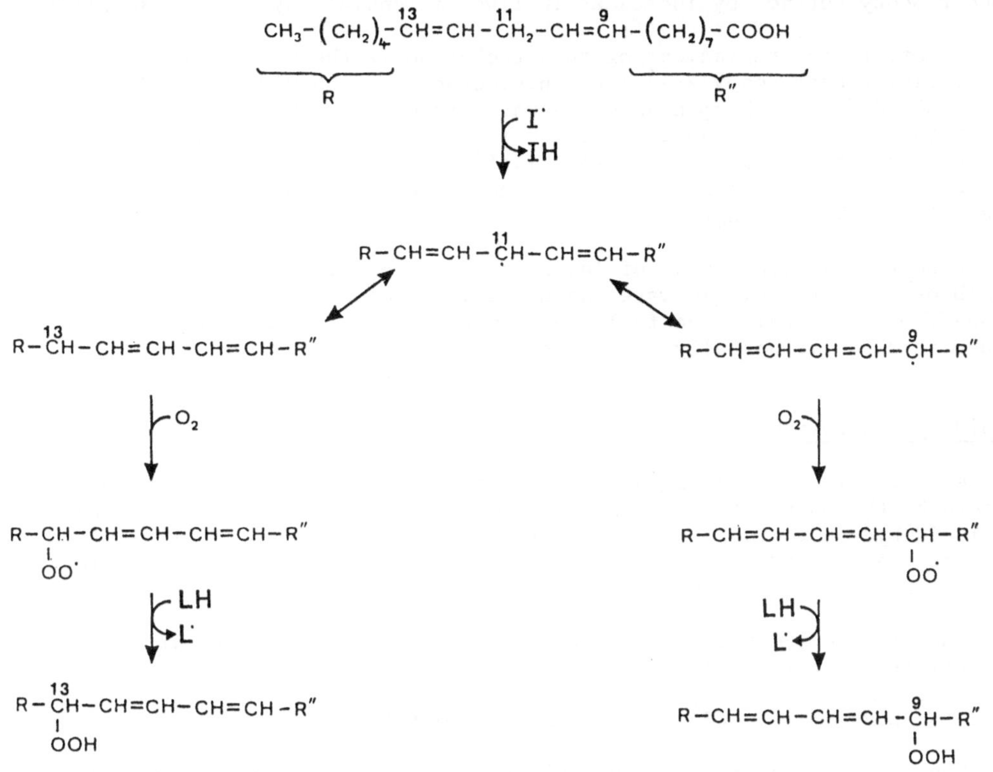

$$CH_3-(CH_2)_4-\overset{13}{C}H=CH-\overset{11}{C}H_2-CH=\overset{9}{C}H-(CH_2)_7-COOH$$

R   R''

$$\overset{I\cdot}{\underset{IH}{\downarrow}}$$

$$R-CH=CH-\overset{11}{\underset{.}{C}}H-CH=CH-R''$$

$$R-\overset{13}{\underset{.}{C}}H-CH=CH-CH=CH-R'' \qquad\qquad R-CH=CH-CH=CH-\overset{9}{\underset{.}{C}}H-R''$$

$$\overset{O_2}{\downarrow} \qquad\qquad\qquad\qquad\qquad\qquad \overset{O_2}{\downarrow}$$

$$R-\underset{OO\cdot}{CH}-CH=CH-CH=CH-R'' \qquad\qquad R-CH=CH-CH=CH-\underset{OO\cdot}{CH}-R''$$

$$\overset{LH}{\underset{L\cdot}{\downarrow}} \qquad\qquad\qquad\qquad\qquad\qquad \overset{LH}{\underset{L\cdot}{\downarrow}}$$

$$R-\overset{13}{\underset{OOH}{C}H}-CH=CH-CH=CH-R'' \qquad\qquad R-CH=CH-CH=CH-\overset{9}{\underset{OOH}{C}H}-R''$$

Fig 1. Formation of conjugated dienes

The arrangement of conjugated double bonds absorbs U.V. light strongly ($\varepsilon=27,000$ $M^{-1}cm^{-1}$) at a peak wavelength of around 233nm. This therefore affords a sensitive method of measuring lipid peroxidation. In essence, the method involves extraction of the lipids in chloroform:methanol; drying down the chloroform phase; and re-dissolution of the lipid in cyclohexane for spectroscopy. Since the diene 'peak' is often just a shoulder on the large absorption due to the lipid itself, difference spectroscopy is usually required using non-peroxidized lipid in the reference cuvette. However, when the investigator requires to compare a group of control samples with a group of test samples (e.g. livers from 4 control rats and from 4 CCl_4-dosed rats) he is faced with the difficulties of preparing an averaged difference spectrum. With the appropriate computer attachment this could be done by averaging the absolute spectra of the tests and subtracting an average spectrum of the controls to give the average difference spectrum. In the absence of computer assistance this must be done manually. The optical density of each sample is measured at each of a series of wavelengths. For each wavelength, the average optical density in the control group and in the test group is calculated and the difference in absorption at each wavelength is calculated. Thus, the average difference spectrum can be plotted. Obviously, this is a time-consuming procedure, but it is sensitive and has been used to good effect. Recknagel used this method for his seminal work on the role of lipid peroxidation in CCl_4-induced liver injury[11]. It is mainly used for in vivo measurements since less tedious assays can be used in vitro but are not often applicable in vivo. With low CCl_4 doses we found that lipid peroxidation in vivo could be detected by conjugated diene spectroscopy but not by increases in liver malondialdehyde, for example[12].

A more recent refinement of this technique is the use of second-derivative spectroscopy[13,14]. By this method, the shoulder absorption can be resolved into a sharp peak or peaks. Often two peaks are found, at 233nm and 245nm; these may represent this cis,trans- and the trans,trans-isomers of the lipid hydroperoxides, respectively[14]. Second-derivative spectroscopy is sensitive and it does not require the measure of a difference spectrum to check the peak wavelength.

Reports sometime appear of 'diene conjugation' measurements based solely on absorption at a single wavelength. This practice is not to be recommended: only recording the full spectrum will reveal if a change in absorbance at 233nm is actually due to conjugated diene formation.

Lipid hydroperoxides

As the primary products of lipid peroxidation, lipid hydroperoxides, should be the most desirable products to measure. However, the measurement of lipid hydroperoxides in biological systems has always been problematic because they are not very stable in the presence of metal ions, reducing agents and certain enzymes.

The classical method for lipid hydroperoxide determination is based on the oxidation of iodide:

$$LOOH + 2I^- + 2H^+ \longrightarrow LOH + H_2O + I_2$$

The iodine is then titrated against thiosulphate. A more recent refinement measures the tri-iodide anion directly by spectrometry:

$$I^- + I_2 \rightleftharpoons I_3^-$$

A high iodine concentration is needed to push the equilibrium to the right and the iodide reagent must be protected from direct oxidation by oxygen. The latter is achieved either by performing the reaction anaerobically throughout[15] or just for the period of the reaction with LOOH followed by addition of cadmium to complex un-reacted iodide[16]. This method is still prone to interference from oxidizing agents other than lipid peroxides and is time-consuming but has recently been automated[17]. The lower limit of detection is around 2nmols.

Various methods exist that utilise peroxidase enzymes. Probably the first example of this approach was that of Heath and Tappel[18] who linked GSH-peroxidase activity to GSSG-reductase activity and followed NADPH oxidation spectrophotometrically. This approach confers a high degree of specificity and sensitivity but the assays are time-consuming and prone to interference.

A very sensitive assay is that described by Cathcart[19] which is based on the oxidation of dichlorofluoroscin to dichlorofluoroscein:

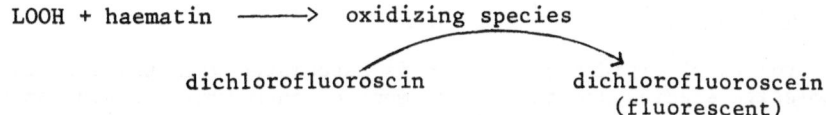

LOOH + haematin ———> oxidizing species

dichlorofluoroscin   dichlorofluoroscein
                                (fluorescent)

The method is sensitive (lower limit 25pmols) and simple to perform but not specific to lipid peroxides and is subject to interference from solvents and antioxidants.

Another simple method is that of Yagi et al.[20,21] who have described a modified version of the classical thiobarbituric acid (TBA) assay. Key elements in the method for measuring lipid peroxides in free lipids or animal tissues is the buffering of the system at around pH4.0 and a prolonged incubation time with TBA reagent (60 min). Because the tissue fraction is still present at this stage, turbidity can result and extraction into butanol is necessary. The coloured product is of the same structure as that formed by reaction of MDA with TBA: apparently MDA is formed from the peroxide breakdown during the long incubation. This method is simple and reliable but there seems potential for peroxidation during the test itself. Yagi has also described a modification of this method for use with plasma or serum[22]. In this assay interfering substances (sugars etc) are removed by precipitation of the lipids with phosphotungistic acid and discarding the supernatant. The TBA-adduct is measured fluorimetrically, which adds sensitivity. Yagi has reviewed the use of this method in human subjects with a range of disorders[23].

Probably the most sensitive assays for lipid peroxides are those based on chemiluminescence, with and without separation of the peroxides chromatographically. Such assays depend on the breakdown of the peroxides, usually by a haem compound, to oxidizing species that react with luminol to produce a chemiluminescent species.

For example, Iwaoka and Tabata[24] describe a method for measuring lipid peroxides in plasma using cytochrome c peptide (also known as microperoxidase) as the catalyst. Plasma is mixed with microperoxidase and luminol and the resultant chemiluminescence is measured in a luminometer. This system was examined and optimised by Belghmi et al.[25] who recommend that the lipid peroxides are first extracted from the plasma since other endogenous compounds interfere with the light emission. Also, catalase can be used to remove hydrogen peroxide. In their hands, the lower detection limit was found to be 1pmol and typical normal plasma values were around 0.2µM.

A development of this method and probably the current method of choice for in vivo (plasma) measurements is the assay based on separation of lipid peroxides according to their lipid class by HPLC, followed by detection using the chemiluminescence reaction described above. This method was developed in parallel in the laboratory of Ames[26] and by Miyazawa.[27] The advantage of this method is that it can yield information on the types of lipid peroxides being measured i.e. cholesterol-peroxides, phospholipid-peroxides etc. It must be said, however, that at the time of writing the method is still not completely characterised in terms of definitive peak identification. Also, investigators must prepare their own standard compounds for use as reference material as they are not commercially available. Nevertheless it is anticipated that these problems will be overcome and this method will be of great value.

Chemiluminescence

The process of lipid peroxidation is associated with the formation of photo-emissive species and the measurement of chemiluminescence can be used to monitor lipid peroxidation. This is not to be confused with the use of chemiluminescence used as a detection method for lipid peroxide measurement as described above. In this section, one refers to the endogenous chemiluminescence associated with the process, in the absence of added amplifiers. The photo-emissive species formed are probably singlet oxygen and excited carbonyl species, each formed during peroxide breakdown. Cadenas, working with Boveris and Chance[28] and with Sies[29] has been foremost in promoting this technique. Chemiluminescence is monitored using a photomultiplier tube and expressed as counts per minute. It is a very versatile technique that has been used with suspensions of cells and cell-fractions, isolated perfused organs and even in exposed organs in situ. It also has the advantages of being sensitive and of enabling continuous monitoring. Thus, as described above for the oxygen uptake technique, the measurement of chemiluminescence is well suited to investigating the effect of antioxidants on the time-course of lipid peroxidation in a given system.

Aldehydes

Lipid peroxidation in biological systems always results in the production of aldehydes, as peroxides degrade. In recent years increasing attention has been paid to the formation of aldehydes, not least because of their biological activity. Foremost in the development of aldehyde analysis has been Esterbauer.[30,31,32,33]

Aldehydes are formed during the scission of lipid hydroperoxides and in biological systems there is a great range of aldehyde products. The most common assay method for aldehydes is that dependent on the formation of dinitrophenylhydrazones by reaction with dinitrophenylhydrazine, followed by chromatographic separation[32,33]. The detection limit is about 1pmol for a single HPLC injection. Because of the large number of products definite identification is not always possible. This assay system was never intended as a routine quantitative index of lipid peroxidation; it is too laborious for that. Its value lies in the qualitative information it can yield about which aldehydes are produced in various systems under various conditions.

Hydroxynonenal is one of the most important aldehyde products, in terms of quantity and biological activity and a direct assay method, without derivatisation, has been developed[34].

In biological systems not only are free aldehydes produced but also aldehydes that remain bound to the phospholipid moiety. Elevated levels of these can be detected in the livers of bromobenzene-poisoned mice and carbon tetrachloride-treated rats.[35,36,37] Comporti's group have also described the detection of aldehydes bound to protein in sections of liver from bromobenzene-poisoned mice, using Schiff's reagent[38]. Thus, these latter methods may prove useful indices of lipid peroxidation in vivo, although they are unlikely to be of great sensitivity.

Returning to the measurement of free aldehydes, Tomita and Yoshino have developed a sensitive detection method based on the formation of fluorescent decahydroacridine derivatives that can be separated by HPLC[39]. This method is more sensitive than the dinitrophenylhydrazine method but less well characterised. Finally, Van Kuijk has developed a GC-MS method for determining hydroxyalkenal levels in animal tissues which is highly sensitive (down to 0.1pmols).

Malondialdehyde

Malondialdehyde (MDA) strictly belongs to the above section but merits special attention. The TBA test for MDA is certainly the most widely used assay for lipid peroxidation[41] and its popularity lies in its simplicity and sensitivity. It is not generally applicable to samples obtained ex vivo as MDA is rapidly metabolised, but as an in vitro test it deserves to be popular and has sometimes attracted unreasonable criticism. Certainly, numerous compounds other than MDA are positive in the TBA test but often these substances can be avoided or removed or have such a low molar extinction coefficient as to be insignificant. In several systems in vitro, it has been demonstrated that all of the TBA-reactive compound produced can be accounted for by MDA[32,42]. This may not always be the case but it can be ascertained by using the direct HPLC method for measuring MDA which is relatively simple to perform[32,42]. Again, it should be stressed that the latter method is not intended as a replacement for the TBA test but as a means to validate (or otherwise) the TBA test as a measure of MDA in a given system.

Ethane and pentane

Ethane and pentane are produced by the scission of ω-3 and ω-6 PUFA hydroperoxides, respectively. Riely first demonstrated that the measurement of these alkanes in the expired air could be used as a non-invasive index of lipid peroxidation in vivo[43]. The method is fully described for use in vivo by Lawrence and Cohen[44] and for use in vitro by Muller and Sies[45]. The essence of the technique is to perform GC separation of those gases either in headspace samples or following an adsorption/desorption process. This remains the only genuinely non-invasive method of measuring lipid peroxidation in vivo and enables the continuous or sequential monitoring of individual animals (including humans) over long time periods. The main disadvantages are the lack of certainty of the precise origin of the gases within the body and the requirement for specialised apparatus.

Fluorescent products

The carbonyl products of lipid peroxidation react with amino groups of (e.g.) proteins to yield Schiff bases with strong optical absorptions and which are also strongly fluorescent. Fluorescent products have been detected in peroxidizing sub-cellular fractions and in vivo, for example in antioxidant-deficient rodent tissues[46]. The spectral properties of these chromophores are practically identical with extracts of the "age-pigment" lipofuschin[47,48]. The measurement of liposoluble pigment extracts is usually taken as a parameter of lipid peroxidation in vivo although it is, of course, rather distanced from the initial event.

Concluding remarks

As outlined in the Introduction, the final choice of method depends on a number of factors including applicability to the system under study and whether one is interested in a particular aspect of lipid peroxidation or not.

Detailed accounts of the methods can be found in the references cited above. A two-volume review of all aspects of lipid peroxidation has recently been published[49,50] and a review of the methods used to measure MDA and other aldehydes is in press[51].

References

1.   H. E. May and P. B. McCay, Reduced triphosphopyridine nucleotide

oxidase-catalysed alterations of membrane phospholipids, J.Biol. Chem. 243:2288 (1968).

2. R. Le Page, K. H. Cheeseman, N. Osman and T. F. Slater, Lipid Peroxidation in purified plasma membrane fractions of rat liver in relation to the hepatotoxicity of carbon tetrachloride, Cell Biochem. Function 6:87 (1988).

3. P. Hochstein and L. Ernster, ADP-activated lipid peroxidation coupled to the TPNH-oxidase system of microsomes, Biochem. Biophys. Res. Commun. 12:388 (1963).

4. P. Hochstein, K. Nordenbrand and L. Ernster, Evidence for the involvement of iron in the ADP-activated peroxidation in lipids in microsomes and mitochondria, Biochem. Biophys. Res. Commun. 14:323 (1964).

5. G. W. Burton, K. H. Cheeseman, T. Doba, K. U. Ingold and T. F. Slater, Vitamin E as an antioxidant in vitro and in vivo, in "Biology of Vitamin E", Pitman Books, London (1983).

6. E. Niki, Antioxidants in relation to lipid peroxidation, Chem. Phys. Lipids 44:227 (1987).

7. K. H. Cheeseman, M. M. Collins, K. P. Proudfoot, T. F. Slater, G. W. Burton, A. C. Webb and K. U. Ingold, Studies on lipid peroxidation in normal and tumour tissue. The Novikoff rat liver tumour, Biochem. J. 235:507 (1986).

8. H. De Groot and T. Noll, The role of physiological oxygen partial pressures in lipid peroxidation. Theoretical considerations and experimental evidence, Chem. Phys. Lipids 44:209 (1987).

9. T. Noll, H. De Groot and P. Wissemann, A computer-supported oxystat system maintaining steady-state O_2 partial pressures and simultaneously monitoring O_2 uptake in biological systems. Biochem. J. 236:765 (1986).

10. M. J. Davies, Applications of electron spin resonance spectroscopy to the identification of radicals produced during lipid peroxidation, Chem. Phys. Lipids 44:149 (1988).

11. K. S. Rao and R. O Recknagel, Early onset of lipoperoxidation in rat liver after CCl_4 administration, Exp. Molec. Pathol. 9:271 (1968).

12. C. J. Reddrop, K. H. Cheeseman and T. F. Slater, Correlations between common tests for assessment of liver damage: indices of the hepatoprotective activity of promethazine in carbon tetrachloride hepatotoxicity, Cell. Biochem. Function 1:55 (1983).

13. F. P. Corongiu and A. Milia, An improved and simple method for determining diene conjugation in autoxidized polyunsaturated fatty acids, Chem. Biol. Interact. 44:289 (1983).

14. F. P. Corongiu, G. Poli, M. U. Dianzani, K. H. Cheeseman and T. F. Slater, Lipid peroxidation and molecular damage to polyunsaturated fatty acids in rat liver. Chem. Biol. Interact. 59:147 (1986).

15. M. Hicks and J. Gebicki, A spectrophotometric method for the determination of lipid hydroperoxides, Anal. Biochem, 99:249 (1979).

16. J. A. Buege and S. D. Aust, Microsomal Lipid Peroxidation, Methods Enzymol. 52:302 (1978).

17. S. M. Thomas, W. Jessup, Jan. M. Gebicki and R. T. Dean, A continuous-flow automated assay for iodometric estimation of hydroperoxides, Anal. Biochem. 176 (1989) In press.

18. R. L. Heath and A. L. Tappel, A new sensitive assay for the measurement of hydroperoxides, Anal. Biochem. 75:184 (1976)

19. R. Cathcart, E. Schwiers and B. N. Ames, Detection of picomole levels of lipid hydroperoxides using a dichlorofluoroscein fluorescent assay, Methods. Enzymol. 105:352 (1984).

20. H. Ohkawa, N. Ohishi and K. Yagi, Reaction of linoleic and hydroperoxide with thiobarbituric acid, J. Lipid. Res. 19:1053 (1978)

21. H. Ohkawa, N. Ohishi and K. Yagi, Assay for lipid peroxides in animal tissues by thiobarbituric acid reaction, Anal. Biochem. 95:351 (1978).

22. K. Yagi, Assay for blood plasma or serum, Methods Enzymol. 105:328 (1984).

23. K. Yagi, Lipid peroxides and human diseases, Chem. Phys. Lipids 45:337 (1987).

24. T Iwaoka and I Tabata, Chemiluminescent assay of lipid peroxide in plasma using cytochrome c peptide, FEBS Lett. 178:47 (1984).

25. K. Belghmi, J-C. Nicolas and A. Crastes de Paulet, Chemiluminescent assay of lipid peroxides, J. Biolum. Chemilum. 2:113 (1988).

26. Y. Yamamoto, M. H. Brodsky, J. C. Baker and B N Ames, Detection and characterisation of lipid hydroperoxides at picomole levels by high performance liquid chromatography, Anal. Biochem. 160:7 (1987).

27. T. Miyazawa, T. Fujimoto and T. Kaneda, Detection of picomole levels of lipid hydroperoxides by a chemiluminescent assay, Agric. Biol. Chem. 51:2569 (1987).

28. A. Boveris, E. Cadenas and B. Chance, Ultra-weak chemiluminescence: a sensitive assay for oxidative radical reactions, Fed. Proc. 40:195 (1981).

29. E. Cadenas, M. Ginsberg, U. Rabe and H. Sies, Estimation of α-tocopherol antioxidant activity in microsomal lipid peroxidation as detected by low-level chemiluminescence, Biochem. J. 223:755 (1984).

30. H. Esterbauer, Aldehydic products of lipid peroxidation, in "Free radicals, Lipid Peroxidation and Cancer", D. C. H. McBrien and T. F. Slater, eds., Academic Press, London (1982).

31. A. Benedetti, M. Comporti and H. Esterbauer, Identification of 4-hydroxy-nonenal as a cytotoxic product originating from the peroxidation of liver microsomal lipids, Biochim. Biophys. Acta 620:281 (1980).

32. H. Esterbauer, K. H. Cheeseman, M. U. Dianzani, G. Poli and T. F. Slater, Separation and characterisation of the aldehydic products of lipid peroxidation stimulated by ADP-Fe^{2+} in rat liver microsomes, Biochem. J. 208:129 (1982).

33. G. Poli, M. U. Dianzani, K. H. Cheeseman, T. F. Slater, J. Lang and H. Esterbauer, Separation and characterisation of the aldehydic products of lipid peroxidation stimulated by carbon tetrachloride or ADP-iron in isolated rat hepatocytes and rat liver microsomal suspensions. Biochem. J. 227:629 (1985).

34. J. Lang, C. Celotto and H. Esterbauer, Quantitative determination of the lipid peroxidation product 4-hydroxynonenal by high performance liquid chromatography, Anal. Biochem. 150:369 (1985)

35. A Pompella, E. Maellaro, A. F. Casini, M. Ferrali, L. Ciccoli and M. Comporti, Measurement of lipid peroxidation in vivo: a comparison of different procedures, Lipids 22:206 (1987).

36. A. Benedetti, R. Fulceri, M. Ferrali, L. Ciccoli, H. Esterbauer and M. Comporti, Detection of carbonyl functions in phospholipids of liver microsomes in CCl$_4$- and BrCCl$_3$-poisoned rats. Biochim. Biophys. Acta 712:628 (1982).

37. A. Benedetti, A. Pompella, R. Fulceri, A. Romani and M. Comporti, Detection of 4-hydroxyonenal and other lipid peroxidation products in the liver of bromobenzene-poisoned mice. Biochim. Biophys. Acta 876:658 (1986).

38. A. Pompella, E. Maellaro, A. F. Casini and Mario Comporti, Histo-chemical detection of lipid peroxidation in the liver bromobenzene-poisoned mice, Am. J. Pathol. 129:295 (1987).

39. K. Yoshino, M. Sano, M. Fujita and I. Tomita, Formation of aliphatic aldehydes in rat plasma and liver due to vitamin E deficiencey, Chem. Pharm. Bull. 34:5184 (1986).

40. F. J. G. M. Van Kuijk, D. W. Thomas, R. J. Stevens and E. A. Dratz, Occurrence of 4-hydroxyalkenals in rat tissues determined as penta-fluorobenzyl oxine derivatives by gas chromatography-mass spectro-scopy. Biochem. Biophys. Res. Commun. 139:144 (1986).

41. T. F. Slater and K. H. Cheeseman, Lipid peroxidation in "Prosta-glandins and Related Substances, A Practical Approach", C. Benedetto S. Nigam, R. McDonald-Gibson and T. F. Slater, IRL Press, Oxford, 243-258 (1987).

42. H. Esterbauer, J. Lang, S Zadravec and T. F. Slater, Detection of malonaldehyde by high-performance liquid chromatography, Methods Enzymol. 105:319 (1984).

43. C. A. Riely, G. Cohen and M. Lieberman, Ethane evolution: a new index of lipid peroxidation, Science 183:208 (1974).

44. G. D. Lawrence and G. Cohen, Concentrating ethane from breath to monitor lipid peroxidation in vivo, Methods Enzymol. 105:305 (1984).

45. A. Muller and H. Sies, Assay of ethane and pentane from isolated organs and cells, Methods Enzymol. 105:311 (1984).

46. C. J. Dillard and A. L. Tappel, Fluorescent damage products of lipid peroxidation, Methods Enzymol. 105:337 (1984).

47. K. Kikugawa and M. Beppu, Involvement of lipid oxidation products in the formation of fluorescent and cross-linked proteins, Chem. Phys. Lipids 44:279 (1987).

48. M. Tsuchida, T. Miura and K. Aibara, Lipofuscin and lipofuscin-like substances, Chem. Phys. Lipids 44:297 (1987).

49. H. Esterbauer and K. H. Cheeseman (Eds) Lipid Peroxidation: Bio-chemical and Biophysical Aspects. Chem. Phys. Lipids 44 (1987).

50. H. Esterbauer and K. H. Cheeseman (Eds) Lipid Peroxidation: Patho-logical Implications. Chem. Phys. Lipids 45 (1987).

51. H. Esterbauer and K. H. Cheeseman, Determination of aldehydic lipid peroxidation products with special attention to malonaldehyde and 4-hydroxynonenal, Methods in Enzymol. In press (1989).

MEASUREMENT OF SUPEROXIDE AND SUPEROXIDE DISMUTASE BY ELECTRON SPIN

RESONANCE AND CHEMILUMINESCENCE ASSAY

Toshikazu Yoshikawa, Yuji Naito, Toru Tanigawa,
Shigeru Sugino, Motoharu Kondo

First Department of Medicine
Kyoto Prefectural University of Medicine
Kamigyo-ku, Kyoto 602, Japan

INTRODUCTION

Spin trapping using dimethyl-1-pyrroline-N-oxide (DMPO) as a spin trapper is a useful method for the detection of superoxide[1, 2]. DMPO-OOH which is formed by the reaction of DMPO and superoxide gives a characteristic electron spin resonance (ESR) signal. This combined with its superoxide dismutase (SOD) inhibitability, is making it possible to give a two fold garantee of specificity to this assay. Because of the relatively short life time of DMPO-OOH, to exactly quantify superoxide is difficult. However it is still possible to semi-quantify in cases where the rate of superoxide generation is relatively constant[3]. These methods may be advantageous when the materials show a strong light absorption or turbidity.

Cypridina luciferin analogs, 2-methyl-6-phenyl-3,7-dihydroimidazo[1,2-a]pyrazin-3-one (CLA) and 2-methyl-6-(p-methoxy-phenyl)3,7-dihydroimidazo[1,2-a]pyrazin-3-one (MCLA) are chemiluminescence probes that are highly sensitive to superoxide[4, 5]. Using the maximal increase in light intensity, the amount of superoxide production can be determined.

By the inhibition of the DMPO-OOH signal or the Cypridana luciferin analog-dependent chemiluminescence, the SOD activity can also be measured. The actual procedure and some examples of its application will be described below.

MEASUREMENT OF SUPEROXIDE GENERATED FROM POLYMORPHONUCLEAR LEUKOCYTES BY ELECTRON SPIN RESONANCE

Polymorphonuclear leukocytes (PMN) were isolated from the peripheral blood of normal human volunteers by dextran sedimentation followed by Ficoll-Paque (Pharmacia AB, Uppsala) separation and the hypotonic lysis of contaminating erythrocytes. PMN were suspended in Hanks' balanced salt solution (HBSS) at pH 7.4 and stored on ice until used.

PMN were stimulated by the addition of phorbol myristate acetate (PMA, Sigma Chemical Co., St. Louis, Mo.) or opsonized zymosan (OZ). PMA was dissolved in dimethylsulfoxide (DMSO), dessicated and stored at -70°C, and diluted to 4 µg/ml in water when needed. The final

Free Radicals, Lipoproteins, and Membrane Lipids
Edited by A. Crastes de Paulet *et al.*
Plenum Press, New York, 1990

concentration in the assay was 400 ng/ml. Zymosan A (Sigma Chemical Co., St. Louis, MO) in 20 mM veronal buffer at pH7.4 was boiled for 1 h and centrifuged at 1,600 g for 10 min. After being washed twice with HBSS, the sediments were suspended in fresh human serum and incubated at 37°C for 30 min. The OZ pellet was washed twice with HBSS, resuspended in the same solution at a concentration of 30 mg/ml, and stored at -70°C. The final concentration in the assay was 3 mg/ml.

ESR spectra were recorded on a JEOL-JES-FR80 ESR spectrometer (JEOL Co., Ltd., Tokyo). The standard reaction mixture contained 2.0×10^6 /ml of PMN in HBSS, 0.1 mM diethylenetriaminepentaacetic acid (DETAPAC, Sigma Chemical Co., St. Louis, Mo), 0.1 M DMPO, and OZ or PMA. The reaction was started by the addition of stimulants, and the reaction mixture was transferred to a flat quartz ESR cell and then placed in the cavity of the ESR spectrometer. The conditions of the ESR spectrometer were as follows: magnetic field 334.8 ± 5.0 mT, microwave power 8.0 mW, modulation frequency 100 kHz, modulation amplitude 0.1 mT, sweep time 5.0 mT/min, and responce time 0.03 sec. The intensity of DMPO-OOH signal was measured as the ratio to the intensity of the Mn^{2+} signal.

DETAPAC increased the relative intensity of the DMPO-OOH signal, and the enhanced effect was maximal at 0.1-0.2 mM of DETAPAC in the PMN+PMA system[3]. The relative intensity of DMPO-OOH signal increased as the DMPO concentration increased up to 0.1 M. DMPO appeared to be toxic to PMN at higher concentrations. The signal of DMPO-OOH generated from stimulated PMN was completely inhibited by the addition of SOD. This confirms that the adduct is of superoxide origin. Catalase and NaN3 had no effect on the signal, indicating that hydrogen peroxide, singlet oxygen, or myeloperoxidase have no effect on the formation or degradation of DMPO-OOH (Fig. 1, 2).

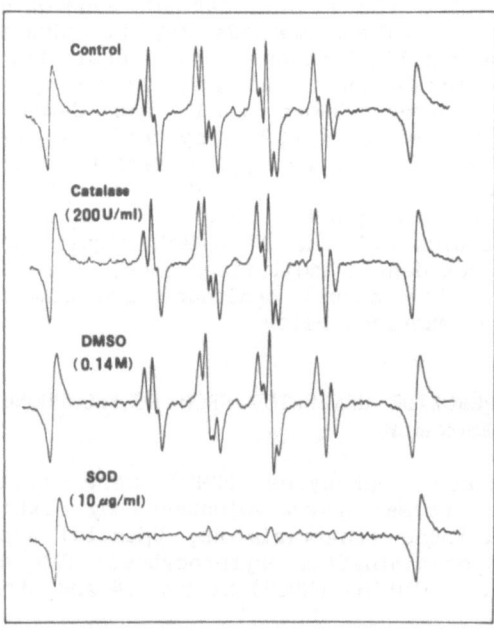

Fig. 1   Effect of scavengers on the ESR spectrum obtained by polymorphonuclear leukocytes stimulated with opsonized zymosan in the presence of DMPO.

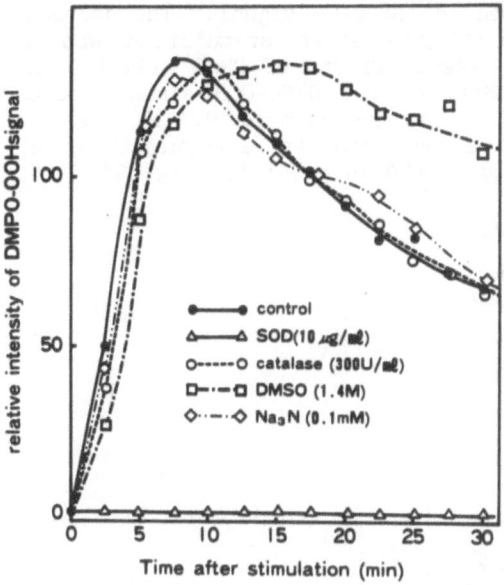

Fig. 2   Effect of scavengers on the relative intensity of DMPO-OOH
         spin adduct generated from polymorphonuclear leukocytes
         stimulated with phorbol myristate acetate.

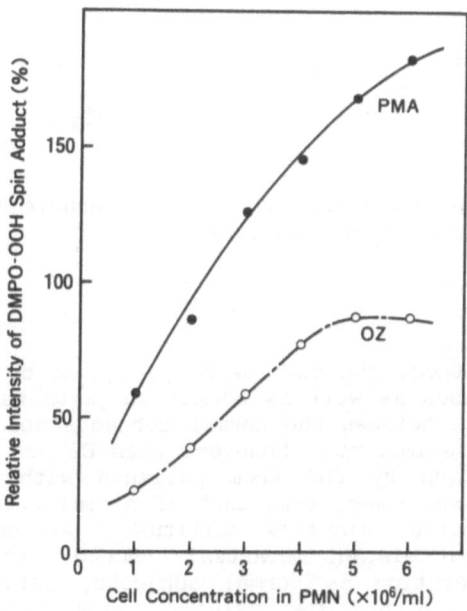

Fig. 3   Relationship between the relative intensity of DMPO-OOH spin
         adduct and the cell concentration in human polymorphonuclear
         leukocytes.

DMSO at a concentration of 0.14 M slightly suppressed the DMPO-OOH signal and increased the DMPO-OH signal. The decrease in the DMPO-OOH signal later than 10 min after stimulation was suppressed by the addition of DMSO. These effects of DMSO will be discussed elsewhere. The relative intensity of the DMPO-OOH signal increased according to the number of PMN in the presence of 0.1 M DMPO and 0.1 mM DETAPAC, indicating that the quantitative determination of the amount of superoxide generated by PMN is possible (Fig. 3).

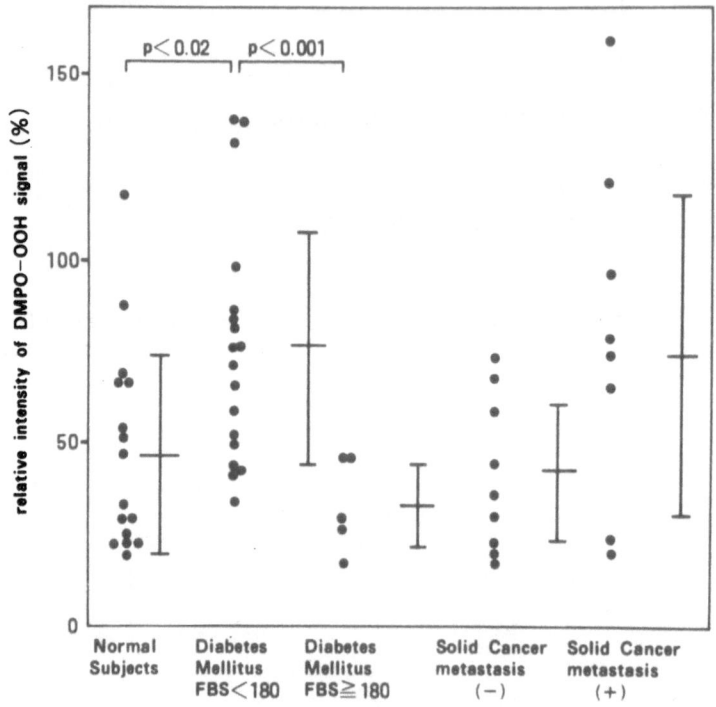

Fig. 4   Superoxide production by polymorphonuclear leukocytes from patients and normal subjects.

Using this method, the superoxide generated by PMN from patients with various diseases as well as normal subjects was measured [3]. No difference was seen between the normal subjects and the patients when PMA was used as a stimulant. However, when OZ was used, the rate of superoxide generation by PMN from patients with fairly controlled diabetes mellitus was higher than that of normal subjects and patients with poorly controlled diabetes mellitus. Although patients with solid cancer without distant metastasis showed almost the same level of superoxide generation as normal subjects, patients with distant metastasis showed a higher mean value of superoxide generation (Fig. 4).

MEASUREMENT OF SUPEROXIDE GENERATED FROM POLYMORPHONUCLEAR LEUKOCYTES
BY CHEMILUMINESCENCE

An Aloka Luminescence Reader (Aloka Co., Ltd., Tokyo) was used for the assay of CLA-dependent chemiluminescence. The standard reaction mixtures contained 2×10^5 /ml, 1 μM CLA, OZ or PMA in a total volume of 2 ml of HBSS. PMN in HBSS were preincubated at 37°C for 3 minutes and the reaction was initiated by the addition of stimulants. The increase in chemiluminescence was completely inhibited by the addition of SOD (Fig. 5), but not affected by the addition of catalase or NaN3. The maximal intensity of CLA-dependent chemiluminescence increased in proportion to the number of PMN up to 5×10^5 /ml, indicating that the quantitative determination of the rate of superoxide generation by PMN is also possible.

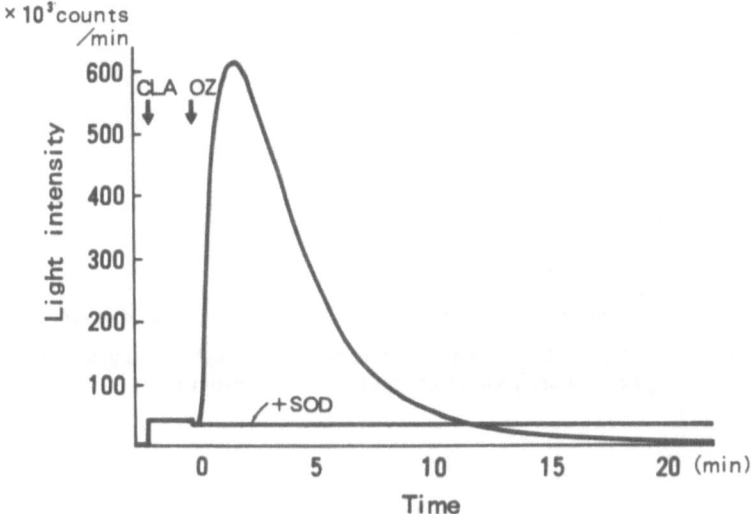

Fig. 5   CLA-dependent chemiluminescence obtained by stimulated human polymorphonuclear leukocytes.

Clinically, the superoxide production by PMN of patients with chronic renal failure treated with chronic hemo dialysis significantly increased compared with that of normal subjects (Fig. 6). The results obtained by ESR assay correlated with those obtained by chemiluminescence assay[3] (Fig. 7). We could not find any common factors in the patients whose results showed a discrepancy among the two methods.

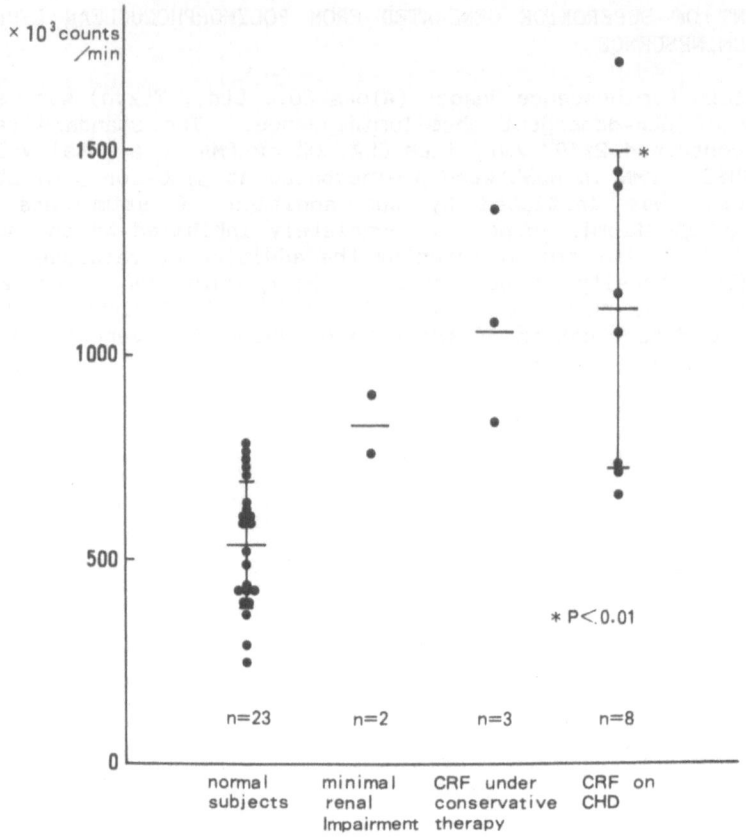

Fig. 6   CLA-dependent chemiluminescence of polymorphonuclear leukocytes from patients and normal subjects.

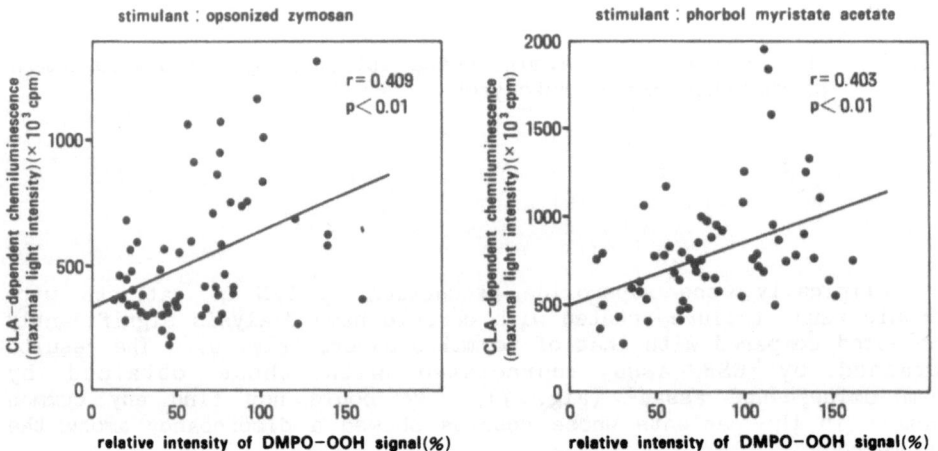

Fig. 7   Correlation between the relative intensity of DMPO-OOH spin adduct and the CLA-dependent chemiluminescence obtained by polymorphonuclear leukocytes from patients.

MEASUREMENT OF SUPEROXIDE DISMUTASE OR SUPEROXIDE DISMUTASE-LIKE ACTIVITY BY ELECTRON SPIN RESONANCE

SOD or SOD-like activity (superoxide scavenging activity) can be measured by the inhibition of the ESR signal of the DMPO-OOH spin adduct generated from the hypoxanthine-xanthine oxidase system in the presence of DMPO[6]. The reaction mixture was previously mixed with phosphate buffer solution (PBS, pH 7.4), 0.5 mM hypoxanthine, 0.1 mM DETAPAC, 0.1 M DMPO and various concentrations of standard SOD or sample in a small tube. The reaction was started by the addition of 5 mU/ml xanthine oxidase. An aliquot of reaction mixture was then transferred into a quartz cell, and the ESR spectrum was determined with a JEOL JES-FR80 spectrometer. The ESR spectra recording was started 2 min after the addition of the xanthine oxidase. The intensity of DMPO-OOH signal was measured as the ratio to the intensity of the Mn^{2+} signal.

The kinetic competition between superoxide dismutase and DMPO was used to determine the activity of superoxide dismutase[2].

$$V/v - 1 = K[SOD] \qquad\qquad (1)$$

> V=rate in the absence of SOD
> v=rate in the presence of SOD
> $K=k_e/k_a \quad k_e \gg k_a$
> k_e=rate constant for the reaction of superoxide and SOD
> k_a=rate constant for the reaction of superoxide and DMPO

Relationship between SOD activity and V/v - 1 is shown in Fig. 8.

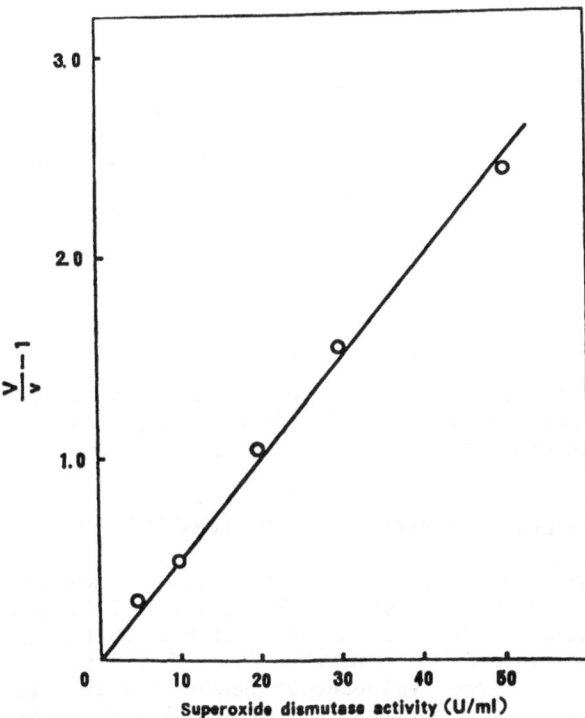

Fig. 8   Relationship between the superoxide dismutase activity and V/v - 1.

Serum superoxide dismutase activity could be measured after the injection of human SOD by three different routes (Fig. 9). After intravenous injection, the SOD activity decreased almost linearly. At 60 min after the injection, the SOD activity returned to near the preinjection level. The half life of the SOD activity was about 6 min. When subcutaneously injected, the SOD activity reached a maximum level 210 min after injection and returned to the preinjection level at 240 min. When intraperitoneally injected, the activity reached a maximum after 90 min and returned to the preinjection level at 210 min after injection.

Superoxide scavenging activity or SOD-like activity in normal human serum was found to be about 9 U/ml. Ascorbic acid and ceruloplasmin which have superoxide scavenging activity were included in the normal serum, indicating that only a little SOD appeared to be present in normal human serum.

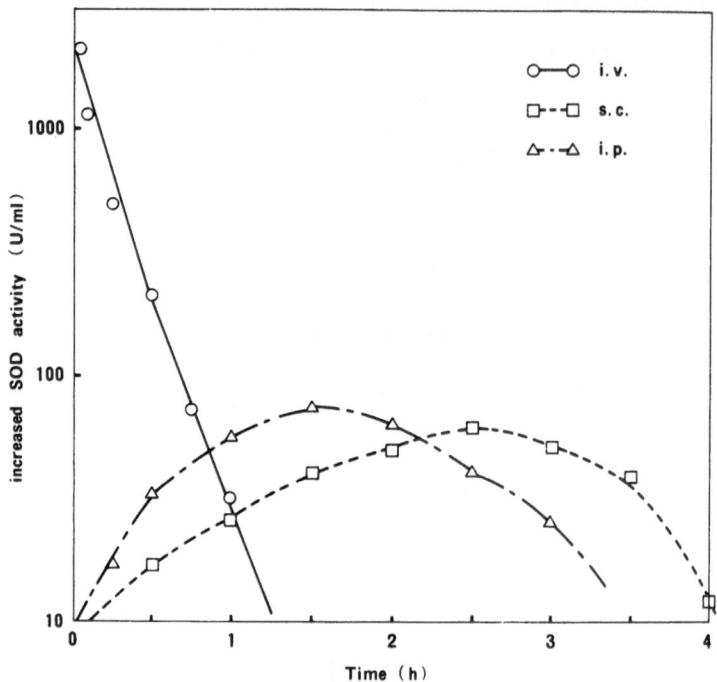

Fig. 9   Increase in the serum superoxide dismutase activity after the injection of human SOD (50,000 U/kg) by three different routes in rats.

MEASUREMENT OF SOD-LIKE ACTIVITY BY CHEMILUMINESCENCE

The standard reaction mixture contained hypoxanthine (50 μM), DETAPAC (100 μM), xanthine oxidase (Sigma grade III, 0.6 mU/ml), MCLA (1 μM) in a total volume 2 ml of Tris-HCl buffer (50 mM). The reaction mixture, except for xanthine oxidase and MCLA, was incubated for 2 min at 37°C in the Luminescence Reader followed by the addition of xanthine oxidase to generate superoxide. After another 1 min, MCLA was added to initiate the chemiluminescence reaction. The background chemiluminescence was measured in the reaction mixture from which xanthine oxidase was ommited. The reaction mixture was agitated by the rotation of the cavity (70 rpm) in the Luminescence Reader. A prompt increase in luminescence was observed after the addition of MCLA followed by a gradual decrease. The maximal intensity of chemiluminescence was corrected by the background intensity.

Sigmoid inhibition curves were obtained using SOD units in logarithmic scales. Under the presence of 2 mM KCN, the inhibition curve for Cu,Zn-SOD shifted to the right and that of Mn-SOD to the left (Fig. 10). The sigmoid curve could be linearized by taking the vertical scale as F/(1-F) and the horizontal scale as SOD units as was done in the ESR assay.

The SOD-like activity by this assay in normal human serum was 6.7±1.1 U/ml (n=10), however, albumin and ascorbic acid also have inhibitory effects on MCLA-dependent chemiluminescence. These SOD-like activities were 3.7 U/ml and 2.7 U/ml at a normal serum concentration, respectively. And they appear to be most of the SOD-like activity in human serum.

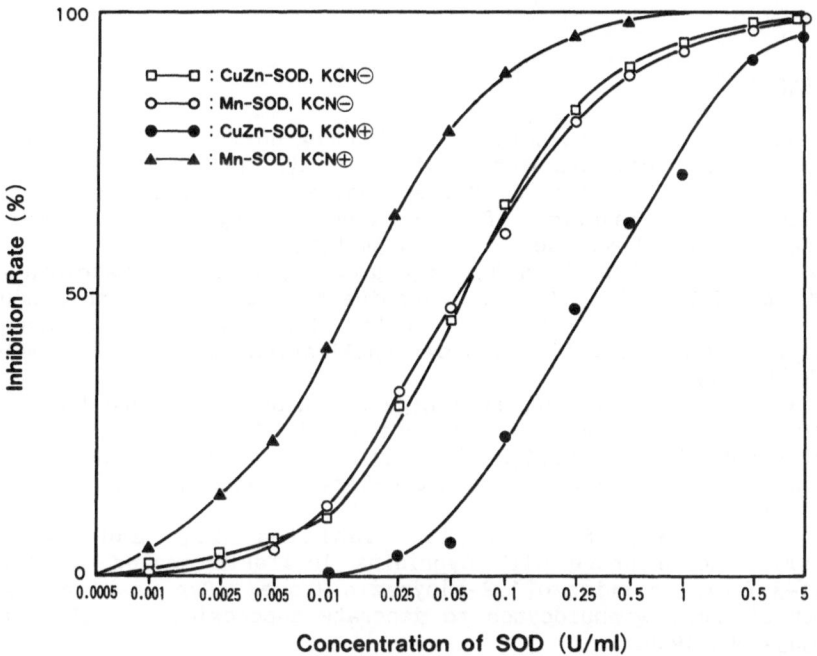

Fig. 10   Inhibition of MCLA-dependent chemiluminescence by SOD in the presence or absence of KCN.

The SOD-like activity in rat organ homogenate was also measured by this assay system and the results are shown in the table 1. They coincided with the reported values. Protein and ascorbic acid concentrations in the homogenate were low enough to neglect their SOD-like activities.

Table 1 SOD or SOD-like activity in rat tissues

Organ (N)	SOD or SOD-like activity	
	U/g wet weight	U/mg protein
Kidney (3)	2669±395	39.4±5.4
Liver (3)	2680±250	35.2±3.6
Adrenal gland	854	28.4
Brain (3)	500±99	17.0±3.9
Lung (3)	386±35	8.9±0.4
Heart (3)	383±21	10.0±0.7
Gastric mucosa (3)	248±53	6.4±0.4

REFERENCES

1) Finkelstein E., Rosen G. M., and Rauckman E. J. : Spin trapping of superoxide. Mol. Pharmacol., 16: 676-685, 1979.
2) Finkelstein E., Rosen G. M., and Rauckman E. J. : Spin trapping; kinetics of the reaction of superoxide and hydroxyl radicals with nitrones. J. Am. Chem. Soc. 102 : 4994-4999, 1980.
3) Tanigawa T., Yoshikawa T., Miyagawa H., Ueda S., Takemura T., Tainaka K., Morita Y., Itani K., Yoshida N., Sugino S., and Kondo M. : Determination of superoxide generated by human polymorphonuclear leukocytes by ESR and its clinical application. Jap. J. Inflammation. 8: 443-447, 1988.
4) Sugioka K., Nakano M., Kurashige S., Akuzawa Y., and Goto T.: A chemiluminescence probe with a Cypridana liciferin analog, 2-methyl-6-phenyl-3,7-dihydroimidazo[1,2-a] pyrazin-3-one, specific and sensitive for superoxide production in phagocytizing manrophages. FEBS Lett. 197: 27-30, 1986.
5) Nakano M., Sugioka K., Ushijima Y., and Goto T.: Chemiluminescence probe with Cypridina luciferin analog, 2-methyl-6-phenyl-3,7-dihydroimidazo[1,2-a]pyrazin-3-one, for estimating the ability of human granulocytes to generate superoxide. Anal. Biochem. 159: 363-369, 1986.
6) Miyagawa H., Yoshikawa T., Tanigawa T., Yoshida N., Sugino S., Kondo M., Nishikawa H., and Kohno M.: Measurement of serum superoxide dismutase activity by electron spin resonance. J. Clin. Biochem. Nutr., 5: 1-7, 1988.

PENTANE MEASUREMENT, AN INDEX OF IN VIVO LIPOPEROXIDATION:

APPLICATIONS AND LIMITS OF THE METHOD

J Pincemail[1], C. Deby[1], M. Lismonde[2], Y. Bertrand[3], G. Camus[4] and M. Lamy[2]

[1]Laboratoire de Biochimie et Radiobiologie, Université de Liège, Institut Chimie, B6, Sart Tilman, 4000 Liège I, Belgium. [2]Service d'Anesthésiologie, CHU, B33, Sart Tilman, 4000 Liège I, Belgium. [3]Clinique Saint-Jean, rue des Marais 104, 1000 Bruxelles, Belgium. [4]Institut Supérieur d'Education physique, B21, Sart Tilman, 4000 Liège I, Belgium

INTRODUCTION

Lipid peroxidation induced by free radicals generation has been implicated in several pathological conditions including adult respiratory distress syndrome[1] and cancer[2], in cardiac surgery such as cardiopulmonary bypass[3], in the ageing process[4], in the ischemic-reperfusion phenomenon[5], etc...

However, the evidence that lipid peroxidation occurs in vivo has not been unequivocally demonstrated. The determination of malonaldehyde (MDA), a decomposition product of lipoperoxides, in blood or plasma by the thiobarbituric acid (TBA) assay or by fluorescence spectroscopy, is the most popular procedure to assess the involvement of in vivo lipid peroxidation processes[6]. Nevertheless, this method is not specific since; 1) other products than MDA give positive reactions with TBA test; 2) MDA may arise from other pathways than lipid peroxidation[7]; 3) the metabolism of MDA is fast. For all these reasons, MDA may not be considered as a marker of in vivo lipid peroxidation.

Other methods such as conjugated dienes[8], 4-hydroxynonenal[9] determination have also been described to quantify in vitro lipid peroxidation, but none of these techniques, like MDA, can be carried out in vivo.

Since a few years, the measurement of hydrocarbon gases such as ethane or pentane (respectively products of peroxidation of n-3 and n-6 fatty acids) has been proposed as a specific and sensitive index of lipid peroxidation in vitro[10]. Then, the determination of hydrocarbon gases in the breathed air has been used to provide evidence for lipid peroxidation in animals and in man. Despite hepatic metabolism[11] and a high solubility in lipid tissues[12], pentane provides a more reliable measure of in vivo lipoperoxidation than ethane (poorly soluble in tissues) since n-6 fatty acids are the predominant polyunsaturated fatty acids in membranes.

Free Radicals, Lipoproteins, and Membrane Lipids
Edited by A. Crastes de Paulet *et al.*
Plenum Press, New York, 1990

By hydrocarbon gases method, evidence for lipoperoxidation was found in animal models : in aged rats[13], during tocopherol or selenenium deficiences[14,15,16], after CCl$_4$[17,18,19], iron[19], iron-dextran[20], cumene hydroperoxide[21] or alchohol administration[17,22], after ozone inhalation[23], after oxygen exposure[24] and in halothane hepatotoxicity[25].

In contrast, only few groups have investigated lipoperoxidation processes in man by this way although the collection of hydrocarbon gases in the breath has the advantage to be a non invasive technique. The most significant studies are related to volunteers treated or non treated with vitamin E and submitted to moderate exercise[26], to volunteers exposed to oxygen[27], to parenteral lipid infusion in newborn infants[28], to home parenteral nutrition patients[29] and to renal transplantation but in a non convincing way[30].

EXPERIMENTAL CONDITIONS

Several systems have been proposed for detecting and quantifying hydrocarbon gases in vivo[26,29,31] and recently, we described in detail the methodology used in our laboratory[32].

In order to improve precisions in the detection of hydrocarbons, some precautions are required. The inspiratory gases (He-O$_2$ mixture) must be hydrocarbon gases free; the ambient air contamination should be detected by monitoring the breathed gases with a nitrogen emission meter; all the internal surfaces of the circuit must be covered with teflon or be in metal instead of plastics which are hydrocarbon gases leaching material; all the expiratory circuit must be maintained at the ebullition point of pentane (37°c) to avoid absorption; two successive washouts of the complete system are necessary before each experiment to eliminate endogenous pentane. At least, a washout of pentane should be made in order to eliminate nonmetabolic and environmental pentane from the lung (fast phase) and then from the viscera, muscle and fats of the subject (slow phase, pentane having a high solubility coefficient in these tissues[12]). A breathing of hydrocarbon free gases for 60-120 minutes has been suggested to reach a point where pentane excretion represents basal metabolic production from lipid peroxidation in a normal volunteer[12,27]. However, a washout of this duration is not particularly suitable to clinical studies. In our experimental conditions, the washout was reduced to 10-15 minutes.

The relative complexity of the equipment and all the care needed for a good sampling collection and for the detection of hydrocarbon gases can partially explain the lack of results in man.

NORMAL VALUES OF EXHALED PENTANE

As indicated in table 1, there is a great variability of results regarding to the establishment of a normal value of exhaled pentane. This discrepancy may arise from differences in the system of gases collection, in the units system (pmol/kg/min, pmol/l, pmol/ml), in the washout time, in the sample treatment and in the detection system.

TABLE 1   Normal values of exhaled pentane

	washout	normal values	n
Dillard et al[26]	0 min	0.71 - 9.28 pmol/kg/min[a]	5
Wade and Van Rij[12]	120 min	120 ± 50 pmol/l	6
Snider et al[31]	60 min	2.58 ± 1.26 pmol/kg/min[b]	11
	90 min	1.99 ± 0.90 pmol/kg/min[b]	11
Morita et al[27]	0-90min	0.25 - 2.25 pmol/kg/min[a]	25
Lemoyne et al[29]	4 min	6.34 ± 0.96 pmol/kg/min[c]	10
Heim et al[30]	non precised	3.06 - 21.50 pmol/ml[a]	26
Pincemail et al[33]	10 min	5.00 - 140 pmol/l[a,d] or 0.40 - 11.25 pmol/kg/min[a,d]	22
Pincemail et al.	10 min	4.41 ± 1.80 pmol/l[b,e] or 0.70 ± 0.28 pmol/kg/min[b,e]	6

a : range values; b : values are mean ± SD; c : values are mean ± SEM;
d : mean weight of the subjects 70 ± 1O kg, time of gases collection : 8 minutes;
e : mean weight of the subjects 70 ± 5 kg, time of gases collection : 8 minutes.

In addition, a wide distribution of pentane values exists in the same group of experiments[26,27,30,32] (table 1). Variations observed in pentane values among individuals may be due to factors such as feeding habits with variation in the intake of n-3 and n-6 fatty acids, smoking, hepatic metabolism[11], muscular activity and age. In 1987, we have found pentane values ranging from 5 to 140 pmol/l in 22 male volunteers (age range 21-57 years) who were smokers and non smokers[33]. Recently, we investigated 6 non smokers and sportsmen volunteers (mean age 21 ± 3 years, mean weight 70 ± 5 kg). In addition, they have all been fasting for 12 hours before the experiment. In these conditions, we found pentane values ranging between only 1.55 to 7.40 pmol/l with a mean value of 4.41 pmol/l (table 1).

APPLICATIONS OF THE METHOD

Despite techinal difficulties and inaccuracy in the concentration of exhaled pentane, significant results can be otained in experiments where a subjet is his own control and when pentane variations are sufficiently marked.

a) Intensive exercise

Exhaustive or moderate excercise is associated with free radicals generation[34] and lipid peroxidation as demonstrated in rats and in man[26] by the measurement of pentane. In order to confirm these results with our system, 6 young sportsmen fasting for 12 hours completed a submaximal cycle ergometer test of graded intensity and of long duration (20 minutes). At rest, the mean value of pentane concentration was of 4.41 ± 1.8 pmol/l. As expected, exercise induced in these volunteers a significant increase of exhaled pentane concentration (15.8 ± 6.9 pmol/l; p <0.01) which was measured 20-25 minutes after the end of exercise (table 2).

TABLE 2   Effect of heavy exercise on pentane concentration (pmol/l) in the exhaled breath of 6 sportsmen volunteers (mean age 21 ± 3 years, mean weight 70 ± 5 kg).

Subject	Rest	20-25 min post exercise
1	5.80	9.15
2	1.55	6.00
3	4.10	13.00
4	4.40	20.00
5	7.40	26.00
6	3.20	20.50
mean ± SD	4.41 ± 1.80	15.80 ± 6.90*

The subjects have all been fasting for 12 hours before the experiment. They went through a graded level of exercise that was 40, 60 and 75% Vo_2 max at 0-10, 10-15, 15-20 minutes respectively. Breath samples (90 liters) were collected for 8 minutes during rest prior exercise and at 20-25 minutes post exercise. * $p < 0.01$ versus rest.

b) Adult Respiratory Distress Syndrome

Lung injury leading to adult respiratory distress syndrome (ARDS) is thought to be associated with toxic species released by activated neutrophils accumulated in the pulmonary capillary bed[1]. Upon stimulation, these cells generate a variety of potentially damaging substances, including oxygen free radicals which can induce lipoperoxidation processes.

In one patient admitted to the Intensive Care Unit of Liege Hospital for an abdominal sepsis and at risk of developing ARDS, we have investigated the evolution of pentane (figure 1). During the stay in ICU, his septic state became more severe as indicated by the increase of white blood cells count (day 3). Diagnosis of ARDS was made at day 4 in the morning on four criteria : a severe hypoxemia with a PaO_2 less than 75 mm Hg or a ratio of arterial to alveolar oxygen tension less than 0.3; a diffusion infiltration of both lungs on chest radiography; a decrease of pulmonary compliance pulmonary wedge < 16 mm Hg.

Evidence that lipoperoxidation phenomena have occured in this patient is given by the spectacular increase of pentane concentration observed at day 4 in the evening. A few hours after the onset of ARDS criteria, the measurement of pentane concentration in the breath gave a value of 171 pmol/l, either a significant 30-fold increase of the value observed at days 1 and 2.

ASSOCIATION PENTANE-VITAMIN E AS INDEX OF IN VIVO LIPOPEROXIDATION

Although hydrocarbon determination as an index of in vivo lipoperoxidation appears very interesting in clinical research because of its specific and non invasive character, this method has also several drawbacks due to variations in metabolic rates of pentane and to its solubility in tissues. Therefore, a reliable evaluation of in vivo lipoperoxidation by hydrocarbon determination in the breath should rely on the simultaneous measurement of vitamin E. Indeed, this vitamin is probably the major natural liposoluble antioxidant which can react with lipidic free radicals and prevent auto-oxidation of polyunsaturated fatty acids[35]. It is therefore expected that plasma vitamin E status can be modified during lipid peroxidation processes. Recently, Lemoyne et al[29] have found a significant negative correlation between breath pentane excretion and plasma vitamin E (r=-0.66) in normal volunteers and in home parenteral nutrition patients. Similarly, we have observed in the patient developing an ARDS (figure 1) that the increase of pentane in the breath was associated with a decrease of plasma vitamin E status (day 4). These two complementary observations give therefore strong evidence that lipoperoxidation processes have effectively occured in this patient.

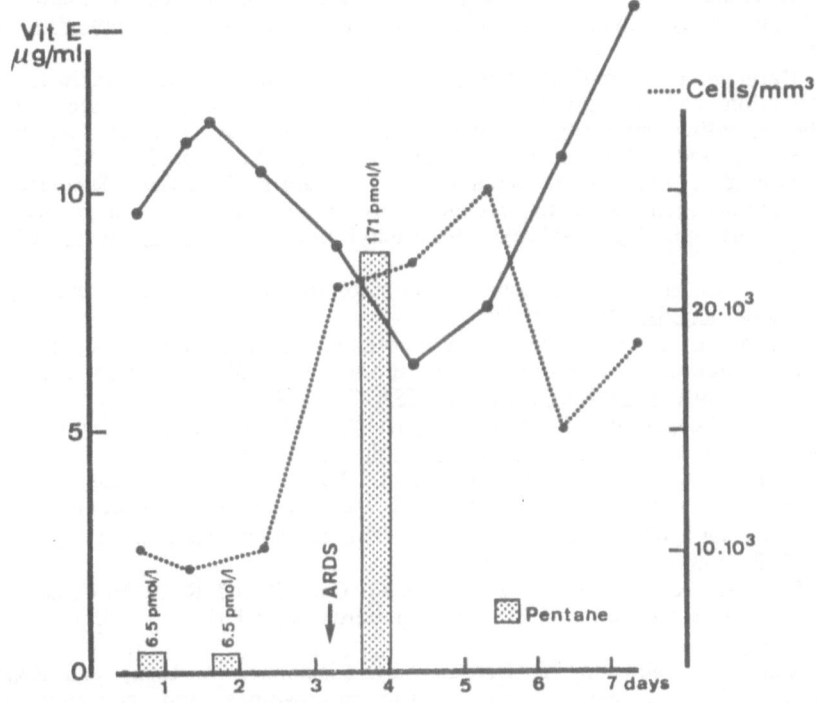

Figure 1   Evolution of white blood cells count, of plasma vitamin E concentration and of exhaled pentane in a patient developing adult respiratory distress syndrome (ARDS)

CONCLUSIONS

Pentane measurement in the breathed air is proposed as a specific index of in vivo lipoperoxidation and can be determined in healthly volunteers as well in hospital patients. Its non invasive character is particularly interesting in clinical research. However, this method requires important safety measures in order to avoid errors during the collection and the detection of pentane and is therefore time-consuming. In addition, in vivo hydrocarbon gases determination has several drawbacks (like the other methods described for in vivo evidence of lipoperoxidation) due to variations in metabolic rate of pentane and to its solubility in tissues, more particularly in fat. For this reason, pentane determination must be considered as a semi-quantitative method for the in vivo evidence of lipoperoxidation phenomena.. Therefore it is strongly suggested that pentane determination as an index of lipoperoxidation must be associated with the simultaneous determination of plasma vitamin E, another biological parameter modified by lipoperoxidation processes.

REFERENCES

1. M. Lamy, G. Deby-Dupont, J. Pincemail, M. Braun, J. Duchateau, C. Deby, J. Van Erck, L. Bodson, P. Damas and P. Franchimont. Biochemical pathways of acute lung injury. Bull. Eur. Physiopathol. Respir. 21:221 (1985)

2. B. N. Ames, Dietary carcinogens and anticarcinogens. Oxygen radicals and degenerative disease, Science 221:1256 (1983)

3. N. C. Cavarocchi, M. D. England, J. F. O'Brien, E. Solis, P. Russo, H. V. Schaff, T. A. Orszulak, J R. Pluth and M. P. Kaye, Superoxide formation during cardiopulmonary bypass. Is there a role for vitamin E, J. Surg Res 40:519 (1986).

4. D. Harman, Free radical theory of aging : role of free radicals in the origination and evolution of life, aging, and disease processes in "Free radicals, aging and degenerative disease". eds, J.E. Johnson, R. Walford, D. Harman and J. Miquel; Alan R. Liss, Inc, New York, pp 3 (1986) .

5. R. Bolli, B. S. Patel, M. O. Jeroudi, E. K. Lai and P. B McCay, Demonstration of free radical generation in "stunned" myocardium of intact dogs with the use of the spin trap -phenyl N-tert-butyl nitrone, J. Clin. Invest. 82:476 (1988)

6. T. Asakawa and S. Matsushita, Thiobarbituric acid test for detecting lipid peroxides, Lipids 14:401 (1979)

7. M. Hamberg, J. Svensson, B. Samuelsson, Prostaglandin endoperoxides. A new concept concerning the mode of action and release of prostaglandins, Proc Natl Acad Sci USA 71:3284 (1974)

8. R. O. Recknagel and E. A Glende Jr, Spectrophotometric detection of lipid conjugated dienes, Methods in Enzymology 105:331 (1984)

9. H. Esterbauer, K. H. Cheeseman, M. U. Dianzani, G. Poli and T. F Slater, Separation and characterization of the aldehydic products of lipid peroxidation stimulated by ADF-Fe^{2+} in rat liver microsomes, Biochem J 208:129 (1982)

10. C. D. Evans, G. R. List, A. Dolev, D. G. Connell and R. L. Hoffman, Pentane from thermal decomposition of lipoperoxidase-derived products, Lipids 2:432 (1967)

11. C. Deby, J. Pincemail, Y. Bertrand, M. Lismonde, M. Lamy and R.Goutier, Consumption of pentane by hepatic microsomes and consequences on pentane measurement in exhaled gases, Arch Int Physio .Biochim. 94:S19 (1986)

12. C. R. Wade and A. M. van Rij, In vivo lipid peroxidation in man. measured by the respiratory excretion of ethane, pentane and other low-molecular-weight-hydrocarbons, Anal. Biochem. 150:1 (1985)

13. M. Sagai and T. Ichinose, Age-related changes in lipid peroxidation as measured by ethane, ethylene, butane and pentane in respired gases of rats, Life Sciences 27:731 (1980)

14. C. J. Dillard, E. E . Dumelin and A. L. Tappel, Effect of dietary vitamin E on expiration of pentane and ethane by the rat, Lipids 12:109 (1977).

15. C. J. Dillard, R. E. Litov and A. L. Tappel, Effects of dietary vitamin E, selenium, and polyunsaturated fats on in vivo lipid peroxidation in the rat as measured by pentane production, Lipids 13:396 (1978)

16. D. G. Hafeman and W. G. Hoekstra, Lipid peroxidation in vivo during vitamin E and selenium deficiency in the rat as monitored by ethane evolution, J. Nutr. 107:666 (1977)

17. U. Köster, D. Albrecht and H. Kappus, Evidence for carbon tetrachloride- and ethanol-induced lipid peroxidation in vivo demonstrated by ethane production in mice and rats, Toxicol. Appl. Pharmacol. 41:639 (1977)

18. M. Sagai and A. L. Tappel, Lipid peroxidation induced by some halomethanes as measured by in vivo pentane production in the rat, Toxicol. Appl. Pharmacol. 49:283 (1979)

19. J. G. Filser, H. M. Bolt, H. Muliawan and H. Kappus, Quantitative evaluation of ethane and n-pentane as indicators of lipid peroxidation in vivo, Arch. Toxicol. 52:135 (1983)

20. C. J. Dillard, J. E. Downey and A.L. Tappel, Effects of antioxidant on lipid peroxidation in iron-loaded rats, Lipids 19:127 (1984)

21. G. Cohen, Lipid peroxidation : detection in vivo and in vitro through the formation of saturated hydrocarbon gases, in "Oxygen free radicals and tissue damage" Ciba Found Symp 65 Excerta Medica Amsterdam, Oxford, New York, pp 177 (1979)

22. R. E. Litov, D. L. Gee, J. E. Downey and A. L. Tappel, The role of lipid peroxidation during chronic and acute exposure to ethanol as determined by pentane expiration in the rat, Lipids 16:52 (1981)

23. E. E. Dumelin, C. J. Dillard and A. L. Tappel, Effects of vitamin E and ozone on pentane and ethane expired by rats, Arch. Environ. Health 33:129 (1978)

24. M. P. Habib, C. Eskelson and M. A. Katz, Ethane production rate in rats exposed to high oxygen concentration, Am. Rev. Respir. Dis. 137:341 (1988)

25. F. O. O'Neal, D. B. Menzel and M. D. Karis, Pentane expiration : a measure of halothane-induced peroxidation, Anesthesiology 51:S255 (1979)

26. C. J. Dillard, R. E. Litov, W. M. Savin, E. E. Dumelin and A. L. Tappel, Effects of exercise, vitamin E, and ozone on pulmonary function and lipid peroxidation. J. Appl. Physiol. : Respirat, Environ. Exercise Physiol. 45(6):927 (1978)

27. S. Morita, M. T. Snider, Y. Inada, Increased N-pentane excretion in humans : a consequence of pulmonary oxygen exposure, Anesthesiology 64:730 (1986).

28. J. R. Wispe, E. F. Bell and R. J. Roberts, Assessment of lipid peroxidation in newborn infants and rabbits by measurements of expired ethane and pentane : influence of parenteral lipid infusion, Pediat. Res. 19:374 (1985)

29. M. Lemoyne, A. Van Gossum, R. Kurian, M. Ostro, J. Axler, K. N. Jeejeebhoy, Breath pentane analysis as an index of lipid peroxidation : a functional test of vitamin E status, Am. J. Clin. Nutr. 46:267 (1987)

30. K.F. Heim, U. M. Makila, R. Leveson, G. S. Ledley, G. Thomas, C. Rackley and P. W. Ramwell, Detection of pentane as a measurement of lipid peroxidation in humans using gas chromatography with a photoionization detector, in: "Lipid Mediators in the Immunology of Shock", M. Paubert-Braquet, ed., NATO ASI series, Plenum Press, New York and London, pp 103 (1987)

31. M. T. Snider, P. O. Balke, K. E. Oerter, N. A. Francalancia, K. A . Pasko, M. E. Robbins, G. S. Gerhard and R. P. Richard, Methods for measuring lipid peroxidation products in the breath of man, Life Chem. Rep. 3:168 (1985)

32. J. Pincemail, C. Deby, A. Dethier, Y. Bertrand, M. Lismonde, and M. Lamy, Pentane measurement in man as an index of lipoperoxidation, Bioelec. and Bioenerg. 18:117 (1987)

33. C. Deby, J. Pincemail, Y. Bertrand and M. Lismonde, The significance of pentane measurements in man, in: "Lipid Mediators in the Immunologof Shock", M. Paubert-Braquet, ed., NATO ASI series, Plenum Press, New York and London, pp 97 (1987)

34. K. J. A. Davies, A. T. Quintanilha, G. A Brooks and L. Packer, Free radicals and tissue damage produced by excercise, Biochem. Biophys. Res. Comm. 107(4):1198 (1982)

35. P. B. McCay and M. M. King, Vitamin E : its role as biological free radical scavenger and its relationship to the mixed-function oxidase system in: " Vitamin E : a comprehensive treatise", eds, L.J. Machlin, New York, Marcel Dekker Inc, pp 289 (1980)

SECTION IV

ACTION ON MODEL SYSTEMS AND CELLULAR STRUCTURES

ACTION OF THE ALDEHYDES DERIVED FROM LIPID PEROXIDATION ON ISOLATED LIVER
PLASMAMEMBRANES

Mario Umberto Dianzani, Luciana Paradisi, Maurizio Parola,
Giuseppina Barrera and Maria Armida Rossi
Department of Experimental Medicine and Oncology, University
of Turin, Corso Raffaello 30, 10125 Turin, Italy

INTRODUCTION

The discovery that lipid peroxidation increases in several pathological
conditions has put forward the problem of its relevance in producing cell
damage, as well as of its eventual mechanisms.

Whereas most authors agree at present on the first point and consider
therefore lipid peroxidation as one of the major changes leading to cell
death (see 1 of references), the discussion is still open about the second
point. In fact lipid peroxidation not only produces damage to the membrane,
but also gives origin to a lot of toxic substances being able to diffuse far
from the production sites and to export the damage. Among the most studied
toxic substances, a special consideration has been devoted to the aldehydes
that are produced in big amounts from peroxidizing fatty acids. Among the
aldehydes, it has been observed that the most toxic are those belonging to
the 4-hydroxy-2,3,-trans-unsaturated series. This is represented by 4-hydro-
xy-nonenal, 4-hydroxy-hexenal, 4-hydroxy-octenal, 4-hydroxy-undecenal,
4,5-dihydroxy-decenal and possibly by other minor products, the first substance in this series being the most represented under a quantitative point
of view[2,3].

The activity of 4-hydroxynonenal (4-HNE) on a big number of cell functions and enzymes has been reported[4]. Most described effects were inhibitory
and took place at relatively high concentrations, i.e. in the range of
0.1-10 mM. The main reason for this need of relatively high concentrations
to get an inhibition of the function "in vitro" is probably the fact that
biological fluids contain a big amounts of masking substances preventing the
contact of the aldehyde with the enzymatic targets. In fact aldehydes react
both with amino- and with sulphydryl groups that are abundantly represented

Free Radicals, Lipoproteins, and Membrane Lipids
Edited by A. Crastes de Paulet *et al.*
Plenum Press, New York, 1990

in biological fluids. That this is an important point, it is shown by the fact that when the aldehydes are assayed on purified enzymes, their active concentration decreases consistently[4,5].

Another reason for the need of high concentrations of the aldehyde may be even the fact that aldehydes are quickly catabolized by cells and cell particles. About 80-90 % of added 4-HNE is removed within 10-15 min from the incubation medium in the case of rat liver isolated hepatocytes[6].

In a few cases, however, the activity of 4-HNE has been observed to occur at concentrations in the order of magnitude of micromolar, that can be found inside the cells as a consequence of local lipid peroxidation[2,3]. One of this function is the stimulation of chemotaxis and chemokinesis displayed by 4-HNE and 4-HOE (4-hydroxyoctenal) at concentrations of 10^{-6}-10^{-7} M (the effect is still demonstrable at 10^{-13} in the case of 4-HOE)[7-9].

It is noteworthy that nonenal, the not-hydroxylated form of HNE, maintains at lower level some stimulatory activity, that is, however, completely missing in the case of several other studied aldehydes, both saturated and unsaturated. 4-HNE seems to act on a specific receptor on polymorphonuclear leucocytes. This receptor shares several characteristics with that of leukotriene B_4[10].

Another function displayed at very low concentration is the specific block of c-myc oncogene expression by a human erythroleukemic cell strain (K 562), cultivated "in vitro"[11,12].

This effect occurs at concentrations of 0.1-1 µM and is accompanied by the increase of the expression of the gamma globin gene. So, 4-HNE displays on these cells not only a block of the oncogene expression, but even a differentiating effect. This result appears very interesting especially if one considers that lipid peroxidation is usually very low in tumours and that in hepatomas it inversely follows the dedifferentiation degree of the cells[13].

A third effect occurring at very low concentrations is finally the stimulation of adenylate cyclase of rat liver purified plasmamembranes. This was seen at concentrations 0.1-1 µM and occurs in very short time after aldehyde addition[14].

During the last few years, the study of the effects of HNE and related aldehydes on isolated plasmamembranes has been intensified. The obtained results are described in this paper.

MATERIAL AND METHODS

Fisher rats weighing 200-240 g were used in this work. They were obtained from Charles River, Italy, and were fed on a chow diet ad libitum. Plasmamembranes were isolated and purified as described elsewhere[14]. The contamination of the samples was studied by electron microscopy, as well as by measurements of the most important markers for mitochondria (cytochrome oxidase), microsomes (glucose-6-phosphatase and cytochrome P_{450}), lysosomes (beta-glucuronidase and acid phosphatase) and cytosol (lactate dehydrogenase). Contamination did not exceed 1%.

When phospholipase C had to be measured, a rough membrane preparation obtained by centrifuging the homogenate deprived of nuclei and cell debris at 25,000 g for 15 min was used[15].

Adenylate cyclase activity was measured as described elsewhere[14].

Guanylate cyclase activity was measured according to Kimura and Murad[16].

Phosphatidylinositol-4,5-diphosphate phospholipase C (PIP_2-phospholipase C) was measured according to Magnaldo et al.[17].

Mg^{2+},Na^+,K^+-ATPase, Na^+,K^+-ATPase and Ca^{2+},Mg^{2+},Na^+,K^+-ATPase were measured as described elsewhere[18,19].

5'-nucleotidase was measured as described by Paradisi et al.[14].

Membrane fluidity of the plasmamembrane preparations was measured by the electron spin resonance method described by Hyde et al.[20].

RESULTS

It is shown in Fig.1 that basal adenylate cyclase activity is strongly stimulated by micromolar 4-HNE, the activity observed in the presence of the aldehyde being about 5 times higher than the basal activity. A consistent

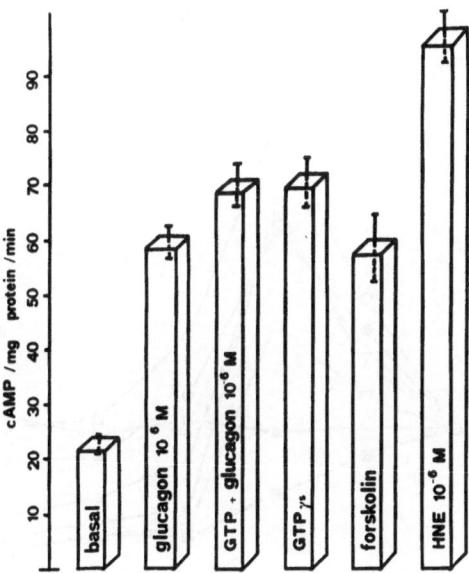

Figure 1. Adenylate cyclase activity in rat liver plasmamembranes after addition of 4-HNE and several activators. Results are given as pmoles cAMP/mg protein/min.

stimulation of the activity is displayed even in the presence of micromolar
glucagon, but its extent is lower than that seen with 4-HNE. The simultaneous addition of 4-HNE and glucagon didn't produce any additive stimulation.
Also the addition of $GTP_\gamma S$ to glucagon did not produce any consistent increase in the degree of stimulation. Forskolin, a well known stimulator of
the catalytic unit of the enzyme[21], produced a stimulation similar to that
provoked by glucagon, whereas the simultaneous addition of both forskolin
and 4-HNE produced the same level of stimulation observed with forskolin
alone. So, the level of stimulation observed in the presence of 4-HNE alone
is cancelled in the presence of forskolin.

The addition of dithiothreitol to plasmamembrane before 4-HNE addition
strongly decreases the effect of 4-HNE, but does not influence those displayed by forskolin or glucagon.

The effect of 4-HNE is not intensified by the simultaneous addition of
GTP S.

Sodium fluoride, a well known stimulator of adenylate cyclase activity[22], thought to act at the level of a stimulatory G protein (Gs), strongly
potentiates the effect of 4-HNE, that becomes evident even at concentraions
of 0.1 and 0.01 micromolar.

Fig.2 describes the kinetics of adenylate cyclase stimulation by several concentrations of 4-HNE. It is clear that maximal stimulatory activity
occurs within five min after the addition of the aldehyde, and declines
therefore on increasing incubation time or aldehyde concentration.

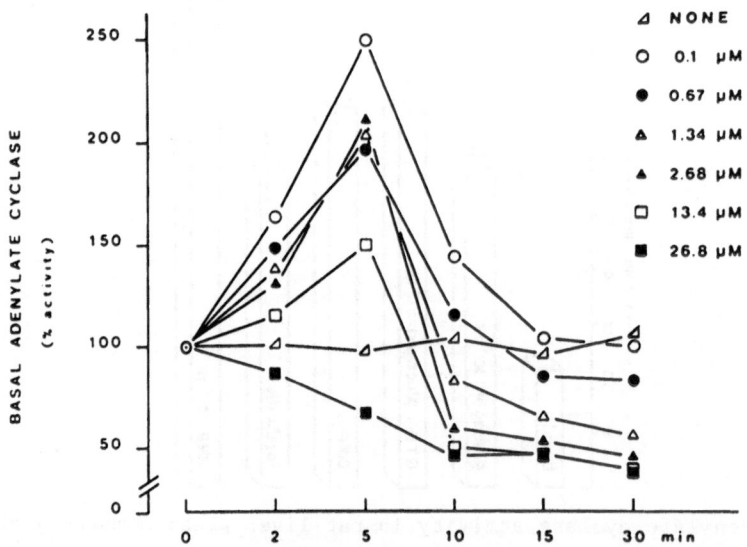

Figure 2. Time-course of basal adenylate cyclase activity (NONE) of rat
liver plasmamembranes in the presence of increasing concentrations of HNE.

174

4-HOE behaves in a very similar way with respect to 4-HNE, but the stimulatory activity appears to be quantitatively less intense than that of 4-HNE. Nonenal and octenal do not stimulate adenylate cyclase, but are able to inhibit the enzymatic activity when used at concentrations in the order of millimolar.

Fig.3 describes the effect of HNE on plasmamembrane guanylate cyclase. The work on this enzyme is rather preliminar, but it is interesting to remark that 0.5 micromolar 4-HNE stimulates the enzymatic activity by about 50 %. It is clear, however, that the extent of the stimulation remains much lower than that seen in the case of adenylate cyclase.

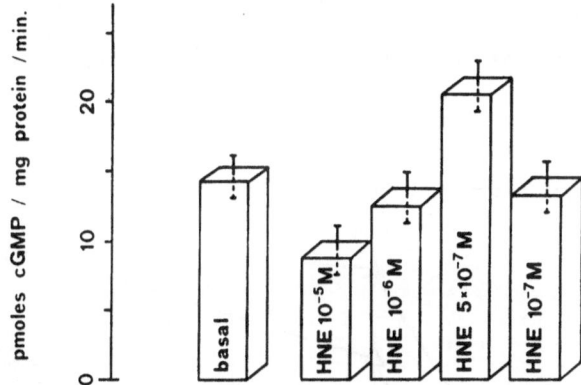

Figure 3. Effect of different concentrations of 4-HNE on guanylate cyclase activity.

It is noteworthy that other plasmamembrane marker enzymes, as for instance 5'-nucleotidase, Mg^{2+},Na^+,K^+-ATPase, Na^+,K^+-ATPase and Ca^{2+},Mg^{2+}-ATPase are unaffected by low 4-HNE concentrations, and become slightly inhibited at concentrations of the order of millimolar (data not shown).

Phospholipase C activity is strongly increased in the presence of 4-HNE at concentrations 0.1-1 micromolar. The stimulation of the activity occurs within 1-3 min after the addition of the aldehyde, and is higher in the presence of GTPgammaS (Fig.4). Maximal stimulation is reached when added Ca^{2+} concentrations is between 100 and 500 nM, i.e. in the physiological range of Calcium concentration (Fig.5).

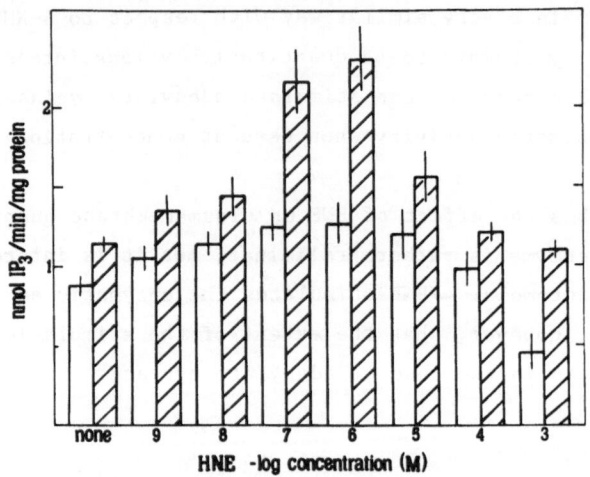

Figure 4. Effect of different concentrations of 4-HNE on phospholipase C activity. Hatched columns refer to samples incubated in the presence of 10 micromolar GTPgammaS.

The data are expressed as nanomoles IP_3 released/min/mg of protein.

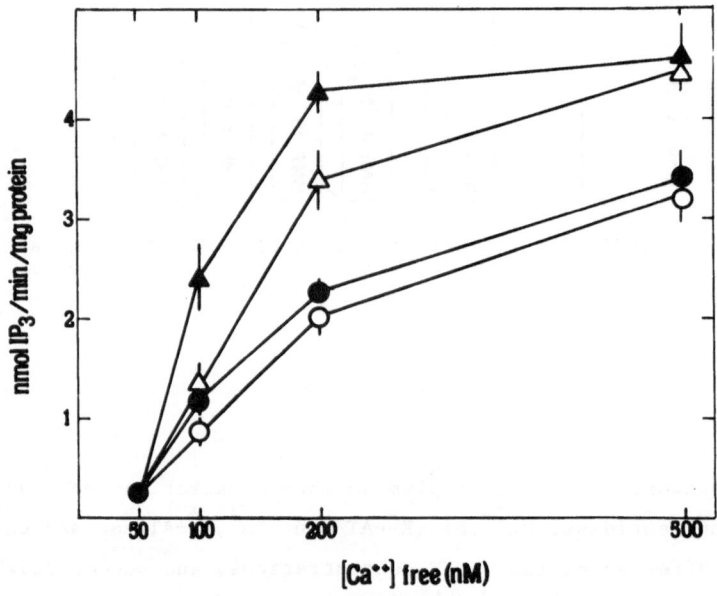

Figure 5. Effect of Ca^{2+} concentrations on phospholipase C activity. Liver membranes were incubated at 37°C under control conditions (O), in the presence of 10 micromolar GTPgammaS (●), 1 micromolar 4-HNE (△) or 4-HNE plus GTPgammaS (▲).

176

ESR-measurement of membrane fluidity showed that a small increase in the rigidity of membranes occurs only in the presence of concentration of 4-HNE of 0.1 mM or higher (Fig.6).

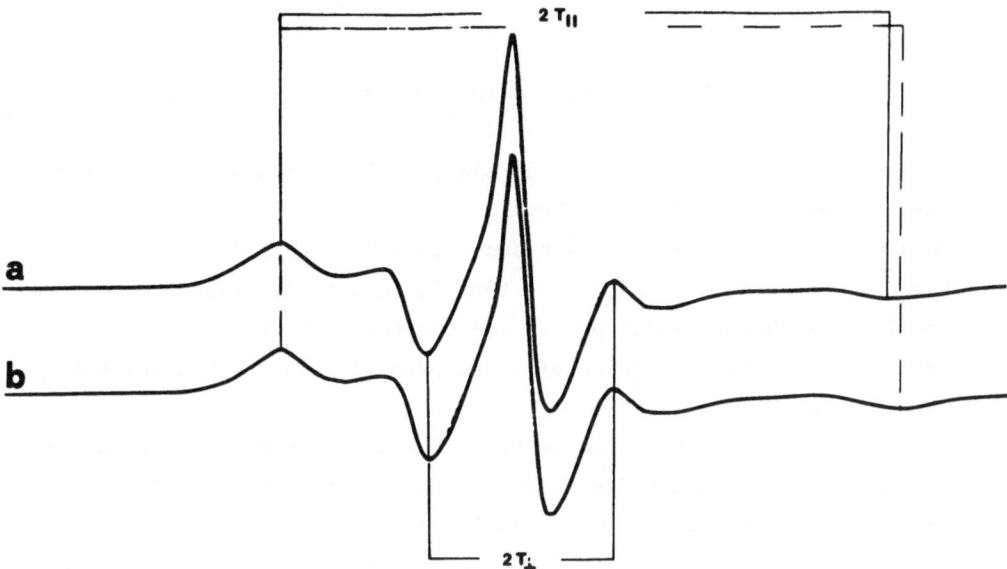

Figure 6. Effect of 4-HNE on the ESR spectra of Spin Label 5-doxylstearic acid added to purified plasmamembranes.
a) control, b) plasmamembranes plus 4-HNE.

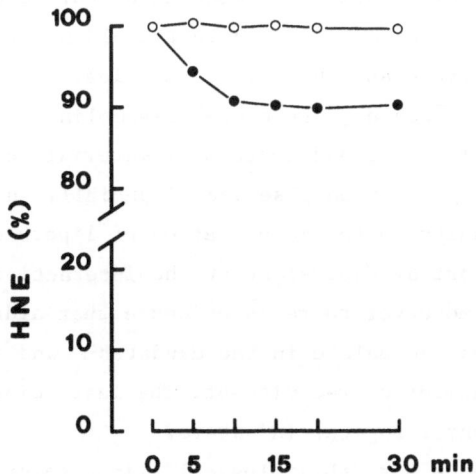

Figure 7. Time-course of decrease of the 4-HNE added in the incubation medium containing purified plasmamembranes. Data are expressed as percentage respect to the blank (only incubation medium plus 4-HNE) values.
Blank (O), plasmamembranes plus 4-HNE (●).

The study of the changes of added 4-HNE to the plasmamembranes showed that, after a small decrease amounting to 5-10 % of the added substance, the concentration of 4-HNE remains pratically constant in the incubation system for 1-2 hours (Fig.7).

DISCUSSION

The demonstration of the effects displayed by 4-HNE on several enzymes at concentrations that can be found inside the cells put forward the problem of a possible function of this aldehyde and of related compounds not only in pathology, but even in cell physiology.

The nature of the affected functions strengthens this hypothesis.

There are, at present, few proofs that lipid peroxidation may occur in normal cells. The most relevant points are the following :

1. the demonstration that products of lipid peroxidation, as MDA and HNE, are present in normal cells;

2. the fact, recently demonstrated by Poli et al.[23] that serum from normal subjects contains these aldehydes firmly bound to both proteins and lipids. The concentration of the aldehydes strongly increases in alcoholic subjects;

3. the demonstration that normal subjects exhale small amounts of pentane and ethane, that increase in the expired air of alcoholics[24]. This fact is, however, discussed as a definite proof for physiological lipid peroxidation, as exhaled pentane and ethane could even have been produced by intestinal bacteria;

4. the knowledge that cell provided with long life-span undergo physiological renewal of their lipid containing subcellular particles.

One of the mechanism used to eliminate the old membraneous structures is their segregation within autophagocytic vacuoles, that finally fuse with lysosomes. Inside such vacuoles, structures resembling those frequently seen during the disintegration of lipid containing material, under lipoperoxidative conditions, can regularly be observed. Especially in nervous cells, the process very often results in the accumulation of lipofuscins, that are considered as the morphological equivalent of the interaction between lipoperoxides and proteins[25]. Moreover there is evidence that aldehydes deriving from lipid peroxidation accumulate in the exudates[26] and can contribute in polymorphonuclear granulocytes recruitment. The last point, however, is pathological and not physiological in nature.

So far, the evidence for "physiological" lipid peroxidation is still feeble. There is, however, no evidence at all against, and the topic can be retained as a working hypothesis. The facts,documented in the present paper, that 4-HNE acts at low concentrations on important physiological functions may be therefore considered under these limits. It is noteworthy, however, that enzymes interested in transmembrane signalling and sited in hepatocyte plasmamembranes are strongly stimulated by 4-HNE at concentrations that do not increase membrane rigidity and are devoid of any influence on other enzymatic markers of the same membranes. The fact that the three enzymes are all concerned with G proteins for their activity adds further interest to the point. The extent of stimulation on the three enzymes is not identical. In fact the stimulation of guanylate cyclase is smaller than those displayed respectively on adenylate cyclase and on phospholipase C. This suggests that the target is probably not G protein itself. Moreover, the conditions for the stimulation of adenylate cyclase and of phospholipase C are also somehow different. So the observation recorded above must be still considered as preliminary, and much more work must be done in order to see if the observed phenomena really correspond to a common mechanism of action.

Another speculation strengthening the working hypothesis reported above is, moreover, that lipid peroxidation is very low or absent in tumours, as well as in other non tumoral tissues also characterized by high multiplication rates. On the contrary, it is usually high in tissues formed by long-living cells, with rare or absent reproductive episodes. The experiment by Baccino et al.[27], who have found an inhibitory effect of 4-HNE on cell multiplication in lymphocytes cultivated "in vitro", acting especially at the level of entering into the S phase, is another point in favour of a possible regulatory effect of lipid peroxidation on cell multiplication.

It is noteworthy that in the experiments by Tessitore et al.[27] the block of cell division appeared to be concerned more with an effect on a surface receptor than on the enzymes (DNA and RNA polymerase) themselves. In fact the effect was seen after a simple contact of the aldehyde on the cells, and resisted to washings after this contact. The experiments by Barrera et al.[11,12] showing the block of c-myc expression by 4-HNE on human erythroleukemic cells is another point in favour of the hypothesis.

To conclude, a final speculation derives from the knowledge that other derivatives of arachidonic acid, as the prostaglandins-prostacyclins, tromboxane group, as well as the leukotriene group, are known already to work in transmembrane signalling. And we know that nature is very often rich in modulation of effects and mechanisms.

REFERENCES
1. G. Poli, E. Albano and M. U. Dianzani, Chemistry and Physics of Lipids, 45: 117 (1987).
2. H. Esterbauer, K. H. Cheeseman, G. Poli, M. U. Dianzani and T.F. Slater, Biochem. J. 208:129 (1982)
3. G. Poli, M. U. Dianzani, K. H. Cheeseman, T. F. Slater, J. Lang and H. Esterbauer, Biochem. J. 227:629 (1985)

4. M. U. Dianzani, Biochemical effects of saturated and unsaturated aldehydes., in:"Free Radicals, Lipid peroxidation and Cancer", D. C. H. McBrien and T. F. Slater eds, Academic Press (1981)

5. E. Gravela et al., unpublished results

6. M. U. Dianzani, R. A. Canuto, M. Ferro, A. M. Bassi, L. Paradisi, M. E. Biocca and G. Muzio, Fourth Sardinian International Meeting on Chemical Carcinogenesis, Alghero (Italy) October 1987

7. M. Curzio, M. V. Torrielli, J. P. Giroud, H. Esterbauer and M. U. Dianzani, Res. Comm. Chem. Pathol. Pharmacol., 36:463 (1982)

8. M. Curzio, H. Esterbauer, C. DiMauro, G. Cecchini and M. U. Dianzani, Biol. Chem. Hoppe Seyler, 367:321 (1986)

9. M. Curzio, C. Di Mauro, H. Esterbauer and M. U. Dianzani, Biomedicine and Pharmacoterapy, 41:304 (1987)

10. M. Curzio et al.,unpublished results

11. G. Barrera, S. Martinotti, V. Fazio, V. Manzari, L. Paradisi, M. Parola, L. Frati and M. U. Dianzani, Toxicol. Pathol., 15:238 (1987)

12. V. Fazio, G. Barrera, S. Martinotti, M. G. Farace, B. Giglioni, O. Gadini, L. Frati, V. Manzari and M. U. Dianzani, Second International Meeting on "Free Radicals in Liver Injury". Torino (Italy) 1988

13. M. A. Rossi and G. Cecchini, Cell Biochem. Function, 1:49 (1983)

14. L. Paradisi, C. Panagini, M. Parola, G. Barrera and M. U. Dianzani, Chem. Biol. Interactions 53:209 (1985)

15. M. A. Rossi and H. Esterbauer, 1988, submitted for publication

16. H. Kimura and F. Murad, J. Biol. Chem., 249:6910 (1974)

17. I. Magnaldo, H. Talwar, W.B. Anderson, J. Poysse'gur, F.E.B.S. Letters, 210:6 (1987)

18. M. Parola, G. Barrera, M. C. Carasso, L. Amoroso, B. Bosia, L. Paradisi and M. U. Dianzani, Boll. Soc. Ital. Biol. Sper, 58:1199 (1982)

19. M. Parola, E. Albano, R. Autelli, G. Barrera, M. E. Biocca, L. Paradisi and M. U. Dianzani, Chem. Biol. Interactions, 73:103 (1990)

20. J. S. Hyde, H. M. Swartz and W. E. Antholine, The spin probe-spin label method, in:"Spin labeling II,theory and applications", L. J. Berliner ed., Acad. Press, N.Y. (1979)

21. K. B. Seamon, W. Padgett and J. W. Daly, Proc. Natl. Acad. Sci. U.S.A., 78:3363 (1981)

22. A. G. Gilman, Ann. Rev. Biochem., 56:615 (1987)

23. R. Carini, R. Mazzanti, F. Biasi, E. Chiarpotto, G. Marno, S. Moscarella, P. Gentilini, M. U. Dianzani and G. Poli, Adv. Biosci., 71:61 (1986)

24. S. Moscarella, G. Laffi, D. Coletta, U. Arena, A. P. Capellini and

P. Gentilini, Volatile hydrocarbons in the breath of patients with
chronic alcoholic liver disease:a possible marker of ethanol-induced
lipoperoxidation., in: "Liver Cirrhosis" P. Gentilini and M. U. Dianzani
eds, P. Rozen, Tel Aviv (1984)

25. W. T. Roubal and A. L. Tappel, Arch. Biochem., 113:5 (1966)

26. M. Curzio, G. Poli, H. Esterbauer, F. Biasi, C. DiMauro and
M. U. Dianzani, IRCS Med. SCI., 14:984 (1986)

27. L. Tessitore, L. Matera, G. Bonelli, F. M. Baccino and M. U. Dianzani,
Chem. Biol. Interactions, 65:217 (1987)

CARBON TETRACHLORIDE-INDUCED OXIDATIVE STRESS AT THE LEVEL OF LIVER GOLGI

APPARATUS: EFFECT ON LIPOPROTEIN SECRETION

Giuseppe Poli, Elena Chiarpotto, Emanuele Albano, Fiorella
Biasi, *Damiano Cottalasso, *Maria A. Pronzato, *Umberto M.
Marinari, *Giorgio Nanni and Mario U. Dianzani

Department of Experimental Medicine and Oncology, University
of Turin, C.so Raffaello 30, 10125 Turin, Italy

*Institute of General Pathology, University of Genoa, via
Alberti 2, 16132 Genoa, Italy

INTRODUCTION

The onset of the liver fatty degeneration which follows rat acute
poisoning with carbon tetrachloride (CCl_4) is provoked by a severe block[1,2,3]
of lipoprotein secretory pathway. Such a complex functional
impairment is already relevant few hours after treatment and probably
involves various cellular compartments. Previous studies employing the
isolated hepatocyte model gave indirect indications for an early
involvement of liver Golgi apparatus in the CCl_4-induced block of
intracellular transport of the lipoprotein micelles.[4,5] More recent
studies with rat hepatocytes in single cell suspension have allowed us to
directly demonstrate a marked impairment of lipoprotein maturation and
transport at the Golgi level soon after CCl_4 treatment.[6] "In vivo"
investigations of the changes occurring in the Golgi apparatus during the
oxidative stress due to CCl_4 are here reported. In addition, the analysis
of the pathogenesis of these Golgi disturbances has been attempted by
using a strong antioxidant treatment represented by rat supplementation
with large doses of alpha-tocopherol prior to CCl_4 dosing.

MATERIALS AND METHODS

Male rats of the Wistar strain (Charles River Italia, Calco, Italy)
weighing 200-250 g, were used. They were maintained one week before using
under controlled feeding with a standard synthetic diet (n. 48, Piccioni,
Gessate Milanese, Italy; vitamin E content: 40 mg/kg) and with water "ad
libitum".

All chemicals were of reagent grade and were obtained from Sigma
Chemical Co., St. Louis, Missouri, USA; Merck, Darmstadt, West Germany;
the radioisotopes from The Radiochemical Centre, Amersham, Bucks, U.K..

Pretreatment of rats with vitamin E, where necessary, was carried out
as described by Pfeifer & McCay:[7] vitamin E was dissolved in 1 vol. of
ethanol and followed by the addition of 9 vol. of 16% (v/v) Tween 80 in
0.9% NaCl. 100 mg/kg b.wt of alpha-tocopherol were injected
intraperitoneally 15 h before sacrifice.

For CCl_4 poisoning, each animal, fasted overnight, received by gastric intubation either a single dose of CCl_4 (250 μl/100 g b.wt as a solution 50% v/v in mineral oil), or mineral oil alone 60 min before sacrifice. For the analysis of palmitate incorporation into the liver Golgi, 100 μCi 5% albumin-bound (9-10-^{3}H)-sodium palmitate (0.2 μmol) were injected into the tail vein of each rat 10 min before sacrifice, according to the procedure previously used.[8] The animals were killed by cervical dislocation, the liver was removed, washed in ice-cold 0.25 M sucrose, forced through a tissue press and then homogenized in 0.25 M sucrose with eight strokes in a Potter-Elvehjem homogenizer to give a 20% (w/v) homogenate.

The isolation of Golgi apparatus was carried out according to the method described by Ehrenreich et al.[9] Three different fractions of purified Golgi membranes were obtained: F_1 and F_2, secretory sides, and F_3, formative side. Morphological characterization of Golgi apparatus sub-fractions was carried out using a Siemens Elmiskop 101 electron microscope. Golgi fractions obtained by ultracentrifugation were fixed in 1.25% glutaraldehyde in 0.1 M Na-cacodylate buffer (pH 7.6) for 30 min, than washed and post-fixed in 1% osmium tetroxide in the aforementioned buffer. Before dehydration in ethanol, fractions were stained for 2 hr with uranyl acetate. Thin sections, stained with lead citrate, were then observed with the electron microscope. For chemical characterization of Golgi membranes, the activity of the marker enzyme UDP-galactose: N-acetyl-glucosamine galactosyl transferase was measured as hereafter reported.

Furthermore, measurement of three other marker enzymes for different subcellular organelles was performed to check the purity of the Golgi fraction preparations. To check for possible microsomal contamination, glucose-6-phosphatase was measured as reported by Poli et al.[10] As a mitochondrial marker enzyme we evaluated cytochrome oxidase activity by the method of de Duve et al.[11] Finally, as a lysosomal marker enzyme N-acetyl-beta-glucosaminidase activity was measured by the method of Sellinger et al.[12] modified according to Baccino et al.[13] using p-nitro-phenyl derivatives as substrate.

In order to follow the dynamics of hepatic lipid transport and secretion through the Golgi, we evaluated the ^{3}H-palmitic acid incorporation into the lipids of the fractions obtained by discontinuous gradient ultracentrifugation, by processing aliquots of these samples for lipid separation and radioactivity detection as described elsewhere.[8] The results were expressed as specific activities after the protein content of the samples was estimated by the method of Lowry as modified by Peterson.[14]

The functional condition of the Golgi apparatus fractions was checked by measuring UDP-galactose: N-acetyl-glucosamine galactosyl transferase activity with the method described by Fleischer et al.[15]

For the evaluation of CCl_4 uptake by the liver, rats treated and untreated with vitamin E were intoxicated 1 hour with 0.25 ml of radioactive CCl_4 (specific activity 25 μCi/ml). After the sacrifice, the livers were homogenized in 0.25 M sucrose (20% w/v) and 0.1 ml aliquots of the homogenates were diluted with 4 ml Instagel scintillation fluid for the determination of the total radioactivity in the livers.

Liver microsomes were prepared as reported by Slater & Saywer;[16] microsomal proteins were precipitated with 10% trichloroacetic acid (TCA) and collected by centrifugation at 2000 g for 10 min. The resulting protein pellets were first washed once with 5% TCA, 4 times with chloroform-methanol-diethyl ether (2:2:1 v/v) mixture and then once with diethyl ether. Finally, the dried protein layers were solubilized with 0.5 ml 1 N NaOH and mixed with 5 ml Pico-Fluor 30 for measurement of radioactivity. The protein content in the solutions was determined before the addition of the scintillation fluid by the Lowry procedure as modified by Peterson.[14]

Aliquots of liver homogenate were taken for analysis of alpha-tocopherol content essentially as described by Burton et al.[17]

184

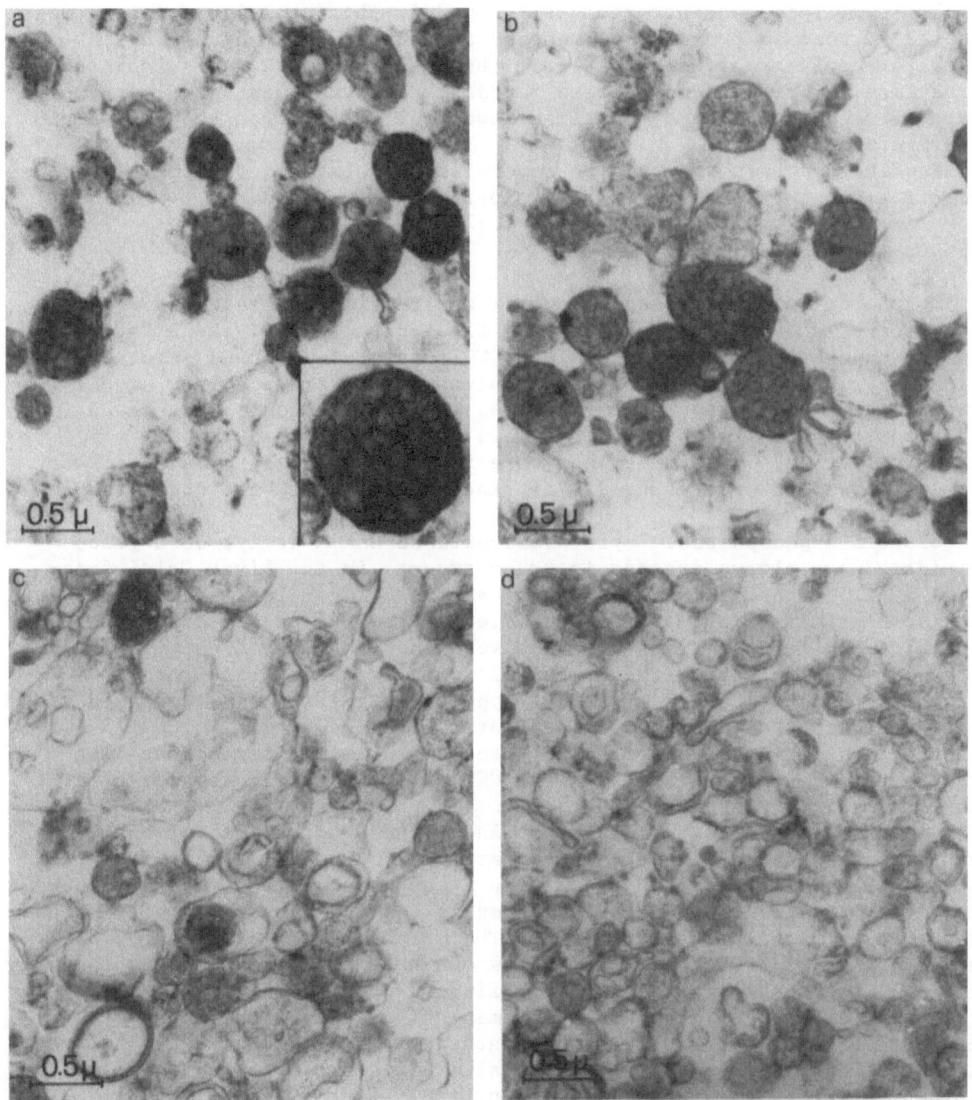

Fig. 1. Transmission electron micrographs of liver Golgi fractions from control animals and rats treated with CCl_4 1 hour before killing.
(a) F_1 secretory fraction from CCl_4 poisoned animals (2.5 ml/Kg b wt) membrane vesicles are filled with lipids (inset);
(b) F_1 fraction from CCl_4 poisoned animals but preloaded with vitamin E (100 mg/Kg b wt);
(c) F_1 Golgi fraction from control animals;
(d) F_1^1 fraction from control rats but preloaded with vitamin E.

The determination of the diene-conjugated signal in liver microsomes obtained from CCl_4-treated rats, preloaded or not with vitamin E, was performed as follows using a Perkin-Elmer 550 S spectrophotometer. The lipid/chloroform extracts of the microsomal preparations were dried down under nitrogen stream then dissolved either in cyclohexane or in ethanol to obtain a lipid concentration of 100 μg/ml. The total lipid extract was eventually scanned from 220-300 nm and the absorbance and the second derivative spectrum recorded. Before the scanning operation a background-correction memorized scan between Suprasil cells containing solvent only was performed in order to counteract spectral differences between sample and reference cells.

Student's t test was used to determine the statistical significance of the difference between control and experimental groups.

RESULTS AND DISCUSSION

Morphological studies on the Golgi subfractions isolated from both CCl_4-treated and control animals were first carried out. The effect of vitamin supplementation on normal and CCl_4-damaged membrane structure was also checked. The Golgi preparations always demonstrated minor or nil contamination with cytoplasm organelles such as lysosomes, mitochondria and ribosomes. Consistent findings were obtained by the biochemical characterization. Golgi membranes were in fact analysed for a series of marker enzymes of other subcellular fractions to check the degree of their homogeneity.

Glucose-6-phosphatase, cytochrome-oxidase and N-acetyl-beta-glucosaminidase were always present in trace amounts in the liver Golgi fractions of all experimental and control animals. Control F_1 and F_2 fractions consisted primarily of membrane vesicles containing or not lipoprotein particles, while in F_3 more frequent were the profiles of Golgi cisternae. After 1 h CCl_4 poisoning the number of vesicles and of lipoprotein particles contained by each vesicle appeared significantly increased as compared to controls (Fig. 1). The characteristic picture in CCl_4 intoxication of vesicles filled up with osmiophilic material was also present but to a minor extent in the CCl_4 plus vitamin E samples (Fig. 1). The intermediate secretory fraction F_2 and the formative F_3 showed, after CCl_4 intoxication, loss of cisternae and vacuolar profiles, probably due to the derangement of these membrane structures induced by the toxin and not significantly protected by vitamin E supplementation (data not shown). The latter treatment did not exert "per se" any deranging activity on the structure of normal Golgi apparatus. The morphological evidence of a marked change of liver Golgi membranes one hour after acute rat poisoning with CCl_4 was then paralleled by the finding of a small but present protective effect of vitamin E supplementation against the block of lipoprotein transit through the apparatus. Both phenomena were confirmed and better characterized by biochemical studies employing a radiolabeled lipid precursor, i.e. ^{3}H-palmitate. In fact, after 60 min from CCl_4 administration, the animals showed an already relevant accumulation of labeled lipids which involved all the apparatus components (Fig. 2). Time course studies elsewhere reported[18] proved that such a lipid accumulation, that clearly demonstrates a diminished efflux of lipoprotein micelles from Golgi apparatus, is significant after 15 min poisoning in the formative (F_3) and intermediate side (F_2), and after only 5 min in the secretory pole (F_1).

In other words, the obstruction of lipid transit through the Golgi takes place rapidly after CCl_4 poisoning, and after a sharp aggravation in F_1 and F_2 during the first 30-60 min.

In order to check if not only the transport, but also the maturation of lipoproteins at the level of Golgi apparatus was affected by CCl_4 acute intoxication, UDP-galactosyl transferase activity was determined in the three liver Golgi fractions after 60 min rat treatment with the halolkane.

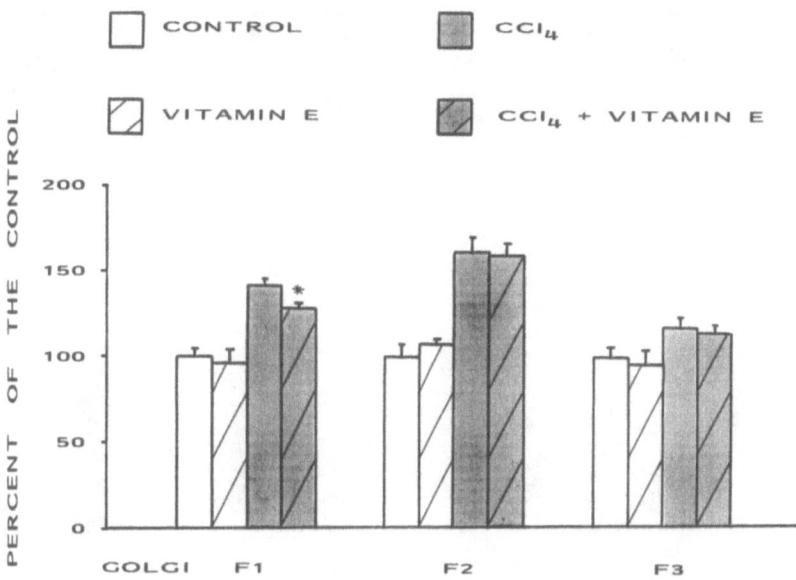

Fig. 2. ^{3}H-palmitic acid incorporation in the lipids of liver Golgi subfractions obtained 1 hour after CCl_4 poisoning in the presence or absence of vitamin E supplementation. The results of 6-8 experiments have been averaged and represent the specific activity (cpm/mg protein $\pm$ S.D.).
* Significant as to CCl_4 group (P < 0.01).

Golgi membranes from the poisoned rats showed a marked inactivation of the key enzyme of glycosylation processes (Fig. 3). Such a decrease of activity appears significant in all Golgi fractions already after 15 min from the poisoning.[18] In a second series of experiments, the pathogenesis of the observed changes was investigated by supplementing the animals with alpha-tocopherol 15 hours before CCl_4 dosing. The rationale of these studies was based on the observation that in isolated rat hepatocytes with an increased vitamin E concentration, the peroxidative damage of membrane lipids due to CCl_4 intoxication was totally prevented without any modification of the covalent binding of CCl_4 metabolites to cell protein and lipid.[19] So, the selective inhibition of lipid peroxidation, i.e. one of the two mechanisms of CCl_4 toxic injury, should further clarify their role in the genesis of the single cellular changes.

Actually, in the whole animal as well as in isolated hepatocytes, the pretreatment with a large dose of alpha-tocopherol achieved a complete protection against CCl_4-prooxidant effect. As illustrated in Fig. 4, the second derivative spectrum of liver microsomal lipids obtained from CCl_4-poisoned rats showed a deep minimum peak at 233 nm which corresponds to the absorption peak of diene conjugation in conventional spectroscopy.[20] Noteworthy is the higher resolution of the second derivative spectrum as to that of UV spectrum. Rat pretreatment with vitamin E, which did not exert any aspecific damage "per se" to any of the parameters so far tested, led to the complete disappearance of the conjugated diene signal due to CCl_4. The protection afforded by this treatment against CCl_4-induced lipid peroxidation is not due to a delay in CCl_4 uptake by the livers of vitamin E supplemented rats.

In fact, as reported in Table 1, the pretreatment with the vitamin did not interfere with the distribution of the toxic compound in the liver.

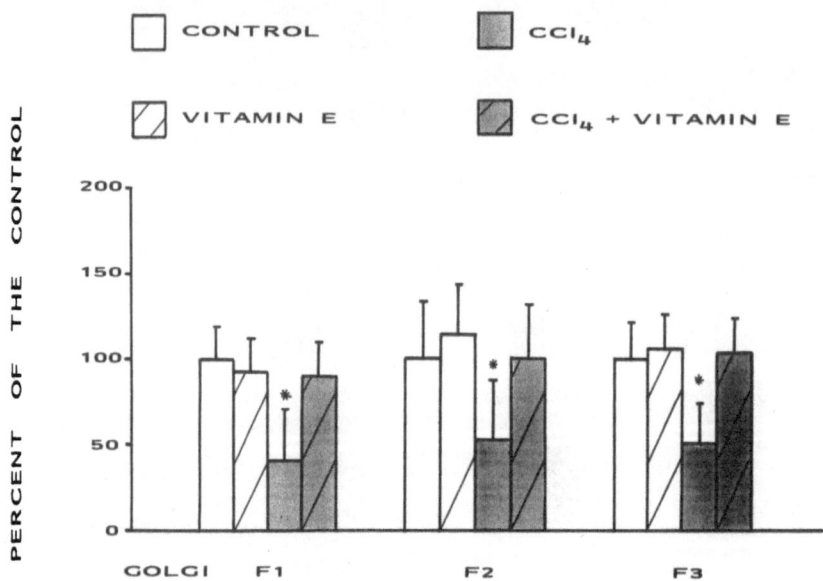

Fig. 3. Galactosyl transferase activity in liver Golgi subfractions
obtained 1 hour after CCl_4 poisoning in the presence or absence of
vitamin E supplmentation. The results of 6-8 experiments have been
averaged and represent the nmol of galactose transferred to
N-acetyl-glucosamine/h/mg protein $\pm$ S.D.
* Significant as to the control group (P < 0.01).

Further, CCl_4 covalent binding to microsomal protein was not affected
by the increased amount of tocopherol in the membranes. Quantitative
h.p.l.c. measurements showed that the vitamin E content in the liver from
supplemented animals was 8-10 times higher than that of normal rats.

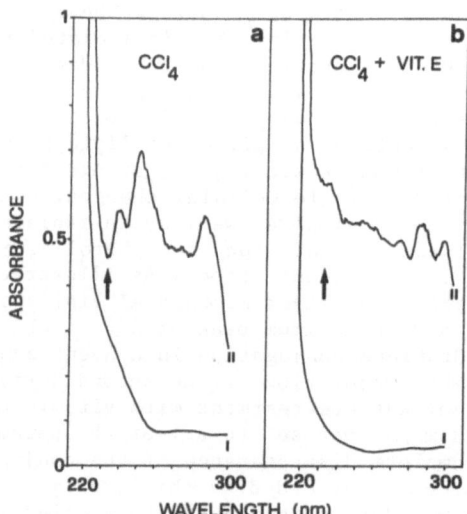

Fig. 4. UV (I) and second derivative (II) spectra of lipid extracted from
microsomes of CCl_4 (2.5 ml/Kg b wt) (a) and CCl_4 + vit. E (b)
treated rats. A typical experiment. The arrows show in the second
derivative spectra the minimum absorption peaks at 233 nm.

Table 1. ^{14}C-CCl$_4$ Recovered in Total Liver Homogenate or Covalently Bound to Microsomal Protein 60 min After Poisoning of Rats Preloaded or Not With Vitamin E.

Experimental groups	homogenates (dpm x 10^{-2} / g liver)	microsomes (pmol/mg protein)
CCl$_4$	113 + 14	12.38 + 1.05
CCl$_4$ + vitamin E	112 + 19	11.13 + 1.26

All values are means + S.D. of two experiments; for each experimental group six animals were used. The total CCl$_4$ amount given to the rats was 2.5 ml/kg b wt.

By pretreating the rat with large amounts of alpha-tocopherol it was then possible to evaluate the actual involvement of prooxidant and haloalkylating effects of CCl$_4$ in the impairment of Golgi lipoprotein transport and hexosylation.

As shown in Fig. 2, despite a complete protection against CCl$_4$-induced lipid peroxidation, vitamin E-pretreatment gave only minor (though statistically significant) protection against lipid accumulation in the F$_1$ fraction and provided no protection at all in the F$_2$ and F$_3$ fractions. This finding is against a major role of lipid peroxidation at least in that early block of lipoprotein Golgi transport. Conversely, the strong inhibition of UDP-galactosyl transferase observed in the livers of the same rats was totally prevented by alpha-tocopherol preloading. In fact, such an enzyme activity was reduced by about 50% in all three Golgi fractions after 1 hour CCl$_4$ poisoning, showed the full normal function when the vitamin E pretreatment was adopted (Fig. 3). In other words the CCl$_4$-induced change of this membrane bound enzyme was clearly shown to be dependent upon peroxidative attack.

In conclusion, two findings of the herereported "in vivo" investigation support haloalkylation as the primary pathogenic mechanism for the early onset of liver fatty degeneration following CCl$_4$ poisoning: i) the only slight, even if significant, reduction of lipoprotein accumulation in the Golgi membranes by vitamin E preloading; ii) the evidence of such an impairment shortly after CCl$_4$ (in F$_1$ after 5 min) when cellular changes proved to be tightly dependent on lipid peroxidation (e.g. inactivation of UDP-galactosyl transferase in F$_1$, F$_2$, F$_3$ after 15 min) are not yet detectable. Of course, this statement does not exclude the implication of lipid peroxidation as well in the genesis of the secretory block, but at a later time and with less emphasis.

In the chronology of CCl$_4$-induced impairment of liver lipoprotein secretion, the first mechanism occurring is most likely the direct binding of reactive free radicals, most likely the trichloromethyl radical, to crucial components of the cell secretory pathway. With the time, this injury is maintained and amplified by the occurrence of various other cell changes partly mediated by the prooxidant action of CCl$_4$, as indirectly suggested by various reports on the valid even if never complete protection exerted against CCl$_4$-induced fatty liver by membrane enrichment in vitamin E.

ACKNOWLEDGMENTS

The authors wish to thank the Consiglio Nazionale delle Ricerche, Gruppo di Gastroenterologia, and the Ministero Pubblica Istruzione, Progetto Cirrosi Epatica, Roma, Italy, for financial support.

REFERENCES

1. T. F. Slater, Hepatotoxicity of carbon tetrachloride: fatty degeneration, in: "Free Radical Mechanisms in Tissue Injury," J. R. Lagnado, ed., Pion Ltd., London (1972).

2. M. U. Dianzani, Biochemical aspects of fatty liver, in: "Biochemical Mechanisms of Liver Injury," T. F. Slater, ed., Academic Press, London (1978).

3. W. J. Brattin, E. A. Glende Jr, and R. O. Recknagel, Pathological mechanisms in carbon tetrachloride hepatotoxicity, J. Free Rad. Biol. Med. 1:27 (1985).

4. E. Gravela, E. Albano, M. U. Dianzani, G. Poli, and T. F. Slater, Effects of carbon tetrachloride on isolated rat hepatocytes, Biochem. J. 178:509 (1979).

5. G. Poli, E. Gravela, E. Albano, and M. U. Dianzani, Studies on fatty liver with isolated hepatocytes, Exp. Mol. Pathol. 30:116 (1979).

6. G. Poli, E. Chiarpotto, E. Albano, D. Cottalasso, G. Nanni, U. M. Marinari, A. M. Bassi, and M. U. Dianzani, Carbon tetrachloride-induced inhibition of hepatocyte lipoprotein secretion: functional impairment of Golgi apparatus in the early phases of such injury, Life Sci. 36:533 (1985).

7. P. M. Pfeifer, and P. B. McCay, Reduced triphosphopyridine nucleotide oxidase-catalyzed alterations of membrane phospholipids, J. Biol. Chem. 246:6401 (1971).

8. U. M. Marinari, D. Cottalasso, G. R. Gambella, M. M. Averame, M. A. Pronzato, and G. Nanni, Effects of acute ethanol intoxication on ^{3}H-palmitic acid transport through hepatocyte Golgi apparatus, FEBS Lett. 86:53 (1978).

9. J. H. Ehrenreich, J. J. M. Bergeron, P. Siekevitz, and G. E. Palade, Golgi fractions prepared from rat liver homogenates, J. Cell Biology 59:45 (1973).

10. G. Poli, K. H. Cheeseman, T. F. Slater, and M. U. Dianzani, The role of lipid peroxidation in CCl_4-induced damage to liver microsomal enzymes: comparative studies in vitro using microsomes and isolated liver cells, Chem.-Biol. Interactions 37:13 (1981).

11. C. de Duve, B. C. Pressman, R. Gianetto, R. Wattiaux, and F. Appelmans F, Tissue fractionation studies. 6. Intracellular distribution patterns of enzymes in rat-liver tissue, Biochem. J. 60:604 (1955).

12. O. Z. Sellinger, H. Beaufay, A. Jacques, A. Doyeu, and C. de Duve, Intracellular distribution and properties of beta-N-acetyl-glucosaminidase and beta-galactosidase in rat liver, Biochem. J. 74:450 (1960).

13. F. M. Baccino, G. A. Rita, and M. F. Zuretti, Studies on the structure-bound sedimentability of some rat liver lysosome hydrolases, Biochem. J. 122:363 (1971).

14. G. L. Peterson, A simplification of the protein assay method of Lowry et al. which is more generally applicable, Anal. Biochem. 83:346 (1977).

15. B. Fleischer, S. Fleischer, and H. Ozawa, Isolation and characterization of Golgi membrane from bovine liver, J. Cell Biology 43:59 (1969).

16. T. F. Slater, and B. C. Sawyer, The stimulatory effects of carbon tetrachloride on peroxidative reactions in rat liver fractions "in vitro", Biochem. J. 123:815 (1971).

17. G. W. Burton, A. Webb, and K. U. Ingold, A mild, rapid, and efficient method of lipid extraction for use in determining vitamin E/lipid ratios, Lipids 20:29 (1985).

18. U. M. Marinari, M. A. Pronzato, D. Cottalasso, A. Zicca-Cadoni, G. Nanni, G. Poli, E. Chiarpotto, E. Albano, F. Biasi, and M. U. Dianzani, CCl_4-induced early functional impairment of rat liver

Golgi apparatus, in: "Free Radicals in Liver Injury," G. Poli, K. H. Cheeseman, M. U. Dianzani, and T. F. Slater, eds., IRL Press, Oxford (1985).

19. G. Poli, E. Albano, F. Biasi, G. Cecchini, R. Carini, G. Bellomo, and M. U. Dianzani, Lipid peroxidation stimulated by carbon tetrachloride or iron and hepatocyte death: protective effect of vitamin E, in: "Free Radicals in Liver Injury," G. Poli, K. H. Cheeseman, M. U. Dianzani, and T. F. Slater, eds., IRL Press, Oxford (1985).

20. F. P. Corongiu, M. Lai, and A. Milia, Carbon tetrachloride, bromotrichloromethane and ethanol acute intoxication, Biochem. J. 212:625 (1983).

OXIDIZED LIPOPROTEINS AND ATHEROGENESIS

ROLE OF LOW DENSITY LIPOPROTEIN OXIDATION IN FOAM CELL FORMATION

Urs P. Steinbrecher

Department of Medicine
University of British Columbia
2211 Wesbrook Mall
Vancouver, British Columbia
V6T 1W5 Canada

INTRODUCTION

The pathophysiology of atherosclerosis is complex, and probably involves multiple interacting mechanisms. The two hypotheses that have been invoked most often to explain atherogenesis are the "response to endothelial injury" hypothesis, and the "lipid infiltration" hypothesis.[1,2] It has become clear that these two seemingly distinct hypotheses are in fact so closely linked that they cannot be considered in isolation.[2] Pathologically, one of the hallmarks of the atherosclerotic lesion is the presence of lipid-laden foam cells in the arterial intima. Several lines of evidence indicate that in early lesions foam cells are derived from macrophages, although in more advanced lesions, smooth muscle cells apparently can also undergo foam cell transformation.[3-8] The mechanism by which these cells accumulate excess lipid in the artery wall <u>in vivo</u> has not been determined. However, studies using cultured macrophages have suggested potential mechanisms by which foam cell formation might occur. When macrophages are exposed to "physiologic" or even to high concentrations or normal lipoproteins <u>in vitro</u>, the rate at which they accumulate cholesterol is usually insufficient to cause foam cell formation.[9] An exception to this is the observation that J774 cells, a mouse macrophage-like cell line that exhibits unusually high acyl-CoA:cholesterol acyltransferase (ACAT) activity, can accumulate cholesterol ester when incubated with normal LDL.[10] This phenomenon apparently does not occur with primary cultured macrophages, and hence its applicability to foam cell formation <u>in vivo</u> remains speculative. Cultured macrophages can also accumulate lipid when exposed to βVLDL, but the reason for this is not entirely clear, as the receptor responsible for the internalization of βVLDL appears to be identical to the classical LDL receptor[11-13] and its expression would therefore be expected to be regulated by cellular cholesterol content. In any case, uptake of βVLDL and related triglyceride-rich lipoproteins remains a potentially important mechanism for lipid accumulation. Macrophages have also been found to possess a high affinity uptake mechanism for certain types of modified lipoproteins.[9,14] This has been termed the scavenger or "acetyl-LDL" receptor. The expression of this receptor is not subject to regulation by cellular cholesterol content, and therefore can lead to massive cholesterol accumulation. Several different chemical modifications of LDL can lead to uptake via this receptor, including acetylation, acetoacetylation, carbamylation, and modification by malondialdehyde.[14-19] These modifications share as a common feature the ability to derivatize lysine residues and result in a neutralization of the positive charge of the lysine epsilon amino group, resulting in a net increase in electrophoretic mobility. In addition to these chemical modifications of LDL, a "biological" modification of LDL has also been described that leads to uptake by the scavenger receptor. This modification is the result of peroxidation of LDL, either initiated by incubation of LDL with cultured cells or by exposure of LDL to redox-active metal ions in the absence of cells.[20-26] Although many

Free Radicals, Lipoproteins, and Membrane Lipids
Edited by A. Crastes de Paulet *et al.*
Plenum Press, New York, 1990

changes to LDL structure and composition result from oxidation, the recognition of oxidized LDL by the scavenger receptor can be explained by the modification of lysine residues.

A number of previous studies have described accelerated uptake of lipoproteins in macrophages by a mechanism that does not involve the scavenger receptor pathway.[27-29] This process probably involves the uptake of aggregated lipoprotein particles by phagocytosis.[30] Insofar as aggregated lipoproteins have been found in fluid isolated from arterial intima[28], and may occur as a consequence of lipoprotein oxidation, this phagocytic pathway may be of physiologic importance.

Activated platelets have also been found to be capable of inducing cholesterol accumulation in cultured macrophages.[31,32] This appears to occur as a result of the release of cholesterol-rich microvesicular structures from the activated platelets, and subsequent rapid internalization of these microvesicles by macrophages.[32] For reasons that have not been defined, the platelet-derived cholesterol leads to a much greater increase in cholesterol ester content than equivalent quantities of cholesterol delivered as opsonized red blood cell membranes. One possible explanation for this might be that platelet products affect the cholesterol ester cycle, for example by stimulating cholesterol ester synthesis (ACAT activity) or by inhibiting cholesterol ester hydrolase. Cholesterol-rich "debris" derived from membranes of other cellular elements has also been reported to cause cholesterol accumulation in cultured macrophages, by an apparently similar mechanism.[33] Finally, increased cholesterol esterification could also occur under circumstances of sustained excessive activity of HMG CoA reductase.

The following discussion will consider several of these mechanisms in further detail. However, it should be kept in mind that while foam cells are characteristic of atherosclerotic lesions, it has not been proven that they are essential in the pathogenesis of atherosclerosis. Indeed, it is conceivable that under some circumstances these cells might even remove cholesterol from extracellular deposits and therefore have a beneficial effect.

PROPERTIES OF OXIDIZED LDL

The potential importance of oxidized LDL in atherosclerosis was first suggested by Chisholm and co-workers, who found that oxidized LDL was cytotoxic to cultured endothelial cells.[34] These investigators also showed that cultured endothelial cells were capable of inducing an oxidative modification of LDL, and that this modification depended on free radical peroxidation, apparently involving superoxide as an intermediate.[35]

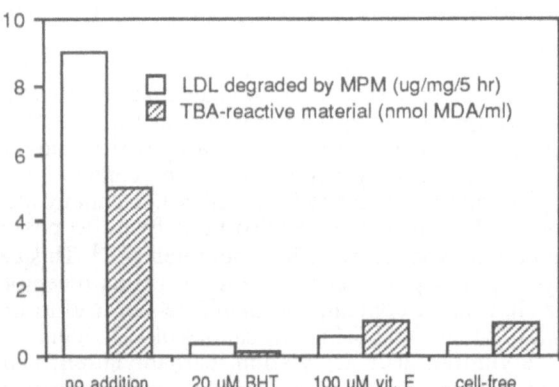

Figure 1. Inhibition of endothelial cell modification of LDL by antioxidants. [125]I LDL (100 ug/ml) was incubated with cultured rabbit aortic endothelial cells and in "cell-free" control dishes for 24 hr in serum-free Ham's F-10 medium with or without antioxidants as indicated. After this incubation, the rate of degradation of the LDL by cultured mouse peritoneal macrophages was determined as trichloroacetic acid-soluble noniodide radioactivity. Extent of oxidation was estimated as thiobarbituric acid-reactive substances, using malondialdehyde (from tetramethoxypropane) as the standard.

Henriksen and colleagues demonstrated that endothelial cell-modified LDL was rapidly internalized and degraded via the scavenger receptor of cultured macrophages. We subsequently showed that antioxidants prevented the modification of LDL by endothelial cells, including the enhanced degradation by macrophages (Figure 1). Chelators of transition metal ions also inhibited modification (Figure 2).

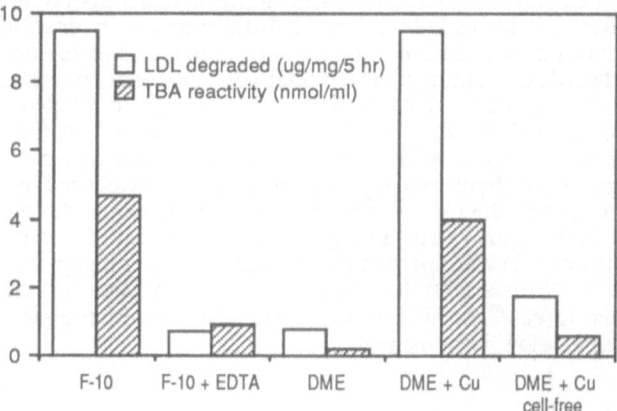

Figure 2. Dependence of endothelial cell modification of LDL on transition metals. ^{125}I LDL (100 ug/ml) was incubated with cultured rabbit aortic endothelial cells and in "cell-free" control dishes for 24 hr. Media were then assayed for rate of LDL degradation by mouse peritoneal macrophages and for TBA reactivity as described in the legend to figure 1. Incubations with endothelial cells were done in Ham's F-10 medium, in Ham's F-10 with 50 uM EDTA, in Dulbecco's modified Eagle's (DME) medium, and in DME supplemented with 5 uM $CuSO_4$.

Furthermore, all of the changes associated with the modification of LDL resulting from incubation with cultured endothelial cells or smooth muscle cells could be reproduced by Cu(II)-dependent oxidation of LDL in the absence of cells[23], and these are summarized in Table I.

Table I. Characteristic changes accompanying LDL oxidation.

- increased electrophoretic mobility
- increased density
- fragmentation of apoB
- fluorescence at excitation 360 nm, emission 430 nm
- decreased content of reactive amino groups
- loss of esterified cholesterol
- loss of polyunsaturated fatty acids
- hydrolysis of phosphatidylcholine to lysophosphatidylcholine
- greater heterogeneity in size, with tendency to form aggregates
- accelerated uptake and degradation by macrophages

A decrease in the number of reactive amino groups was noted during LDL oxidation, and this decrease in amino group reactivity was accompanied by the derivatization of lysine amino groups (Figure 3). Amino acid analysis showed that this decrease in amino group reactivity was due to a decrease in lysine residues, apparently due to covalent binding of lipid peroxidation products to the epsilon amino groups.[36] To determine if the derivatization of lysine groups was a necessary and sufficient feature to permit recognition by the scavenger receptor, we performed an experiment in which the lysine groups were modified by lipid peroxidation products under conditions where the LDL particle itself was not oxidized, and therefore none of the other changes that accompany LDL oxidation were present. This was accomplished by isolating reactive products from autoxidized polyunsaturated fatty acids, and then incubating these products with intact LDL under conditions where the LDL itself could not undergo oxidation. The reactive products derived from polyunsaturated fatty acids were indeed capable of modifying LDL, and a significant increase in electrophoretic mobility occurred, comparable to that seen with Cu(II)-induced oxidation of LDL. The LDL modified by oxidation products showed a dramatic increase in degradation by macrophages, and this degradation was inhibited by unlabeled acetyl LDL as well as polyinosinic acid, confirming that the uptake was largely due to the scavenger receptor pathway. It was also possible to modify bovine serum albumin with the oxidation products, thus proving that the recognition of the modified proteins by the scavenger receptor was not dependent on structural domains intrinsic to apolipoprotein B. Analysis of the reactive products by derivatization with 2,4-dinitrophenylhydrazine and thin layer chromatography revealed several bands which migrated in the same region as derivatives of medium-chain aldehydes, including crotonaldehyde, pentenal, heptenal, and nonenal. However, when LDL was derivatized with these 2-unsaturated aldehydes, there was no increase in the uptake or degradation of the LDL via the scavenger receptor of cultured macrophages even though lysine residues were extensively modified. Some aldehydes, for example nonenal, produced significant aggregation of LDL as evidenced by the development of turbidity and retention of the LDL at the origin of the agarose gel during electrophoresis. These aggregated LDLs were taken up significantly more rapidly than native LDL by macrophages, but the uptake was not mediated by the scavenger receptor pathway, and was probably due to phagocytosis. Similar results have recently been reported for LDL after aggregation by shear forces[30], or by LDL incubated with 4-hydroxynonenal, which results in a very similar aggregation phenomenon.

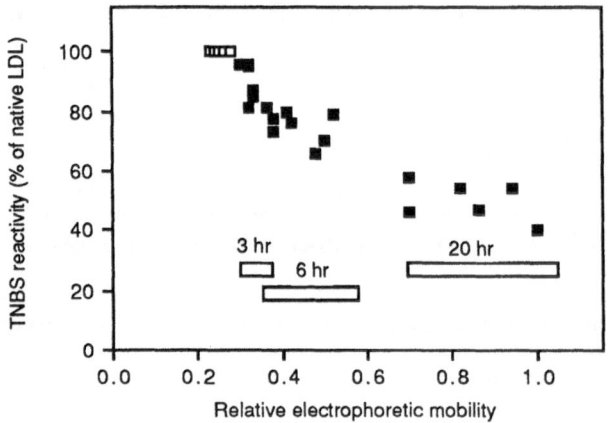

Figure 3. Decrease in reactive amino groups during LDL oxidation. LDL (100-200 ug/ml) was incubated at 37°C in PBS containing 5uM $CuSO_4$ for 3, 6, or 20 hours. Aliquots were then taken for agarose gel electrophoresis and estimation of reactive amino groups with trinitrobenzenesulfonic acid (TNBS). Electrophoretic mobility is expressed relative to that of bovine albumin (2.5 cm in this system). Native LDL (□), oxidized LDL (■).

MECHANISMS BY WHICH CULTURED CELLS MODIFY LDL

Recent studies have elucidated the mechanism by which cultured cells promote LDL oxidation. Several different cell lines have been tested for their ability to modify LDL, and it has been found that cultured endothelial cells, as well as smooth muscle cells from several species, have potent activity in modifying LDL.[20-26,38] Stimulated macrophages are also relatively active in modifying LDL.[38] Human skin fibroblasts are intermediate, and a line of bovine aortic endothelial cells has been found not to modify LDL.[25] We found that the ability of a given cell type to modify LDL was directly correlated with the rate of secretion of superoxide by these cells (figure 4). Removal of O_2 inhibited modification, as did addition of superoxide dismutase (figure 5). The apparent inhibition by catalase was probably due to a heat-stable antioxidant contaminant and not to catalase itself. Heinecke and colleagues have shown that the secretion of superoxide by cultured monkey smooth muscle cells is dependent on the presence of cystine in the medium.[24] From their studies it seems likely that superoxide anion is not secreted directly by the cells, but that reducing equivalents are generated in the form of cysteine, which is the released into the medium. Through catalysis by redox-active metal ions such as copper, the cysteine leads to the reduction of oxygen to superoxide.

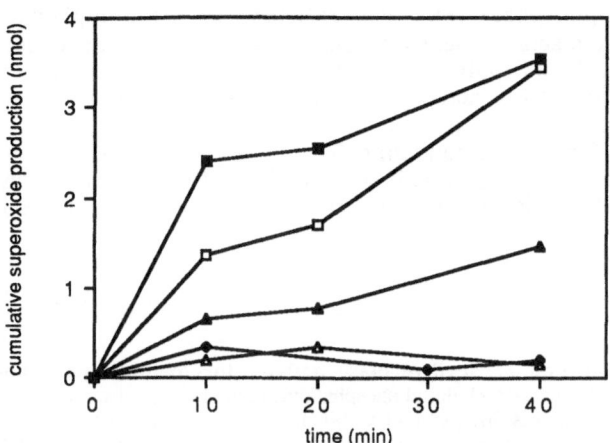

Figure 4. Correlation of superoxide production and LDL modification by cultured cells. Confluent 60mm diameter dishes of rabbit aortic smooth muscle cells (■), rabbit aortic endothelial cells (□), human skin fibroblasts (▲), bovine aortic endothelial cells (△), or cell-free control dishes (◆) were incubated in 2 ml phenol red-free Ham's F-10 medium containing 20 uM ferricytochrome c. Superoxide accumulation in the medium was calculated from the superoxide dismutase-inhibitable component of cytochrome c reduction. Parallel dishes were incubated for 24 hr with 200 ug/ml LDL (without cytochrome c) to assess ability of the cells to modify LDL. The electrophoretic mobility of LDL after incubation with rabbit smooth muscle cells was 2.1 cm, with rabbit endothelial cells 2.0 cm, with fibroblasts 1.4 cm, with bovine endothelial cells 0.7 cm, and with cell-free control dishes, 0.7 cm.

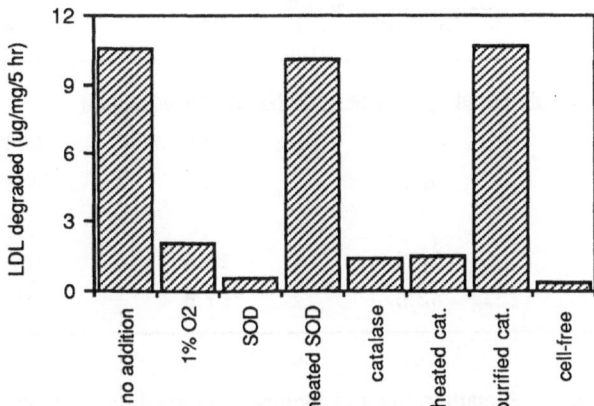

Figure 5. Effect of oxygen concentration, superoxide dismutase, and catalase on LDL modification by endothelial cells. Rabbit aortic endothelial cells were incubated for 20 hr with ^{125}I LDL (200 ug/ml) in serum-free Ham's F-10 medium with the indicated additions. Oxygen concentration was 19% except where otherwise indicated. Superoxide dismutase (SOD) was added at 100 ug/ml, and catalase at 200 ug/ml. Heat-inactivation was shown to destroy more than 90% of enzyme activity.

197

ENZYMATIC HYDROLYSIS OF PHOSPHATIDYLCHOLINE DURING LDL OXIDATION

As shown in Table I, degradation of phosphatidylcholine to lysophosphatidylcholine occurs during oxidative modification of LDL. We recently showed that this phospholipid hydrolysis was brought about by an LDL–associated phospholipase A_2 that acted only on oxidized LDL phospholipids.[22,39] The enzyme was calcium-independent, heat-stable, and was inhibited by the serine esterase inhibitors phenylmethylsulfonylfluoride and diisopropylfluorophosphate, but was resistant to p-bromophenacylbromide and dithiobisnitrobenzoic acid.[39] Nearly all the activity in EDTA-anticoagulated plasma was physically associated with apoB-containing lipoproteins. However, apoB was not essential for enzyme activity, as full activity was present in plasma from abetalipoproteinemic patients.[39] The properties of this enzyme were very similar to those which have been recently reported for platelet activating factor acetylhydrolase.[40,41] We found that the activity against oxidized phospholipids copurified with PAF acetylhydrolase (Table II), and that oxidized phospholipids and PAF acted as competitors for hydrolysis by this enzyme (Figure 6).

TABLE II. Copurification of acylhydrolase activity for oxidized phospholipids with PAF acylhydrolase. Autoxidized 2-arachidonyl phosphatidylcholine (APC) substrate was prepared as follows: 10 mg phosphatidylcholine was dried in a 13 x 100 mm screw-top borosilicate test tube, to which 500 ul benzene were added and then vortexed for 1 min. to yield reverse-phase micelles. The tube was flushed with O_2, sealed, and heated to 100°C for 60 min, resulting in a brown solution. The benzene was removed under reduced pressure, and then the oxidized phospholipid was dispersed in PBS and briefly sonicated to yield an optically clear solution. Assays were done with 250 nmol of oxidized APC and 1 hr incubation at 37°C. The hydrolysis product lysophosphatidylcholine was assayed by thin layer chromatography and phosphorus analysis. 1-0-alkyl 2-[acetyl-^{3}H]-sn-glycero-3-phosphocholine was synthesized by reacting L-2-lyso PAF with [^{3}H] acetic anhydride. The assay mixture contained 50 uM [^{3}H-acetyl] PAF (2-5 Ci/mole), and plasma or lipoprotein fractions in a final volume of 0.4 ml Dulbecco's PBS. After incubation for 10 minutes at 37°C, the lipids were extracted by the method of Folch et al., and the amount of ^{3}H recovered in the aqueous phase was determined by liquid scintillation counting. The assays were linear for up to 15 minutes using 10 ul of human plasma as the enzyme source. Assays were done with 5-10 ul plasma, 20 ug LDL, and 3-5 ug partially purified enzyme. Purification steps included sequential ultracentrifugation (d=1.019-1.063), detergent solubilization and chromatography over Sephacryl S-300, followed by ion-exchange chromatography on DEAE Sepharose 4B. Values shown are means of duplicates that varied less than ± 10%. The percentage of total enzyme activity recovered is indicated in parentheses.

	Substrate		
	Oxidized APC	PAF	recovery
	phosphatidylcholine hydrolyzed (nmol/mg/min)		
plasma	.12	.39	100%
LDL	6.5	12.3	39%
S-300 peak	36.3	24.3	37%
DEAE peak	43.6	57.8	11%

A recent report by Quinn, Parsatharathy and Steinberg showed that lysophosphatidylcholine was a potent chemoattractant for macrophages.[42] The significance of this observation is that during LDL oxidation, oxidized phospholipids would be rapidly hydrolyzed by platelet activating factor acetylhydrolase yielding lysophosphatidylcholine, which could then lead to the recruitment of macrophages. As noted above, macrophages can themselves modify LDL, and hence this might lead to amplification of the peroxidation process.

Enzymatic assays of cholesterol and cholesterol ester suggest a substantial decrease in the content of these sterols, particularly of cholesterol ester. However, when the fate of cholesterol was traced using ^{3}H cholesterol linoleate, it was found that virtually all of the radioactive sterol nucleus remained associated with the LDL. We hypothesized that this discrepancy was due to the presence of oxidized sterols in LDL that were not detected by the enzymatic assay. Analysis of sterols in oxidized LDL by thin layer chromatography revealed numerous spots a few of which had Rf values and color (after spraying with sulfuric acid) corresponding to oxidized sterols, including 25-hydroxycholesterol, and 7-ketocholesterol. There were also several unidentified oxysterols. Because a number of oxysterols, including 25-hydroxycholesterol, have been shown to be potent stimulators of ACAT, it is possible that the oxysterols in oxidized LDL might increase the actitvity of ACAT, and hence promote cholesterol esterification, analagous to the relative overactivity of ACAT reported with J774 macrophages. To test this, we measured cholesterol synthesis rates, and ACAT activity in cultured fibroblasts and in cultured macrophages, and found that somewhat surprisingly, both ACAT activity and cholesterol synthesis were suppressed by oxidized LDL. These findings could not be explained on the basis of cytotoxicity, as cell morphology, and total proteins synthesis as assessed by [^{14}C]leucine incorporation were not affected by oxidized LDL. This suggests that ACAT and HMG CoA reductase are under separate regulation, and that of the oxysterols present in oxidized LDL, at least one is capable of suppressing ACAT activity. Although the inhibitory oxysterol in oxidized has not yet been identified, previous studies by Bates and colleagues have shown that 22-hydroxycholesterol is capable of suppressing ACAT.[43] As well, several other closely related compounds have been developed as potential therapeutic agents to inhibit ACAT <u>in</u>

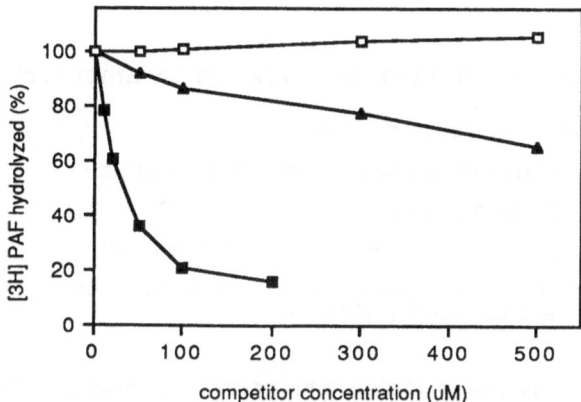

Figure 6. Substrate competition for acylhydrolase. 20 nmol of [^{3}H]PAF was added to assay tubes containing 20 ug LDL and varying amounts of unlabelled PAF (■), oxidized 2-arachidonyl phosphatidylcholine (▲), or dimyristoylphosphatidylcholine (□), and after 20 min the amount of PAF hydrolyzed was determined. The value in the absence of competitor was 0.32 nmol/min. The lower apparent potency of oxidized LDL as a competitor compared to PAF probably reflects the fact that the oxidized APC preparation was highly heterogeneous, and therefore only a fraction of the PC molecules were actually capable of interacting with the enzyme. The amount of competition observed is close to what one would predict from the hydrolysis rates of the two substrates shown in table II, in that the substrate concentration for APC was 5 times that of PAF for comparable hydrolysis rates.

LDL OXIDATION <u>IN VIVO</u>

The conditions under which LDL oxidation was achieved in the experiments described above may not be comparable to those which are present <u>in vivo</u>, in that a variety of potent antioxidant defences are normally present in the bloodstream and in other tissues. These defences include lipophilic antioxidants such as vitamin E and vitamin A, as well as enzymatic defences such as glutathione peroxidase, catalase, superoxide dismutase, and ceruloplasmin. Because of these antioxidant defences, it is unlikely that significant LDL oxidation would occur in plasma. In fact, the addition of as little of 1 mg/ml of lipoprotein-deficient plasma can completely inhibit LDL oxidation in vitro.[25] Furthermore, even if oxidation of LDL were to occur in the plasma compartment, studies on the turnover and tissue sites of catabolism of oxidized LDL in guinea pigs showed that most of the intravenously injected oxidized LDL was cleared by the liver, probably by sinusoidal endothelial cells.[23] On the other hand, oxidation might be selectively favored in regions such as the aortic intima,where the concentration of antioxidant defences would be expected to be lower than in whole blood, and where there is intimate contact with a large surface area of endothelial cells and smooth muscle cells that have been shown in vitro to be capable of promoting LDL oxidation. There are several lines of evidence that suggest that lipid peroxidation occurs in the region of atherosclerotic plaques, as summarized in Table III. However, these studies do not prove that the oxidation is of pathogenetic importance in the formation of the atherosclerotic plaque. Studies of the properties of LDL extracted from aortic intima are of considerable interest in this regard. Reports by Clevidence et al. have suggested that aortic LDL has increased electrophoretic mobility, and fragmentation of apoB compatible with oxidative change.[27] In addition, it has been found that aortic LDL undergoes accelerated degradation by cultured macrophages.[27-29] Some of these studies have been performed on homogenized aortic tissue, and therefore are rather difficult to interpret in view of the findings of Khoo and coworkers showing that even vortexing of LDL can cause aggregation, and lead to accelerated uptake via phagocytic mechanisms.[30] Recent studies obtained in our laboratory and in a study reported by Yla-Herttuala et al. with aortic LDL extracted without homogenization, reveal evidence of a mild and somewhat variable degree of oxidation, probably insufficient to lead to scavenger receptor uptake.[28] However, the aortic LDL did cause a significant stimulation of oleate incorporation into cholesterol esters in cultured macrophages. It has not yet been determined if this is due to the uptake of lipid or lipoprotein aggregates, or to some other mechanism.

TABLE III. Evidence for lipid peroxidation in atherosclerotic lesions.

- presence of ceroid pigments
- increased electrophoretic mobility of aortic LDL
- partial fragmentation of apoB in aortic LDL
- decreased content of linoleic acid in aortic LDL
- immunohistochemical detection of malondialdehyde adducts in WHHL rabbit aorta

Further evidence for LDL oxidation in the artery wall was obtained by Haberland and colleagues in immunohistochemical studies of aorta from WHHL animals.[44] Monoclonal antibodies against malondialdehyde-modified LDL were used to identify immunoreactive material, mainly in the aortic intima, that colocalized almost exactly with apoB immunoreactivity.

The most direct evidence that oxidation may play a pathogenetic role in atherosclerosis has been obtained in two recent studies in Watanabe heritable hyperlipidemic (WHHL) rabbits.[45,46] In these experiments, the rate of development of atherosclerosis in control WHHL rabbits was compared with that of WHHL rabbits treated with probucol, which is a potent lipophilic antioxidant that is transported in the LDL core, and hence is highly effective in preventing LDL oxidation in vitro.[47] The lesion area in aortas of treated animals was significantly less than that in control animals, and this could not be explained by a reduction in LDL cholesterol. Furthermore, it was shown that intimal macrophages in the treated animals degraded less LDL than intimal macrophages in the controls.[46] Although there may be explanations unrelated to inhibition of oxidation to account for the observed effect of probucol in these experiments, for example an effect on HDL metabolism or on macrophages[48], at present this experimental approach has provided the most convincing evidence that LDL oxidation may play a pathogenetic role in atherogenesis.

SUMMARY AND CONCLUSIONS

The results of studies described above have provided a good understanding of the physical and biochemical changes that accompany LDL oxidation, and of the mechanism by which cultured arterial wall cells can promote LDL oxidation. Oxidized LDL has been shown to have a number of properties that might contribute to atherogenesis, including accelerated uptake by macrophages by both scavenger receptor-dependent and non-receptor processes, cytotoxicity to endothelial cells, chemotactic activity toward monocytes, and a tendency to self-aggregate. Analyses of LDL in the aortic intima have revealed changes that may be consistent with oxidative change, and recent experiments in WHHL rabbits demonstrated that the antioxidant drug probucol caused a reduction in atherosclerosis. However, much additional work in required to determine what role lipoprotein oxidation plays in the generation of atherosclerosis in humans. Because there is a clear potential for therapeutic intervention, such studies will likely receive a high priority.

ACKNOWLEDGEMENTS

The author is indebted to the following individuals for valuable discussions and advice: Dr. Joseph L Witztum, Dr. Daniel Steinberg, Dr. Alan Chait, Dr. Dennis Vance, Dr. Sam Parthasarathy, Dr. Peter Kwan, and Dr. Jay Heinecke. Studies from the author's laboratory were done with the expert technical assistance of Marilee Lougheed, Harkamal Basra, and Diane Thorpe, and were supported by the Medical Research Council of Canada, The Canadian Heart Foundation, and the B.C. Health Care Research Foundation.

REFERENCES

1. Ross, R. (1981) Arteriosclerosis 1, 293-311.
2. Steinberg, D. (1983) Arteriosclerosis 3, 283-301.
3. Fowler, S., Shio, H., and Haley, W.J. (1979) Lab. Invest. 4, 372-378.
4. Schaffner, T., Taylor, K., Bantucci, E.J., Fischer-Dzoga, K., Beenson, J.H., Glagov, S., and Wissler, R. (1980) Am. J. Pathol. 100, 57-80.
5. Gerrity, R.G. (1981) Am. J. Pathol. 103, 181-190.
6. Gerrity, R.G. (1981) Am. J. Pathol. 103, 191-200.
7. Faggiotto, A., Ross, R., and Harker, L. (1984) Arteriosclerosis 4, 323-340.
8. Rosenfeld, M.E., Tsukuda, T., Gown, A.M., and Ross, R. (1987) Arteriosclerosis 7, 9-23.
9. Goldstein, J.L., Ho, Y.K., Basu, S.K., and Brown, M.S. (1979) Proc. Natl. Acad. Sci. USA 76, 333-337.
10. Tabas, I., Boykow, G.C., and Tall, A.R. (1987) J. Clin. Invest. 79, 418-426.
11. Koo, C., Wernette-Hammond, M. E., and Innerarity, T.L. (1986) J. Biol. Chem. 261, 11194-11201.
12. Koo, C., Wernette-Hammond, M. E., Garcia, Z., Malloy, M., Uauy, R., East, C., Bilheimer, D.W., Mahley, R.W., and Innerarity, T.L. (1988) J. Clin. Invest. 81, 1332-1340.
13. Ellsworth, J.L., Kraemer, F.B., and Cooper, A.D. (1987) J. Biol. Chem. 262, 2316-2325.
14. Brown, M.S.,Basu, S.K., Falck, Y.K., Ho, Y.K., and Goldstein, J.L. (1980) J. Supramol. Struct. 13, 67-81.
15. Brown, M.S., and Goldstein, J.L. (1983) Annu. Rev. Biochem. 52, 223-261.
16. Mahley, R.W., Innerarity, T.L., Weisgraber, K.H., and Oh, S.Y. (1979) J. Clin. Invest. 64, 743-750.
17. Gonen, B., Cole, T., and Hahm, K-S. (1983) Biochim. Biophys. Acta 754, 201-207.
18. Fogelman, A.M., Schecter, I.S., Seager, J., Hokom, M., Child, J.S., and Edwards, P.A. (1980) Proc. Natl. Acad. Sci. USA 77, 2214-2218.
19. Haberland, M.E., Fogelman, A.M., and Edwards, P. A. (1982) Proc. Natl. Acad. Sci. USA 79, 1712-1716.
20. Henriksen, T., Mahoney, E.M., and Steinberg, D. (1981) Proc. Natl. Acad. Sci. USA 78, 6499-6503.
21. Henriksen, T., Mahoney, E.M., and Steinberg, D. (1983) Arteriosclerosis 3, 149-159.

22. Steinbrecher, U.P., Parthasarathy, S., Leake, D.S., Witztum, J.L., and Steinberg, D. (1984) Proc. Natl. Acad. Sci. USA 81, 3883-3887.
23. Steinbrecher, U.P., Parthasarathy, S., Witztum, J.L., and Steinberg, D. (1987) Arteriosclerosis 7, 135-143.
24. Heinecke, J.W., Baker, L., Rosen, H. and Chait, A. (1986) J. Clin. Invest. 77, 757-761.
25. Steinbrecher, U.P. (1988) Biochim. Biophys. Acta 959, 20-30.
26. Hiramatsu, K., Rosen, H., Heinecke, J.W., Wolfbauer, G., and Chait, A. (1987) Arteriosclerosis 7, 55-60.
27. Clevidence, B.A., Morton, R.E., West, G., Dusek, D.M., and Hoff, H.F. (1984) Arteriosclerosis 4, 196-207.
28. Yla-Herttuala, S., Jaakkola, O., Ehnholm, C., Tikkanen, M., Solakivi, T., Sarkioja, T., and Nikkari, T. (1988) J. Lipid Res. 29, 563-572.
29. Morton, R.E., West, G.A., and Hoff, H.F. (1986) J. Lipid Res. 27, 1124-1134.
30. Khoo, J.C., Miller, E., McLoughlin, P., and Steinberg, D. (1988) Arteriosclerosis 8, 348-358.
31 Kruth, H.S. (1985) Science 227, 1243-1245.
32. Curtiss, L.K., Black, A.S., Takagi, Y., and Plow, E.F. (1987) J. Clin. Invest. 80, 367-373.
33. Wong, H., and Hashimoto, S. (1987) Arteriosclerosis 7, 185-190.
34. Morel, D.W., Hessler, J.R., and Chisolm, G.M. (1983) J. Lipid Res. 24, 1070-1076.
35. Morel, D.W., DiCorleto, P.E., and Chisolm, G.M. (1984) Arteriosclerosis 4, 357-364.
36. Steinbrecher, U.P. (1987) J. Biol. Chem. 262, 3603-3608.
37. Hoff, H.F., Morel, D.W., Juergens, G., Esterbauer, H., and Chisolm, G.M. (1987) Arteriosclerosis 7, 523a.
38. Parthasarathy, S., Printz, D.J., Boyd, D., Joy, L., and Steinberg, D. (1986) Arteriosclerosis 6, 505-510.
39. Steinbrecher, U.P., and Pritchard, P.H. (1989) J. Lipid Res. (in press).
40. Stafforini, D.M., McIntyre, T.M., Carter, M.E., and Prescott, S.M. (1987) J. Biol. Chem. 262, 4215-4222.
41. Stafforini, D.M., Prescott, S.M., McIntyre, T.M. (1987) J. Biol. Chem. 262, 4223-4230.
42. Quinn, M.T., Parthasarathy, S., and Steinberg, D. (1988) Proc Natl Acad Sci USA 85, 2995-2998.
43. Bates, S.R., Jett, C.M., and Miller, J.E. (1983) Biochim. Biophys. Acta 753, 281-293.
44. Haberland, M.E., Fong, D., and Cheng, L. (1988) Science 241, 215-218.
45. Kita, T., Nagano, Y., Yokode, M., Ishii, K., Kume, N., Ooshimi, A., Yoshida, H. and Kawai, C. (1987) Proc. Natl. Acad. Sci. USA 84, 5928-5931.
46. Carew, T.E., Schwenke, D.C. and Steinberg, D. (1987) Proc. Natl. Acad. Sci. USA 84, 7725-7729.
47. Parthasarathy, S., Young, S.G., Witztum, J.L., Pittman, R.C., and Steinberg, D. (1986) J. Clin. Invest. 77, 641-644.
48. Yamamoto, A., Takaichi, S., Hara, H., Nishikawa, O., Yokoyama, S., Yamamura, T., Yamaguchi, T. (1986) Atherosclerosis 62, 209-217.

IN VITRO OXIDATION OF LOW DENSITY LIPOPROTEINS

Martina Rotheneder, Georg Striegl and Hermann Esterbauer

Institute of Biochemistry, University of Graz
Schubertstrasse 1, A-8010 Graz
Austria

INTRODUCTION

Oxidized low density lipoproteins (o-LDL) are considered to be causally involved in the formation of early atherosclerotic leasons[1,2,3]. Lipid peroxidation is an extremely complex process and the consequences for LDL are manifold (for review see [4]).

It has been shown that copper(II)-chloride is a very potent pro-oxidant for LDL[4,5]. If trace amounts of $CuCl_2$ are added to an LDL solution the LDL becomes heavily oxidized within a few hours. Figure 1 shows in a simplified scheme the main events which probably take place during this in vitro oxidation process.

The left part of the scheme shows the initiation process, the right part the chain breaking effect of antioxidants, the central part indicates the propagation phase or chain reaction of the lipid peroxidation process and the lower part shows secondary and tertiary reactions evolving from the lipid hydroperoxides.

The initiation process is not well investigated. We think that the initiation depends on preexisting lipid hydroperoxides (LOOH) present in the LDL, we found in fact in all LDL preparations a trace amount of peroxides equivalent to about 10 LOOH molecules per LDL particle. Copper (II) ions probably decompose these preexisting LOOH to lipid peroxy-radicals ($LOOH + Cu^{2+} \rightarrow LOO^{\cdot} + Cu^{+} + H^{+}$). The Cu^{+}-ion formed in this reaction could then react in a Fenton type reaction with other preexisting LOOH to give the highly reactive lipid alkoxy radical (LOOH + $Cu^+ \rightarrow LO^. + Cu^{2+} + OH^-$). Both copper driven initial reactions leading to the decomposition of the preexisting hydroperoxides can formally be combined and written as gross reaction : $2\ LOOH \rightarrow LO^. + LOO^. + H_2O$. The LO$^.$ radical has a similar reactivity as the hydroxyl radical and it is likely to assume that it is the lipid alkoxy radical LO$^.$ which attacks neighbouring polyunsaturated fatty acids by abstracting a hydrogen atom and generating the initial lipid radical L$^.$ (LH + LO$^. \rightarrow$ L$^.$+ LOH). It is known from other studies that the lipid radical L$^.$ reacts very rapidly diffusion controlled with molecular oxygen to give LOO$^.$ radical. If α-tocopherol is present in LDL, the LOO$^.$ radical is immediately scavenged and converted to a lipid hydroperoxide. This prevents that the lipid peroxidation process can enter the self-sustaining chain reaction shown in the central part of the scheme in Fig.1. One vitamin E molecule can scavenge two LOO$^.$ radicals and becomes thereby oxidized to an inactive form. Thus the LOO$^.$ radical scavenging is for the vitamin E a "suicide" and as a consequence the protection of LDL against oxidation progressively decreases as more and more LOO$^.$ radicals are formed by the initiation reactions. Since 10 vitamin E molecules are contained in a LDL molecule the generation of about 20 LOO$^.$ radicals is sufficient to deplete LDL from vitamin E.

We have shown [5] that LDL contains in addition to vitamin E a last barrier against oxidation i.e. carotenoids and retenoids. It is not well known how these compounds act, but it was clearly shown that they act as antioxidants at a phase when the LDL has lost already its vitamin E. Finally also this last protective barrier breaks down and the lag phase is over and each additional lipid radical L$^.$ formed would lead to a self sustaining chain reaction in which the PUFAs are now rapidly converted to lipid hydroperoxides as shown in the central part of the scheme in Fig.1.

We have in fact found in kinetics studies that a rapid increase of lipid hydroperoxides in LDL only occurs when the LDL has lost all antioxidants [5].

Lipide hydroperoxides are labile compounds and can subsequently undergo a number of secondary and tertiary reactions. In the presence of copper ions most important probably is the formation of LOO$^.$ and LO$^.$ radicals, both of them can lead to a chain branching i.e. they can initiate new cycles of the central chain reaction. Of particular importance in respect to the oxidative modification of apo-B probably are the lipid alkoxyradicals LO . Firstly these radicals could directly attack the protein and lead to cleavage of peptide bonds; a heavy fragmentation of the apo-B to smaller peptides has in fact been found to occur during LDL oxidation [2]. Secondly the LO˙ radical can undergo a β-cleavage reaction which leads to the formation of aldehydes within the LDL particle. Some of these aldehydes such as malonaldehyde, 4-hydroxy-nonenal or 2,4-alkadienals are highly reactive and modify LDL in several ways as shown in the lower part of the scheme in Fig.1.

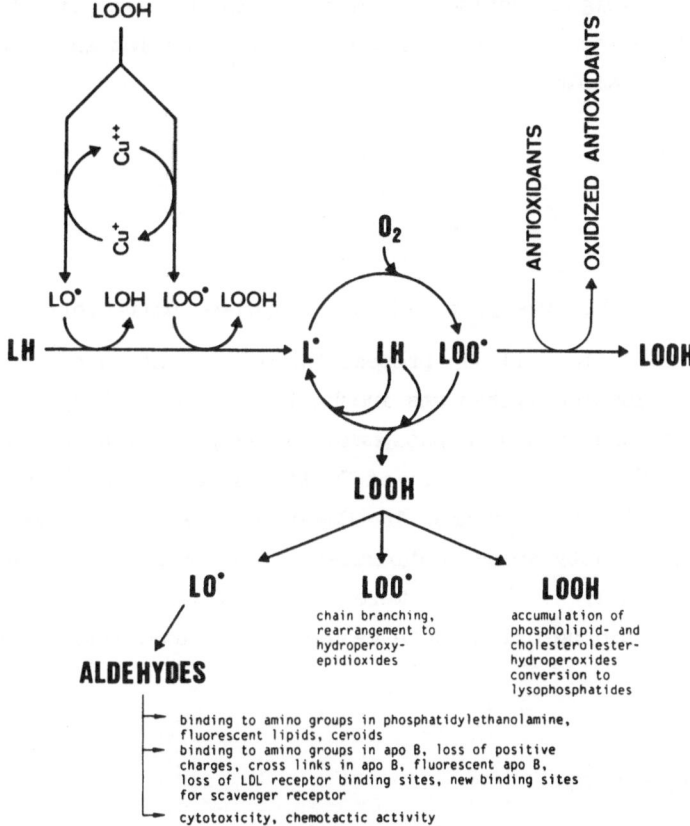

Fig. 1. Oxidation of LDL (description in the text). LH = polyunsaturated fatty acids, LOOH = lipid hydroperoxides, LO˙ = alkoxy-radicals, LOO˙ = peroxy-radicals, LOH = hydroxylated fatty acid

Reports from several laboratories suggest but of course do not prove that aldehydic lipid peroxidation products are causally involved in the alteration of the functional and structural properties of LDL i.e. cytotoxicity, chemotactic activity, loss of positive surface charges in the apo-B region recognized by the LDL-receptor, formation of new binding sites recognized by the scavenger-receptors of macrophages and others.

The scheme of Fig.1 also shows which parameters could be used to measure the rate and extent of LDL oxidation in vitro. These parameters are : Decrease of fatty acids, consumption of oxygen, loss of antioxidants, accumulation of lipid hydroperoxides or dienes and formation of secundary and tertiary products of lipid hydroperoxides such as MDA and other aldehydes like 4-hydroxy-nonenal. Moreover it is known that products of lipid peroxidation interact with the apolipoprotein-B (apo-B) and therefore changes occurring in apo-B can be used to follow oxidation of LDL. Such changes are : Increase of the fluorescence, increase of carbonyl-functions of the apo-B and its fragmentation into lower molecular weight peptides. We have studied most of these above mentioned alterations occurring in LDL during oxidation and describe here a few selected examples.

METHODS

Human plasma LDL was prepared by stepwise ultracentrifugation as described [4]. EDTA (1mg/ml) was present through all steps of preparation. Our in vitro oxidation system was incubation of LDL (1.5 mg/ml or lower) in an oxygen saturated 0.01 M phosphate buffer pH 7.4, 0.16 M NaCl with a freshly prepared aqueous solution of $CuCl_2$. The ratio LDL:Cu was in all experiments 1:16, that means 16 Cu-atoms per LDL molecule. The determination of fatty acids, antioxidants and aldehydes was done as described previously [4]. Dienes were determined from the 234 nm UV absorption [5]. Lipid peroxides were measured by a iodometric assay [6].

RESULTS AND DISCUSSION

Table 1 gives the fatty acid composition of LDL prior and after oxidation. Native LDL contains a rather high amount of polyunsaturated fatty acids (PUFA). The mean value is 422 nmol PUFA per mg LDL which corresponds to about 1000 PUFA molecules per LDL particle. The PUFA are mainly linoleic acid (18:2) and arachidonic acid (20:4). LDL incubated for three hours with 10 μM copper has a significantly different fatty acid composition, the number of PUFA-molecules decreases from 1000 to 260 per LDL-particle.

Table 1. Fatty acids in native and oxidized LDL

	nanomole/mg LDL	
	native mean +/- s.dev.	oxidized single value
Fatty acids (n=8)		
14:0	28 +/- 11	14
16:0	278 +/- 36	267
16:1	20 +/- 7	20
18:0	49 +/- 10	54
18:1	161 +/- 36	185
18:2	384 +/- 79	104
20:4	36 +/- 17	2
Total FA	930 +/-152	646
Total PUFA	422 +/- 95	106

Table 2. Antioxidants and oxidation products in native and oxidized LDL

	nanomole/mg LDL	
	native mean +/- s.dev.	oxidized single value
Antioxidants (n=16)		
Alpha-tocopherol	2.61 +/- 0.62	0
Gamma-tocopherol	0.20 +/- 0.07	0
Beta-carotene	0,13 +/- 0.07	0
Lycopine	0.07 +/- 0.04	0
Retinylstearate	0.77 +/- 0.25	0
Antioxidants/PUFAs	1:110	0
Peroxides (n=6)	4.10 +/- 1.60	110
Dienes (n=6)	traces	134
rel.Fluorescence Ex.360/Em.430 nm	50	250

Table 2 shows the lipophilic antioxidants and various oxidation products in native and oxidized LDL. The antioxidants are alpha-tocopherol, gamma-tocopherol, beta-carotene, lycopine and retinylstearate. On the concentration base the most important are alpha-tocopherol and retinylstearate. Oxidized LDL is completly depleated from its lipophilic antioxidants and does not contain even traces of them. All fresh LDL-preparations which we have investigated so far contained a low but clearly detectable level of about 4 nanomol of lipid hydroperoxides per mg LDL. Important is that oxidized LDL (o-LDL) contains huge amounts of hydroperoxides. We determined the peroxide-content with a simple method previously published by El-Saadani et al [6]. Another physico-chemical parameter that is significantly changed during oxidation is the fluorescence in the visible range which increases from 50 to 250 units during oxidation.

An excellent correlation (r = 0.99) was found between the amount of conjugated dienes and the amount of MDA or peroxides. The amount of conjugated dienes was determined by a method previously published by our laboratory[5], where the change of the absorbance of an LDL solution during oxidation was followed at 234 nm and the diene content was calculated (ε = 29500). MDA and peroxides were measured as described [6,7].

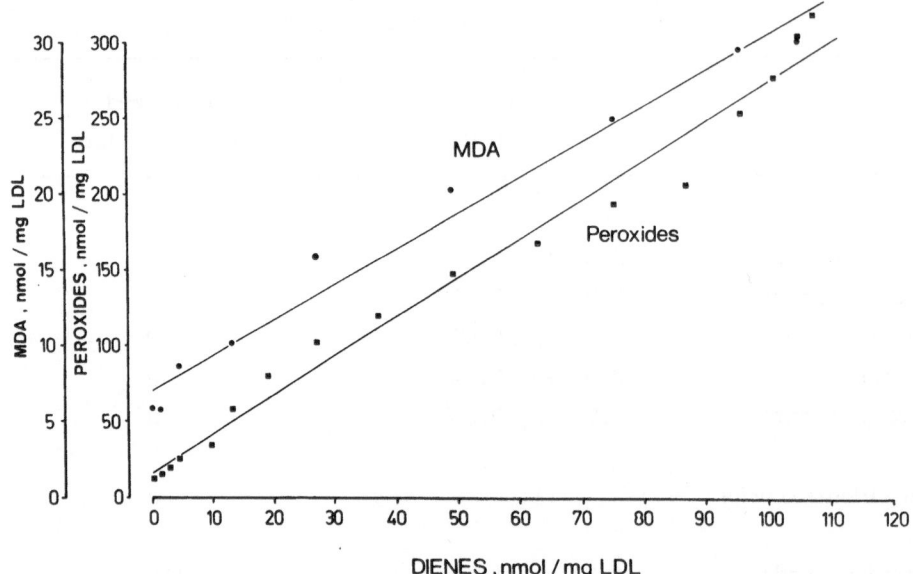

Fig. 2. Correlation between the amount of conjugated dienes, lipid peroxides and MDA in o-LDL. LDL (0.25 mg/ml) in 0.01M phosphate buffer pH 7.4, 0.16M NaCl was oxidized by addition of 1.66 μM CuCl₂. The amount of dienes was calculated from the increase of the absorbance at 234 nm (ε=29500), MDA was determined by the TBA assay[7] and peroxides by the iodometric method as described in [8].

Table 3. Aldehydes in native and oxidized LDL

| | nanomole/mg LDL | |
	native mean +/- s.dev.	oxidized single value
Propanal	0.24 +/- 0.14	1.03
Butanal	0.33 +/- 0.17	0.65
Pentanal	0.09 +/- 0.08	0.89
Hexanal	0.00	9.97
2.4-Heptadienal	0.00	0.88
4-Hydroxyhexenal	0.00	1.53
4-Hydroxyoctenal	0.00	1.42
4-Hydroxynonenal	0.14 +/- 0.17	4.97
Malonaldehyde	0.46 +/- 0.13	16.51

Table 3 shows peroxidic aldehydes that means aldehydic lipid peroxidation products. These aldehydes are formed by the degradation of the lipid hydroperoxides. It is interesting to note that some of these aldehydes are already present in native LDL e.g. propanal, butanal, malonaldehyde and trace amounts of pentanal and 4-hydroxynonenal. We think that these aldehydes can be considered as remnants of in vivo lipid peroxidation. We have intentionally given only single values of the aldehydes in o-LDL (values for one particular LDL preparation). If other LDL-preparations from other donors are investigated in principle the same trend is found. However, the absolute values vary significantly from LDL to LDL preparation. This variation in the oxidizibility most likely depends on the variation of the lipophilic antioxidants present in LDL.

Figure 3 shows a chromatogram of the determination of the hydroxyalkenals formed during lipid peroxidation. The most frequently used procedure is that of derivatizing with 2,4-dinitrophenylhydrazine to the corresponding hydrazones and separation of the hydrazones by HPLC [4]. The identified hydroxy-aldehydes are 4-hydroxy-hexenal, 4-hydroxy-octenal and 4-hydroxy-nonenal as mentioned above.

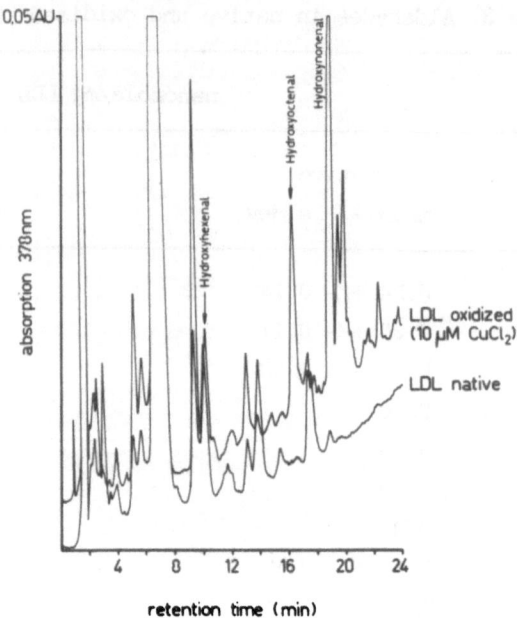

Fig. 3. HPLC separation of 4-hydroxyalkenals in native and o-LDL (10 μM Cu, 3h). Gradient: 77,5% methanol to 100% methanol in 30 min, flow: 1 ml /min.

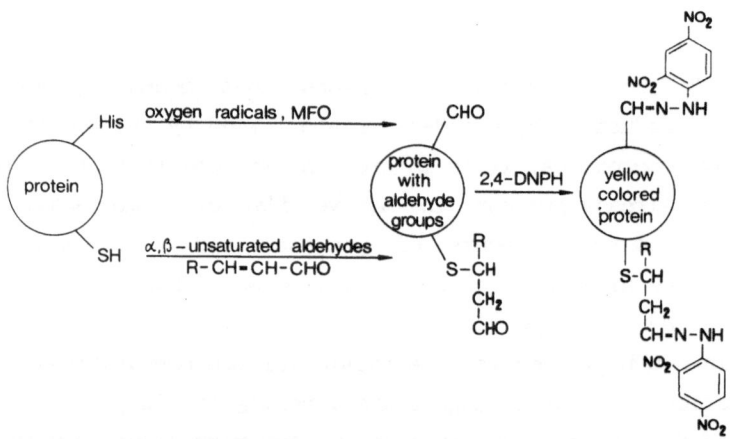

Fig. 4. Determination of aldehydes covalently bound to protein (apo-B) with 2,4-dinitro-phenylhydrazine (2,4-DNPH).

Can aldehydes formed during lipidperoxidation modify apo-B? Unsaturated aldehydes like 4-hydroxyalkenals, 2,4-alkadienals or 2-alkenals are able to bind to thiol groups or amino groups of the protein in a Michael-type addition reaction. Through this reaction aldehydes become covalently bound to the protein and can be determined photometricly after derivatisation with 2,4-dinitro-phenylhydrazine (Fig.4).

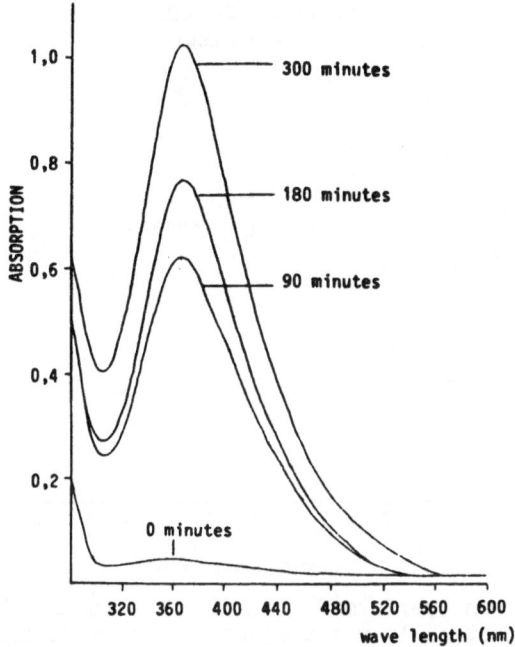

Fig. 5. Absorption spectra of the 2,4-dinitro-phenylhydrazones of apoB isolated from LDL oxidized for different times.

The experiment shown in Fig.5 clearly shows that the amount of protein-bound carbonyls determined as 2,4-dinitro-phenylhydrazones strongly increase during oxidation which supports the assumption that aldehydic lipid peroxidation products can alter the apo-B. From the maximum absorbance reached after 300 minutes of oxidation it results that about 50 aldehyde groups are present in the apo-B of one LDL molecule.

From the development of lipid peroxide values shown in Fig.6 one can see that LDL oxidation is a dynamic process which does not at all lead to a stationary end stage. Based on the development of the peroxide values the oxidation process can be divided in principle into three phases. A first phase where nearly no oxidation occurs (lag-phase), a second phase where oxidation proceeds very rapidly (propagation phase) and a third phase where the peroxide values decrease again. If this is now compared with the malonaldehyde (MDA) values one can clearly see that if only MDA would be determined one would get a completly wrong impression of the stage of oxidation. The MDA development also shows a lag phase and a kind of propagation but then remains more or less at a constant end value although the peroxide values strongly decrease.

211

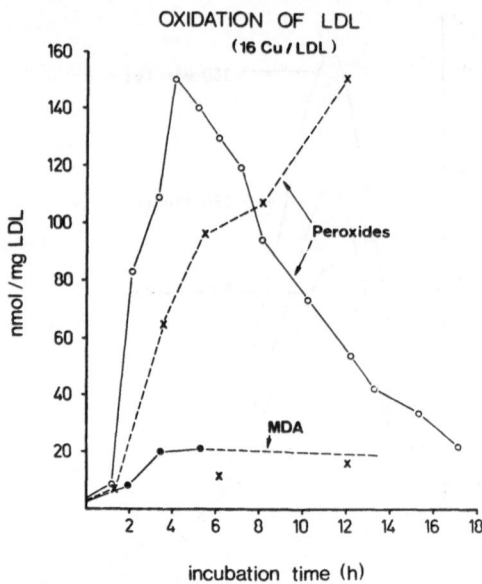

Fig. 6. Oxidation was performed with two different LDL preparations (o = preparation 1, x = preparation 2). Experimental conditions as described in Fig.2.

Big variations in the oxidizibility were found in different LDL-preparations. These variations probably depend on the amount of antioxidants, preformed lipid hydroperoxides and heterogeneity of the lipoproteins itself. Each LDL has its own characteristic oxidation behavior in respect to the length of the lag phase, the maximum amount of peroxides or other oxidation products and other parameters. As example Fig.6 shows the development of peroxides in two different LDL preparations. The one reached a maximum peroxide content of about 150 nmol per mg LDL already after four hours oxidation whereas the other LDL reached the same value only after 13 hours oxidation. Both LDL preparations showed after 4 hours about the same MDA content of 20 nmol per mg LDL. It is reasonable to assume that the biological properties (binding to receptors, cytotocicity, chemotactic activity and others) of LDL change in parallel with the degree of oxidation. The above example clearly shows that the malonaldehyde value alone is not a sufficient index to define the degree of oxidation. This might also be the reason why different laboratories have reported biological results with significantly quantitative differences. We think that it is important to determine several parameters at different time points or even better to follow oxidation of LDL continuously [5].

The investigation of oxidatively modified LDL has become an

important topic in atherosclerosis research in the recent few years. MDA-modified protein could be detected in atherosclerotic plaques [8]. The reactive species of the complex mixture of substances generated in the course of lipid peroxidation have to be further investigated to better understand their role in oxidative modification of LDL both in vitro and in vivo.

ACKNOWLEDGEMENTS

Our work has been supported by the Association for International Cancer Research (U.K.) and by the Austrian Science Foundation, project no. P 6176 B.

REFERENCES

1. U. P. Steinbrecher, S. Parthasaraty, D. S. Leake, J. L. Witztum and D. Steinberg, Modification of low density lipoprotein by endothelial cells involves lipid peroxidation and degradation of low density lipoprotein phospholipids, Proc. Natl. Acad. Sci. USA 81:3883 (1984).

2. J. W. Heinecke, Free radical modification of low density lipoprotein: Mechanisms and biological consequences. Free Radical Biology and Medicine 3:65 (1987).

3. G. Jürgens, H. F. Hoff, G. M. Chisolm III and H. Esterbauer, Modification of human serum low density lipoprotein by oxidation -characterization and pathophysiological implications, Chem. Phys. Lipids 45:315 (1987).

4. H. Esterbauer, O. Quehenberger and G. Jürgens, Oxidation of human low density lipoprotein with special attention to aldehydic lipid peroxidation products, in: "Free Radicals: Methodology and Concepts, C. Rice-Evans and B. Halliwell, eds., The Richelieu Press, London, 243 (1988).

5. H. Esterbauer, G. Striegl, H. Puhl and M. Rotheneder, Continuous monitoring of in vitro oxidation of human low density lipoprotein Free Rad. Res. Comms. : in press (1989).

6. M. El-Saadani, H. Esterbauer, M. El-Sayed, M. Goher, A. Y. Nassar and G. Jürgens, A spectrophotometric assay for lipid peroxides in serum lipoproteins using a comercially available reagent, J. Lip. Res. : in press (1989).

7. K. H. Cheeseman, A. Beavis and H. Esterbauer, Hydroxyl-radical-induced iron-catalysed degradation of 2-deoxyribose. Quantitative determination of malonaldehyde, Biochem. J. 252:649 (1988).

8. M. E. Haberland, D. Fong and L. Cheng, Malondialdehyde-altered protein occurs in atheroma of Watanabe heritable hyperlipidemic rabbits, Science 241: 215 (1988).

PARADOXICAL EFFECTS OF VITAMIN E: OXIDIZED LIPOPROTEINS, PROSTANOIDS AND THE PATHOGENESIS OF ATHEROSCLEROSIS

David G. Cornwell, Hanfang Zhang, W. Bruce Davis
Ronald L. Whisler and Rao V. Panganamala

Department of Physiological Chemistry
and Internal Medicine
The Ohio State University
Columbus, Ohio, U.S.A.

INTRODUCTION

Sixty years ago Michel Macheboeuf published his Doctor of Science Thesis and a year later his first paper on the isolation of a lipoprotein from horse serum. His outstanding studies[1] showed for the first time that a lipoprotein of constant composition was obtained by repeated precipitation from an aqueous solution. Shortly after Macheboeuf's work appeared, Sorensen[2] questioned whether isolated lipoproteins truly represented the molecular complexes found in serum. Sorensen was most concerned with the association and dissociation of solutes during lipoprotein isolation but his paper, as Chargaff[3] has noted, identified the problem of true lipoprotein composition in vivo. Some twenty years later, Oncley and Gurd[4] reported that isolated serum low density lipoproteins (LDL) were highly unstable complexes readily susceptible to chemical modification by processes such as oxidation (LDL_{OXID}). Lipoprotein instability raised again the question first posed by Sorensen that purified lipoproteins might be significantly different from native lipoproteins.

Several chemical and physical changes were found in LDL_{OXID}.[4,5] LDL contain β-carotene, lycopene and lutein[6] and absorbance of LDL_{OXID} in the 410 to 540 nm region decreased as the result of carotenoid destruction.[4,5] Absorbance of LDL_{OXID} increased in the 240 to 410 nm region and this effect was attributed by Oncley and Gurd[4] to the oxidation of unsaturated fatty acid residues in phospholipids and cholesteryl esters. LDL$_{OXID}$ also had different physical properties which included a decreased flotation rate with increased boundary spreading, and increased electrophoretic mobility.[4,5]. Physical properties[7] were altered dramatically when oxidation was catalyzed by metal ions such as Cu^{2+}. Recent work with LDL$_{OXID}$ is sometimes the rediscovery of these early observations.

The natural occurrence of LDL$_{OXID}$ was suggested by a number of observations. Lipids peroxides were identified as thiobarbituric acid reactants (TBAR) in plaques that had probably been infiltrated by serum components[8] and hydroxy fatty acids containing conjugated diene groups, classic per-oxidation products of polyunsaturated fatty acids, were found in the cholesteryl esters of severe plaques.[9,9a] However, other investigators suggested that the peroxides detected in lipid extracts from plaques were actually artifacts.[10] TBAR were found in serum and isolated LDL[11-13] and although these may have been artifacts formed during aging or isolation[14] very recent experiments in which LDL$_{OXID}$ were identified after dialysis under nitrogen in the presence of EDTA provided strong support for LDL$_{OXID}$ in vivo.[15]

We began to investigate LDL$_{OXID}$ because it appeared that the formation of LDL$_{OXID}$ in vivo might explain what we have come to call the vitamin E paradox.[16,17] Different anti-oxidants had three distinct effects on fatty acid metabolism in tissues culture.[18] Some antioxidants such as dipyridamole decreased TBAR and increased prostanoid levels probably by protecting cyclooxygenase from oxidative denaturation.[19,20] Other antioxidants such as α-naphthol, propyl gallate and butylated hydroxytoluene (BHT) decreased both TBAR and the prostanoid level probably by neutralizing oxygen centered radicals that catalyzed cyclooxygenase.[16-21] Finally, vitamin E was an effective inhibitor of intracellular TBAR but this agent had no effect on the prostanoid level in tissue culture.[19-21] Paradoxically, vitamin E administration to deficient animals blocked both plasma TBAR and prostanoid synthesis.[22-25] Since tissue culture media does not contain large amounts of LDL and a number of investigators found that LDL and LDL$_{OXID}$ had both stimulatory and inhibitory effects on prostanoid synthesis,[12,13,26,27] we proposed[17] that the paradoxical effect of vitamin E in tissue culture and animals depended on the formation of LDL_{OXID} and the different effects of exogenous lipid peroxides represented by LDL_{OXID} and endogenous lipid peroxides (represented by intracellular TBAR) on prostanoid synthesis (Fig. 1). If LDL_{OXID} but not endogenous lipid peroxides promoted prostanoid synthesis in cell cultures, then it was reasonable to assume that LDL_{OXID} was formed _in vivo_ and that vitamin E acted indirectly on prostanoid synthesis by regulating LDL_{OXID}.

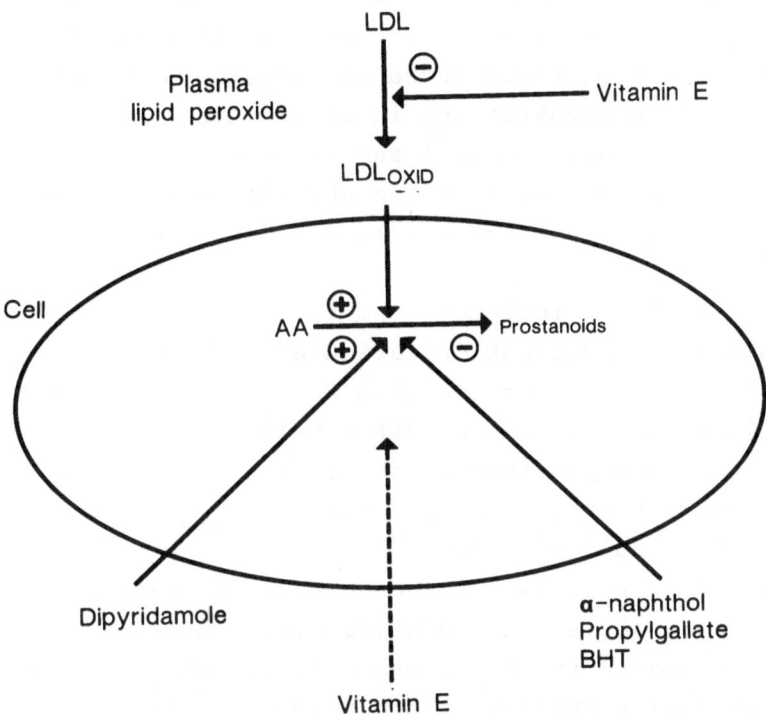

Fig. 1. Hypothetical scheme explaining the paradoxical effect of vitamin E on prostanoid synthesis in animals and in tissue culture. Vitamin E in plasma blocks extracellular lipid peroxidation and indirectly diminishes prostanoid synthesis by abolishing the stimulatory effects of extracellular lipid peroxides. Vitamin E also blocks intracellular lipid peroxidation (dashed line) but vitamin E, unlike other antioxidants in tissue culture, has no direct effect on prostanoid synthesis.

CONTROLLED OXIDATION OF LDL

Recent studies have reported that freshly isolated LDL contained some TBAR unless BHT or EDTA were added at the end of the ultracentrifugation flotation step and before dialysis against 0.15 M NaCl.[17,28-30] These observations confirmed the susceptibility of LDL to oxidation that was reported in early studies.[4,5,7] We found that it was necessary to add BHT before dialysis in order to prepare unoxidized LDL_{BHT} for comparison with LDL_{OXID}.[17,28]

LDL oxidation in the absence of BHT was a slow, controlled process when LDL solutions in 0.15 M NaCl were incubated at 37°C under 96% air-4% CO_2.[17,28] Oxidation was proportional to the decrease in carotenoid color and it was possible to obtain LDL_{OXID} preparations with the same TBAR levels by adjusting incubation times based on carotenoid disappearance.[17,28] LDL concentrations in these studies were obtained by a cholesterol measurement[31] and lipid peroxidation was estimated as TBAR reported as the absorbance at 532 nm or the conversion of absorbance to a malondialdehyde (MDA) concentration.[19-21,32]

LDL were always isolated from individual sera and these LDL preparations underwent oxidation at different rates.[17,28] Data for the oxidation of LDL from three different sera are reported in Figure 2. Since TBAR levels depended on incubation times, it was possible to obtain reproducible LDL_{OXID} preparations with much less TBAR than LDL_{OXID} prepared by incubation with Cu^{++} or cells.[30,33-37]

Physical properties were not altered significantly in LDL_{OXID}. For example, the relative electrophoretic mobility of LDL_{OXID} compared to LDL_{BHT} was only 1.1 when the TBAR reached 1.9 nmoles MDA/200 µg cholesterol.[17,28] LDL_{OXID} were re-isolated by ultracentrifugal flotation when the density was adjusted to 1.1.[28] Flotation at lower densities was not examined.

The chemical modification of LDL was examined by extracting lipids from both LDL_{BHT} and LDL_{OXID}, separating lipid classes by thin-layer chromatography (TLC), and staining both for lipid and lipid peroxides.[28,32] These experiments which are summarized in Table 1, showed that LDL_{BHT} contained cholesteryl esters (CE), triglycerides (TG), cholesterol (C), phosphatidylcholine (PC), lysophosphatidylcholine (LPC), and sphingomyelin (SPH). No LDL_{BHT} fractions stained for lipid peroxides. While LDL_{OXID} contained the same lipid components as LDL_{BHT}, the PC spot decreased and the LPC spot increased in these preparations. This observations confirmed other studies which showed that LPC was found in LDL_{OXID}.[30,33,38,39]

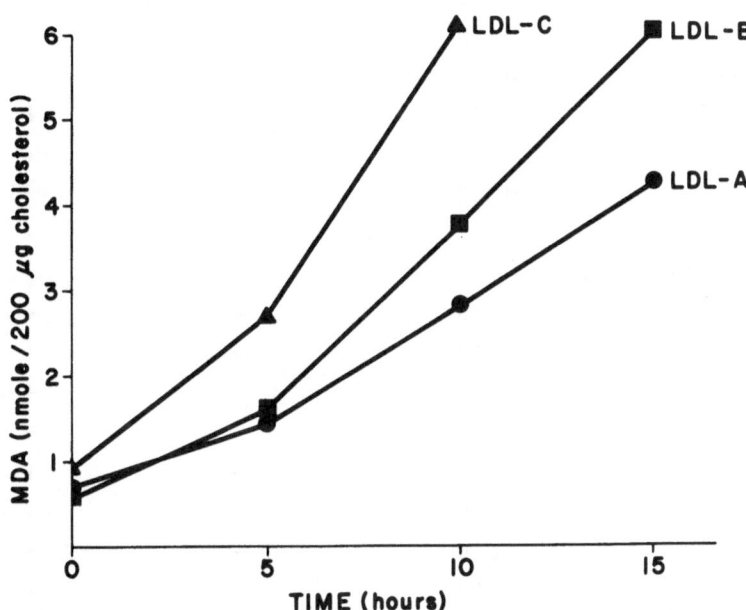

Fig. 2.   LDL from individual sera are oxidized at different rates.   Three LDL preparations (A,B and C) were isolated at the same time from three different sera, dialyzed against 0.15 M NaCl, sterilized, and then incubated at 37°C in 0.15 M NaCl.   TBAR were measured at incubation times of 5, 10 and 15 hr and are reported as nmol MDA/200 µg cholesterol (reprinted from 28).

Free fatty acid (FFA) could not be identified in lipid extracts from either LDL_{BHT} or LDL_{OXID} (Table 1).   A number of complex oxidation products were identified as broad bands on high performance liquid chromatography (HPLC) when LDL was co-oxidized with free arachidonic acid (AA).   These experiments showed that FFA released during LDL oxidation could be

Table 1. TLC Separation of Lipid Classes from LDL_{BHT} and LDL_{OXID}

Reference Compounds	R_f	LDL_{BHT}	LDL_{OXID}
A. Polar lipid classes[a]			
NL	0.96	+[c]	+ Oxid
FFA	0.75	−	−
15-HPETE	0.68	−	−
PC	0.60	+	↓
SPH	0.45	+	+
LPC	0.32	+	↑
B. Neutral lipid classes[b]			
CE	0.90	+	+
TG	0.56	+	+
	0.42	−	+ Oxid
	0.40	−	+ Oxid
	0.37	−	+ Oxid
	0.26	−	+ Oxid
FFA	0.20	−	−
C	0.19	+	+
15-HPETE	0.14	−	−
PL	0	+	+

[a]Whatman LK5D plates, chloroform-methanol-40% methylamine (60:20:5, v/v).

[b]What LK6D plates, hexane-ether-acetic acid (80:20:1,v/v).

[c]Lipid class present (+), absent (−), increased (↑), decreased (↓), contained lipid peroxides (Oxid). Data are summarized from reference 28.

converted to oxidation products that would not separate as FFA on TLC.[28]

Neutral lipid (NL) and phospholipid fractions isolated from LDL_{OXID} did not stain for lipid peroxides (Table 1). However, 4 new spots with R_f values between TG and C were identified in extracts from LDL_{OXID} and these sports all stained for lipid peroxides. The new spots had R_f values that coincided with spots from autoxidized trilinolenin and cholesteryl arachidonate. 15-Hydroperoxy-eicosatetraenoic acid (15-HPETE) had a much lower R_f value. The TLC data indicated that TBAR in LDL_{OXID} was derived from oxidized triglycerides and cholesteryl esters, a possibility first suggested by Oncley and Gurd.[4]

PROSTANOID ARTIFACTS IN LDL_{OXID}

Polyunsaturated fatty acid oxidation generates a large number of products including acyclic and cyclic peroxides.[40,41] Since some oxidation products resemble prostanoids that are synthesized through cyclic endoperoxide intermediates, we examined LDL_{OXID} for cross-reactivity in the radioimmunoassay (RIA) of prostanoids.[17,28] LDL_{BHT} did not cross-react in the RIA of prostanoids. However, the data in Figure 3 showed that material in LDL_{OXID} cross-reacted in the RIA for PGE_2 but not 6-keto-$PGF_{1\alpha}$. Cross-reacting materials were not formed through prostanoid bio-synthesis since indomethacin (IM) had no effect on the formation of the PGE_2 artifact when it was added to LDL before oxidation.

LDL_{OXID} did not destroy antibodies to PGE_2 since they had no effect when they were added to solutions of authentic PGE_2 prior to RIA.[28] The PGE_2 artifact was not a part of the LDL_{OXID} molecular complex since it was found in the infranatant when LDL_{OXID} was re-isolated by ultracentrifugal flotation.[28] Finally, the formation of the PGE_2 artifact was specific for LDL_{OXID} since cross-reacting material was not increased in the synthetic intravenous fat emulsion, Intralipid[R], by oxidation.[28,42] However, cross-reactivity was increased by the co-oxidation of Intralipid[R] and free AA.[28] These data are explained by the synthesis of PGE_2 artifacts from the oxidation of FFA released in the formation of LDL_{OXID}. The synthesis of these complex oxidation products explains why neither FFA nor 15-HPETE were found in lipid extracted from LDL_{OXID}.

LDL AND PROSTANOID BIOSYNTHESIS

Any effect of LDL on PGE_2 biosynthesis is difficult to measure by RIA since PGE_2 artifacts are formed during LDL oxidation (Fig. 3). This problem was overcome in tissue cultures by two RIA measurements. If total or apparent PGE_2 was determined in cultures incubated without IM, the PGE_2

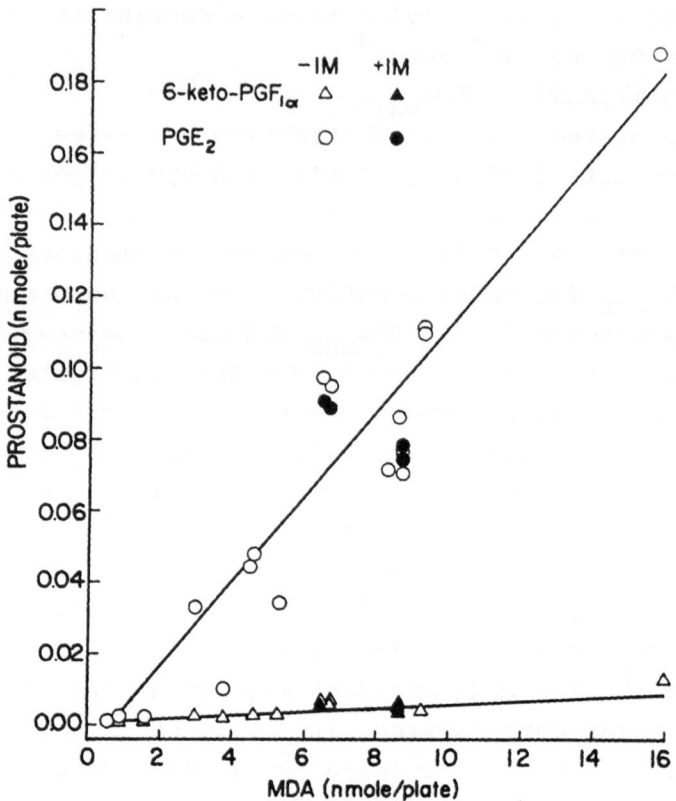

Fig. 3. Cross-reactivity of LDL_{OXID} to PGE_2 antibodies
varies directly with lipid peroxide content and
is not inhibited by IM. LDL_{OXID} does not cross-
react significantly with 6-keto-$PGF_{1\alpha}$ antibodies.
LDL from 17 individual sera were prepared with
different levels of lipid peroxidation. LDL concentrations were adjusted to 800 µg cholesterol/ml
and the LDL were incubated in tissue culture media
alone for 24 hr at 37°C in the absence or presence
of 10 µM IM. RIA was then used to estimate prostanoid levels. PGE_2 in the graph refers to immunoreactive PGE_2-like material (reprinted from 28).

artifact was then determined as the cross-reacting material found when prostanoid biosynthesis in cell cultures was blocked by IM.[17] The difference between these values represented de novo PGE_2 biosynthesis. RIA of the stable PGI_2 metabolite 6-keto-$PGF_{1\alpha}$ was used directly in experiments on prostanoid biosynthesis since material cross-reacting with 6-keto-$PGF_{1\alpha}$ was not formed in LDL_{OXID} preparations (Fig. 3).

Smooth muscle cells (SMC) from guinea pig aorta were chosen for tissue culture experiments since these cell cultures have been studied extensively in our laboratory.[16-21,32,42] Initial studies, which will be reported elsewhere in detail, showed that LDL_{BHT} had no effect on cellular prostanoid synthesis. For example, confluent cultures of SMC incubated for 24 hr with media alone contained 0.049 nmole PGE_2/plate and cultures from the same primary incubated for 24 hr with LDL_{BHT} (800 µg cholesterol/ml) also contained 0.049 nmole PGE_2/plate.[17]

LDL_{OXID} has in contrast to LDL_{BHT} profound effects on prostanoid metabolism. Preliminary studies[17] which will be reported elsewhere in detail showed that LDL_{OXID} stimulated PGI_2 synthesis and appeared to stimulate PGE_2 synthesis (Fig. 4). Two LDL preparations containing different amounts of TBAR were examined in these studies. The first preparation was designated LDL_{CON} because it was a "control" preparation consisting of freshly isolated LDL after dialysis against 0.15 M NaCl. LDL_{CON} retained the yellow carotenoid color and contained 1.1 nmole MDA/200 µg cholesterol. The second LDL preparation, LDL_{OXID}, was obtained when LDL_{CON} was incubated at 37°C under 96% air-4% CO_2. LDL_{OXID} appeared colorless and contained 1.7 nmole MDA/200 µg cholesterol.

PGI_2 synthesis was enhanced and then diminished by increasing concentrations of both LDL_{CON} and LDL_{OXID}. LDL_{CON}, the lipoprotein containing the lower amount of TBAR, had a greater stimulatory effect than LDL_{OXID}, the lipoprotein containing the greater amount of TBAR. These data showed both that oxidized lipoproteins, in contrast to unoxidized lipoproteins, stimulated prostanoid synthesis and that the stimulatory effect was diminished when the degree of lipid peroxidation exceeded optimal levels.

The RIA data reported in Figure 4 suggested that PGE_2 synthesis varied directly with LDL concentration and did not differ for LDL_{CON} and LDL_{OXID}. However, the PGE_2 artifact

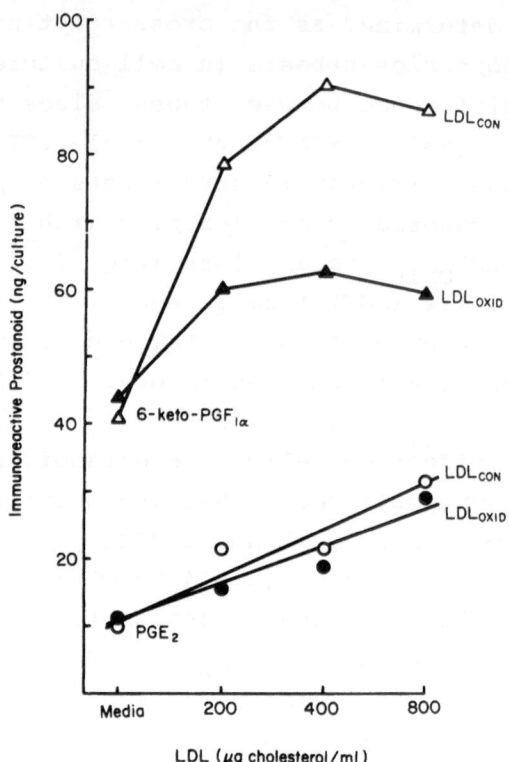

Fig. 4. Release of prostanoid immunoreactive compounds during
the incubation of confluent SMC with LDL_{CON} and
LDL_{OXID} (reprinted from 17).

(Fig. 3) interferred with the interpretation of RIA data for
this prostanoid. In fact, unpublished studies with IM showed
that true PGE_2 obtained by _de novo_ synthesis and PGI_2 synthesis
responded to LDL in a similar manner.

EXOGENOUS AND ENDOGENOUS LIPID PEROXIDES COMPARED

Enhanced and diminished prostanoid synthesis with low-TBAR
and high-TBAR-LDL are explained by well-known effects of lipid
peroxides on prostanoid synthesis in several tissues and cell
lines. These data show that fatty acid hydroperoxides in low
concentrations activate cyclooxygenase[16,18,43,44] while fatty
acid hydroperoxides at high concentrations inactivate cyclooxygenase and PGI_2 synthase.[12,13,16,18,43,45,46] However, it

has not been established whether the effects of hydroperoxy fatty acids are general properties that are shared by other lipid peroxides. We have approached this question by comparing the effects of 15-HPETE and 15-hydroperoxy-eicosapentaenoic acid (15-HPEPE) with the effects of oxidized lipoproteins on prostanoid synthesis in SMC.[42]

Table 2. Exogenous Lipid Peroxides Supplied as 15-HPETE and 15-HPEPE have both Stimulatory and Inhibitory Effects on Prostanoid Synthesis in SMC

Lipid Peroxide (μM)	6-keto-PGF$_{1\alpha}$ (nmole/plate)	
	No AA	120 μM AA
A. 15-HPETE		
0	0.24 ± 0.02	0.74 ± 0.04
5	0.37 ± 0.05	0.91 ± 0.05
25	0.43 ± 0.02	1.17 ± 0.01
50	0.42 ± 0.03	1.47 ± 0
B. 15-HPEPE		
0	0.39 ± 0.02	1.46 ± 0.10
5	0.45 ± 0.02	2.29 ± 0.16
25	0.30 ± 0.03	1.95 ± 0.07
50	0.16 ± 0.0	1.99 ± 0.03

[a]Confluent SMC cultures were incubated for 24 hr in media alone or media containing 120 μM AA with and without different concentrations of hydroxperoxy fatty acids. 6-keto-PGF$_{1\alpha}$ was estimated by RIA. Data are adapted from reference 42.

Two hydroperoxy fatty acids, 15-HPETE and 15-HPEPE, had effects similar to oxidized lipoproteins (Fig. 4) on PGI$_2$ synthesis in SMC (Table 2). 15-HPETE stimulated PGI$_2$ synthesis in the presence and absence of free AA while 15-HPEPE had stimulatory effects at low concentrations and inhibitory effects at high concentrations. The hydroperoxy fatty acids probably acted at the cyclooxygenase step in the AA cascade since they enhanced PGI$_2$ synthesis in the presence of a large

amount of free AA.[47] Any effect on AA release would have
been overwhelmed by the addition of a large amount of free AA.

A number of studies from our laboratory have shown that
lipid peroxides formed in the cell had, in contrast to exogenous lipid peroxides, no discernible effects on prostanoid
synthesis.[16-20,42,48] Although SMC cultures synthesized
bound lipid peroxides and prostanoids when they were challenged with increasing amounts of free AA, endogenous lipid
peroxides did not interact in prostanoid synthesis since both
a low concentration of BHT (non-inhibitory) and vitamin E
diminished lipid peroxide levels significantly without having
any effect on prostanoid synthesis in these cultures (Table 3).

Table 3. Endogenous Lipid Peroxides (TBAR) Generated by the
          Addition of Free AA have no Effect on Prostanoid
          Synthesis in SMC

Treatment[a]	TBAR	6-keto-PGF$_{1\alpha}$
	(nmole/culture)	
Media alone	1.00	0.48
120 µM AA	18.8	2.97
" + 10 µM BHT	3.3	2.90
120 µM AA	12.50	2.45
" + 50 µM vitamin E	0.50	2.10

[a]Confluent SMC cultures were incubated for 24 hr in media
alone or media containing 120 µM AA with and without an
antioxidant. Lipid peroxides were estimated as TBAR and
6-keto-PGF$_{1\alpha}$ was estimated by RIA. Data are adapted from
references 19 and 21.

OXIDIZED LDL AND THE VITAMIN E PARADOX

Polyunsaturated fatty acid metabolism in cultured cells
can be manipulated in various ways to increase, decrease, or
leave unchanged prostanoids and lipid peroxides. As a consequence, there are eight possible combinations between lipid
peroxides and prostanoids in which either group or both groups
are altered. We[18] have classified many agents by dividing
them on the basis of their effects on prostanoid synthesis

Table 4. Class I agents in a wide concentration range all
increase prostanoid levels while Class II agents in a wide
concentration range leave prostanoid levels unchanged and Class
II agents at high concentrations all decrease prostanoid levels.
Agents in each class may then be subdivided by their effect
on lipid peroxides. All antioxidants belong to subclass C in
that they diminish endogenous lipid peroxide levels. Dipyridamole, a Class I-C antioxidant, increases prostanoid levels
 while the Class III-C antioxidants α-naphthol, propyl gallate
and BHT decrease prostanoid levels.

Table 4. Prostanoid and Lipid Peroxide Combinations Generated
by Manipulating Fatty Acid Metabolism in SMC[a]

	Prostanoids	Lipid Peroxides
Class I		
A	Up	Up
B	Up	Unchanged
C	Up	Down
Class II		
A	Unchanged	Up
B	Unchanged	Down
Class III		
A	Down	Up
B	Down	Unchanged
C	Down	Down

[a]Reprinted from reference 18.

Vitamin E, a Class II-C antioxidant, is clearly different from other antioxidants. In tissue culture, vitamin E
functions as a highly effective inhibitor of endogenous lipid
peroxides without having any effect on prostanoid synthesis
and yet vitamin E appears to block prostanoid synthesis in
whole animals. We have called the different effects of vitamin E in cultured cells and whole animals the Vitamin E Paradox
(see INTRODUCTION).

The Vitamin E Paradox is readily explained as suggested
in Figure 1 by the differences we have observed between exogenous an endogenous lipid peroxides. Endogenous lipid peroxides

have little effect on prostanoid synthesis and therefore vitamin E which blocks endogenous lipid peroxide formation has little effect on prostanoid levels in cells. Exogenous lipid peroxides in relatively low concentrations stimulate prostanoid synthesis and therefore vitamin E which blocks exogenous lipid peroxide formation diminishes prostanoid levels in animals. The available data do not distinguish between hydroperoxy fatty acids and oxidized lipoproteins as the source of these effects. However, cells appear more sensitive to oxidized lipoproteins than hydroperoxy fatty acids. For example, a small change in TBAR content alters the effect of oxidized lipoproteins (Fig. 4) while a ten-fold increase does not change the effect of 15-HPETE (Table 2). Plasma contains only about 0.5 M total peroxide[49] and it appears unlikely that sufficient hydroperoxy fatty acid alone would be available in plasma to affect cellular prostanoid synthesis. In any event, the Vitamin E Paradox lends strong support to the physiologic significance of extracellular lipid peroxides present in plasma as oxidized lipoprotein and/or hydroperoxy fatty acid.

ROLE OF EXTRACELLULAR LIPID PEROXIDES IN CELL VIABILITY AND PROLIFERATION

The SMC cultures used in our studies have been described in a number of publications.[16-21,28,42,48] Cells were obtained from the dissected medial layer of guinea pig aorta from prepubertal males. Cell cultures contained only SMC[18] which were identified by their reactivity to antibodies against human umbilical cord F-actin which had been shown by others[50] to react specifically with muscle actin isoforms. Trypan blue exclusion and a new staining procedure[51] employing fluorescein diacetate and propidium iodide (FDA-PI) showed that the cultures contained only viable cells.[17,42] In confluent cultures, cells were elongated with a few cytoplasmic processes.[52] The cells had large nuclei, 2 to 6 nucleoli and a granular perinuclear area. Electron-microscopy suggested phenotypic modulation in that cells contained a few thin filaments and dense bodies, but no thick filaments, and they had more organelles than unmodulated cells.[52] A small number of cells had very large nuclei suggesting polyploidy. Mitotic figures were evident and the mitotic index was estimated at 3 to 4%.[17] ^{3}H-thymidine incorporation showed that DNA synthesis occurred in both confluent and proliferating cultures.[42]

The effects of different LDL preparations on confluent SMC cultures are summarized in Table 5. Cells treated with LDL_{BHT} had the same morphology, viability and mitotic index as untreated cells.[17] Low-TBAR-LDL_{OXID} did not affect morphology, viability, mitotic index and ^{3}H-thymidine incorporation.[17,42] Viability, mitotic index and ^{3}H-thymidine incorporation decreased as both lipoprotein concentration and the TBAR level increased. High-TBAR-LDL_{OXID} had a number of dramatic effects on SMC in confluent cultures. Cellular outlines were retained but the cytoplasm was shrunken and the refractive index changed. FDA-PI staining showed that very few cells were viable even though cells remained attached to the culture dish. These cells incorporated very little ^{3}H-thymidine.[42] Finally, the antioxidant BHT blocked many of the effects of high-TBAR-LDL_{OXID} although the mitotic index was not always restored.[17]

Table 5. The Effects of LDL_{BHT} and LDL_{OXID} on the Morphology, viability and Proliferation of Confluent SMC[a]

Property	Media alone	LDL_{BHT}	LDL_{OXID}	
			Low-TBAR	High-TBAR
Morphology	Normal	Unchanged	Unchanged	Greatly altered
Viability	100%	Unchanged	Unchanged[b] ▼ decreased	0
Mitotic index	3-4%	Unchanged	Unchanged[b] ▼ decreased	0
Thymidine uptake	100%	Unchanged	Unchanged[b] ▼ decreased	10%

[a]Properties are summarized from references 17,42 and 52.
[b]Property changed by an increase in either lipoprotein concentration or TBAR level.

Endogenous lipid peroxides, unlike exogenous lipid peroxides, had little effect on the properties of cells in confluent cultures. For example, high concentrations of free AA did not block thymidine incorporation in confluent cells even though large amounts of endogenous lipid peroxides were formed.[42] SCM were very sensitive to the inhibitory effects of lipid peroxides during clonal growth and even endogenous lipids peroxides blocked proliferation in these cells.[16-21,53]

Prostanoids in low concentrantions enhance clonal growth in medial SMC.[16,18-20] PGI_2 also exhibits cell-cycle dependent inhibition of DNA synthesis in medial but not intimal SMC.[54,55] Oxidized LDL depending on the TBAR level either stimulate or inhibit PGI_2 synthesis. Thus oxidized LDL may modulate cell proliferation through the direct effect of lipid peroxides and an indirect effect that is the result of variations in the level of PGI_2. TBAR levels may explain why some investigators have reported that LDL are cytotoxic [29,37,56,57] and other investigators have reported that LDL are mitogenic in cell cultures.[58-61]

LIPID PEROXIDES, VITAMIN E AND ATHEROSCLEROSIS

Many reasons have been advanced to implicate lipid peroxides in the pathogenesis of atherosclerosis.[16,18,20,24,62,63] Relatively high levels of lipid peroxides certainly cause endothelial cell injury and greatly diminish PGI_2 synthesis. Both these effects result in platelet aggregation and the elevation of the platelet derived growth factor which in turn initiate disease and promote cell proliferation. Cell proliferation in response to oxidative injury is complex since lipid peroxides themselves inhibit mitogenesis. However, factors elaborated in response to oxidative injury may overwhelm any modulating effects of lipid peroxides on cell proliferation (low vitamin E/high TBAR state in Table 6). Since vitamin E blocks lipid peroxidation, in vivo, vitamin E and other antioxidants should significantly inhibit the initiation and progression of atherosclerosis. Yet dietary vitamin E is disappointing as a therapeutic agent in most pathological conditions including atherosclerosis.[64,65]

The rationale for antioxidant therapy is found in the negative effects of excess lipid peroxides leading to atherosclerosis through oxidant injury. A very different state may obtain when injury has already occurred or is initiated by a process other than lipid peroxidation. As with many redox systems, oxygen centered radicals have a number of conflicting biological properties. Cell proliferation may be controlled by the direct effect of lipid peroxides them-

Table 6. Hypothesis for the Paradoxical Effects of Vitamin E on the Pathogenesis of Atherosclerosis

Vitamin E	TBAR	PGI$_2$	Proliferation	Atherosclerosis
Low	High	↓	↑ (injury)	Initiation
Intermediate	Low	↑	↓	Controlled
High	Absent	↓	↑	Progression

selves and the indirect effect of PGI$_2$ levels enhanced when prostanoid synthesis is stimulated by lipid peroxides (intermediate vitamin E/low TBAR state in Table 6). Finally, high levels of vitamin E and other antioxidants might be considered as mitogenic agents[16,18,20,24,42] that enhance the progression of atherosclerotic disease (high vitamin E/absent TBAR state in Table 6).

The antiaggregatory agent dipyridamole is an example of the antioxidant dilemma. We have shown that dipyridamole is an antioxidant that promotes SMC proliferation in cell cultures.[19,20] Dipyridamole also enhances the proliferative activity of capillary wall cells,[66] prolongs the survival of experimental skin flaps,[67] and enhances plaque formation (a process that involes intimal SMC that are less susceptible to PGI$_2$ than medial SMC[55]) in animals fed atherogenic diets.[68-71] Furthermore, dipyridamole has little effect on metastasis even though antiaggregatory agents generally diminish this process.[72,73] We suggest that the paradoxical effects of dipyridamole are best understood when both the antiaggregatory and mitogenic effects of this agent are considered.[16,19,20]

Similar problems may be encountered when dietary vitamin E is used indiscriminately as a therapeutic agent in atherosclerosis. It may be necessary to monitor plasma for vitamin E, TBAR, and oxidative susceptibility since further oxidation may occur when lipoproteins come in contact with intimal cells. These analyses could be particularly important with dietary regimens that include significant amounts of the eicosapentaenoic and docosahexaenoic fatty acids from fish oils.[74] Only intermediate vitamin E levels (see box in Table 6) may allow the formation of low-TBAR-LDL$_{OXID}$ which provide exogenous lipid peroxides with direct or indirect (elevated PGI$_2$) inhibitory effects on cell proliferation. For example, an appropriate dietary vitamin E level may be important in converting the diabetic animal from a high TBAR state to a low TBAR state thus diminishing the progression of atherosclerosis in this disease.[24,25,75]

ACKNOWLEDGMENT

This paper is based on studies which were supported in part by a Research Challenge Grant from the Ohio State University.

REFERENCES

1. M. Macheboeuf, "Etat des lipides de la matière vivante: Les cénapses et leur importante biologique," Herman and Co., Paris (1936).
2. S.P.L. Sorensen, Die Konstitution der loclichen Proteinstoffe als reversibel dissoziable komponentensysteme Kolloid-Z. 53:306 (1930).
3. E. Chargaff, Lipoproteins, Adv. Protein Chem. 1:1 (1944).
4. J. L. Oncley and F. R. N. Gurd, The lipoprotein of human plasma, in: "Blood Cells and Plasma Proteins, Their state in Nature," J. L. Tullis, ed., Academic Press, New York (1953).
5. J. L. Oncley, K. W. Walton and D. G. Cornwell, A rapid method for the bulk isolation of β-lipoproteins from human plasma, J. Am. Chem. Soc. 79:4666 (1957).
6. N. I. Krinsky, D. G. Cornwell and J. L. Oncley, The transport of vitamin A and carotenoids in human plasma, Arch. Biochem. Biophys. 73:233 (1958).
7. B. R. Ray, E. O. Davisson and H. L. Crespi, Experiments on the degradation of lipoproteins from serum, J. Phys. Chem. 58:841 (1954)
8. J. Glavind, S. Hartmann, J. Clemmesen, K. E. Jessen and H. Dam, Studies of the role of lipid peroxides in human pathology. II. the presence of peroxidized lipids in the atherosclerotic aorta, Acta Pathol. Microbiol. Scand. 30:1 (1952).
9. C. J. W. Brooks, G. Steel, J. D. Gilbert and W. A. Harland, Lipids of human atheroma. Part 4. Characterization of a new group of polar sterol estes from human atherosclerotic plaques, Atherosclerosis 13:233 (1971).
9a. H.-O. Mowri and T. Takano, Significance of lipid peroxides in atherosclerosis, in: "Clinical and Nutritional Aspects of Vitamin E," O. Hayaishi and M. Mino, eds., Elsevier, Amsterdam (1987).
10. F. P. Woodford, C. J. F. Bottcher, K. Oette and E. H. Ahrens, Jr., The artifactural nature of lipid peroxides detected in extracts of human aorta, J. Atheroscler. Res. 5:311 (1965).
11. K. Yagi, Assay for serum lipid peroxide level and its clinical significance, in: "Lipid Peroxides in Biology

and Medicine," K. Yagi, ed., Academic Press, New York (1982).

12. A. Szczeklik and R. J. Gryglewski, Low density lipo-proteins (LDL) are carriers for lipid peroxides and inhibit prostacyclin (PGI$_2$) biosynthesis in arteries, Artery 7:488 (1980).

13. J. Beitz, M. Panse, S. Fischer, C. Hora and W. Forster, Inhibition of prostaglandin I$_2$ (PGI$_2$) formation by LDL-cholesterol or LDL-peroxides? Prostaglandins 26:885 (1983).

14. D. M. Lee, Malondialdehyde formation in stored plasma, Biochem. Biophys. Res. Commun. 95:1663 (1980).

15. G. Bittolo-Bon, G. Cazzolato, M. Saccardi and P. Avogaro, Presence of modified LDL in humans: Effect of vitamin E, in: "Clinical and nutritional Aspects of Vitamin E," O. Hayaishi and M. Mino, eds., Elsevier Science Publish-ers, Amsterdam (1987).

16. D. G. Cornwell and N. Morisaki, Fatty acid paradoxes in the control of cell proliferation: prostaglandins, lipid peroxides, and cooxidation reactions, in: "Free Radicals In Biology," Vol. VI, W. A. Pryor, ed., Academic Press, New York, (1984).

17. H. Zhang, K. H. Jones, W. B. Davis, R. L. Whisler, R. V. Panganamala and D. G. Cornwell, Oxidized low density lipoproteins in smooth muscle cell cultures: differential effects on prostanoid synthesis and via-bility, in: "Clinical and Nutritional Aspects of Vitamin E, "O. Hayaishi and M. Mino, eds., Elsevier Science Publishers, Amsterdam (1987).

18. D. G. Cornwell and H. Zhang, Fatty acid metabolism and cell proliferation, in: "Lipid Peroxidation in Biologi-cal Systems," A. Sevanian, ed., Am. Oil Chemists' Soc., Champaign, IL (1988).

19. N. Morisaki, J. M. Stitts, L. Bartels-Tomei, G. E. Milo, R. V. Panganamala and D. G. Cornwell, Dipyridamole: an antioxidant that promotes the proliferation of aorta smooth muscle cells, Artery 11:88 (1982).

20. D. G. Cornwell, J. A. Lindsey, H. Zhang and N. Morisaki, Fatty acid metabolism and cell proliferation. VI. Properties of antithrombotic agents that influence metastasis, in: "Icosanoids and Cancer," H. Thaler-Dao, A. Crastes de Paulet and R. Paoletti, eds., Raven Press, New York (1984).

21. N. Morisaki, J. A. Lindsey, J. M. Stitts, H. Zhang and D. G. Cornwell, Fatty acid metabolism and cell prolif-eration. V. Evaluation of pathways for the generation of lipid peroxides, Lipids 19:381 (1984).

22. W. C. Hope, C. Dalton, L. J. Machlin, R. J. Filipski and F. M. Vane, Influence of dietary vitamin E on prostaglandin biosynthesis in rat blood, Prostaglandins 10:557 (1975).

23. A. E. Kitabchi, Hormonal status in vitamin E deficiency in: "Vitamin E, A Comprehensive Treatise," L. J. Machlin ed., Marcel Dekker, New York (1980).

24. R. V. Panganamala and D. G. Cornwell, The effects of vitamin E on arachidonic acid metabolism, Ann. N. Y. Acad. Sci. 393:376 (1982).

25. C. W. Karpen, K. A. Pritchard, J. H. Arnold, D. G. Cornwell and R. V. Panganamala, Restoration of prosta-cyclin/thromboxane A$_2$ balance in the diabetic rat, Diabetes 31:947 (1982).

26. K. B. Pomerantz, A. R. Tall, S. J. Feinmark and P. J. Cannon, Stimulation of vascular smooth muscle cell prostacyclin and prostaglandin E_2 synthesis by plasma high and low density lipoproteins, Circ. Res. 54:554 (1984).

27. M. Yokode, T. Kita, Y. Kikawa, T. Ogorochi, S. Narumiya and C. Kawai, Stimulated arachidonate metabolism during foam cell transformation of mouse peritoneal macrophages with oxidized low density lipoprotein, J. Clin. Invest. 81:720 (1988).

28. H. Zhang, W. B. Davis, X Chen, R. L. Whisler and D. G. Cornwell, Studies on oxidized low density lipoproteins. Controlled oxidation and a prostaglandin artifact, J. Lipid Res. (in press).

29. D. W. Morel, J. R. Hessler and G. M. Chisolm, Low density lipoprotein cytotoxicity induced by free radical peroxidation in lipid, J. Lipid Res. 24:1070 (1983).

30. U. P. Steinbrecher, S. Parthasarathy, D. S. Leake, J. L. Witztum and D. Steinberg, Modification of low density lipoprotein by endothelial cells involves lipid peroxidation and degradation of low density lipoprotein phospholipids, Proc. Natl. Acad. Sci. USA 81:3883 (1984).

31. D. L. Whitte, D. A. Barrett and D. A. Wycoff, Evaluation of an enzymatic procedure for determination of serum 20:1282 (1974).

32. V. C. Gavino, J. S. Miller, S. O. Ikharebha, G. E. Milo and D. G. Cornwell, Effect of polyunsaturated fatty acids and antioxidants on lipid peroxidation in tissue cultures, J. Lipid Res. 22:763 (1981).

33. T. Henricksen, E. M. Mahoney and D. Steinberg, Enhanced macrophage degradation of biologically modified low density lipoprotein, Arteriosclerosis 3:149 (1983).

34. J. F. Nagelkerke, L. Havekes, V. W. M. Van Hinsberg and T. J. C. Van Berkel, In vivo catabolism of biologically modified LDL, Arteriosclerosis 4:256 (1984).

35. J. W. Heinecke, L. Baker, H. Rosen and A. Chait, Superoxide-mediated modification of low density lipoprotein by arterial smooth muscle cells, J. Clin. Invest. 77:757 (1986).

36. V. O. Ivanov, S. N. Preobrazhensky, V. P. Tsibulsky, V. R. Bakaev, V. S. Repin and V. N. Smirnov, Liposome uptake by cultured macrophages mediated by modified low-density lipoproteins, Biochim. Biophys. Acta 846:76 (1985).

37. D. W. Morel, P. E. DiCorleto and G. M. Chisolm, Endothelial and smooth muscle cells alter low density lipoprotein in vitro by free radical oxidation, Arteriosclerosis 4:357 (1984).

38. S. Parthasarathy, U. P. Steinbrecher, J. Barnett, J. L. Witztum and D. Steinberg, Essential role of phospholipase A_2 activity in endothelial cell-induced modification of low density lipoprotein, Proc. Natl. Acad. Sci. USA 82:3000 (1985).

39. U. P. Steinbrecher, J. L. Witztum, S. Parthasarathy and D. Steinberg, Decrease in reactive amino groups during oxidation or endothelial cell modification of LDL, Arteriosclerosis 7:135 (1987).

40. W. A. Pryor, J. P. Stanley and E. Blair, Autoxidation of polyunsaturated fatty acid. II. A suggested mechanism for the formation of TBA-reactive materials from prostaglandin-like endoperoxides, Lipids 11:370 (1976).

41. N. A. Porter, Mechanisms for autoxidation of polyunsaturated lipids, Acc. Chem. Res. 19:262 (1986).
42. H. Zhang. K. H. Jones, W. B. Davis, R. L. Whisler, R. V. Panganamala and D. G. Cornwell, Heterogeneity in lipid peroxides: cellular arachidonic acid metabolism and DNA synthesis, in: "Pharmacologic Effects of Lipids," Vol. III, J. J. Kabara, ed., Am. Oil Chemists' Soc., Champaign, Il. (in press).
43. P. H. Gale and R. W. Egan, Prostaglandin endoperoxide synthase-catalyzed oxidation reactions, in: "Free Radicals in Biology," Vol. VI, W. A. Pryor, ed., Academic Press, New York (1984).
44. W. E. M. Lands, R. J. Kulmacz and P. J. Marshall, Lipid peroxide actions in the regulation of prostaglandin biosynthesis, in: "Free Radicals in Biology," Vol. VI,. W. A. Pryor , ed., Academic Press, New York (1984).
45. J. L. Humes, E. E. Opas, M. Galavage, D. Soderman and R. J. Bonney, Regulation of macrophage eicosanoid production of hydroperoxy- and hydroxy-eicosatetraenoic acids, Biochem. J. 233:199 (1986).
46. B. Mayer, R. Moser, H. Gleispach and W. R. Kukovetz, Possible inhibitory function of endogenous 15-hydroperoxy-eicosatetraenoic acid on prostacyclin formation in bovine aortic endothelial cells, Biochim. Biophys. Acta 875:641 (1986).
47. H. Zhang, H. Kaseki, W. B. Davis, R. L. Whisler and D. G. Cornwell, Mechanisms for the stimulation of prostanoid synthesis by cyclosporine A and bacterial lipopolysaccharide, Transplantation (in press).
48. N. Morisaki, H. Sprecher, G. E. Milo and D. G. Cornwell, Fatty acid specificity in the inhibition of cell proliferation and its relationship to lipid peroxidation and prostaglandin biosynthesis, Lipids 17:893 (1982).
49. M. A. Warso and W. E. M. Lands, Presence of lipid hydroperoxide in human plasma, J. Clin. Invest. 75:667 (1985).
50. J. L. Lessard, Two monoclonal antibodies to actin: one muscle selective and one generally reactive, Cell Motility Cytoskeleton 10:349 (1988).
51. K. H. Jones and J. A. Senft, An improved method to determine cell viability by simultaneous staining with fluorescein diacetate-propidium iodide, J. Histochem. Cytochem. 33:77 (1985).
52. J. S. Miller, V. C. Gavino, G. A. Ackerman, H. M. Sharma, G. E. Milo, J. C. Geer and D. G. Cornwell, Triglycerides, lipid droplets, and lysosomes in aorta smooth muscle cells during the control of cell proliferation with polyunsaturated fatty acids and vitamin E, Lab. Invest. 42:495 (1980).
53. V. C. Gavino, G. E. Milo and D. G. Cornwell, Image analysis for the automated estimation of clonal growth and its application to the growth of smooth muscle cells, Cell Tissue Kinet. 15:225 (1982).
54. N. Morisaki, T. Kanzaki, N. Motoyama, Y. Sato and S. Yoshida, Cell cycle-dependent inhibition of DNA synthesis by prostaglandin I_2 in cultured rabbit aortic smooth muscle cells, Atherosclerosis 71:165 (1988).
55. N. Morisaki, T. Kanzaki, Y. Sato and S. Yoshida, Lack of inhibition of DNA synthesis by prostaglandin I_2 in cultured intimal smooth muscle cells from rabbits, Atherosclerosis 73:67 (1988).

56. J. R. Hessler, A. L. Robertson and G. M. Chisolm, LDL-induced cytotoxicity and its inhibition by HDL in human vascular smooth muscle and endothelial cells in culture, Atherosclerosis 32:213 (1979).

57. T. Henriksen, S. A. Evensen and B. Carlander, Injury to human endothelial cells in culture induced by low density lipoproteins, Scand. J. Clin. Lab. Invest. 39:361 (1979).

58. B. G. Brown, R. Mahley and G. Assman, Swine aortic smooth muscle in tissue culture: some effects of purified swine lipoproteins on cell growth and morphology, Circ. Res. 39:415 (1976).

59. K. Fischer-Dzoga, R. A. Fraser and R. W. Wissler, Stimulation of proliferation in stationary primary cultures of monkey and rabbit aortic smooth muscle cells. I. Effects of lipoprotein fractions of hyperlipemic serum and lymph, Exp. Mol. Pathol. 24:346 (1976).

60. D. L. Layman, B. L. Jelen and D. R. Illingworth, Inability of serum from abetalipoproteinemic subjects to stimulate proliferation of human smooth muscle cells and dermal fibroblasts in vitro, Proc. Natl. Acad. USA 77:1511 (1980).

61. J. P. Tauber, J. Cheng and D. Gospodarowicz, Effects of high and low density lipoproteins on proliferation of cultured bovine vascular endothelial cells, J. Clin. Invest. 66:696 (1980).

62. D. G. Cornwell and R. V. Panganamala, Atherosclerosis: an intracellular deficiency in essential fatty acids, Prog. Lipid Res. 20:365 (1981).

63. D. Steinberg, Metabolism of lipoproteins and their role in the pathogenesis of atherosclerosis, Atherosclerosis Rev. 18:1 (1988).

64. M. L. Scott, Vitamin E, in: "Handbook of Lipid Research. 2. The Fat-Soluble Vitamins," H. F. DeLuca, ed., Plenum Press, New York (1978).

65. P. M. Farrell, Deficiency states, pharmacological effects, and nutrient requirements, in: "Vitamin E, A Comprehensive Treatise," L. J. Machlin, ed., Marcel Dekker, Inc., New York (1980).

66. G. Tornling, G. Unge, L. Skoog, A. Ljunggvist, S. Carlsson and J. Adolfsson, Proliferative activity of myocardial capillary wall cells in dipyridamole-treated rats, Cardiovascular Res. 12:692 (1978).

67. C. Arnander, G. Jurell, G. Tornling and G. Unge, Effect of dipyridamole on the survival of experimental critical skin flaps, Scand. J. Plast. Reconstruct. Surg. 13:261 (1979).

68. A. Dembinska-Kieč, W. Rücker and P. S. Schönhöfer, Effects of dipyridamole in experimental atherosclerosis, Atherosclerosis 33:315 (1979).

69. A. Dembinska-Kieč, W. Rücker and P. S. Schönhöfer, Effects of dipyridamole in vivo on ATP and cAMP content in platelets and arterial walls and on atherosclerotic plaque formation, Naunyn-Schmiedeberg's Arch. Pharmacol. 309:59 (1979).

70. J. K. Koster, Jr., A. F. Tryka, P. H'Doubler and J. J. Collins, Jr., The effect of low-dose aspirin and dipyridamole upon atherosclerosis in the rabbit, Artery 9:405 (1981).

71. W. Hollander, b. Kirkpatrick, J. Paddock, M. Colombo, S. Nagraj and S. Prusty, Studies on the progression and regression of coronary and peripheral atherosclerosis in the cynomolgus monkey. I. Effects of dipyridamole and aspirin, Exp. Mol. Pathol. 30:55 (1979).

72. P. Hilgard, Blood platelets and tumor dissemination in: "Interaction of Platelets and Tumor Cells," G. A. Jamieson, Alan R. Liss, Inc., New York (1982).

73. K. V. Honn, R. S. Bockman and L. J. Marnett, Prostaglandins and Cancer: a review of tumor initiation through tumor metastasis, Prostaglandins 21:833 (1981).

74. F. Hirahara and S. Kimura, Effects of different dietary oil levels and E/PUFA ratio on tocopherol contents and lipid peroxidative values in serum and tissues of rats, in: "Clinical and Nutritional Aspects of Vitamin E," O. Hayaishi and M. Mino, eds., Elsevier, Amsterdam (1987).

75. C. W. Karpen, A. J. Merola, R. W. Trewyn, D. G. Cornwell and R. V. Panganamala, Modulation of platelet thromboxane A_2 and arterial prostacyclin by dietary vitamin E, Prostaglandins 22:651 (1981).

THE ROLE OF OXIDATIVE METABOLISM AND ANTIOXIDANTS IN LOW-DENSITY LIPOPROTEIN

STRUCTURE AND METABOLISM

Wendy Jessup, Stephen Bedwell and [*]Roger T. Dean

Cell Biology Research Group,
Department of Biology and Biochemistry,
Brunel University, U.K.
and [*]Heart Research Institute,
Sydney, Australia

INTRODUCTION

The mechanisms by which free radicals mediate alterations in the structure and function of proteins, both in aqueous and lipid-containing (e.g. membranes and lipoproteins) systems, are of relevance in many normal and pathological conditions, since proteins are frequently the components of biological systems which are responsible for functional activity and specificity. We describe here studies of these processes in human low-density lipoprotein, and their likely relevance to atherogenesis.

Low-density lipoprotein is normally cleared from the circulation by endocytosis, mediated largely via cell-surface LDL receptors, which recognize binding domains on the Apolipoprotein B component of the LDL particle. Several types of chemical alterations to ApoB can reduce its affinity for the LDL receptor. Some, such as acetylation, can also promote binding of LDL to receptors on macrophages, of which the acetyl-LDL or 'scavenger' receptor is the best characterized.[1] This can lead to the accumulation of large amounts of cholesterol by these cells. Recently, cell-mediated modification of LDL by certain cell types, such as arterial endothelial and smooth muscle cells and macrophages, has also been shown to produce an altered form of LDL which is apparently recognized by binding sites (including the scavenger receptor) on macrophages.[2-4]

Cell-mediated modification is believed to be dependent upon oxidation of LDL lipids. This view is supported by observations that oxidation occurs in all productive cell-mediated modifications reported, and that the degree of modification achieved is roughly related to the extent to which LDL oxidation has occurred.[3-5] Furthermore, if oxidation is suppressed by the addition of chain-breaking antioxidants (BHT, probucol, Vitamin E) or metal chelators (EDTA, desferrioxamine) to the cell mediation system, then modification does not occur.[4-6] Cell-free oxidative modification producing LDL with apparently similar properties to cell-oxidized material can be achieved if LDL is incubated with relatively high concentrations of copper or iron.[4]

The precise nature of the change in LDL which produces a particle capable of binding to macrophage receptors as a result of oxidative modification is not known, although there is some evidence that it resides in the Apo B molecule.[7] It may involve conformational changes with or without cleavage of the molecule which expose previously buried domains, excision of parts of the polypeptide, the formation of protein peroxides, or binding of products of lipid peroxidation to essential residues.

The complexity of the cell-mediated modification system makes identification of the initiating species difficult. The <u>in vitro</u> modification of LDL in the presence of transition metals or copper probably occurs via catalytic decomposition of endogenous peroxides in the LDL. The inhibition of the cell-mediated system by antioxidants and metal chelators suggests that an analogous process is also occurring here, accelerated by the production of active oxygen-centred free radicals by the cells which stimulate oxidation of LDL lipids. It has been suggested that the initiating species is superoxide,[8,9] based on sensitivity of the system to superoxide dismutase, but this has not been reproduced by some other groups.[10,11] An alternative possibility is that a primary site of peroxidation is the cell plasma membrane, and that products of this peroxidation subsequently attack the LDL particle.

We have addressed two main questions pertinent to oxidative modification of LDL. Firstly, to what extent are biologically relevant free radicals capable of stimulating LDL oxidation? In particular, is superoxide radical, which is believed to be incapable of initiating lipid peroxidation, able to modify LDL? Secondly, what role do endogenous antioxidants, such as Vitamin E, play in oxidative modification, and how effectively do exogenously supplied natural and synthetic antioxidants protect LDL against oxidative attack?

MATERIALS AND METHODS

Human LDL was isolated from fresh plasma by differential density ultracentrifugation in the density range 1.019-1.050, using KBr solutions for density adjustments. The isolated LDL was dialysed extensively against phosphate-buffered saline, sterilized by ultrafiltration and stored in the dark under nitrogen at $4^{o}C$. All solutions and dialysis buffers contained 1.0mg/ml EDTA and 0.1mg/ml chloramphenicol and were deoxygenated before use, and all dialyses were performed under essentially anaerobic conditions in stoppered bottles at $4^{o}C$. LDL was iodinated by the iodine monochloride method.[12]

Radical species were generated in solution by irradiation of LDL (1.0mg protein/ml) using the Brunel Biochemistry ^{60}Co source [13] at a dose rate of 30-50 Gy/min. LDL was pre-saturated with the appropriate gas by top-gassing for 15min before and continuously during irradiation. For generation of predominantly hydroxyl radical, the gas was N_2O or N_2O/O_2. Substitution of air and the addition of 10mM formate permitted selective generation of either superoxide radical (pH 7.4) or hydroperoxyl radical (pH 4.0). Macrophage-mediated LDL oxidation was performed as described by Rankin and Leake [6], and LDL 'autoxidation' as described by Esterbauer et al. [14]

The lipid hydroperoxide content of samples of LDL was determined by an automated version of the triiodide assay. [15] Vitamin E was extracted by the SDS method [16] and measured by HPLC using a Merck Lichrosorb CN column (250 x 4mm; 7um particle size) with hexane/ isopropanol (99:1) as mobile phase and a fluorescence detector set at 295nm excitation and 325nm emission. Agarose electrophoresis was performed in 1% agarose gels in barbitone buffer at pH 8.6; SDS PAGE was in 3-20% discontinuous linear gradient gels. [17] Binding of LDL to the Apo B/E and macrophage receptors was determined as described by [18] and [19] respectively.

RESULTS AND DISCUSSION

In vitro oxidation of LDL by defined free radicals

LDL was exposed in vitro to several oxygen-centred free radicals which are believed to occur in vivo, and which may be involved in cell-mediated modification of LDL. We studied the ability of hydroxyl, superoxide and hydroperoxyl radicals to initiate LDL oxidation, and the effects of these radicals on Apolipoprotein B structure and functional properties.

LDL peroxidation was monitored by direct measurement of hydroperoxide content, which we have found to be a more sensitive and convenient assay than the commonly used TBA test. Both hydroxyl and hydroperoxyl radicals were able to induce dose-dependent oxidation of LDL. In contrast, superoxide radicals were much less effective (Table 1.). This is consistent with previous studies which have established the inability of superoxide to initiate lipid peroxidation. The peroxidation which did occur may be attributed to the small amount of hydroperoxyl radical (its conjugate acid) which is in equilibrium with superoxide at pH 7.4 (pK= 4.8). This limited peroxidation consequent on superoxide exposure could be greatly amplified during subsequent incubation of LDL with Cu^{2+} (data not shown).

LDL is a major transporter of plasma Vitamin E (mainly alpha-tocopherol). This lipid-soluble antioxidant may protect LDL against oxidative stress as it does in various natural and synthetic membranes.[21,22]

Table 1. Oxidation and alpha-tocopherol consumption following exposure of LDL to oxygen-centred free radicals

Radical species	Radical/protein molar ratio				
	0	90	180	270	360
HYDROPEROXIDE (nmol/mg. LDL protein)					
$HO_2{}^{\cdot}/O_2$	n.d.	1.2	14.5	22.9	43.4
$O_2{}^{\cdot -}/O_2$	1.0	3.3	10.1	5.5	8.4
$OH^{\cdot -}/O_2$	2.1	6.9	26.9	59.2	89.2
$OH^{\cdot -}/anoxic$	6.7	5.9	6.9	6.9	8.2
ALPHA-TOCOPHEROL (nmol/mg. LDL protein)					
$HO_2{}^{\cdot -}/O_2$	7.5	0.2	n.d.	n.d.	n.d.
$O_2{}^{\cdot -}/O_2$	6.9	4.4	2.3	0.4	n.d.
$OH^{\cdot -}/O_2$	7.0	1.2	0.3	0.2	0.2
$OH^{\cdot -}/anoxic$	7.2	6.7	7.0	7.2	7.5

n.d.; not detectable. Data are from a single experiment, but are representative of several separate experiments using different preparations.

Exposure of LDL to hydroxyl or hydroperoxyl radicals led to a dose-dependent consumption of endogenous alpha-tocopherol. Interestingly, LDL peroxidation was largely suppressed until most of this antioxidant was consumed (Table 1.) Superoxide was also capable of lowering the alpha-tocopherol content of LDL. Thus the increased sensitivity of superoxide-treated LDL to subsequent redox-active metal-catalysed oxidation may result either from the formation of low amounts of hydroperoxides in the LDL (as we have observed), or from depletion of the content of endogenous antioxidants such as Vitamin E, or to a combination of both.

Cell-mediated oxidation of LDL is accompanied by an increase in anodic mobility in agarose and extensive fragmentation of Apolipoprotein B.[3,5] In vitro radical exposure had similar effects on LDL electrophoretic mobility (in a dose- and radical-dependent manner)(Table 2.). Loss of intact Apolipoprotein B was detected by SDS-PAGE, when the 514kDa band of Apo B gradually diminished with increasing radical dose (data not shown). This was accompanied by the appearance of lower molecular weight fragments for superoxide, hydroperoxyl and hydroxyl radical-treated LDLs. Some specificity in cleavage was also evident from the appearance of discrete lower molecular weight bands against a backgound smear of Coomassie-staining material. Hydroxyl radical-treated LDL also showed insoluble material in the sample wells, which we presume derived from intermolecular cross-linking induced by this radical species, as noted previously for other proteins in aqueous and lipid environments.[13,23-25] The efficiency with which the various radicals fragmented Apolipoprotein B was related to their ability to peroxidize LDL, which is consistent with the view that hydroperoxides can promote protein cleavage.[26]

Treatment of LDL with these free radicals also caused a decline in the affinity of LDL for the fibroblst Apo B/E receptor, but did not lead to the formation of species of modified LDL which could be rapidly endocytosed and

Table 2. Oxygen-centred free radicals alter LDL electrophoretic mobility and affinities for fibroblast and macrophage receptors.

| Property | Radical species (radical/protein molar ratio = 300) | | | |
	$HO_2^{\bullet}/O_2$	$O_2^{\bullet -}/O_2$	$OH^{\bullet -}/O_2$	$OH^{\bullet -}/anoxic$
Relative mobility (*)	1.38	1.04	1.70	1.57
Fibroblast Apo B/E receptor affinity (%, **)	78	89	75	69
Macrophage uptake (***)	0.19	0.29	0.21	0.19

(*) relative to native LDL in 1% agarose gels, as described in Methods.
(**) Compares the ability of irradiated LDLs to compete with labelled native LDL for high affinity binding to fibroblast surface receptors at $4^{\circ}C$. All LDLs were used at 20ug/ml; results expressed as a percentage of the affinity of native LDL measured under identical conditions.
(***) Estimated as the ability of various LDLs to stimulate [14-C]-oleate incorporation into cholesteryl oleate when incubated at 25ug/ml with mouse peritoneal macrophages at $37^{\circ}C$; results expressed as nmol [14-C]-oleate incorporated/mg cell protein/24 hr. (native LDL, 0.21; acetylated LDL, 20.9)

degraded by macrophages (Table 2). Thus, several of the alterations in LDL structure which are normally observed during cell-mediated oxidation and which were also found in these in vitro experiments, are not predictive of the generation of high-uptake forms of LDL.

From these simple studies we conclude that cell-derived free radicals, such as hydroperoxyl/superoxide, may be capable of acting as initiators of LDL oxidation, and of producing several of the changes in LDL structure associated with this process. Radical attack alone (and the range of doses used here enclosed those measured in the cell-mediated systems) is not sufficient to produce forms of LDL with high affinity for macrophage surface receptors, such as the 'scavenger' receptor. Perhaps amplification of this oxidation, catalysed by the essential redox-active metals in the culture medium, is the other component of cell-mediated modification. Endogenous antioxidants (such as alpha-tocopherol) may protect LDL against oxidative attack in vitro, and may also be a significant component of the prevention of oxidative damage to these particles in vivo.

Exogenous antioxidants also suppress LDL oxidation in vitro

If endogenous lipid-soluble antioxidants protect LDL against oxidation, then conditions which either preserve or supplement these agents should increase the resistance of LDL to oxidative attack. We therefore studied the effects of exogenously supplied antioxidants on the consumption of alpha-tocopherol and the onset of lipid peroxidation during 'autoxidation' of LDL.[14] This system depends on the presence of trace amounts of redox-active metals and an oxygen-saturated buffer to promote oxidation of LDL. Table 3 shows that during a 24hr incubation, alpha-tocopherol was progressively consumed and peroxidation increased. This process was accelerated if Cu^{2+} was added to the system, and completely blocked by EDTA. Vitamin E analogues such as butylated hydroxytoluene and probucol (not shown) retarded the rate of alpha-tocopherol consumption and the onset of peroxidation. Ascorbate can protect Vitamin E in model lipid systems by reducing the tocopheroxyl radical to native Vitamin E [22]; in the presence of modest concentrations of redox-active metals it can also act as a pro-oxidant (see [27]).

Table 3. Modulation of the progress of peroxidation and alpha-tocopherol consumption during LDL autoxidation.

Hydroperoxide [alpha-tocopherol] nmol/mg LDL protein.

Additions	Incubation time (hr.)							
	3		6		12		24	
None	19	[10.0]	27	[9.7]	55	[5.7]	101	[n.d.]
CuSO$_4$ (10uM)	62	[11.2]	119	[n.d.]	378	[n.d.]	328	[n.d.]
EDTA (1mg/ml)	16	[11.2]	16	[11.1]	15	[9.6]	18	[10.8]
BHT (2uM)	19	[10.5]	16	[9.0]	41	[6.0]	92	[1.5]
Ascorbate (0.01mM)	17	[11.5]	17	[11.3]	34	[8.2]	84	[n.d.]
(0.10mM)	16	[10.9]	14	[10.2]	15	[10.2]	66	[3.1]
(1.00mM)	14	[10.9]	15	[11.8]	17	[10.1]	19	[10.5]

Autoxidation was performed as described [14]. At the start of the experiment, the levels were; hydroperoxide, 15nmol/mg protein; alpha-tocopherol, 11.5 nmol/mg protein. n.d.; not detectable.

In this autoxidation system, ascorbate functioned as a net antioxidant and induced a lag-period during which no net consumption of alpha-tocopherol was observed, and no lipid peroxidation was detected. The duration of the lag period was related to the initial concentration of ascorbate in the system (Table 3). Since ascorbate is a major water-soluble antioxidant in plasma, it may be important in the in vivo maintenance Vitamin E in many lipid-containing systems, including LDL.

Alpha-tocopherol consumption and peroxidation during macrophage-mediated LDL modification

Previous studies [3,6] have shown that incubation of LDL with cultured macrophages in the presence of micromolar concentrations of Fe^{2+} leads to the formation of a modified species of LDL which is endocytosed and degraded 10-20 times more rapidly than native LDL by a second set of macrophages. This can lead to relatively uncontrolled accumulation of cholesterol by these cells and may mimic the events which occur in the developing atherosclerotic lesion. Rankin and Leake [6] have shown that there is an induction period of several hours before any high-uptake LDL is formed in this system. In collaboration with this group, we have studied the kinetics of alpha-tocopherol consumption and LDL peroxidation during similar incubations of LDL with macrophages. We observed that alpha-tocopherol was rapidly and completely consumed by 4hr. This was dependent on the presence of macrophages in the incubations. As the tocopherol levels declined, so the rate of LDL peroxidation accelerated. Most interestingly, the formation of high-uptake modified forms of LDL was apparently completely suppressed until all the alpha-tocopherol in the system was exhausted. We conclude, as previously suggested, that cell-mediated modification is dependent on LDL oxidation, and that the endogenous antioxidant content of LDL may be a significant factor in determining its susceptibility to oxidation. Variations in the antioxidant content of LDL between individuals may explain why some groups have found differences in the ease with which different preparations of LDL can be oxidatively modified.[10]

Micromolar concentrations of flavonoids such as morin, myricetin or quercetin added to macrophage incubations resulted in potent inhibition of the progress of alpha-tocopherol consumption, LDL peroxidation and the formation of high-uptake LDL. Since these compounds also prevented cell-free oxidation of LDL by micromolar concentrations of Cu^{2+} [28], their suppression of macrophage-mediated modification of LDL is likely to be largely due to their properties as antioxidants [29] rather than to any direct effects on the activities of the macrophages. Flavone, which is structurally related to these flavonoids but lacks phenolic groups, does not inhibit macrophage-mediated oxidation. Exogenous antioxidants can therefore protect LDL against cell-mediated oxidation.

SUMMARY

In vitro oxidation of LDL by defined oxygen-centred free radicals produces many of the alterations in LDL associated with cell-mediated modification of LDL. Superoxide radical is unable to induce much peroxidation in LDL but can react with and deplete alpha-tocopherol. LDL which has been exposed to superoxide radical is more sensitive to copper-catalysed peroxidation than native LDL.

During in vitro oxidation of LDL by defined free radicals, or by cultured macrophages, alpha-tocopherol consumption precedes the formation of high-uptake, potentially atherogenic, LDL. Similarly, protection of LDL by the addition of exogenous antioxidants such as flavonoids and ascorbate can delay the onset of oxidation and prevent endogenous alpha-tocopherol consumption and the formation of high-uptake LDL.

We conclude that the endogenous antioxidant content of LDL is a significant determinant of its susceptibility to oxidation in vitro, and probably also in vivo. A number of antioxidants have been identified in LDL (see Rotheneder et al., this volume) and it is likely that there are others yet to be detected.

REFERENCES

1. J.L. Goldstein, Y.K. Ho, S.K. Basu and M.S. Brown, Binding site on macrophages that mediates uptake and degradation of acetylated low-density lipoprotein, producing massive cholesterol deposition., Proc. Natl. Acad. Sci. USA., 76:333 (1979)

2. T. Henricksen, E.M. Mahoney and D. Steinberg, Enhanced macrophage degradation of low density lipoprotein previously incubated with cultured endothelial cells: recognition by receptors for acetylated low density lipoproteins., Proc. Natl. Acad. Sci. USA., 78:6499 (1981)

3. S. Parthasarathy, D.J. Printz, D. Boyd, L. Joy and D. Steinberg, Macrophage oxidation of low density lipoprotein generates a modified form recognized by the scavenger receptor., Arteriosclerosis, 6:505 (1986).

4. J.W. Heinecke, H. Rosen and A. Chait, Iron and copper promote modification of low density lipoprotein by human arterial smooth muscle cells in culture., J. Clin. Invest., 74:1890 (1984)

5. U.P. Steinbrecher, S. Parathasarathy, D.S. Leake, J.L. Witzum and D. Steinberg, Modification of low density lipoprotein by endothelial cells involves lipid peroxidation and degradation of low-density lipoprotein phospholipids.,Proc. Natl. Acad. Sci. USA. 81:3883 (1984)

6. S.M. Rankin and D.S. Leake, The modification of low density lipoproteins by macrophages by oxidation or proteolysis., Agents Actions, in press (1988).

7. S. Parathasarathy, L. Fong, D. Otero and D. Steinberg, Regognition of solubilized apoproteins from delipidated, oxidized low density lipoprotein (LDL) by the acetyl-LDL receptor., Proc. Natl. Acad. Sci. USA. 84:537 (1987).

8. U.P. Steinbrecher, Role of superoxide in endothelial-cell modification of low-density lipoproteins., Biochim. Biophys. Acta 959:20 (1988).

9. J.W. Heinecke, L. Baker, H. Rosen and A. Chait, Superoxide-mediated modification of low density lipoprotein by arterial smooth muscle cells., J. Clin. Invest., 77:757 (1986).

10. V.W.M. van Hinsbergh, M. Scheffer, L. Havekes and H.J. van Kempen, Role of endothelial cells and their products in the modification of low density lipoproteins., Biochim. Biophys. Acta, 878:49 (1986).

11. R. Montgomery, C.F. Nathan and Z.A. Cohn, Effect of reagent and cell-generated hydrogen peroxide on the properties of low density lipoprotein., Proc. Natl. Acad. Sci. USA., 83: 6631 (1986).

12. D.W.S. Bilheimer, S. Eisenberg and R.I. Levy, The metabolism of very low-density lipoproteins. Biochim. Biophys. Acta 60:212 (1972).

13. R.T. Dean, C.R. Roberts and W. Jessup, Fragmentation of intracellular and extracellular polypeptides by free radicals, in: "Intracellular protein catabolism," E.A. Khairallah, J.S. Bond and J.W.C. Bird, eds., A.R. Liss, New York (1985).

14. H. Esterbauer, G. Jurgens, O. Quehenberger and E. Koller, Autoxidation of human low density lipoprotein: loss of polyunsaturated fatty acids and Vitamin E and generation of aldehydes., J. Lipid Res., 28:495 (1987).

15. S.M. Thomas, W. Jessup, J.M. Gebicki and R.T. Dean, A continuous-flow automated assay for iodometric estimation of hydroperoxides., Anal. Biochem., 176:in press (1988).

16. G.W. Burton, A. Webb and K.U. Ingold, A mild, rapid, and efficient method of lipid extraction for use in determining Vitamin E/lipid ratios, Lipids 20:29 (1985).

17. U.K. Laemmli, Cleavage of structural proteins during the assembly of the head of bacteriophage T4., Nature 227:680 (1970).

18. W. Jessup, G. Jurgens, J. Lang, H. Esterbauer and R.T. Dean, Interaction of 4-hydroxynonenal-modified lipoproteins with the fibroblast apolipoprotein B/E receptor., Biochem. J. 234:245 (1986).

19. M.S. Brown, J.L. Goldstein, M. Kreiger, Y.K. Ho and R.G.W. Anderson, Reversible accumulation of cholesteryl esters in macrophages incubated with acetylated lipoproteins., J. Cell Biol. 82:597 (1979).

20. J.M. Gebicki and B.H.J. Bielski, Comparison of the capacities of the perhydroxyl and the superoide radicals to initiate chain oxidation of linoleic acid., J. Am. Chem. Soc., 103:7020 (1981).

21. R.T. Dean and K.H. Cheeseman, Vitamin E protects proteins against free radical damage in lipid environments. Biochem. Biophys. Res. Commun., 148:1277 (1987).

22. E. Niki, Antioxidants in relation to lipid peroxidation., Chem. Phys. Lipids, 44:227 (1987).

23. S.P. Wolff and R.T. Dean, Fragmentation of polypeptides by free radicals and its effect on susceptibility to proteolysis., Biochem. J. 234:399 (1986).

24. R.T. Dean, S.M. Thomas and A.C. Garner, Free radical-mediated fragmentation of monoamine oxidase in the mitochondrial membrane, Biochem. J. 240:489 (1986).

25. R.T. Dean, S.M. Thomas, G. Vince and S.P. Wolff, Oxidation induced proteolysis and its possible restriction by some secondary protein modifications., Biomed. Biochim. Acta 45:1563 (1986).

26. J.V. Hunt, J.A. Simpson and R.T. Dean., Hydroperoxide-mediated fragmentation of proteins., Biochem. J. 250:87 (1988).

27. B. Halliwell and J.M.C. Gutteridge, "Free radicals in biology and medicine," Oxford University Press, Oxford (1985).

28. S.M. Rankin, J.R.S. Hoult and D.S. Leake, Effects of flavonoids on the oxidative modification of low density lipoproteins by macrophages, Brit. J. Immunol., in press (1988).

29. W. Bors, H. Heller, C. Michel and M. Saran, Flavonoids as antioxidants. Determination of their radical scavenging efficiencies. Methods in Enzymology, in press. (1988)

A NEW MODEL SYSTEM FOR STUDYING THE CYTOTOXICITY OF PEROXIDIZED LIPOPROTEINS IN CULTURED CELLS

R. Salvayre[1], A. Nègre[1], M. Lopez[1], N. Dousset[1], T. Levade[1],
A. Maret[1], M.T. Pieraggi[2] and L. Douste-Blazy[1]

[1] Laboratoire de Biochimie, INSERM 101, Faculté de Médecine,
37 Allées J. Guesde, 31073 Toulouse Cedex, France
[2] Anatomie Pathologique, CHU Rangueil, Toulouse, France

INTRODUCTION

Free radicals can be generated by physical agents, e.g. ionizing radiations[1,2] as well as by a wide variety of biochemical reactions occuring in living organisms[3,4,5]. Transition metals seem to play a major role in the interconversion of oxygen free radical species, for example in the Haber-Weiss reaction[6,7]. Lipid peroxidation induced by free radical attack is amplified by autocatalytic cycles generating in turn lipid peroxide free radicals; this "propagation" step leads to the "termination" step which is characterized by the formation of non-radical compounds such as short chain aldehydes, ketones, alkanes, carboxylic acids, lipid dimers[6].

Peroxidation of lipoproteins can occur in vitro, as well as in vivo[8,9,10] In vitro, during lipoprotein isolation, peroxidation is catalysed by iron ions[11] and can be prevented by chelators and antioxidants[12]. In biological systems, LDL peroxidation can be induced by oxygen free radicals produced by endothelial and smooth muscle cells[4,11,13,14] and by neutrophils and monocytes[15].

LDL attack by free radicals generated by iron ions or by cells induces several modifications of the apo B (modifications of charge and of molecular weight) and of the lipids (decrease of polyunsaturated fatty acids, increase of aldehyde compounds and of thiobarbituric reactive substance[9,10,14] or TBARS). The "modified" apo B is no more recognized by the apo B/E receptor but only by the scavenger receptor of macrophages[13,16,17,18]. In addition, the oxidized lipoproteins exhibit a strong cytotoxicity against various types of cultured cells[8,15,19].

These elegant studies are considerably hindered by the impossibility to control and quantify the intensity of the peroxidizing stress and to study the cytotoxicity subsequent to the uptake of the peroxidized LDL through the apo B/E receptor pathways.

This prompted us to define new experimental systems in which lipoproteins contain peroxidized lipids and unmodified (or only slightly modified) apo B. We report here the study of the chemical and biological properties of lipoproteins treated by UV light (at 254 nm) and the effect of the replacement of native neutral lipids by peroxidized triacylglycerols or cholesteryl esters.

MATERIALS AND METHODS

Preparation and UV-treatment of the lipoproteins

Lipoproteins were isolated by sequential ultracentrifugation according to Havel et al.[20,12] under previously used conditions[21]. After careful
dialysis[12], in buffer containing 10 mmol/l EDTA, the isolated lipoproteins
were irradiated by short-UV (254 nm) for variable period of time and we
monitored the formation of conjugated dienes[22] and TBARS[23] and analysed
the fatty acid[24] composition and the apoprotein structure[25].

Cell culture

The lymphoid cell lines (LCL) were established by Epstein-Barr virus
transformation of blood B-lymphocytes as previously described[26].
Cells were grown in a standard culture medium RPMI 1640 containing
10% fetal calf serum and glutamine. 3 days before the experiments, the
standard culture medium was substituted by a lipoprotein-free medium and
supplemented with 2% Ultroser HY (IBF, Villeneuve-la Garenne, F).
The lipoprotein uptake was followed by using the fluorescent
carbocyanine labelling according to Via and Smith[27].

Study of the cytotoxicity

The cytotoxic effect of the UV-irradiated lipoproteins on the cultured
LCL was investigated by 3 tests simultaneously performed: trypan-blue test,
[3]H-Thymidine incorporation into cells and release of the cellular LDH
(lactate dehydrogenase) in the culture medium according to Jürgens et al.[10].

RESULTS AND DISCUSSION

Lipoperoxidation induced by UV-treatment (Table 1)

UV-irradiation of the purified LDL and VLDL induced the formation of
conjugated dienes which reach a plateau after 20 h of UV-irradiation, under
the used experimental conditions, and also of the TBARS which were
continuously increasing with time, during all the UV-irradiation period.
In contrast, under the same experimental conditions, UV-irradiation
affected neither the conjugated diene nor the TBARS levels in HDL.

Table 1. Conjugated dienes and TBARS during UV-irradiation of the various
classes of purified human serum lipoproteins.

UV-irradiation time (hours)	conjugated dienes[a]			TBARS[b]		
	VLDL	LDL	HDL	VLDL	LDL	HDL
0	100	100	100	0.1	0.1	0.06
5	210	140	102	0.2	0.3	0.07
20	300	180	102	2.1	2.5	0.07
30	310	185	97	3.1	3.5	0.09
60	320	190	100	5.2	6.0	0.08

[a] Conjugated diene (measured at 233 nm) contents are expressed as percent
of the initial level (i.e. initial absorbance at 233 nm).
[b] TBARS content is expressed as μmol per g of apoprotein.

The study of the fatty acid composition of lipoproteins showed a significant decrease in polyunsaturated fatty acid (PUFA) content in UV-irradiated LDL, but not in HDL. In comparison, the treatment of lipoproteins by Fe^{++} induced a significant decrease of PUFA content in both LDL and HDL (Table 2).

SDS-PAGE showed that the molecular weights of the apo B and apo A1 from the respective UV-irradiated LDL and HDL were not affected by UV, under the used experimental conditions (Figure 1). These results are different from those observed in the Cu- or Fe-treated[28] LDL which exhibit a fragmentation of the apo B after peroxidation catalyzed by transition metals.

Table 2. Effect of UV-irradiation on the polyunsaturated fatty acids (PUFA) and apoproteins of lipoproteins.

		Controls	UV	Fe^{++}
LDL				
PUFA %	TG	8	4	5
	CE	-	-	-
	PL	26	14	8
	FFA	15	13	1
ApoB		unbroken	unbroken	fragments
HDL				
PUFA %	TG	22	18	11
	CE	24	20	8
	PL	-	-	-
ApoA$_1$		unbroken	unbroken	unbroken

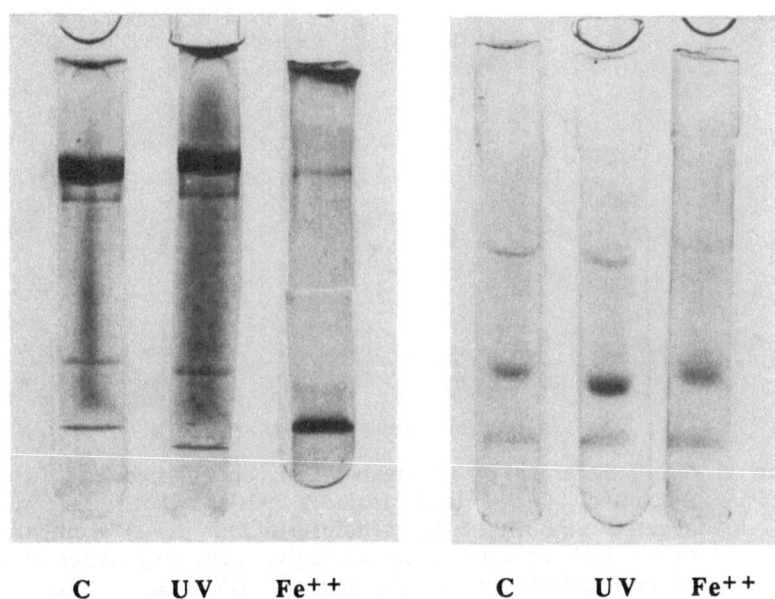

C   UV Fe^{++} C   UV Fe^{++}

Figure 1. SDS-PAGE of the apoproteins from the native (C), UV- or Fe-treated lipoproteins, LDL (left) and HDL (right).

The study of the amino-acid composition showed a significant decrease of His content (-23%) and only a slight decrease of that of Lys (-5%) in UV-irradiated apoB.

Cytotoxic effect of the UV-irradiated LDL

As shown in Figure 2, the cytotoxicity was dependent on the dose of the UV-irradiated LDL taken up by the cells. Low doses (50 µg apoprotein/ml) of UV-treated LDL were cytotoxic for normal cells but not for cells from Familial Hypercholesterolemia (Receptor-negative); in these cells the cytotoxicity appeared only at high doses (> 200 µg apoprotein/ml under the used experimental conditions). In order to confirm the role of the LDL-receptors in the cytotoxic mechanism, we examined the uptake of the UV-irradiated LDL labelled by carbocyanine. The native and the UV-treated LDL were internalized in normal lymphoid cells at a similar rate and by a saturable mechanism, in agreement with the receptor-dependent pathway reported by Ho et al.[29].

In the receptor-negative cells, the lack of cytotoxicity of low doses is probably due to the complete defect of the receptor-mediated uptake and to the very low efficiency of receptor-independent uptake (at these low concentrations); the cytotoxic effect appeared only when the UV-treated LDL (internalized through the receptor-independent pathway) reached a critical threshold.

It is noteworthy that UV-treated HDL were not cytotoxic at all, even after a very long UV-treatment and at very high doses.

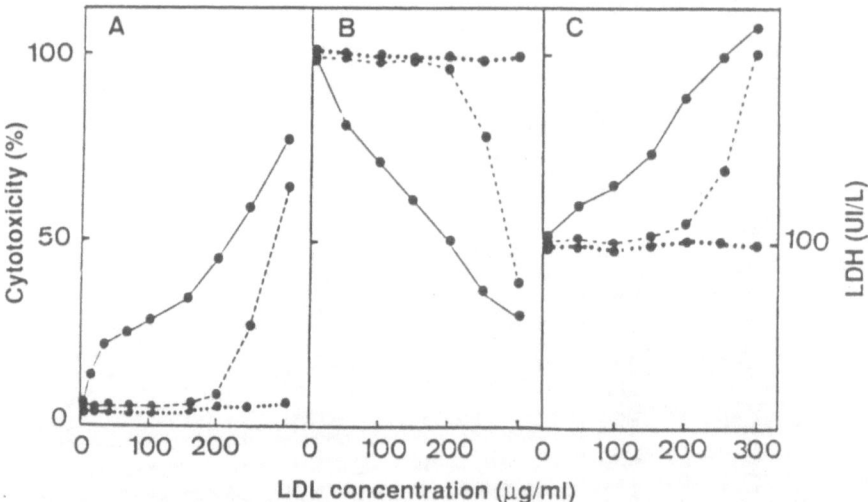

Figure 2. Cytotoxic effect of UV-treated LDL on cultured LCL from a control (——) and from a patient with receptor-negative Familial Hypercholesterolemia (---) compared to the lack of cytotoxicity of native LDL (···). After 48 h of cell culture in the presence of UV-treated LDL, the cytotoxicity was determined by using the Trypan-Blue Test (A), the [³H]-Thymidine incorporation in the cells (B) and the release of LDH in the culture medium (C).

Relative cytotoxicity of the lipid classes from UV-treated LDL (Table 3)

Using native and UV-treated LDL partially delipidated and relipidated according to the procedure of Krieger[30] we have tested the relative cytotoxicity of LDL reconstituted with various UV-treated lipid fractions. The LDL reconstituted with triglycerides or cholesteryl esters extracted and purified from UV-LDL were highly cytotoxic. The "envelopes" from UV-LDL (partially delipidated LDL constituted by phospholipids and apo B), utilized without relipidation or with relipidation by non irradiated neutral lipids, were less cytotoxic than the non-delipidated UV-treated LDL. These data suggest that the cytotoxicity of the UV-LDL is mainly due to the neutral lipids contained in the "core" of the LDL.

Table 3. Relative cytotoxicity of LDL reconstituted with various lipid fractions from UV-LDL.

Composition of the reconstituted LDL			Cytotoxicity (%)
"envelope"		Core lipids	
Native	+	UV-triglycerides	70
Native	+	UV-Cholesteryl Esters	50
UV-treated	+	Native neutral lipids	30
Native	+	Native neutral lipids (controls)	15

* The relative cytotoxicity was determined by trypan blue-test.

Electron microscopy features (Figure 3)

As shown by scanning electron microscopy (Figure 3-right) the UV-LDL induced the formation of numerous vesicles or "blebs" in the plasma membrane of the cells. Transmission electron microscopy showed the presence of structural alterations of the mitochondria in the cells treated by UV-LDL (data not shown).

These morphological changes show that the peroxidizing stress used in these experiments induces several pleiotropic effects, but we do not know yet the nature of the cytotoxic molecules nor the molecular mechanism of the cytotoxicity.

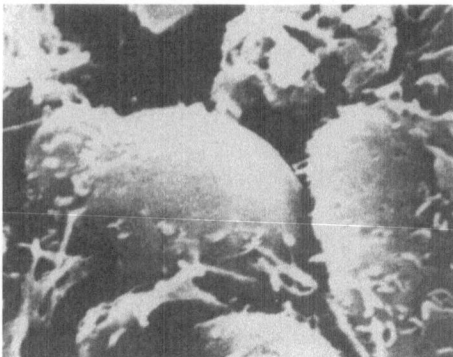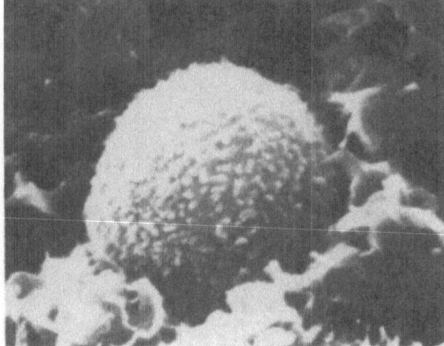

Figure 3. Scanning electron microscopy of the lymphoid cell before (left) and after (right) culture in the presence of UV-irradiated LDL.

Protective effect of antioxidants (Table 4)

The cultured cells could be completely protected against the cytotoxicity of the UV-LDL by antioxidants introduced in the cultured medium simultaneously with the UV-LDL. 1 µg/ml catechin or 100 µg/ml vitamin E induced a complete inhibition of the cytotoxicity in the used experimental conditions. These antioxidants could either block a secondary peroxidation step induced by the lipid peroxides of the UV-LDL or could take part in the reduction of the peroxide to form a much less toxic alcohol group.

Table 4. Protective effect of antioxidants against the cytotoxic effect of UV-LDL

Antioxidant	Antioxidant concentration (µg/ml)					
	0	0.01	0.1	1	10	100
Catechin	100	80	35	0	0	n d
Tocopherol	100	100	90	50	15	0

* The relative cytotoxic effect was determined by 3 tests (trypan blue, [3H] Thymidine and release of LDH) and calculated by comparison with the maximal cytotoxic effect (in the presence of UV-LDL).

In conclusion, we can emphasize the interest of the experimental model system described in this paper:
- the intensity of the peroxidizing physical agents, i.e., UV, is easy to modulate,
- under the used conditions, the lipidic part of the LDL is cytotoxic without major structural and functional modification of the apoprotein part.
- this experimental system seems to be a good model for testing the cytotoxicity of lipid peroxides which are potentially involved in various pathological processes such as atherosclerosis.

ACKNOWLEDGEMENTS

The authors wish to thank C. Vieu, J.C. Thiers, M. Troly and J. Dumoulin for their excellent technical assistance.

REFERENCES

1. R. Latarjet, Les peroxydes organiques en radiologie, in "Action chimique et biologique des radiations" (4° serie), Masson, Paris (1958).

2. J. B. Mudd, The role of free radicals in toxicity of air polluants, in "Free radicals in biology", W.A. Pryor Ed., vol. 2, Academic Press, New-York, 203-211 (1976).

3. R. Mason, Free-radical intermediates in the metabolism of toxic chemicals, in "Free radicals in biology", W.A. Pryor Ed., vol. 5, Academic Press, New-York, 161-221 (1982).

4. J. W. Heinecke, L. Baker, H. Rosen, A. Chait, Superoxide-mediated modification of low density lipoprotein by arterial smooth muscle cells, J. Clin. Invest., 77: 757-761 (1986).

5. P. Gorog, J. D. Pearson, V. V. Kakkar, Generation of reactive oxygen metabolism by phagocytosing endothelial cells, Atherosclerosis, 72: 19-27 (1988).

6. S. D. Aust, B. A. Svingen, The role of iron in enzymatic lipid peroxidation, in "Free radicals in biology", W.A. Pryor Ed., vol. 5, Academic Press, New-York, 1-28 (1982).

7. J. M. Braughler, L. A. Duncan, R. L. Chase, The involvement of iron in lipid peroxidation. Importance of ferric to ferrous ratios in initiation, J. Biol. Chem., 261: 10282-10289 (1986).

8. J. R. Hessler, D. W. Morel, L.J. Lewis, G. M. Chisolm, Lipoprotein oxidation and lipoprotein-induced cytotoxicity, Arteriosclerosis, 3: 215-222 (1983).

9. H. Esterbauer, G. Jürgens, O. Quehenberger, E. Koller, Autoxidation of human low density lipoprotein: loss of polyunsaturated fatty acids and vitamine E and generation of aldehydes, J. Lip. Res., 28: 495-509 (1987) .

10. G. Jürgens, H. Hoff, G. M. Chisolm, H. Esterbauer, Modification of human serum low density lipoprotein by oxidation: characterization and pathophysiological implications, Chem. Phys. Lipids, 45: 315-336 (1987).

11. J. W. Heinecke, H. Rosen, A. Chait, Iron and copper promote modification of low density lipoprotein by human arterial smooth muscle cells in culture, J. Clin. Invest., 74: 1890-1894 (1984).

12. V. N. Schumaker and D.L. Puppione, Sequential flotation ultracentrifugation, Meth. Enzymol., 128: 155-170 (1986).

13. T. Henriksen, E. M. Mahoney, D. Steinberg, Enhanced macrophage degradation of low density lipoprotein previously incubated with cultured endothelial cells: recognition by receptors for acetylated low density lipoproteins, Proc. Natl. Acad. Sci., 78: 6499-6503 (1981) .

14. U. P. Steinbrecher, S. Parthasarathy, D. Leake, J. L. Witztum, D. Steinberg, Modification of low density lipoprotein by endothelial cells involves lipid peroxidation and degradation of low density lipoprotein phospholipids, Proc. Natl. Acad. Sci., 81: 3883-3887 (1984).

15. M. K. Cathcart, D. W. Morel, and M. G. Chisholm, Monocytes and neutrophils oxidise low density lipoprotein making it cytotoxic, J. Leukocyte Biol., 38: 341-350(1985).

16. J. L. Goldstein, Y. K. Ho, S. K., Basu, M. S. Brown, Binding site on macrophages that mediates uptake and degradation of acetylated low density lipoprotein, producing massive cholesterol deposition, Proc. Nat. Acad. Sci., 76: 333-337 (1979).

17. S. Parthasarathy, Oxidation of low density lipoprotein by thiol compounds leads to its recognition by the acetyl LDL receptor, Biochim. Biophys. Acta, 917: 337-340 (1987).

18. U. P. Steinbrecher, J. L. Witztum, S. Parthasarathy, D. Steinberg, Decreased in reactive amino groups during oxidation on endothelial cell modification of LDL. Correlation with changes in receptor-mediated catabolism, <u>Arteriosclerosis</u>, 7: 135-143 (1987).

19. J. R. Hessler, A. Lazzarini-Robertson, G. M. Chisolm, LDL-induced cyto-toxicity and its inhibition by HDL in human vascular smooth muscle and endothelial cells in culture, <u>Atherosclerosis</u>, 32: 213-229 (1979).

20. R. I. Havel, H. A. Eder, J. H. Braigon, The distribution and chemical composition of ultracentrifugally separated lipoproteins in human serum, <u>J. Clin. Invest.</u>, 39: 1345-1363 (1955).

21. R. Salvayre, A. Nègre, A. Maret, J. Radom, L. Douste-Blazy, Extracellular origin of the lipid lysosomal storage in cultured fibroblasts from Wolman's disease, <u>Eur. J. Biochem.</u>, 170: 453-458 (1987).

22. R. O. Recknagel, and E. A. Glende, Spectrophotometric detection of lipid conjugated dienes, <u>Meth. Enzymol</u>. 105: 331-337 (1984).

23. K. Yagi, A simple fluorimetric assay for lipoperoxides in blood plasma, <u>Biochem. Med.</u>, 15: 212-216 (1976).

24. J. M. Lillington, D. J. Trafford and H. L. Makin, A rapid and simple method for the esterification of the fatty acids and steroid carboxylic acids prior to gas-liquid chromatography, <u>Clin. Chim. Acta</u>, 111: 91-98 (1981).

25. A. Lespine, N. Dousset, B. Perret, M. De Forny, H. Chap, L. Douste-Blazy, Accumulation of large VLDL in cyclophosphamide-treated rabbits. Relationship with lipoprotein lipase deficiency, <u>Biochem. Biophys. Res. Comm.</u>, 154: 633-640 (1988).

26. R. Salvayre, A. Nègre, A. Maret, G. Lenoir, L. Douste-Blazy, Separation and properties of molecular forms of alpha-galactosidase and alpha-N-acetylgalactosaminidase from blood lymphocytes and lymphoid cell lines transformed by Epstein-Barr virus. <u>Biochim. Biophys. Acta</u> 659: 445-456 (1981).

27. D. P. Via, L. C. Smith, Fluorescent labelling of lipoproteins, <u>Meth. Enzymol.</u>, 129: 848-857 (1986).

28. S. Parthasarathy, U. P. Steinbrecher, J. Barnett, J. L. Witztum, D. Steinberg, Essential role of phospholipase A2 activity in endothelial cell induced modification of low density lipoprotein, <u>Proc. Natl. Acad. Sci.</u>, 82: 3000-3004 (1985).

29. Y. K. Ho, M. S. Brown, A. J. Kayden and J. L. Goldstein, Binding, internalization and hydrolysis of LDL in long-term lymphoid cell lines from a normal subject and a patient with homozygous familial hypercholesterolemia, <u>J. Exp. Med.</u>, 114: 444-455 (1976).

30. M. Krieger, M.S. Brown, J.R. Faust, J.L. Goldstein, Replacement of endogenous cholesteryl esters of low density lipoprotein with exogenous cholesteryl linoleate, <u>J. Biol. Chem.</u>, 252: 4093-4101 (1978).

BIOLOGICAL EFFECTS

MOLECULAR MECHANISMS OF OXIDATIVE CELL DAMAGE

Sten Orrenius

Department of Toxicology, Karolinska Institutet

S-104 01 Stockholm, Sweden

INTRODUCTION

Tissue necrosis has long been known to be associated with the accumulation of calcium in the necrotic tissue, and it has been proposed that the calcium ion may play a critical role in cell killing.[1] Although the generality of this hypothesis has been questioned,[2] the involvement of Ca^{2+} in the development of lethal cell injury is now supported by a number of observations. For example, Shanne and associates[3] found that removal of Ca^{2+} from the medium protected cultured hepatocytes from the toxicity of a variety of agents. This and similar observations led to the proposal that an influx of extracellular Ca^{2+} could play an important role in the development of irreversible cell damage.[4]

Studies in this and other laboratories have demonstrated that oxidative stress in hepatocytes is associated with the inhibition of Ca^{2+} transport and the disruption of intracellular Ca^{2+} homeostasis.[5] This may lead to a sustained elevation of cytosolic Ca^{2+} concentration which, in turn, can cause the activation of various Ca^{2+}-dependent degradative enzymes, including phospholipases, proteases, and endonucleases. This paper will focus on the mechanisms by which a perturbation of intracellular Ca^{2+} homeostasis can cause cytotoxicity and cell death.

Free Radicals, Lipoproteins, and Membrane Lipids
Edited by A. Crastes de Paulet *et al.*
Plenum Press, New York, 1990

The low resting concentration of Ca^{2+} in the cytosol of hepatocytes ($\approx 0,1$ μM) is maintained by active compartmentation processes. The mitochondria and the endoplasmic reticulum represent the predominant sites of intracellular Ca^{2+} sequestration, whereas the plasma membrane Ca^{2+} pump plays a major role in the maintenance of the concentration gradient existing between the extra- and intracellular environments by actively extruding Ca^{2+} from the cell.

Mitochondrial Ca^{2+} homeostasis is regulated by a cyclic mechanism, involving Ca^{2+} uptake by an energy-dependent pathway, and Ca^{2+} release which, in the liver, is probably mediated by a Ca^{2+}/H^+ antiporter. The latter appears to be regulated by the redox level of intramitochondrial pyridine nucleotides,[6] although membrane-bound protein thiols may also be important in modulating mitochondrial Ca^{2+} fluxes. The active transport of calcium ions through the endoplasmic reticular and plasma membranes is mediated by Ca^{2+}-stimulated, Mg^{2+}-dependent ATPases which appear to be dependent on free sulfhydryl groups for their activity.[7,8]

Thus, it seems that the major mechanisms responsible for Ca^{2+} transport across cellular membranes are all dependent on protein thiol groups. It is therefore conceivable that oxidation and/or arylation of these critical thiol groups may result in the inactivation of the translocases involved in the regulation of intracellular Ca^{2+} compartmentation, leading to an alteration of the level of cytosolic Ca^{2+}.

GLUTATHIONE OXIDATION AND PROTEIN THIOL MODIFICATION DURING OXIDATIVE STRESS

Damage to biological systems caused by generation of active oxygen species is often referred to as oxidative stress.[9] Since aerobic cells possess antioxidant systems to trap and/or inactivate such reactive species, oxidative stress is the result of an imbalance between the generation of active oxygen species and their inactivation by the protective systems.

Glutathione (GSH) plays a unique role in cellular defence against reactive intermediates, inasmuch as it functions as a reductant in the metabolism of hydrogen peroxide and organic hydroperoxides and can also bind electrophilic metabolites.[10] During the glutathione peroxidase-catalyzed metabolism of hydroperoxides, GSH serves as an electron donor and the glutathione disulfide (GSSG) formed in the reaction is subsequently reduced back to GSH by glutathione reductase at the expense of NADPH. Under conditions of oxidative stress, when the cell must cope with large amounts of H_2O_2 and/or organic hydroperoxides, the glutathione reductase is unable to match the rate of glutathione oxidation, and GSSG accumulates. In an apparent effort to avoid the detrimental effects of increased intracellular levels of GSSG, e.g. formation of mixed disulfides with protein thiols, the cell actively excretes the disulfide which can lead to a depletion of the intracellular glutathione pool.

Other thiol groups of toxicological significance are those present in cellular proteins. In general, sulfhydryl groups are highly reactive and can participate in several different types of reactions, such as arylation, oxidation, and thiol-disulfide exchange. All of these reactions may be involved in the modification of protein thiols as a consequence of their interaction with reactive intermediates formed during the metabolism of toxic chemicals. Such modifications of protein thiols may be particularly important during oxidative stress.[11]

QUINONE-INDUCED OXIDATIVE STRESS

In recent years our laboratory has been actively engaged in the study of possible mechanisms of cell killing by oxidative stress. Menadione (2-methyl-1,4-naphthoquinone) is a model compound we have used extensively in these studies. In mammalian cells menadione metabolism involves both one- and two-electron reduction pathways which differ in their contribution to cytotoxicity.[12] One-electron reduction of quinones results in the formation of semiquinone radicals which can rapidly reduce dioxygen, forming the superoxide anion radical ($O_2^{\cdot-}$), and regenerating the parent quinone. Dismutation of $O_2^{\cdot-}$ to hydrogen peroxide, and the production of other highly reactive species, quickly lead to a condition of oxidative stress as redox cycling of the quinone continues.

Exposure of freshly isolated rat hepatocytes to toxic concentrations of menadione results in the extensive formation of $O_2^{\bar{}}$ and H_2O_2 and the oxidation of GSH and pyridine nucleotides. Following GSH depletion, a dose- and time-dependent loss of protein thiols precedes cell death in hepatocytes exposed to menadione.[11] Although a general decrease in protein thiols has been observed in cells exposed to cytotoxic levels of menadione, this does not imply that all protein thiols are equally important for cell survival. In fact, it appears that the fraction of critical thiols may be quite small. As mentioned above, the proper functioning of intracellular Ca^{2+} translocases is dependent on thiol groups in specific proteins, and the disruption of Ca^{2+} homeostasis resulting from their modification appears to be ultimately related to the onset of cytotoxicity. Other critical thiols affected by menadione metabolism are located on microfilament proteins, whose modification may result in the disruption of the structural integrity of the plasma membrane and the onset of widespread bleb formation.

APPEARANCE OF SURFACE BLEBS IN HEPATOCYTES EXPOSED TO TOXIC AGENTS

Incubation of hepatocytes with cytotoxic levels of menadione results in a loss of microvilli and the appearance of multiple blebs on the surface of the hepatocytes.[12] Many other toxic agents cause similar alterations of surface structure, indicating that plasma membrane blebbing is a common event in the progression of toxic injury.[13] Blebs usually appear before any sign of increased plasma membrane permeability is observed and seem to be initially reversible, i.e. they often disappear when the toxic agent is removed from the incubation.

Cell surface morphology is thought to be determined by the organization of cortical microfilaments associated with the plasma membrane. This assumption is supported by the finding that two classes of compounds, the cytochalasins and phalloidins which disrupt cortical microfilament structure, cause bleb formation on the surface of hepatocytes similar to that observed with other toxic agents.

As will be discussed below, during oxidative stress a perturbation of intracellular Ca^{2+} homeostasis is intimately related to a modification of the intracellular thiol status. Disruption of intracellular thiol and Ca^{2+} homeostasis may both result in alterations of cytoskeletal structure. Thiol oxidation can perturb microfilament organization through S-S-cross-linking of cytoskeletal polypeptides.[14] A change in intracellular Ca^{2+} distribution can affect cytoskeletal structure because the calcium ion and its associated binding proteins play a pivotal role in regulating the cytoskeletal network.[15] Alternatively, Ca^{2+}-activated catabolic enzymes, such as cytosolic proteases, can cleave cytoskeletally-associated proteins and thus cause alteration of the microfilament network and blebbing.[16-18] It is therefore conceivable that both thiol modification and Ca^{2+}-dependent processes may combine to produce cytoskeletal alterations during oxidative stress.

DISRUPTION OF INTRACELLULAR CALCIUM HOMEOSTASIS DURING OXIDATIVE STRESS

In studies with isolated hepatocytes we have been able to demonstrate that menadione-induced oxidative stress causes the mobilization of Ca^{2+} from both the mitochondria and the endoplasmic reticulum.[12] Further studies with subcellular fractions have shown that menadione metabolism impairs the ability of mitochondria to take up and retain Ca^{2+} by causing the oxidation of intramitochondrial pyridine nucleotides.[19] Menadione also inhibits Ca^{2+} uptake by rat liver microsomes through a mechanism which seems to involve oxidation/arylation of thiol group(s) required for Ca^{2+}-ATPase activity.[20] The incubation of hepatocytes with menadione therefore causes the release into the cytosol of Ca^{2+} which cannot be resequestered. Normally, this would result in only a transient rise in the cytosolic Ca^{2+} concentration, because the plasma membrane Ca^{2+}-ATPase would remove this Ca^{2+} from the cell, and the Ca^{2+} concentration would return to its usual, very low level. However, recent studies in our laboratory have shown that the hepatic plasma membrane Ca^{2+}-ATPase is inhibited by agents, including menadione, which oxidize membrane protein thiols.[21] The redox cycling of menadione can therefore inhibit the Ca^{2+} translocases present in the mitochondria, endoplasmic reticulum, and plasma membrane of hepatocytes, leading to a sustained rise in cytosolic Ca^{2+} which, in turn, appears to be critically related to the development of cytotoxicity.[22]

MECHANISMS OF CALCIUM-MEDIATED CYTOTOXICITY

The discovery of a relationship between a sustained increase in cytosolic Ca^{2+} concentration and the toxicity of menadione and several other agents in hepatocytes has led to a search for mechanism(s) by which the increased cytosolic Ca^{2+} could trigger cytotoxicity. For example, it has been proposed that an increase in intracellular Ca^{2+} concentration may cause abnormal stimulation of various Ca^{2+}-dependent degradative processes.

Ca^{2+}-activated phospholipases are widely distributed in mammalian cells, and it has been suggested that an enhanced rate of phospholipid hydrolysis may result in irreversible cell damage. This assumption is supported by the observation that inhibitors of phospholipase activity can prevent ischemic cell death in liver and heart, and by the finding that phospholipid breakdown is enhanced during tissue injury.[23] We have recently investigated whether phospholipase activation may contribute to the toxicity of the reactive disulfide cystamine in hepatocytes.[17] Indeed, exposure to cystamine did increase the rate of phospholipid hydrolysis, and this effect followed the elevation of cytosolic Ca^{2+} and preceded loss of cell viability. However, in our study, phospholipase inhibitors failed to prevent cystamine toxicity.

Intracellular protein degradation is also known to be stimulated by a rise in Ca^{2+}, and a group of cytosolic cysteinyl proteases, whose activity depends on micromolar concentrations of Ca^{2+}, seems to be involved in this process.[24] These proteases function in the cleavage of membrane receptors, cytoskeletal proteins, and protein kinase C. In our study, exposure of hepatocytes to toxic levels of cystamine resulted in enhanced proteolytic activity which was prevented by inhibitors of the Ca^{2+}-dependent proteases; the same inhibitors also protected the cells from cystamine toxicity.[17]

Another activity found to be stimulated by a sustained elevation of cytosolic Ca^{2+} concentrations is that of an endogenous endonuclease, which has been associated with "apoptosis", or programmed cell death.[25] Thus, both glucocorticoid hormones and the calcium ionophore A23187 stimulate apoptosis in isolated mouse thymocytes.[26,27] Characteristic morphological changes accompany this process, including plasma membrane blebbing and extensive chromatin condensation.

As noted above, we have found that plasma membrane blebbing can often be linked to a disruption of Ca^{2+} homeostasis, and that the blebbing observed in hepatocytes treated with ionophore A23187 is associated with the activation of Ca^{2+}-dependent proteases.[18] The appearance of blebs on apoptotic cells could therefore be an indication that a perturbation of Ca^{2+} homeostasis accompanies this process. Since menadione metabolism creates such a disruption, we have recently studied its effects on endogenous endonuclease activity in hepatocytes.[28] In the presence of moderately toxic concentrations of menadione, endonuclease activation was apparent after 2-4 hours of incubation. Moreover, activation of the endonuclease preceded cell death by an appreciable period of time and was associated with the appearance of plasma membrane blebbing, characteristic of menadione-induced oxidative stress.

Interestingly, morphological alterations associated with menadione treatment of hepatocytes follow those observed with apoptotic thymocytes quite closely. The formation of surface blebs has already been mentioned, but chromatin condensation is also seen and a close similarity exists between DNA cleavage products isolated from apoptotic thymocytes and from menadione-treated hepatocytes. Gel electrophoresis of these products has shown that host chromatin is preferentially cleaved into oligonucleosome-length fragments by the endonuclease in each case.[25,28] Thus, it appears that endonuclease activation is not only a critical step in the sequence of events involved in programmed cell death but can also occur during the development of oxidative injury in hepatocytes.

CONCLUDING REMARKS

A series of potentially toxic mechanisms are activated when a cell is exposed to agents which cause oxidative stress. They include a perturbation of the normal redox balance, thiol depletion, and disruption of intracellular Ca^{2+} homeostasis. The pathways through which a perturbation of intracellular Ca^{2+} homeostasis may cause irreversible damage appear to involve the activation of degradative enzymes such as phospholipases, proteases, and endonucleases. Although additional mechanisms may contribute to cell killing, our findings suggest that a disruption of Ca^{2+} homeostasis is a critical event in the development of lethal cell injury.

REFERENCES

1. A. K. Campbell, "Intracellular Calcium: Its Universal Role as Regulator," Wiley and Sons, Chichester (1983).
2. P. E. Starke, J. B. Hoek, and J. L. Farber, Calcium-dependent and calcium-independent mechanisms of irreversible cell injury in cultured hepatocytes, J. Biol. Chem. 261:3006 (1986).
3. F. A. Shanne, A. B. Kane, E. E. Young, and J. L. Farber, Calcium dependence of toxic cell death: a final common pathway, Science 206:700 (1979).
4. J. L. Farber, The role of Ca^{2+} in cell death, Life Sciences 29:1289 (1981).
5. S. Orrenius, and G. Bellomo, Toxicological implications of perturbation of Ca^{2+} homeostasis in hepatocytes, in: "Calcium and Cell Function," Vol. VI, W. Y. Cheung, ed., Academic Press, Orlando (1986).
6. A. L. Lehninger, A. Vercesi, and E. Bababunmi, Regulation of Ca^{2+} release from mitochondria by the oxidation-reduction state of pyridine nucleotides, Proc. Natl.

Acad. Sci. USA 75:1690 (1978).

7. L. Moore, T. Chen, H. R. Knapp, and E. Landon, Energy-dependent Ca^{2+} sequestration activity in rat liver microsomes, J. Biol. Chem. 250:4562 (1975).

8. G. Bellomo, F. Mirabelli, P. Richelmi, and S. Orrenius, Critical role of sulfhydryl group(s) in the ATP-dependent Ca^{2+}-sequestration by the plasma membrane fraction from rat liver, FEBS Letters 163:136 (1983).

9. H. Sies, Biochemistry of oxidative stress, Angewandte Chemie International Editions English 25:1058 (1986).

10. S. Orrenius, and P. Moldéus, The multiple roles of gluta-thione in drug metabolism, Trends Pharmacol. Sci. 5:432 (1984).

11. D. Di Monte, G. Bellomo, H. Thor, P. Nicotera, and S. Orrenius, Menadione-induced cytotoxicity is associated with protein thiol oxidation and alteration in intra-cellular Ca^{2+} homeostasis. Arch. Biochem. Biophys. 235:343 (1984).

12. H. Thor, M. Smith, P. Hartzell, G. Bellomo, S. A. Jewell, and S. Orrenius, The metabolism of menadione (2-methyl-1,4-naphthoquinone) by isolated hepatocytes - A study of implications of oxidative stress in intact cells, J. Biol. Chem. 257:12419 (1982).

13. S. A. Jewell, G. Bellomo, H. Thor, S. Orrenius, and M. T. Smith, Bleb formation in hepatocytes during drug metabolism is caused by disturbances in thiol and calcium ion homeostasis, Science 214:1257 (1982).

14. F. Mirabelli, A. Salis, V. Marinoni, G. Finardi, G. Bellomo, H. Thor, and S. Orrenius, Menadione-induced bleb formation in hepatocytes is associated with the oxidation of thiol groups in actin, Arch. Biochem. Biophys. 264:261 (1988).

15. W. Y. Cheung, Calmodulin plays a pivotal role in cellular regulation, Science 217:1257 (1980).

16. N. L. Collier, and K. Wang, Purification and properties of human platelets P235, J. Biol. Chem. 257:6937 (1982).

17. P. Nicotera, P. Hartzell, C. Baldi, S.-Å. Svensson, G. Bellomo, and S. Orrenius, Cystamine induces toxicity in hepatocytes through the elevation of cytosolic Ca^{2+} and the stimulation of a non-lysosomal proteolytic system, J. Biol. Chem. 261:14628 (1986).

18. P. Nicotera, P. Hartzell, G. Davis, and S. Orrenius, The formation of plasma membrane blebs in hepatocytes exposed to agents that increase cytosolic Ca^{2+} is mediated by the activation of a non-lysosomal proteolytic system, FEBS Letters 209:139 (1986).

19. G. A. Moore, J. P. O'Brien, and S. Orrenius, Menadione (2-methyl-1,4-naphthoquinone)-induced Ca^{2+} release from rat-liver mitochondria is caused by NAD(P)H oxidation, Xenobiotica 16:873 (1986).

20. H. Thor, P. Hartzell, S.-Å. Svensson, S. Orrenius, F. Mirabelli, V. Marinoni, and G. Bellomo, On the role of thiol groups in the inhibition of liver microsomal Ca^{2+}-sequestration by toxic agents, Biochem. Pharmacol. 34:3717 (1985).

21. P. Nicotera, M. Moore, F. Mirabelli, G. Bellomo, and S. Orrenius, Inhibition of hepatocyte plasma membrane Ca^{2+}-ATPase activity by menadione metabolism and its restoration by thiols, FEBS Letters 181:149 (1985).

22. P. Nicotera, D. McConkey, S.-Å. Svensson, G. Bellomo, and S. Orrenius, Correlation between cytosolic Ca^{2+} concentration and cytotoxicity in hepatocytes exposed to oxidative stress, Toxicology 52:55 (1988).

23. K. R. Chien, R. G. Pfau, and J. L. Farber, Ischemic myocardial cell injury. Prevention by chlorpromazine of an accelerated phospholipid degradation and associated membrane dysfunction, Am. J. Pathol. 97:505 (1979).

24. T. Murachi, Intracellular Ca^{2+} proteases and its inhibitor protein: Calpain and calpastatin, in: "Calcium and Cell Function," Vol. IV, W. Y. Cheung, ed., Academic Press, Orlando (1983).

25. A. H. Wyllie, Glucocorticoid-induced thymocyte apoptosis is associated with endogenous endonuclease activation, Nature 284:555 (1980).

26. J. J. Cohen, and R. L. Duke, Glucocorticoid-activation of a calcium-dependent endonuclease in thymocyte nuclei leads to cell death, J. Immunol. 132:38 (1984).

27. A. H. Wyllie, R. G. Morris, A. C. Smith, and D. Dunlop, Chromatin cleavage in apoptosis: association with condensed chromatin morphology and dependence on macromolecular synthesis, J. Pathol. 14:67 (1984).

28. D. McConkey, P. Hartzell, P. Nicotera, A. H. Wyllie, and

S. Orrenius, Stimulation of endogenous endonuclease activity in hepatocytes exposed to oxidative stress, <u>Toxicol. Lett.</u> 42:123 (1988).

DECOMPARTMENTALISED IRON, MICROBLEEDING AND MEMBRANE OXIDATION

Catherine Rice-Evans

Department of Biochemistry and Chemistry
Royal Free Hospital School of Medicine
London NW3 2PF

INTRODUCTION

In pathological states, current models which account for the phenomenon of oxidative stress in cells and tissues include: increased generation of oxygen radicals, modified antioxidant defences and decompartmentalisation of iron proteins and complexes, all of which may be mutually interactive (Fig 1).

The iron in the human body is normally compartmentalised into its functional locations in haem-containg proteins and iron-binding proteins. The majority of the iron is in the divalent state in haemoglobin and myoglobin. The rest is distributed between the storage sites bound to ferritin, predominantly present in the liver, spleen and the bone marrow, in the low molecular weight iron pool, available for the synthesis of iron-containing proteins and enzymes, and bound to plasma transferrin for transport. Normally, therefore, iron is not available for catalysing damaging radical reactions.

Situations may arise in diseased states whereby iron proteins may be released from their normal functional compartments. For example (Fig 1), delocalisation of iron proteins can occur in rheumatoid arthritis, in which ferritin and haemosiderin have been reported to be deposited in the synovial membrane[1] as well as microbleeding in the joint[2]; microbleeding in certain tumours has been observed[3]; in thalassaemia major haemoglobin precipitation occurs in the erythrocytes[4,5]; immediately after an acute myocardial infarction there is the rapid appearance of myoglobin in the blood. Iron can be mobilised from ferritin by reducing agents such as superoxide radical, ascorbate[6-8] etc. Iron can be delocalised from haemoglobin[9,10] and from myoglobin under the influence of excess hydrogen peroxide[11]. As a result of the increased iron availability, the

Free Radicals, Lipoproteins, and Membrane Lipids
Edited by A. Crastes de Paulet *et al.*
Plenum Press, New York, 1990

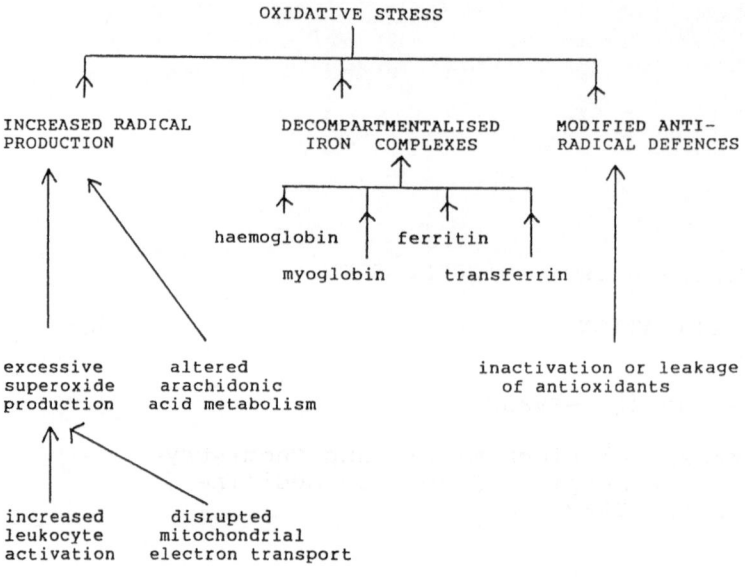

Fig 1

Factors influencing oxidative stress in cells and tissues

possibility arises for the initiation of free radical damage mediated by hydroxyl radical formation[12] via the iron-catalysed Haber-Weiss reaction or the formation of iron-ferryl radicals or other iron-oxygen complexes[13,14]. The propagation of oxidative events is also amplified by the decomposition of lipid hydroperoxides by iron complexes[15,16], such as haem, methaemoglobin, metmyoglobin, haemin in such reactions as:

$$LOOH \quad + \quad Fe(II) \quad ---> \quad LO. \quad + \quad Fe(III) \quad + \quad OH-$$

$$LOOH \quad + \quad Fe(III) \quad ---> \quad LOO. \quad + \quad Fe(II) \quad + \quad H+$$

In this presentation I should like to discuss the importance and significance of the dual role of haemoglobin: as protective a device against oxidative damage on the one hand, and its involvement in toxic reactions when decompartmentalised as membrane-bound complexes or as a result of microbleeding in the tissues, on the other hand.

DELOCALISATION OF HAEMOGLOBIN INTRACELLULARLY AND EXTRACELLULARLY

Evidence is accumulating that when haemoglobin is released during microbleeding processes it becomes toxic: specific observations have been made in the brain[17], in retinal degeneration in the eye[18] and in bleeding at the sites of inflammation[2]. In beta-thalassaemia major, increased iron decompartmentalisation is observed, as mentioned above, both intracellularly and

extracellularly (Fig 2). Thalassaemic erythrocytes are under
increased oxidative stress and inside the erythrocyte denaturation products of unpaired alpha globin chains precipitate,
forming inclusion bodies. Furthermore, the regular blood
transfusion to combat the severe anaemia in this disorder, as
well as the premature erythrocyte destruction, causes iron
overload. The consequences are increased iron in the blood,
extra-erythrocytically, accompanying saturation of the iron
binding capacity of the serum transferrin and increased ferritin
levels, as well as the deposition of iron in the tisues
especially the liver, the reticuloendithelial system and the
endocrine glands.

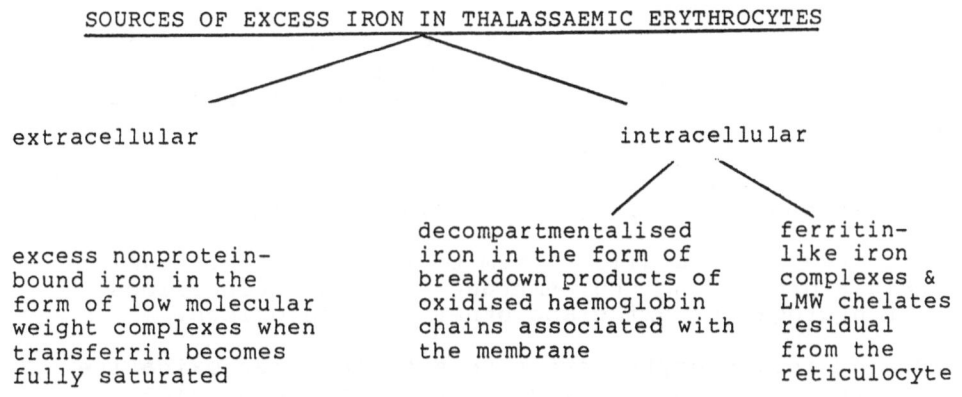

SOURCES OF EXCESS IRON IN THALASSAEMIC ERYTHROCYTES

extracellular

intracellular

excess nonprotein-
bound iron in the
form of low molecular
weight complexes when
transferrin becomes
fully saturated

decompartmentalised
iron in the form of
breakdown products of
oxidised haemoglobin
chains associated with
the membrane

ferritin-
like iron
complexes &
LMW chelates
residual
from the
reticulocyte

Fig 2

HAEMOGLOBIN AS A PROTECTIVE DEVICE

    Haemoglobin and myoglobin are normally safely compartment-
alised within the cytoplasm of the erythrocyte and myocyte
respectively. They are highly susceptible to autoxidation. In the
erythrocyte, for example, there normally develops a balance
between the spontaneous formation of methaemoglobin and
superoxide radicals and the restoration of this oxidised
haemoglobin to its normal functional state, controlled by the
antioxidant defences (Fig 3).

    Depending on the circumstances the presence of haemoglobin
and myoglobin in cells may be important components in the
protection against oxidative damage, rather than the initiators
or propagators of such damage. The ability of the iron-containing

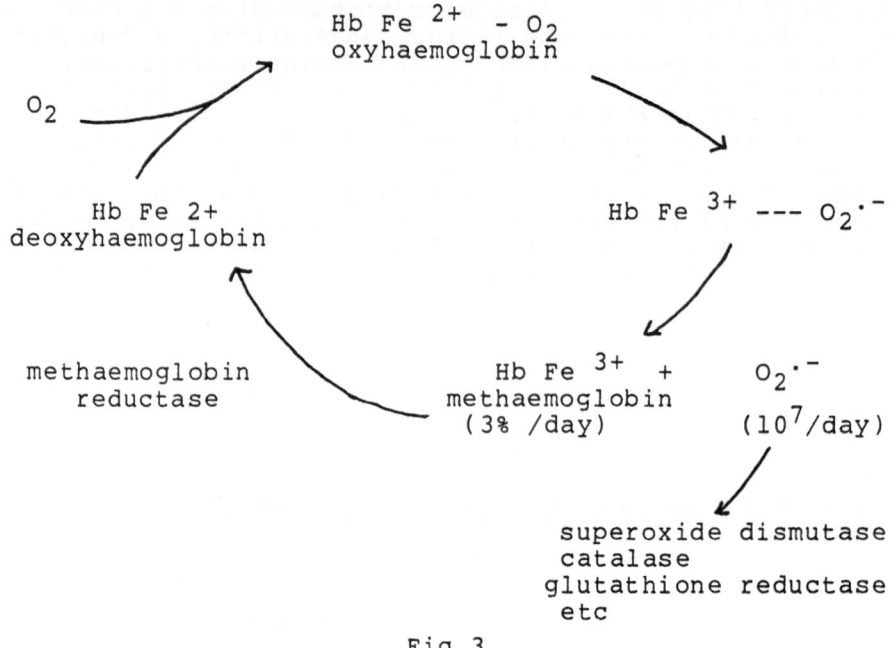

Fig 3

Oxidation-reduction cycle of normal haemoglobin

proteins to redox cycle allows them to serve as additional
oxidant sinks protecting the membrane from the deleterious
effects of hydrogen peroxide and other hydroperoxides.

Exposing intact erythrocytes to extracellular oxidative
stress in vitro induces high levels of haemoglobin oxidation
intracellularly but membrane peroxidation only occurs to a small
extent[19,20](Table 1). Haemoglobin seems to be acting as an
oxidant sink in protecting the membrane from radical attack and
in the process itself being converted to methemoglobin. If during
the course of exposing the cells to iron-mediated oxidative
stress the energy requirements of the cell are maintained,
thereby substantially suppressing the formation of and any
secondary effects due to methaemoglobin, the peroxidation of the
membrane is considerably enhanced (Table 1). This implies that
when methaemoglobin production is minimised under conditions of
oxidative stress, the membranes become more vulnerable to
oxidative damage and increased lipid peroxidation ensues. This
interpretation is confirmed by exposure to oxidative stress under
similar conditions of, firstly, haemoglobin-free membranes and
secondly, membrane-free haemolysates (Table 1): in the former
case high levels of membrane peroxidation are observed in the
absence of the haemoglobin; in the latter case, the induction of
haemoglobin oxidation was considerably less extensive than in the
intact erythrocyte. These observations support the role of
haemoglobin in intercepting the mediator of the oxidative damage
to the membrane in the intact erythrocyte exposed to oxygen
radicals. In other words, the mechanism of scavenging propagating
oxidative species in the membrane is balanced by the oxidation of

TABLE 1

INFLUENCE OF THE OXIDATION STATE OF HAEMOGLOBIN ON THE RESPONSE OF THE MEMBRANE LIPIDS TO OXIDATIVE STRESS

| | ERYTHROCYTES | | HAEMOGLOBIN-FREE | MEMBRANE-FREE |
	-ATP	+ATP	MEMBRANES	HAEMOLYSATE
METHAEMOGLOBIN (%)				
5h:	6 ± 2	0		2 ± 1
24h:	74 ± 8	15 ± 4	-	43 ± 7
	(n=5)			(n=13)
THIOBARBITURIC ACID-REACTIVE COMPOUNDS				
5h:	6.2 ± 1.0 (4)	22.3 ± 3.0 (3)	11.1 ± 2.0 (6)	
24h:	9.4 ± 3.0 (7)	41.3 ± 8.0 (3)		
	n mole/10^{10} cells		n mole/mg protein	

haemoglobin to methaemoglobin within the erythrocyte in its capacity to act as an oxidant sink.

HAEMOGLOBIN AND MYOGLOBIN AS PROPAGATORS OF OXIDATIVE DAMAGE

The role of haemoglobin in the propagation of oxidative damage may be expressed intracellularly in some pathological states in which intracellular delocalisation of haemoglobin occurs or extracellularly in microbleeding states. In situations in which haemoglobin becomes oxidised over and above the compensatory antioxidant mechanisms, the oxidation progresses and oxidative denaturation eventually leads to the formation of membrane-bound breakdown products of haemoglobin (Fig 4)[21]. During the oxidative denaturation of haemoglobin, methaemoglobin is converted to reversible haemichromes which are reducible back to functional haemoglobin. In due course reversible haemichromes are oxidatively denatured to irreversible haemichromes. These may bind to the membrane to form inclusion bodies, or they may release the globin from the haem, from which iron may be released. The formation of inclusion bodies gives rise to further oxidative stress as a result of the increased production of oxygen radicals which may be involved in membrane disruption and the formation of lipid hydroperoxides. For example, decompartmentalised haemoglobin complexes or iron released from

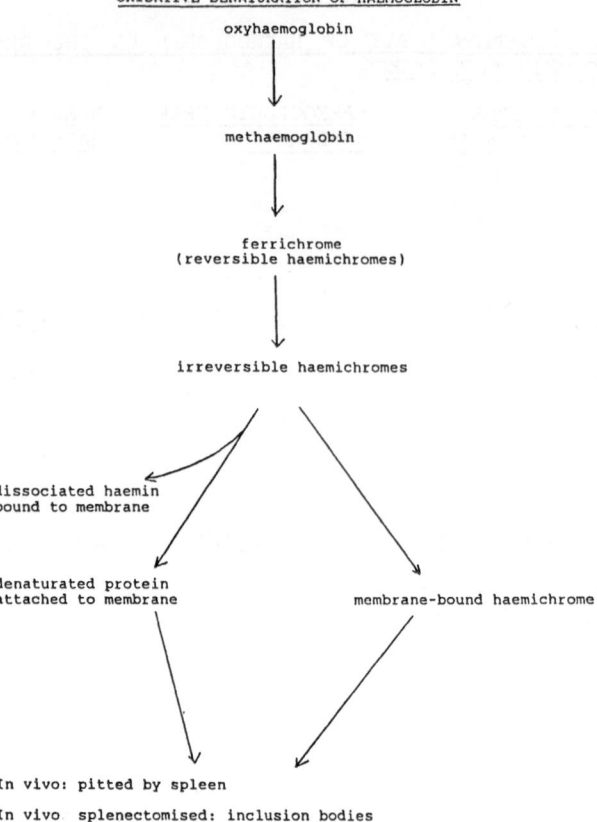

OXIDATIVE DENATURATION OF HAEMOGLOBIN

oxyhaemoglobin

methaemoglobin

ferrichrome
(reversible haemichromes)

irreversible haemichromes

dissociated haemin
bound to membrane

denatured protein
attached to membrane

membrane-bound haemichrome

In vivo: pitted by spleen

In vivo splenectomised: inclusion bodies

Fig 4

haem may be involved in the decomposition of lipid hydroperoxides to alkoxy and peroxy radicals, which can initiate
new cycles of peroxidative breakdown and exacerbate the
initial damage.

The presence of delocalised iron and/or iron-containing
metabolites of haemoglobin denaturation in or close to the
erythrocyte membrane of patients with thalassaemia or sickle
cell anaemia provide a source of continued oxidative stress,
amplifying the initial lesion and contributing towards premature erythrocyte destruction[22-25]. The toxic response to
membrane-bound iron complexes was assessed in sickle cell
membranes by incubation for 5h at 37oC, demonstrating the
increase in the level of the secondary metabolites of lipid
hydroperoxides compared to the endogenous levels; total levels
of membrane-bound iron species were determined by atomic
absorption spectroscopy (Table 2).

TABLE 2

BIOAVAILABILITY OF MEMBRANE-BOUND IRON COMPLEXES IN ABNORMAL

ERYTHROCYTE MEMBRANES

MEMBRANES	IRON LEVELS ug iron/ mg protein	LIPID PEROXIDATON tbar-products A_{532}/mg protein
CONTROL	0.15+0.1 (6)	0.07+0.04 (8)
SICKLE - endogenous levels	2.00+1 (16)	0.15+0.03 (19)
- incubation stress		0.23+0.04 (9)
THALASSAEMIC		
- tBH stress	0.81+0.6 (5)	0.36+0.25 (12)

In thalassaemic erythrocyte membranes, the bioavailability of membrane-bound iron complexes was assessed by studying their ability to stimulate lipid peroxidation via t-butyl hydroperoxide production (Table 2). A correlation is shown between membranes containing elevated membrane-bound iron and the ability to propagate peroxidative damage to the membrane lipids mediated by the organic hydroperoxide.

The presence of iron-containg breakdown products of haemoglobin or decompartmentalised iron copmplexes in or close to the sickle or thalassaemic membrane provide a source of continued oxidative stress which possibly underlies the membrane abnormalities and contributes towards the premature erythrocyte destruction.

EXTRACELLULAR DELOCALISATION OF HAEMOGLOBIN AND MICROBLEEDING PROCESSES

When haemoglobin is released via microbleeding processes and the antioxidant components are diluted out, it becomes toxic. In particular, at sites of tissue damage microbleeding may occur across the damaged endothelium and excessive superoxide radical is produced. The generation of hydrogen peroxide from the superoxide formed at such sites may eventually mobilise iron from the haemoglobin depending on their relative concentrations.

In oxyhaemoglobin, iron is in a nonpolar haem crevice. When methaemoglobin forms and water enters, the haem pocket becomes more polar and consequently hydrogen peroxide has access to the iron binding site. Such an interaction between methaemoglobin or metmyoglobin and hydrogen peroxide forms a complex activating iron to a ferryl species. In this iron IV state of haemoglobin, the iron is in an oxidation state one higher than that of the iron III in methaemoglobin.

The intermediate oxidation state of iron within these proteins to a higher valency state may be an important protective device and this concept is enhanced by experiments which illustrate the capacity of a variety of electron donors to reduce rapidly the iron of these haem proteins, eg the reduction of ferryl myoglobin to metmyoglobin by reduced glutathione[26]. It has been proposed that the functional site of myoglobin, for example, is protected from oxidative damage by the proximity of several aromatic groups. The transfer of the radical site to a tyrosine or histidine on the surface of the globin allows it to be reduced by local electron-donating substrates. Higher concentrations of hydrogen peroxide can interact with ferryl haemoglobin or ferryl myoglobin as shown[27] in Fig 5. The subsequent generation of the hydroxyl radical within the haem pocket will damage the porphyrin ring and, after several rounds of attack, leads to the release of iron from haemoglobin or myoglobin. It is this released iron that can stimulate hydroxyl radical formation away from the haem group and which, in addition to the ferryl haemoglobin radical, has the potential to induce membrane, cell and tissue damage[11,12].

INTERACTION OF HAEMOGLOBIN WITH HYDROGEN PEROXIDE

H_2O_2/[haemoglobin] high

$$\text{ferrihaemoglobin} \xrightarrow[]{+ H_2O_2} \text{ferrylhaemoglobin}$$

    stable   reactive

dissociable $+ H_2O_2$

haemichrome   ·OH generated in the haem pocket and reacts inducing release of iron

Fig 5

REFERENCES

1. D.R. Blake, P.J. Gallagher, A.R. Potter, M.J. Bell, P.A. Bacon, (1984). The effect of synovial iron on the progression of rheumatoid disease. A histologic assessment of patients with early rheumatoid synovitis. Arthritis Rheum. 27, 495 (1984).
2. S. Yoshino, D.R. Blake, S. Hewitt, C. Morris, P.A. Bacon, Effect of blood on the activity and persistence of antigen reduced inflammation in the rat air pouch. Ann.Rheum. Dis. 44, 485 (1985).
3. I.J. Rowland and M.C.R. Symons, "Tumours and iron: the use of electron spin resonance", in: Free Radical Methodology and Concepts," C. Rice-Evans and B. Halliwell, eds, Richelieu Press, London (1988).
4. E.A. Rachmilewitz, E. Shinar, O. Shalev, U. Galili, S.L. Schrier, Erythrocyte membrane alteration alterations in beta-Thalassaemia. Clin. Haematol. 14, 163 (1985)
5. C. Rice-Evans, "Iron chelators and the suppression of oxidative damage in erythrocytes: extracellular and intracellular responses", in: Free Radicals, Metal ions and Biopolymers, P. Beaumont, D. Deeble, B. Parsons, C. Rice-Evans, eds, Richelieu Press, London (1989) In press.
6. J.F. Koster and R.G. Slee, Ferritin, a physiological iron donor for microsomal lipid peroxidation. FEBS Lett 199, 85 (1986).
7. P. Biemond, H. Van Eijk, A.J.G. Swaak, J.F. Koster, J. Iron mobilisation from ferritin by a superoxide derived from stimulated polymorphonuclear leukocytes. Clin. Invest. 73, 1576 (1984).

8. P. Biemond, A.J. G.Swaak, C.M. Beindorff, J. Koster, Biochem J. 239, 169 (1986). On the superoxide-dependent and independent mechanism of iron mobilisation from ferritin by Xanthine oxidase its implications for oxygen free radical induced tissue destruction during ischaemia and inflammation.

9. J.M.C. Gutteridge, Iron promoters of the Fenton reaction and lipid peroxidation can be released from haemoglobin by peroxides. FEBS Lett. 201, 291 (1986).

10. J.M.C. Gutteridge, The antioxidant activity of haptoglobin towards haemoglobin- stimulated lipid peroxidation. Biochem. Biophys. Acta 917, 219 (1987).

11. C. Rice-Evans, R. Khan, G. Okunade: Myoglobin oxidation, hydrogen peroxide and iron release. Free Rad. Res. Comm. Submitted (1989).

12. A. Puppo and B. Halliwell. Formation of hydroxyl radicals from hydrogen peroxide in the presence of iron. Biochem J. 249, 185 (1988).

13. G. Minotti and S.D. Aust, The requirement for iron (IV) in the initiation of lipid peroxidation by iron (II) and hydrogen peroxide. J. Biol. Chem 262, 1098 (1987).

14. G. Minotti and S.D. Aust, Role of iron in the initiation of lipid peroxidation. Chem. Phys. Lipids 44, 191 (1987).

15. P.J. O'Brien, Intracellular mechanisms for the decomposition of a lipid hydroperoxide. Can.J. Biochem, 47, 485 (1969).

16. R. Labeque and L. Marrett, Reaction of haematin with allylic fatty acid hydroperoxides. Biochemistry, 27. 7060 (1988).

17. S. Scott Panter, S.M. Sadrzadeh, P.E. Hallaway, J.L. Hanes, V.E. Anderson, J.W. Eaton. Hypohaptoglobinemia associated with familial epilepsy. J. Exp. Med. 161, 748 (1985).

18. M. Doly, B. Bonhomme, J.C. Vennat, Experimental study of the retinal toxicity of haemoglobinic iron. Ophthalmis Res. 18, 21 (1986).

19. C. Rice-Evans and E. Baysal, Iron-mediated oxidative stress in erythrocytes. Biochem.J. 244, 191.

20. E. Baysal and C. Rice-Evans, Modulation of iron-mediated oxidative damage in erythrocytes by cellular energy levels. Free Rad. Res. Comm. 3,227.

21. C. Rice-Evans and A. Hartley, "Free radicals, erythrocyte disorders and iron decompartmentalisation", in: Medical, Biochemical and Chemical Aspects of Free Radicals, E. Nikki and Y. Toshikawa, eds. Elsevier, Amsterdam (1989) In press.

22. C. Rice-Evans, S.C. Omorphos, E. Baysal, Sickle cell membranes and oxidative damage. Biochem J. 237, 265.

23. C. Rice-Evans, Sickle Cell Patholgy: is the membrane important, in: Free Radicals, Cell Damage and Disease, C. Rice-Evans, ed. Richelieu Press, London (1986).

24. A. Hartley and C. Rice-Evans. Membrane-associated iron species and membrane oxidation in sickle cell disease. Biochem. Soc. Trans. In press (1989a).

25. A. Hartley and C. Rice-Evans. The nature of membrane-bound iron-species involved in radical-mediated damage to sickle erythrocytes. Biochem. Soc. Trans. In press (1989b).

26. D. Galaris, E. Cadenas, P. Hochstein, Glutathione-dependent reduction of peroxides during ferryl-and met-myoglobin interconversion. Free Rad. Biol. Med. In Press (1989).

27. M.C.R. Symons, Application of radiation and ESR spectroscopy to the study of ferryl haemoglobin and myoglobin. J Chem Soc Faraday Trans. In press (1989).

MUTAGENIC EFFECTS OF OXIDIZED LIPIDS

Leland L. Smith and Sontin Mossanda Kensese

Department of Human Biological Chemistry and Genetics
University of Texas Medical Branch
Galveston,TX 77550

INTRODUCTION

Oxidized lipids are suspected as being mutagenic, the mutagenicity being associated in concept with carcinogenicity as well. However, although mutagenic responses for more highly oxidized lipid degradation products have been obtained in a variety of bioassays, convincing demonstration of mutagenicity of early formed products of lipid oxidation has not been readily achieved. In many of the studies conducted todate mutagenicity has been assessed using test strains of Salmonella typhimurium in the Ames test, the bioassay emphasized in the present report.

As examples of degraded oxidized lipid mutagens, the fecapentaenes (polyunsaturated glycerol alkyl ethers formed by intestinal microflora) and the corresponding polyunsaturated aldehydes are directly acting mutagens towards S. typhimurium, as are also hydroxylated unsaturated aldehydes derived by lipid peroxidation and malonaldehyde, the ultimate lipid peroxidation product. However, it is not these degraded products that are of present interest. Rather, interest lies in those oxidation products retaining the two oxygen atoms of the peroxide bond that is formed initially as result of lipid peroxidation or autoxidation of unsaturated lipids.

Demonstration of mutagenicity of the early formed lipid peroxides has been elusive, there being reports of weak or questionable mutagenicity and of nonmutagenicity.[1-7] As peroxides are notoriously toxic to bacteria and other cultured cells, one suspects that the test systems employed were not altogether suitable and that special bioassays for peroxides must be utilized in order to disclose the more subtle mutagenic effects.

To counter these limitations several modified strains of S. typhimurium have been devised for work with mutagenic oxidants, including strains sensitive to forward mutations. We have approached the matter differently, using standard test strains of S. typhimurium but employing a liquid preincubation patterned after that of Mitchell[8] prior to plating on agar in the usual manner, thereby allowing toxic effects of the peroxide analytes to be adjusted. The liquid incubation protocol also involves a longer incubation period that allows recovery of bacterial growth. By such manipulation of incubation conditions it has proven possible to obtain measures of mutagenicity of sterol hydroperoxides that would otherwise be undetected.[9] There is always the question of whether such contrived

Free Radicals, Lipoproteins, and Membrane Lipids
Edited by A. Crastes de Paulet *et al.*
Plenum Press. New York, 1990

bioassays reflect any physiological meaning and whether the weakly mutagenic oxidized lipids detected pose any threat to human health. These
questions remain unanswered.

BIOASSAY OF EDIBLE OILS AND PEROXIDES

The bioassay of mutagens in the presence of massive amounts of lipids
with the standard agar plate protocol is probably without meaning unless
the validity of the test protocol for detecting known mutagens in the
presence of such lipids is established. The insoluble lipid mass must pose
physical barrier problems affecting bacterial respiration and survival as
well as diffusion of analytes from the lipid mass to the test organism.
Negative tests in such cases should be supported by evidence that known
mutagens if present in a nonmutagenic lipid mass would be detected. However, these restraints have not been imposed on the standard agar plate
bioassays, and heated fats have generally been found nonmutagenic using
agar plate bioassays with reversion test strains of S. typhimurium.[5,10-13]

The same limitations hold for liquid incubation bioassays where lipid
analytes may still pose phase separation problems that interfere with the
test. Some of these difficulties are evinced in our prior demonstration
that the mutagenicity of benzo[a]pyrene in 1000-fold amounts of human
aortal lipids is detected using a liquid incubation protocol prior to
routine agar plating, but detection of the mutagenicity of weakly mutagenic
autoxidized cholesterol required much higher levels of added analyte.[14]
Moreover, in bioassay of real foods using liquid incubation protocols
positive mutagenicity has been demonstrated for aged corn oil using the
forward mutation S. typhimurium SV50 strain[13], and we report here detection
of mutagens in commercially available edible palm oils from Zaire using S.
typhimurium TA1537.

Edible Oils

The edible palm oils (unrefined palm oil used for cooking, several
refined palm oil products, and margarines made from palm oil) aged for at
least one year in their original containers at room temperature were weakly
mutagenic towards several S. typhimurium strains, with strain TA1537
providing significant results reproducibly using our liquid incubation
protocol.[9] Further examination by thin-layer chromatography (hexane-diethyl
ether-acetic acid, 70:30:1) of the oils in a bioassay-directed search for
the specific mutagens present revealed the presence of three mutagenic
peroxide-containing components in each preparation, together with nonmutagenic triacylglycerols, fatty acids, sterols, and other components.

Material recovered from each peroxide zone was weakly mutagenic, with
a dose-response rise to maximum mutagenicity followed by decline. Bioassay
data using S. typhimurium TA1537 for the three peroxide components of unrefined edible palm oil and of margarine made from palm oil are presented
in Fig.1. Mutagenicity is expressed by the Increase in Revertants, defined:

$$\text{Increase in Revertants} = \frac{\text{Induced Revertants} - \text{Spontaneous Revertants}}{\text{Spontaneous Revertants}}$$

By this measure positive mutagenicity is indicated by dose responses with
an Increase in Revertants greater than 3.0, with weak mutagenicity being
assessed for values between 2.0 and 3.0 and questionable mutagenicity for
values between 1.5 and 2.0.[15] The two-fold Increase in Revertants
indicating weak mutagenicity corresponds to a count of induced revertants

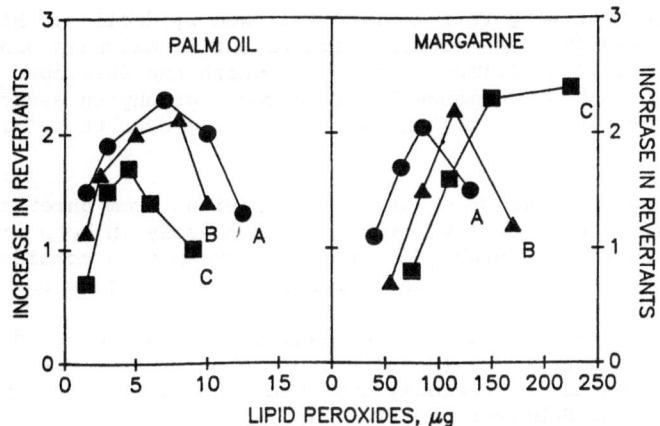

Fig.1. Liquid incubation bioassay with S. typhimurium TA1537
of lipid peroxide fractions A, B, and C (in order of
decreasing mobility) from thin layer chromatograms of
unrefined edible palm oil (left) and margarine made
from palm oil (right).

three-fold that of the spontaneous count. Additionally, mutagenicity was
evaluated with the Q value of Mitchell, defined:

$$Q = \frac{F_e - F_c + F_a}{F_a}$$

where F_e is the frequency of mutation with the analyte, F_c the frequency of
spontaneous mutation, and F_a is the historical (average) frequency of
spontaneous mutation. Significant mutagenicity is indicated by dose-
responses and Q value greater than 1.5.[8]

Peroxide fractions A and B from unrefined palm oil and all three peroxide fractions from margarine were weakly mutagenic by the Increase in
Revertants measure, but unrefined palm oil peroxide fraction C was only
questionably mutagenic. However, by Mitchell's Q values, peroxide fractions
A, B, and C were all mutagenic (1.9, 1.8, 1.7 respectively). The amount of
margarine required for maximum mutagenicity responses was ten-fold that of
liquid unrefined edible palm oil. These same relationships were confirmed
using a second preincubation protocol, that of Ames in which a short 20 min
incubation preceeds plating.[16,17] For example, peroxide fractions A and C
of unrefined edible palm oil were weakly mutagenic, with fraction B being
but questionably mutagenic in the 20 min preincubation test. Although both
incubation bioassays gave confirmatory results, the short 20 min incubation
protocol was less sensitive, requiring approximately 100-fold greater
amounts of lipid peroxide fractions for maximum dose responses.

Further fractionation of each mutagenic initial lipid peroxide
fraction A, B, C using high performance liquid chromatography (Zorbax SIL,
hexane-diethyl ether-acetic acid, 1000:200:1)[18] afforded several resolved
components from each, and most of these subfractions were weakly mutagenic,

with dose-responses rising to a maximum followed by decline. Here the mutagenicity is expressed as material equivalent to malondialdehyde, using a thiobarbituric acid colorimetric assay in which the developed chromogen was recovered by high performance liquid chromatography in association with an array detector recording the chromogen spectrum over the visible range.[19,20]

For instance, four major peroxide subfractions from unrefined palm oil peroxide fraction A were weakly mutagenic on bioassay in both liquid incubation and 20 min preincubation protocols. Despite a considerable concentration into subfractions of components of fraction A there were no accompanying increases in mutagenicity; the four subfractions were characterized by dose-responses rising to maximum with subsequent decline, with Increase in Revertants of 3.0 ($Q = 2.2$), 2.8 ($Q = 1.8$), 2.8 ($Q = 2.2$), and 3.7 ($Q = 2.4$), thus not appreciably improved over data for the parent peroxide fraction A. Subfractions from lipid peroxide fractions B and C from unrefined palm oil were similarly weakly mutagenic.

Continued fractionation of these subfractions was unfruitful, as many additional subfractions were obtained but no improvement in bioassay potency resulted for any subfraction. These results were very much like those previously obtained in the extensive fractionation of very polar mutagenic oxidized cholesterol derivatives where weak mutagenicity was found in most fractions but with no increase in potency of any given fraction.[21]

As the mutagenicity responses of unrefined palm oil, its peroxide fractions A, B, and C, and subfractions from fractions A, B, C were reminiscent of those of sterol hydroperoxides in which reduced oxygen species were implicated, bioassays were also conducted in the presence of catalase or superoxide dismutase now a standard means of testing such systems for the participation of hydrogen peroxide (H_2O_2) and superoxide (O_2^-) respectively. Exogenously added catalase abolished the observed mutagenicity of the mutagenic palm oil peroxide fractions, but added superoxide dismutase gave variable results, with no change in response in some tests, with increased responses in others. We conclude that the observed mutagenicity abolished by catalase is caused by H_2O_2 formed during bioassay. The variable results with superoxide dismutase pose more difficult matters of interpretation, but where increased responses were noted the disproportionation of O_2^- to H_2O_2 may have occured.

These results taken together suggest that the palm oil peroxides served as source of H_2O_2 in the bioassay system or stimulated the reduction of dioxygen during bioassay, yielding O_2^- and H_2O_2, thus in the same manner previously recognized in the case of mutagenic sterol hydroperoxides.[9]

Bioassay of Organic Hydroperoxides

Polyunsaturated fatty acid hydroperoxides have not been found to be mutagenic to S. typhimurium strains heretofore, but methyl esters of linoleic and linolenic acid hydroperoxides exhibit weak mutagenicity to S. typhimurium strains TA97, TA98, TA100, and TA102, generally on agar plates incorporating liver S-9 enzyme preparations. However, negative results are also recorded.[1-7] Furthermore, the corresponding reduced alcohol and other nonperoxide derivatives of methyl linoleate and linolenate are also nonmutagenic in these tests,[6,7] suggesting that destruction of the peroxide bond lead to inactivation. Different results are had in bioassay of commercially available organic hydroperoxides such as cumene hydroperoxide where weak mutagenicity is uniformly revealed with both the common strains and the special test strains advanced for detection of mutagens that are also oxidants.[3,4,22-24]

Our present results evince the weak mutagenicity of linoleic and linolenic acid hydroperoxides in these test systems. The fatty acyl hydroperoxides tested were prepared by air oxidation of pure substrate and consisted of the several recognized product isomeric hydroperoxides. The linoleic acid hydroperoxide preparation contained 13-hydroperoxy-(9Z,11E, 13RS)- and (9E,11E,13RS)-octadeca-9,11-dienoic acids and 9-hydroperoxy-(9RS,10E,12Z)- and (9RS,10E,12E)-octadeca-10,12-dienoic acids; the linolenic acid hydroperoxide preparation contained 13-hydroperoxy-(9Z,11E,13RS, 15Z)-octadeca-9,11,15-trienoic, 16-hydroperoxy-(9Z,12Z,14E,16RS)-octadeca-9,12,14-trienoic, 12-hydroperoxy-(9Z,12RS,13E,15Z)-octadeca-9,13,15-trienoic, and 9-hydroperoxy-(9RS,10E,12Z,15Z)-octadeca-10,12,15-trienoic acids, thus only five of the eight possible hydroperoxides.

The linoleic acid hydroperoxides were weakly mutagenic toward $\underline{S.}$ $\underline{typhimurium}$ TA1537 in the liquid incubation and 20 min preincubation bioassays (maximum Increase in Revertants 2.1 (Q = 1.7) and 3.4 respectively). Linolenic acid hydroperoxides were likewise weakly mutagenic in both protocols (maximum Increase in Revertants 2.3 (Q = 2.2) and 5.6 respectively). The individual isomeric linolenic acid hydroperoxides were bioassayed in a few cases with similar results; the 13-hydroperoxy-(9Z,11E,13RS,15Z)-octadeca-9,11,15-trienoic, 16-hydroperoxy-(9Z,12Z,14E, 16RS)-octadeca-9,12,14-trienoic, and 12-hydroperoxy-(9Z,12RS,13E,15Z)-octadeca-9,13,15-trienoic acids being characterized by maximum Increase in Revertants 2.6, 2.6, and 2.4 respectively. All these dose-responses maxima were abolished by exogenously added catalase.

The mutagenicity of cumene hydroperoxide was confirmed with several $\underline{S.}$ $\underline{typhimurium}$ strains (Table 1). In this case the agar plate protocol was the more effective in showing the effect, and only with strain TA1537 is there weak mutagenicity exhibited in the liquid incubation bioassay. Incorporation of catalase in the agar plate bioassay led to erratic and unreproducible results, but in the case of the liquid incubation bioassay with strain TA1537 catalase abolished the observed mutagenicity, as did also superoxide dismutase.

Bioassay of Sterol Hydroperoxides

The mutagenicity of two sterol hydroperoxides 3β-hydroxy-5α-cholest-6-ene-5-hydroperoxide and cholesterol 7α-hydroperoxide towards $\underline{S.\ typhimurium}$ TA1537 and other strains was previously established using the liquid incubation bioassay protocol. In the case of the 5α-hydroperoxide an unprecedented multiphasic dose-response was obtained with two maxima, one at low dose corresponding to the single maximum response obtained with the isomeric 7α-hydroperoxide, the second maximum corresponding to that of H_2O_2 (Table 2). Mutagenic responses were abolished by exogenous catalase, but added superoxide dismutase increased response.[9]

Table 1. Mutagenicity of Cumene Hydroperoxide

Bioassay[a]	TA97	TA98	TA100	TA102	TA1537	TA1538
Agar plate	18	1.7	2.8	6	13	1.1
Liquid incubation	1.7	0.9	1.2	1.1	2.1	1.9
with catalase	-	0.8	1.1	-	0.7	1.1
with SOD	-	1.0	1.4	-	1.0	2.3

a Abbreviation: SOD, superoxide dismutase

Table 2. Mutagenicity of Sterol Hydroperoxides

ROOH[a]	Maximum Increase in Revertants (µmol)				Q(µmol)
	TA98	TA100	TA1538	TA1537	TA1537
5α-OOH	2.5 (1.2)	5.3(0.3)	2.0(0.1)	1.3(0.1-0.3) 2.5(3.5)	1.2-3.4(0.3-0.9) 2.0-4.4(2.9-3.6)
7α-OOH	-	-	-	3.3(0.2-0.3)	-

[a] Abbreviations: 5α-OOH, 3β-hydroxy-5α-cholest-6-ene-5-hydroperoxide; 7α-OOH, 3β-hydroxycholest-5-ene-7α-hydroperoxide.

These are very strange results and evince dynamics within the bioassay system, including the participation of reduced oxygen species in the mutagenicity responses. It was our practice to recover analytes from bioassay media in order to establish the integrity of the analyte during bioassay and to discover any metabolism of the analyte. Accordingly, the metabolism of the sterol hydroperoxides by strain TA1537 was directly noted.

The 5α-hydroperoxide was formally reduced to the corresponding alcohol 5α-cholest-6-ene-3β,5-diol but also transformed to 3β-hydroxycholest-5-en-7-one, this transformation requiring allylic rearrangement of some sterol derivative in the process. Also occurring was the allylic rearrangement of the 5α-hydroperoxide to the 7α-hydroperoxide, an unprecedented in vivo transformation. The 7α-hydroperoxide was also reduced to the alcohol cholest-5-ene-3β,7α-diol and formally dehydrated to the same 7-ketone, these reactions of the 7α-hydroperoxide being recognized as typically those following peroxide bond homolysis to the corresponding 7-oxyl intermediate 3β-hydroxycholest-5-ene-7α-oxyl radical. Subsequent hydrogen abstraction by the 7-oxyl species yields cholest-5-ene-3β,7α-diol; subsequent β-scission yields 3β-hydroxycholest-5-en-7-one.[9,25]

These transformations were enzyme-catalyzed; at least, cultures that were heat-inactivated under identical incubation conditions did not cause these transformations. It is, of course, uncertain whether substrate-enzyme interactions occurred or whether some product formed by the living bacteria caused the transformations.

The metabolism of the sterol 5α-hydroperoxide to the related mutagenic sterol 7α-hydroperoxide but also to nonmutagenic metabolites as well as the involvement of H_2O_2 as ultimate mutagen appear to be factors in accounting for some of the complexity of the dose-response curves of the 5α-hydroperoxide.

Bioassay of Hydrogen Peroxide

Some uncertainty about the mutagenicity of H_2O_2, the ultimate in peroxides, towards S. typhimurium persists.[26,27] Weak mutagenicity has been observed with some strains[9,15,22,24,28-33] but not in others.[3,4,31,34] The mutagenicity of H_2O_2 is also inferred by the action of exogenous catalase in abolishing observed mutagenicity.[35-38] As H_2O_2 is implicated in the observed mutagenicity of sterol hydroperoxides, linoleic and linolenic acid hydroperoxides, cumene hydroperoxide, and edible palm oil products, and in the absence of bioassay data obtained with standardized bioassay protocols, we evaluated the effects of H2O2 with several S. typhimurium strains under comparable bioassay conditions.

Our data (Table 3) evince weak mutagenicity of H2O2 on the standard agar plate bioassay with strains TA97, TA98, TA102, and TA1537; in the 20 min preincubation bioassay weak mutagenicity was demonstrated with the same strains but also with strains TA100 and TA1538. Our more lengthy liquid incubation protocol was less effective, strain TA1537 being the only one with a weak response.

Maximum responses were obtained with 0.9-4.8 mmol H2O2 in the agar plate bioassay, 0.8-2.8 μmol in the liquid incubation bioassay. For the less sensitive 20 min preincubation bioassay 170-340 μmol were required for maximum responses.

Incorporation of catalase in the standard agar plate bioassay gave erratic bacterial growth and did not accord us reproducible bioassay results. However, catalase abolished or reduced responses otherwise, as did also S-9 enzymes. Exogenous superoxide dismutase had no effect in some cases, decreased responses in others, and stimulated responses for strains TA97 and TA1537.

LIPID PEROXIDES OF EDIBLE OILS

The mutagenic lipid peroxide-containing components present in edible palm oils were more polar than polyunsaturated fatty acids and sterols on thin-layer chromatograms and exhibited mobility and color test behavior of preparations of polyunsaturated fatty acyl hydroperoxides. Sterol hydroperoxides were not detected in the edible oils. In our attempt to discover the nature of these weak mutagens the several mutagenic fractions were further fractionated by high performance liquid chromatographic and sub-fractions therefrom were subjected to additional chromatography, chemical reactions, and spectral examination. Lipid peroxides with chromatographic properties very much like those of linoleic and linolenic acid hydroperoxides were present in each mutagenic fraction.

Each fraction contained material giving a positive color test on thin-layer chromatograms for peroxides using N,N-dimethyl-p-phenylenediamine,

Table 3. Mutagenicity of Hydrogen Peroxide

Bioassay[a]	Maximum Increase in Revertants					
	TA97	TA98	TA100	TA102	TA1537	TA1538
Agar plate	3.2	2.2	1.2	3.1	4.8	1.2
Preincubation	2.4	2.4	3.8	2.8	2.7	6.0
with catalase	1.7	-	-	1.5	1.7	2.4
with SOD	1.8	-	-	2.0	2.7	1.1
Liquid incubation	1.7	1.0	1.5	1.3	2.1	1.3
with catalase	0.8	0.9	1.2	0.8	1.0	1.2
with SOD	3.0	1.0	1.9	1.5	3.5	1.5
with S-9	0.9	1.0	1.3	0.8	0.8	0.9

a Abbreviations: SOD, superoxide dismutase; S-9, rat liver S-9 enzymes.

charring following spraying with 50% sulfuric acid and heating, with mobilities very similar to those of linoleic and linolenic acid hydroperoxides. High performance liquid chromatography of the mutagenic fractions yielded the mutagenic subfractions containing lipid peroxides that were characterized by retention data essentially the same as those of the identified reference linoleic and linolenic acid hydroperoxides. These subfractions reacted with thiobarbituric acid to give chromogens that were characterized by the same spectral properties obtained from reference linoleic acid hydroperoxides, thus a red absorption band at 529 nm with associated maximum or inflecion at 540 nm, an orange band at 490-496 nm, and a yellow band at 452-460 nm. These three absorption bands have been associated with malonaldehyde, dienals, and saturated aldehydes respectively, present in the analyte or formed from lipid peroxides during the analysis.[39]

Additionally, attempted recovery of the mutagenic peroxide fractions by further high performance liquid chromatography invariably resulted in alteration of the sample. A defined component with chemical and retention properties of an individual linoleic acid hydroperoxide, for example, upon reanalysis contained up to four lipid peroxides with retention properties of the four recognized linoleic acid hydroperoxides. This sample instability mimics the established isomerization of linoleic acid hydroperoxides during manipulations.[18]

These data suggest but do not establish the presence of linoleic and linolenic acid hydroperoxides as probable components of the edible palm oil preparations. In view of the problems associated with recovery of pure lipid peroxide mutagens and the involvement of H_2O_2 in the observed mutagenicty responses attempts to identify the mutagens in edible palm oil were suspended at this point.

CONCLUSIONS

Our data evince the presence of lipid peroxide mutagens in edible palm oil foods that have aged in their original containers in contact with air. The chemical nature of the mutagens remains uncertain, but the presence in the samples of lipid peroxides with chemical and chromatographic properties of linoleic and linolenic acid hydroperoxides suggest that these hydroperoxides here demonstrated to be mutagenic in the bioassay contribute significantly to the observed mutagenicities. Moreover, the abolition of the observed mutagenicities of the edible oils and of reference linoleic and linolenic acid hydroperoxides by catalase further support the suggestion and implicate H_2O_2 or a species derived from H_2O_2 as ultimate mutagen. These results are very much like those previously obtained with sterol hydroperoxides[9] and suggest that H_2O_2 be a common agent for the weak mutagenicity observed.

The mutagenicity responses of the edible palm oil fractions, reference linoleic and linolenic acid hydroperoxides, sterol hydroperoxides, and H_2O_2 mimic in detail the dose-responses of toxic substances in general, with an initial rise to a maximum followed by decline. The dose-response declines have been attributed to uncompensated toxicity of analyte or of substances formed during bioassay, by loss of deactivation of mutagen during bioassay, and to unidentified processes. However, it is now recognized that exposure of S. typhimurium to low levels of H_2O_2 induces catalase and related protective proteins that in turn destroy H_2O_2.[30,40-44] The induction of catalase provides an acceptable explanation of the decline in dose-response of the bacteria to higher levels of H_2O_2 and suggests a similar possibility for the dose-response declines observed for sterol and polyunsaturated fatty acyl hydroperoxides. Bacterial metabolism of the fatty acyl hydroperoxides to nonmutagenic metabolites may also contribute to the dose-response declines, as demonstrated for sterol hydroperoxides,[9,25] but we have not investigated the integrity of fatty acyl hydroperoxides during bioassay.

As it appears that H_2O_2 or a species derived from H_2O_2 be the mutagen in these instances, it follows that sterol and fatty acyl hydroperoxides and the mutagens of edible palm oil must stimulate the reduction of oxygen in the test system. These results pose the question whether lipid peroxides exhibit biological activity in other test systems via stimulated reduction of oxygen as well.

The mutagenicity observed here in edible oils, attributed to H_2O_2, does not appear to be retained in heated oils,[5,10-12] a matter rationalized by heat destruction of lipid hydroperoxides and H_2O_2. The mutagenicity present in aged but unheated edible oils thus may be of no more concern as a dietary problem than is the mutagenicity of brewed tea and coffee also attributed to H_2O_2.[32,36-38]

ACKNOWLEDGEMENT

The generous support for these studies of the Council for the International Exchange of Scholars, Washington City,DC and the Robert A. Welch Foundation, Houston,TX is gratefully acknowledged.

REFERENCES

1. T. Yamaguchi and Y. Yamashita, Mutagenic Activity of Autoxidized Linolenic and Linoleic Acid, Ag. Biol. Chem. 43:2225 (1979).
2. T. Yamaguchi, Y. Yamashita, and T. Abe, Desmutagenic Activity of Peroxidase on Autoxidized Linolenic Acid, Ag. Biol. Chem. 44:959 (1980).
3. T. Yamaguchi and Y. Yamashita, Mutagenicity of Hydroperoxides of Fatty Acids and Some Hydrocarbons, Ag. Biol. Chem. 44:1675 (1980).
4. T. Yamaguchi, Activation with Catalase of Mutagenicity of Hydroperoxides of Some Fatty Acids and Hydrocarbons, Ag. Biol. Chem. 44:1989 (1980).
5. M. Scheutwinkel-Reich, G. Ingerowski, and H.-J. Stan, Microbiological Studies Investigating Mutagenicity of Deep Frying Fat Fractions and Some of Their Components, Lipids 15:849 (1980).
6. H. W. Gardner, C. G. Crawford, and J. T. MacGregor, Negative Ames Tests of Epoxide Fatty Methyl Esters Derived from Homolysis of Linoleic Acid Hydroperoxides, Food Chem. Toxicol. 21:175 (1983).
7. J. T. MacGregor, R. E. Wilson, W. E. Neff, and E. N. Frankel, Mutagenicity Tests of Lipid Oxidation Products in Salmonella typhimurium: Monohydroperoxides and Secondary Oxidation Products of Methyl Linoleate and Methyl Linolenate, Food Chem. Toxicol. 23:1041 (1985).
8. I. deG. Mitchell, Microbial Assays for Mutagenicity: A Modified Liquid Culture Method Compared with the Agar Plate System for Precision and Sensitivity, Mutation Res. 54:1 (1978).
9. L. L. Smith, V. B. Smart, and N. M. Made Gowda, Mutagenic Sterol Hydroperoxides, Mutation Res. 161:39 (1986).
10. S. L. Taylor, C. M. Berg, N. H. Shoptaugh, and V. N. Scott, Lack of Mutagens in Deep-Fat-Fried Foods Obtained at the Retail Level, Food Chem. Toxicol. 20:209 (1982).
11. S. L. Taylor, C. M. Berg, N. H. Shoptaugh, and E. Traisman, Mutagen Formation in Deep-Fat Fried Foods as a Function of Frying Conditions, J. Am. Oil Chem. Soc. 60:576 (1983).
12. A. Van Gastel, R. Mathur, V. V. Roy, and C. Rukmini, Ames Mutagenicity Tests of Repeatedly Heated Edible Oils, Food Chem. Toxicol. 22:403 (1984).

13. P. L. Stapleton, C. L. Hansen, K. A. Marley, R. A. Larson, and M. J. Plewa, Mutagenicity and Toxicity of Aged Corn Oil, <u>Environm. Mutagenesis</u> 7(Suppl 3):42 (1985).

14. B. H. Johnson and L. L. Smith, A Search for Mutagens in Human Aortal Lipid Extracts, <u>Atherosclerosis</u> 53:331 (1984).

15. S. De Flora, A. Camoirano, P. Zanacchi, and C. Bennicelli, Mutagenicity Testing with TA97 and TA102 of 30 DNA-damaging Compounds, Negative with Other Salmonella Strains, <u>Mutation Res.</u> 134:159 (1984).

16. B. N. Ames, J. McCann, and E. Yamasaki, Methods for Detecting Carcinogens and Mutagens with the Salmonella/Mammalian-Microsome Mutagenicity Test, <u>Mutation Res.</u> 31:347 (1975).

17. D. M. Maron and B. N. Ames, Revised Methods for the Salmonella Mutagenicity Test, <u>Mutation Res.</u> 113:173 (1983).

18. J. I. Teng and L. L. Smith, High-performance Liquid Chromatography of Linoleic Acid Hydroperoxides and their Corresponding Alcohol Derivatives, <u>J. Chromatog.</u> 350:445 (1985).

19. R. L. Bertholf, J. R. P. Nicholson, M. R. Wills, and J. Savory, Measurement of Lipid Peroxidation Products in Rabbit Brain and Organs (Response to Aluminum Exposure), <u>Ann. Clin. Lab. Sci.</u> 17:418 (1987).

20. H. Y. Wong, J. A. Knight, S. M. Hopfer, O. Zaharia, C. N. Leach, and W. Sunderman, Lipoperoxides in Plasma as Measured by Liquid-Chromatographic Separation of Malondialdehyde-Thiobarbituric Acid Adduct, <u>Clin. Chem.</u> 33:214 (1987).

21. G. A. S. Ansari, R. D. Walker, V. B. Smart, and L. L. Smith, Further Investigations of Mutagenic Cholesterol Preparations, <u>Food Chem. Toxicol.</u> 20:35 (1982).

22. D. E. Levin, M. Hollstein, M. F. Christman, E. A. Schwiers, and B. N. Ames, A New <u>Salmonella</u> Tester Strain (TA102) with A.T Base Pairs at the Site of Mutation Detects Oxidative Mutagens, <u>Proc. Natl. Acad. Sci.</u> 79:7445 (1982).

23. D. E. Levin, L. J. Marnett, and B. N. Ames, Spontaneous and Mutagen-Induced Deletions: Mechanistic Studies in <u>Salmonella</u> Tester Strain TA102, <u>Proc. Natl. Acad. Sci.</u> 81:4457 (1984).

24. M. Ruiz-Rubio, E. Alejandre-Durán, and C. Pueyo, Oxidative Mutagens Specific for A.T Base Pairs Induce Forward Mutations to L-Arabinose Resistance in <u>Salmonella typhimurium</u>, <u>Mutation Res.</u> 147:153 (1985)

25. L. L. Smith, N. M. Made Gowda, and J. I. Teng, Sterol Hydroperoxide Metabolism by <u>Salmonella typhimurium</u>, <u>J. Steroid Biochem.</u> 26:259 (1987).

26. L. E. Kier, D. J. Brusick, A. E. Auletta, E. S. Von Halle, M. M. Brown, V. F. Simmon, V. Dunkel, J. McCann, K. Mortelmans, M. Prival, T. K. Rao, and V. Ray, The <u>Salmonella typhimurium</u>/Mammalian Microsomal Assay. A Report of the U. S. Environmental Protection Agency Gene-Tox Program, <u>Mutation Res.</u> 168:69 (1986).

27. W. von der Hude, C. Behm, R. Gürtler, and A. Basler, Evaluation of the SOS Chromotest, <u>Mutation Res.</u> 203:81 (1988).

28. I. deG. Mitchell, Forward Mutation in <u>Escherichia coli</u> and Gene Conversion in <u>Saccharomyces cerevisiae</u> Compared Quantitatively with Reversion in <u>Salmonella typhimurium</u>, <u>Agents Actions</u> 10:287 (1980).

29. B. N. Ames, M. C. Hollstein, and R. Cathcart, Lipid Peroxidation and Oxidative Damage to DNA. in: "Lipid Peroxides in Biology and Medicine", K. Yagi, ed., Academic Press, New York (1982), pp.339-351.

30. L. Winquist, U. Rannug, A. Rannug, and C. Ramel, Protection from Toxic and Mutagenic Effects of H_2O_2 by Catalase Induction in <u>Salmonella typhimurium</u>, <u>Mutation Res.</u> 141:145 (1984).

31. J. Xu, W.-Z. Whong, and T.-m. Ong, Validation of the Salmonella (SV50)/Arabinose-Resistant Forward Mutation Assay System with 26 Compounds, <u>Mutation Res.</u> 130:79 (1984).

32. Y. Fujita, K. Wakabayashi, M. Nagao, and T. Sugimura, Implication of Hydrogen Peroxide in the Mutagenicity of Coffee, <u>Mutation Res.</u> 144:227 (1985).

288

33. E. H. Berglin and J. Carlsson, Effect of Hydrogen Sulfide on the Mutagenicity of Hydrogen Peroxide in Salmonella typhimurium Strain TA102, Mutation Res. 175:5 (1986).

34. H. F. Stich, L. Wei, and P. Lam, The Need for a Mammalian Test System for Mutagens: Action of Some Reducing Agents, Cancer Lett. 5:199 (1978).

35. W. H. Kalus, W. G. Filby, and R. Münzner, Chemical Aspects of the Mutagenic Activity of the Ascorbic Acid Autoxidation System, Z. Naturforsch. 37c:40 (1982).

36. M. Nagao, Y. Suwa, H. Yoshizumi, and T. Sugimura, Mutagens in Coffee, Banbury Report 17:69 (1984).

37. Y. Fujita, K. Wakabayashi, M. Nagao, and T. Sugimura, Characterization of Major Mutagens of Instant Coffee, Mutation Res. 142:145 (1985).

38. E. Alejandre-Durán, A. Alonso-Moraga, and C. Pueyo, Implication of Active Oxygen Species in the Direct-Acting Mutagenicity of Tea, Mutation Res. 188:251 (1987).

39. H. Kosugi and K. Kikugawa, Thiobarbituric Acid Reaction of Aldehydes and Oxidized Lipids in Glacial Acetic Acid, Lipids 29:915 (1985).

40. J. A. Watson and J. Schubert, Action of Hydrogen Peroxide on Growth Inhibition of Salmonella typhimurium, J. Gen. Microbiol. 57:25 (1969).

41. G. J. Finn and S. Condon, Regulation of Catalase Synthesis in Salmonella typhimurium, J. Bacteriol. 123:570 (1975).

42. P. C. Lee, B. R. Bochner, and B. N. Ames, AppppA, Heat-shock Stress, and Cell Oxidation, Proc. Natl. Acad. Sci. 80:7496 (1983).

43. M. F. Christman, R. W. Morgan, F. S. Jacobson, and B. N. Ames, Positive Control of a Regulon for Defenses against Oxidative Stress and Some Heat-Shock Proteins in Salmonella typhimurium, Cell 41:753 (1985).

44. R. W. Morgan, M. F. Christman, F. S. Jacobson, G. Storz, and B. N. Ames, Hydrogen Peroxide-Inducible Proteins in Salmonella typhimurium Overlap with Heat Shock and Other Stress Proteins, Proc. Natl. Acad. Sci. 83:8059 (1986).

BIOLOGICAL EFFECTS OF OXYSTEROLS

A. CRASTES de PAULET, M.E. ASTRUC, J. BASCOUL et R. DEFAY

INSERM Unité 58
60, rue de Navacelles
34090 MONTPELLIER France

The oxysterol family is very extensive : more than fifty members have been presently identified from either natural sources (plasma, tissue extracts) or artificial systems. These oxysterols are characterized by the presence of one or more oxygenated functions on the cholesterol molecule : primary, secondary or tertiary alcohol or hydroperoxide, aldehyde, ketone, epoxides, endoperoxide. The positions of the main functionalized carbon atoms are shown in Fig. 1.

Some oxysterols are formed by enzymatic reactions (involving Cyt P_{450} species) and are intermediates in bile acids or steroid hormone biosynthesis. But cholesterol can readily result in a large variety of oxysterols by autoxidation, peroxidation or action of $^1O_2$, as clearly explained by SMITH (1). The main problem, when an oxysterol is isolated from a biological medium, is to answer the following question : is it an "endogenous" oxysterol or an artefact formed during the isolation process ?

The first oxysterols were identified in the forties by BERGSTRÖM and WINTERSTEINER (2a,2b) in plasma (7α-OH, 7β-OH, 7-ketocholesterol). Later, in the sixties, it was shown that these oxysterols were mainly in an esterified form (CRASTES de PAULET et al. 3,4). At this time, oxysterols were considered as curiosities, probably artefacts. They were very difficult to study since their concentrations were very low, and at that time analytical methods were not advanced enough to go further.

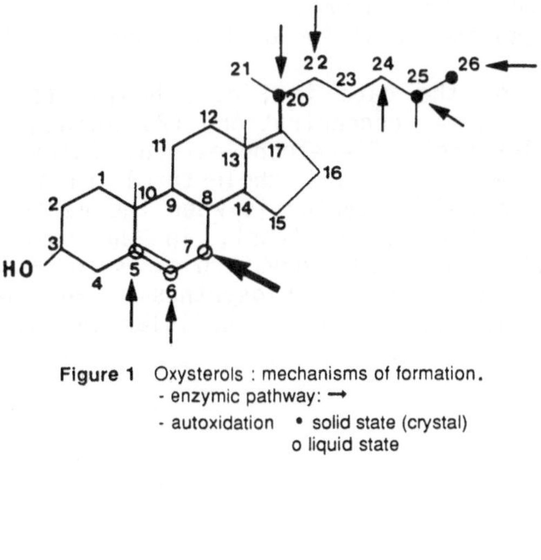

Figure 1 Oxysterols : mechanisms of formation.
- enzymic pathway: →
- autoxidation • solid state (crystal)
o liquid state

The interest in oxysterols arose in 1973 with the discovery by CHEN and KANDUTSCH (5) that some oxysterols (7α-OH, 7β-OH, 7 keto-cholesterol) were strong inhibitors of HMG CoA reductase activity, a key enzyme in cholesterol biosynthesis, in L cells and mouse hepatocytes. Since then the interest in this new class of sterols has grown exponentially with the discovery of other biological activities, such as inhibition of cell division, mutagenicity, cytotoxicity, immunosuppressive effects, complement activation, angiotoxicity and atherogenicity.

If we consider the problem of oxysterol biological activites generally, the situation seems very confusing. However, it must be kept in mind that this family is very large, properties may differ depending on the type of oxysterol, the concentration used and the target. The same situation exists for the biological properties of steroid hormones and their metabolites.

In the seventies our laboratory was interested in steroid hormone metabolisms and mechanisms of action, and in cellular cholesterol homeostasis, particularly with respect to cellular division.

We thus decided to work on oxysterol action and mechanisms of action using different cellular systems and a vast range of oxysterols, either synthesized in our laboratory, or gifts from colleagues, such as VAN LIER (Sherbrooke), OURISSON (Strasbourg), or the ROUSSEL UCLAF Group.

Our first assays were performed on human lymphocytes. When lymphocytes are stimulated by lectins, such as phytohemagglutinin (PHA), they behave as synchronized cells by entering the same cell cycle stage (G$_1$) together, in so called "lymphoblastic transformation". We first used 25-OH cholesterol, a very efficient and readily available oxysterol, which could be obtained in tritiated form with very high specific activity.

When PHA was added to human lymphocytes cultured in RPMI 1640 + 20 % AB serum, we got successively the following results (Fig. 2a) : 1) an enhancement of the HMG CoA reductase activity, 2) then a strong stimulation of cholesterogenesis, 3) finally, 20 hours later these two events, the so called "blastic transformation", as indicated by the incorporation of [methyl-3H] thymidine in DNA. When 25-OH cholesterol is added simultaneously, all these phenomena were completely suppressed (Fig. 2b).

This first observation gave rise to two important questions :
1°) Is the suppression of cell division a consequence of the suppression of cholesterol biosynthesis ?
2°) What is the mechanism of action of this oxysterol ?

An analysis of the data in Fig. 3 begins to answer these two questions. At the same concentration (25 µg/ml), some oxysterols such as 7-ketocholesterol, 22-ketocholesterol almost completely suppressed reductase activity (and cholesterol biosynthesis), without having any effect on DNA biosynthesis when the medium is supplemented with cholesterol or LDL. By contrast, in the same conditions, DNA synthesis is blocked by other oxysterols, such as 25-OH or 26-OH cholesterol. Thus, cholesterol biosynthesis and DNA biosynthesis are two independent mechanisms in this cellular system.

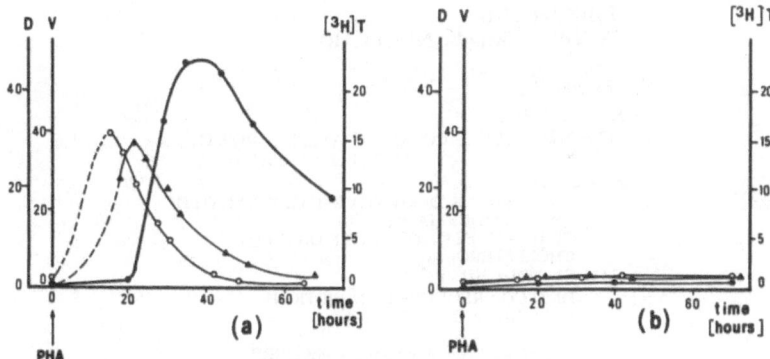

Figure 2 Time course of cholesterol biosynthesis and [³H] thymidine incorporation into DNA in human lymphocytes stimulated by PHA in the absence (a) or in the presence (b) of 25-OH cholesterol (25 µg/ml).

Legend : Experimental conditions as in ref. 6.V(o) : HMG CoA reductase (pmoles/min/mg P). D (▲)3 : [¹⁴C] digitonin-precipitated sterols (dpm x 10⁻³).

• : [³H] thymidine incorporated into DNA (dpm x 10⁻³).

Very small differences in oxysterol structure lead to large differences in the efficiency of HMG CoA reductase inhibition : compare for instance the effects of 24(R)- and 24(S)-OH cholesterol epimers, or 23(R)- and 23(S)-OH cholesterol epimers. Hence, considering the sharp specificity in the action of these oxysterols, the answer to the 2nd question (What is the mechanism of action of the 25-OH cholesterol on the reductase activity and DNA biosynthesis ?) should be : oxysterols act by a mechanism involving some "specific receptor". We therefore decided to work on this hypothesis and to try to characterize this "specific receptor". We were encouraged by the idea that cell mechanisms were involved in this phenomenon through the following observation, made first by CAVENEE et al. (8) : reductase inhibition by oxysterols was observed only in intact cells no effect was observed with microsomes.

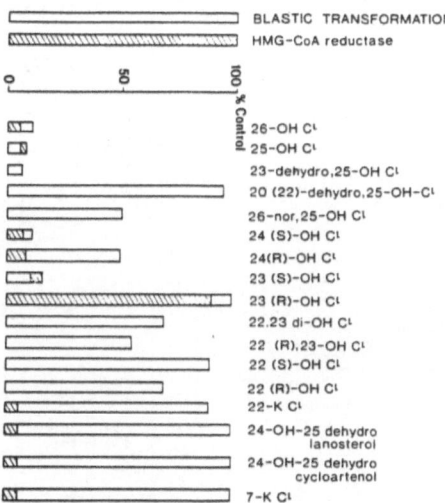

Figure 3 Effect of several oxysterols (25 µg/ml) on blastic transformation and HMG CoA reductase activity of human lymphocytes stimulated by PHA.

Legend : Experimental conditions as in ref. 7.

- **UBIQUITOUS**
- **CYTOSOL (Maybe NUCLEAR)**
- **8 S**
- **165 000 Da**
- **Kd : 6.10⁻⁹M** $6.10^{-9}M$
- **SITE NUMBER** : ACCORDING TO CELL GROWTH
                     AND CELL CYCLE PHASIS
- **SPECIFICITY**
    - CHOLESTEROL DERIVATIVES HYDROXYLATED
      ON THE SIDE CHAIN (+++)
    - DERIVATIVES OXYGENATED ON C-7 (±)
    - CHOLESTEROL : 0
- **ROLE ~ "RECEPTOR"**
    - HMG COA REDUCTASE INHIBITION
    - CELL GROWTH INHIBITION

Figure 4   Main characteristics of OSBP.

After five years of extensive work, first on a human lympho-cyte cell system, then on fibroblasts, HTC cells and rat liver, using [³H] 25 OH cholesterol to track the postulated receptor, we characte-rized a so called "oxysterol binding protein" (OSBP) in the "cytosol" of all the cell types we investigated. This OSBP had the same physico-chimical characteristics in all cell types studied : sedimentation constant, apparent MW of the OS/OSBP complex (fig. 4). It also had all of the characteristics of a true receptor : limited number of sites, high affinity for the ligand, and specificity as clearly shown in Fig. 5, obtained by OSBP isolated from cytosol rat liver.

What is the biological role of this OSBP ?

There is some evidence that this protein is involved in the control of HMG CoA reductase level. Indeed, a recent work of TAYLOR (10) on OSBP of L cells shows a statistically significant positive

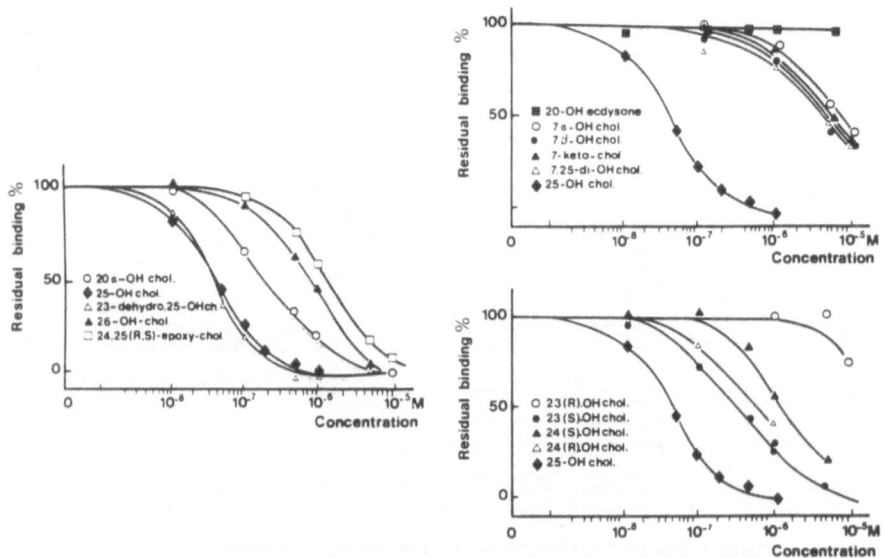

Figure 5   Competition of several oxygenated sterols with [³H] 25-OH cholesterol in rat liver cytosol.
Legend :   Experimental conditions as in ref. 9.

correlation between the logs of the binding affinities of about 45 different sterols and their reductase repression efficiency. But the same reductase inhibition is sometimes obtained with oxysterols having different binding affinities !

We observed the same phenomenon with the rat fibroblast OSBP. But the situation is completely different if we consider the correlation between binding and DNA inhibition, there is a clear linear correlation (Fig. 6).

The strong discrepancy, when we consider the effect on reductase activity and DNA biosynthesis, could be related to some difference(s) in the affinities of the OS/OSBP complex for the "nuclear acceptors", involved respectively in reductase synthesis and the synthesis of some proteins controlling cell division. Currently, we do not know what the "nuclear acceptors" of OS/OSBP complex are. However, if we consider the cell mechanisms, there is other indirect evidence that OS/OSBP complex could be involved in genome control : we have identified a nuclear OS/OSBP complex in rat fibroblast (Fig. 7). Furthermore, in synchronized rat fibroblasts, the number of binding sites depends on the cell cycle phase (Fig. 8).

Returning to HMG CoA reductase inhibition, the questions which now arise are the following :

1) "Are endogenous oxysterols involved in cellular cholesterol homeostasis ?"
- The question of whether oxysterols formed during lanosterol demethylation could be endogenous regulators of cholesterol biosynthesis has been raised by GIBBONS et al. (12), and our laboratory (13). Oxysterols such as those shown in Fig. 9 are found in human lymphocytes and other cells, but their concentrations and their affinities for OSBP are low. Hence, until present, they do not seem to be good candidates.

- The by-products of oxidosqualene cyclisation could be alternative candidates. Indeed, the 24,25-oxidocholesterol has a high affinity for OSBP and is a very efficient HMG-CoA reductase suppressor. But its cellular concentration is normally very low. However, it can be raised to high levels using oxidosqualene-lanosterol cyclase inhibitors, such as azadecalines, and thus contribute to the lowering effect of this drug on cholesterol biosynthesis, as recently suggested by GERST et al. (14).
- 7α-OH cholesterol could also be involved in cholesterol homeostasis. This oxysterol is enzymatically formed during the first step of bile acid production. But its IC_{50} for OSBP is in the µM range, probably too high to be efficient in physiological situations.

2) "Is (or are) some oxysterol(s) produced by the liver involved in cholesterol homeostasis in extrahepatic tissue" ? :

26-OH cholesterol could be a good candidate since it is present in plasma : JAVITT et al. (15). It has an IC_{50} for OSBP in the 10^{-8}M range, and it is very efficient in HMG CoA reductase inhibition (Fig. 3). Moreover, it, and LDL cholesterol, have another effect on the internalized LDL : they block the binding of LDL, and subsequently block the internalization and the degradation of LDL, as recently shown by LORENZO et al. (16). This effect could be interpretated as resulting from the inhibition of Apo B receptor biosynthesis, exactly as internalized cholesterol is supposed to act, but at much higher concentrations.

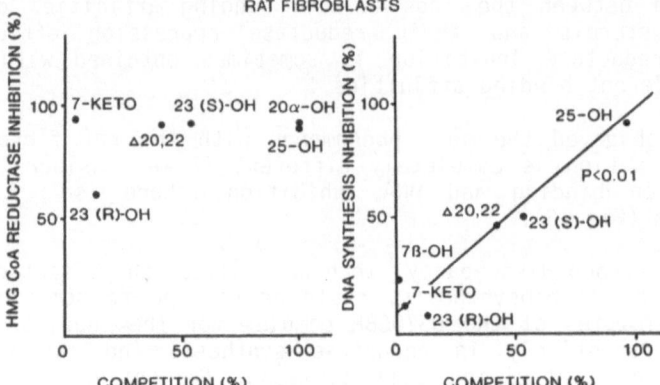

Figure 6 Relationship between OSBP.affinity for several oxysterols and inhibitory activity of these OS on HMG CoA reductase activity (a) or DNA synthesis inhibition (b).

Legend : Experimental conditions as in ref. 11.

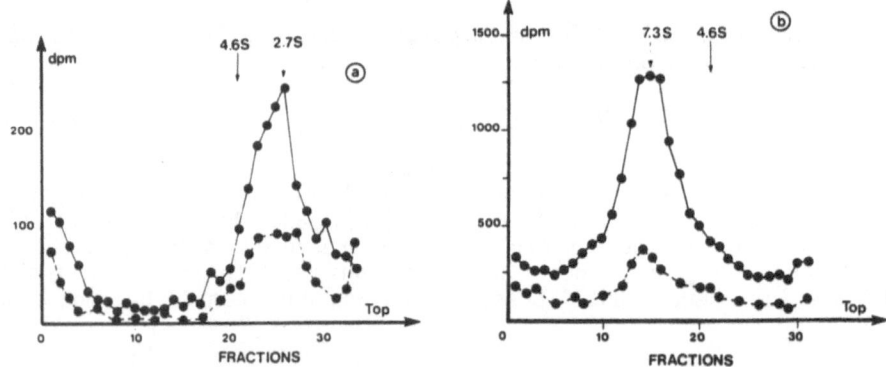

Figure 7 Binding pattern of [³H]-25-OH cholesterol to rat fibroblast nucleic macromolecules (from ref. 11).

Legend : a) Supernatant S1 obtained after digestion of labelled nuclei with DNAse I in the presence of isotonic buffer.
b) Supernatant S2 obtained after extraction of pellet N1 by hypotonic buffer.

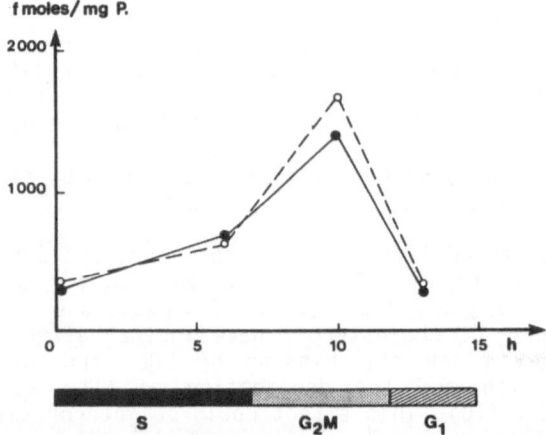

Figure 8 25-OH cholesterol binding variations during the rat embryo fibroblast cell cycle (from ref. 11).

296

Thus, the final question arises : "is cholesterol really involved in cholesterol homeostasis ?"
Another challenging question could be asked : "is cholesterol really involved in atherogenenis ?"

Concerning this question, we observed that as clearly shown in Table I, 26 OH cholesterol is the most important oxysterol in atheroma plates, as previously shown by SMITH and VAN LIER (17). Where does it come from ? What is the significance of this observation in the mechanism of atherogenesis. These are two questions without any clear answers, until present.

To conclude, we present some of our recent results on the metabolism of 7α-OH cholesterol in humans treated with cholestyramine.

Two cationic polymers, cholestyramine and colestipol, are given orally, mainly in association with drugs inhibiting cholesterol biosynthesis (fibrates, HMG CoA reductase inhibitors belonging to the "Compactin" series), to lower circulating cholesterol in hypercholesterolemic patients. These polymers strongly bind the bile acids in the gut. Thus, they completely block the entero-hepatic cycle of these bile acids, and bile acid concentrations in the liver are lowered. In this situation, the 7α-hydroxylase activity of the hepatocyte increases, thus lowering the cholesterol level in the cell. This results in stimulation of Apo B receptor synthesis and LDL clearance improvement, which finally leads to a hypocholesterolemic effect.

Table 1

EVALUATION OF THE MOST REPRESENTATIVE OXYSTEROLS IN HUMAN
ATHEROMATOUS AORTA BY THIN LAYER CHROMATOGRAPHY
AND GAS LIPID CHROMATOGRAPHY
(Results in µg/g of tissue)

	TRIOL	7α(OH)C^L	7β	7 OXO	C^L EPOXYDES β	α	20(OH)C^L	25(OH)C^L	26(OH)C^L
ATH. 02.A	12	44	45	20	15	15	0	0	360
ATH. 03.G	40	-	50	40	18.5	-	0	0	147
ATH. 04.C	15	10	20	20	35	-	0	0	90
ATH. 05.B	17	90	70	-	50	-	0	7	300
ATH. 06.B	100	120	200	-	-	-	0	11	1000
ATH. 07.P	33	7	8	20	13.5	-	0	0	25
ATH. 08.P	71	80	80	-	45	45	0	0	150
ATH. 09.D	58	17	22	-	50	-	0	0	220
ATH. 10.B	52	36	30	19	96	92	0	8	219
ATH. 11.G	37	45	60	10	53	61	0	0	436
ATH. 12.M	42	72	93	8	47	24	0	7	738

Triol	:	3β-5-6β-trihydroxy-5α-cholestane triol
7α(OH)C^L	:	3β-7α-dihydroxy-cholesta-5-ene
7β(OH)C^L	:	3β-7β-dihydroxy-cholesta-5-ene
7-oxo C^L	:	3β-hydroxy-7-oxo-cholesta-5-ene
Epox-α	:	5-6α-epoxy-5α-cholesta-3β-ol
Epox-β	:	5-6β-epoxy-5β-cholesta-3β-ol
20(OH)C^L	:	3β-20-dihydroxy-(20 S)-cholesta-5-ene
25(OH)C^L	:	3β-25-dihydroxy-cholesta-5-ene
26(OH)C^L	:	3β-26-dihydroxy-(20 R)-cholesta-5-ene

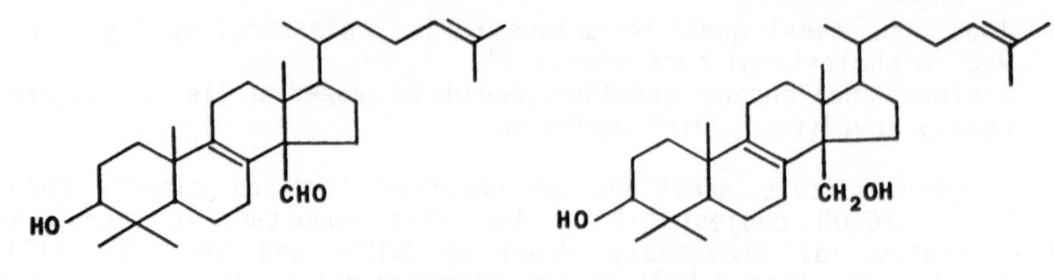

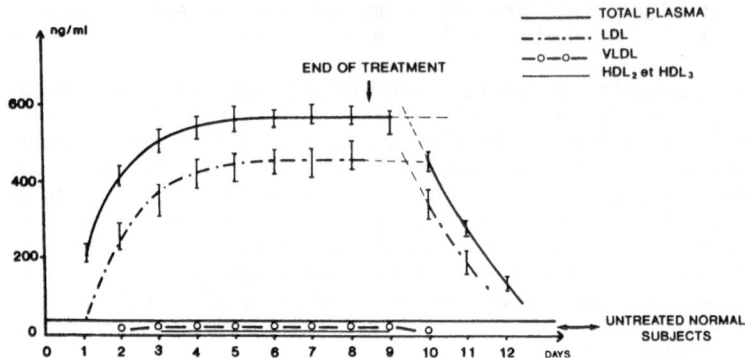

3-hydroxy-5-lanosta-8,24-diene-32-al   5-lanosta-8,24,diene-3,32-diol

Figure 9

Figure 10 7α-OH cholesterol in plasma of a normal adult treated with cholestyramine 0,15 g/kg/day.

In normal situations, 7α-OH cholesterol levels in plasma are low (< 100 μg/l). High levels are observed (up to 2 mg/l) after cholestyramine administration. They are not correlated with concentrations of total cholesterol, HDL cholesterol, Apo A_1, or Apo B. In normal subjects, the half-life of total plasma 7α-OH cholesterol is $\simeq$ 30-50 h, close to the approximative half-life of LDL (Fig. 10). It is mainly transported by LDL (up to 80 %). Hence, 7α-OH cholesterol is probably internalized with LDL. Since high concentrations ($\simeq$ 5 μM) are reached in plasma of subjects treated by cholestyramine, the following question arises : since the IC_{50} of this oxysterol for OSBP is $\simeq$ 2 μM, and since this oxysterol is a very efficient reductase suppressor, could this 7α-OH cholesterol participate in cholesterol cellular homeostasis in these subjects ?

If the answer is positive, this very interesting indirect effect of cholestyramine could perhaps contribute to the lowering effect of this polymer on circulating cholesterol. This question is currently under investigation.

Many other questions arise when we consider the biological effects of oxysterols. Some of the observed effects, depending on concentrations reached, targets, and the oxysterol structure, are probably the result of a direct action on membrane organisation and signal transduction. These include alterations of membrane permeability (18), decreased rates of endocytosis (19), changes in cell shape (20), in calcium permeability (21), potentiation of arachidonic acid release and prostaglandin biosynthesis (22) ; some of which are probably involved in the cytotoxicity of oxysterols. These must be taken into account for the study of the pathophysiological role of oxysterols, or for prospective investigations into their therapeutic use.

REFERENCES

1. L.L. Smith. Mechanisms of formation of oxysterols : a general survey. in "Nato Advanced Research Workshop. Action of free radicals and active forms of oxygen on lipoproteins and membrane lipids : cellular interactions and atherogenesis". A. Crastes de Paulet, L. Douste-Blazy and R. Paoletti Editors, Plenum Press (1990).

2a. S. Bergström and O. Wintersteiner. Autoxydation of sterols in colloïdal aqueous solution. III : quantitative studies on cholesterol. J. Biol. Chem., 145:309-326 (1942).

2b. S. Bergström and O. Wintersteiner. Autoxydation of sterols in colloïdal aqueous solution. IV : influence of esterification and of constitutional factors. J. Biol. Chem., 145:327-333 (1942).

3. A. Crastes de Paulet et P. Crastes de Paulet. Hétérogénéité de la fraction "ester du cholestérol" obtenue par chromatographie sur acide silicique des lipides totaux du sérum humain. in "The enzymes of lipid metabolism", P. Desnuelle Ed., pages 106-109, Pergamon Press (1961).

4. A. Crastes de Paulet, P. Crastes de Paulet et L. Bardou. Autoxydation einiger 3β-hydroxysterin esters. 1st International Congress on Fat Research, Hambourg, 84-85, (1964).

5. A.A. Kandutsch, H.W. Chen. Inhibition of sterol synthesis in cultured mouse cells by 7α-hydroxycholesterol, 7β-hydroxycholesterol and 7-ketocholesterol. J. Biol. Chem., 248:8408-8417 (1973).

6. C. Tabacik, M. Astruc, M. Laporte, B. Descomps, A. Crastes de Paulet and B. Serrou. Comparative study of the kinetics of 3-hydroxy-3-methylglutaryl Coenzyme A reductase and [^{14}C]-acetate incorporation into cholesterol in human lymphocytes stimulated by phytohemagglutinin or sterol efflux. Biomedecine, 34:128-132 (1981).

7. R. Defay, M.E. Astruc, S. Roussillon, B. Descomps and A. Crastes de Paulet. DNA synthesis and 3-hydroxy-3-methylglutaryl CoA reductase activity in PHA-stimulated human lymphocytes : a comparative study of the inhibitory effects of some oxysterols with special reference to side chain hydroxylated derivatives. Biochem. Biophys. Res. Comm., 106:362-372 (1982).

8. W.K. Cavenee, H.W. Chen and A.A. Kandutsch. Regulation of cholesterol biosynthesis in enucleated cells. J. Biol. Chem., 256:2675-2681 (1981).

9. F. Besème, M.E. Astruc, R. Defay and A. Crastes de Paulet. Rat liver oxysterol-binding protein : characterization and comparison with the HTC cell protein. FEBS Letters, 210:97-103 (1987).

10. F.R. Taylor, S.E. Saucier, E.P. Shown, E.J. Parish, A.A. Kandutsch. Correlation between oxysterol binding to a cytosolic binding protein and potency in the repression of 3-hydroxy-3-methylglutaryl CoA reductase. J. Biol. Chem., 259:12382-12387 (1984).

11. F. Besème, M.E. Astruc, R. Defay, B. Descomps and A. Crastes de Paulet. Characterization of oxysterol-binding protein in rat embryo fibroblasts and variations as a function of the cell cycle. Biochim. Biophys. Acta, 886:96-108 (1986).

12. G.F. Gibbons, C.R. Pullinger, H.W. Chen, W.K. Cavenee and A.A. Kandutsch. Regulation of cholesterol biosynthesis in cultured cells by probable natural precursor sterols. J. Biol. Chem., 255:395-400 (1980).

13. C. Tabacik, S. Aliau, B. Serrou and A. Crastes de Paulet. Post HMG-CoA reductase regulation of cholesterol biosynthesis in normal human lymphocytes : lanosten-3β-ol-32-al, a natural inhibitor. Biochem. Biophys. Res. Comm., 101:1087-1095 (1981).

14. N. Gerst, A. Duriatti, F. Schuber, M. Taton, P. Benveniste et A. Rahier. Potent inhibition of cholesterol biosynthesis in 3T3 fibroblasts by N-[1,5,9)-trimethyldecyl]-4α, 10-dimethyl-8-aza-trans-decal-3β-ol, a new 2,3-oxidosqualene cyclase inhibitor. Biochem. Pharmacol., 37:1955-1964 (1988).

15. N.R. Javitt, E. Kok, S. Burstein, B. Cohen and J. Kutscher. 26-hydroxycholesterol : identification and quantitation in human serum. J. Biol. Chem., 256:12644-12646 (1981).

16. J.L. Lorenzo, M. Allorio, F. Bernini, A. Corsini and R. Fumagalli. Regulation of low density lipoprotein metabolism by 26-hydroxy-cholesterol in human fibroblasts. FEBS Letters, 218:77-80 (1987).

17. L.L. Smith and J.E. Van Lier. 26-hydroxycholesterol levels in human aorta. Atherosclerosis, 12:1-14 (1970).

18. H.W. Chen, H.J. Heiniger and A.A. Kandutsch. Alteration of ^{86}Rb^{+} influx and efflux following depletion of membrane sterol in L-cells. J. Biol. Chem., 253:3180-3185, (1978).

19. H.J. Heiniger, A.A. Kandutsch and H.W. Chen. Depletion of L-cell sterol depresses endocytosis. Nature, 263:515-517 (1976).

20. S. Yachnin, R.A. Streuli, L.I. Gordon and R.C. Hsu. Alteration of peripheral blood cell membrane function and morphology by oxygenated sterols ; a membrane insertion hypothesis. Curr. Topics Hematol., 2:245-271 (1979).

21. G.A. Boissonault and H.J. Heiniger. 25-hydroxycholesterol indu-ced elevation in ^{45}Ca uptake : correlation with depressed DNA synthesis. J. Cell Physiol., 120:151-156 (1984).

22. Z. Lahoua, M.E. Astruc and A. Crastes de Paulet. Serum induced arachidonic acid release and prostaglandin biosynthesis are potentiated by oxygenated sterols in NRK49 F cells. Biochim. Biophys. Acta, 958:396-404 (1988).

MECHANISMS OF HYDROPEROXIDE-INDUCED BRONCHO- AND VASOCONSTRICTION IN THE

PERFUSED RAT LUNG

Kristin Olafsdottir, Åke Ryrfeldt, Margareta Berggren and
Peter Moldéus

Department of Toxicology, Karolinska Institutet
S-104 01 Stockholm, Sweden

INTRODUCTION

Active forms of oxygen are now recognized as important factors in the
etiology of lung diseases such as chronic bronchitis and emphysema. Reactive
species such as the superoxide anion radical (O_2^-), hydrogen peroxide (H_2O_2)
and the hydroxyl radical (·OH) may be formed as a consequence of inflammation
which is generally a result of excessive cigarette smoking and environmental
factors. Oxidative stress may also be caused directly by components present
in cigarette smoke such as polymeric phenoxy radicals, hydroperoxides and
other oxidative reactive agents. Experimentally, active oxygen species
generated enzymatically in the air-passages have been shown to cause pulmonary toxicity.[1] Pulmonary toxicity has also been observed to be caused by
stimulated granulocytes.[2] This damage has been postulated to be due to direct
oxidative damage by reactive oxidants produced by the granulocytes.

In addition to direct oxidative damage by reactive oxygen species they
may stimulate production of endogenous mediators and thus also physiological
responses. This has been demonstrated using an organic hydroperoxide, ter-[3,4]
tiary butylhydroperoxide (TBH), and various isolated perfused lung models.
In these models TBH may cause both broncho- and vasoconstriction. The mechanism of these hydroperoxide effects are not yet elucidated but the vasoconstriction has been suggested to be, at least partly, due to stimulated
production of thromboxane.[5]

The present manuscript summarizes some recent findings regarding hydroperoxide induced broncho- and vasoconstriction in an isolated perfused rat
lung model.

MATERIALS AND METHODS

Lungs were isolated from male Wistar rats and perfused essentially as
described,[6] using Krebs-ringer buffer pH 7.4 (composition in mM: NaCl, 118.0;
KCl, 4.7; $CaCl_2$, 2.5; $MgSO_4$, 1.2; $NaHCO_3$, 24.9; KH_2PO_4, 1.2) with the addition of 2% BSA, 0.1% glucose and 12.5 mM HEPES. The perfusion flow was 15-20
ml/min. The lungs were ventilated at 80 breaths/min by creating an alternative negative pressure (-1 to -8 cm H_2O) inside the thoracic chamber relative
to the ambient atmosphere. The pressure changes and the tracheal air flow
were recorded simultaneously on a computer where calculations of lung con-

Free Radicals, Lipoproteins, and Membrane Lipids
Edited by A. Crastes de Paulet *et al.*
Plenum Press, New York, 1990

ductance and dynamic compliance were performed.[7,8] The perfusion flow rate was measured with a drop counter, as well as manually, and recorded with time. Lastly the pH of the buffer influent and the pH and pO_2 in the effluent were recorded. A scheme of the perfused lung model is shown in Figure 1.

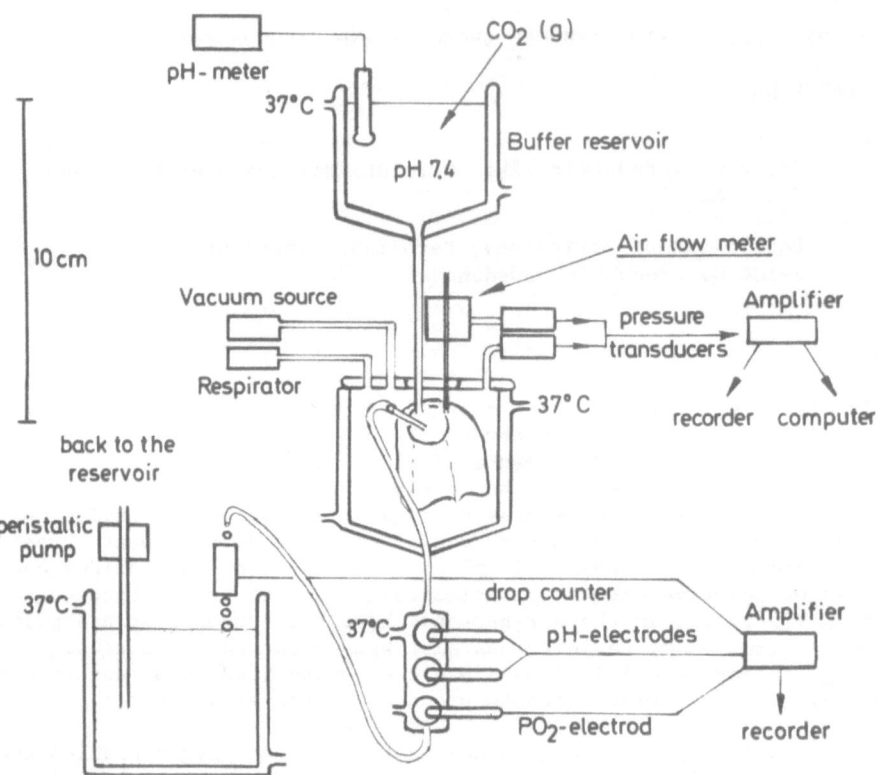

Figure 1. Schematic drawing of the isolated perfused lung set-up.

After 30 min equilibration, TBH (500 µM) was infused via the pulmonary circulation for 10 min in a recirculating system. This dosing was repeated after a 10-20 min recovery period in control buffer. Augmented breaths were induced during this period to recover lung mechanics. Indomethacin and nordihydroguaiaretic acid (NDGA) were added during the recovery period, via the pulmonary circulation and these were also present during the second dosing with TBH. At the end of the perfusions a part of the lung was used to estimate wet/dry weight and the rest was homogenized in 7% perchloric acid. Glutathione (GSH) was analyzed by HPLC according to Reed et al.[9]

TBH, Indomethacin and NDGA were obtained from Sigma Chemical Co. and other chemicals were of the highest grade of purity available from local commercial sources.

RESULTS

   Infusion of either TBH (500 μM) (Figure 2) or H_2O_2 (500 μM) (Figure 3)
for 10 min caused both bronchoconstriction expressed as decreased conductance
and compliance and vasoconstriction expressed as decreased pulmonary circulation. Maximal effects were obtained after 1 to 3 min after the start of
hydroperoxide infusion. The effect diminished during the later part of the
infusion period and a slight decrease in sensitivity of the preparation was
noted during repeated infusions (Figure 2 and 3).

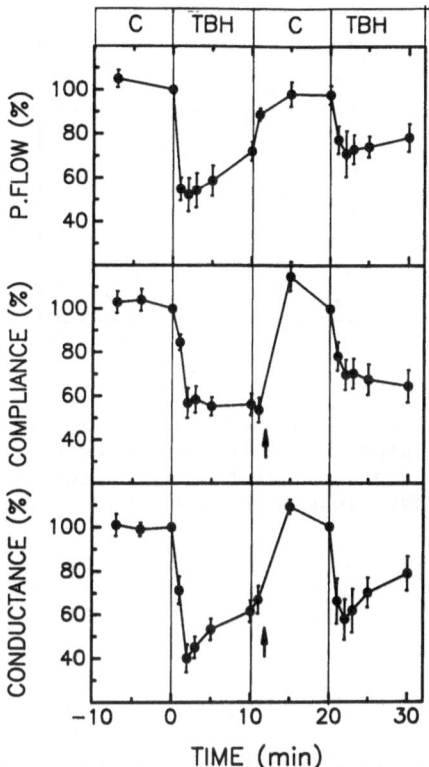

Figure 2. Effect of TBH (500 μM) on perfusion flow and lung mechanics in the
perfused rat lung. Infusion regiments are indicated on the figure. An augmented breath was induced at the onset of the recovery period between the two
TBH-infusions, to recover lung mechanics (↑). Values are means ± SEM, n=4.

   GSH is considered to be of vital importance to metabolize hydroperoxides
through the glutathione peroxidase/reductase system and thus protect the lung
from hydroperoxide dependent oxidative stress. As would be expected TBH
infused via the pulmonary circulation decreased intracellular GSH significantly (Table 1). In addition when lung GSH was depleted by DEM treatment the
lung was more sensitive towards the TBH induced bronchoconstriction (Table 1).

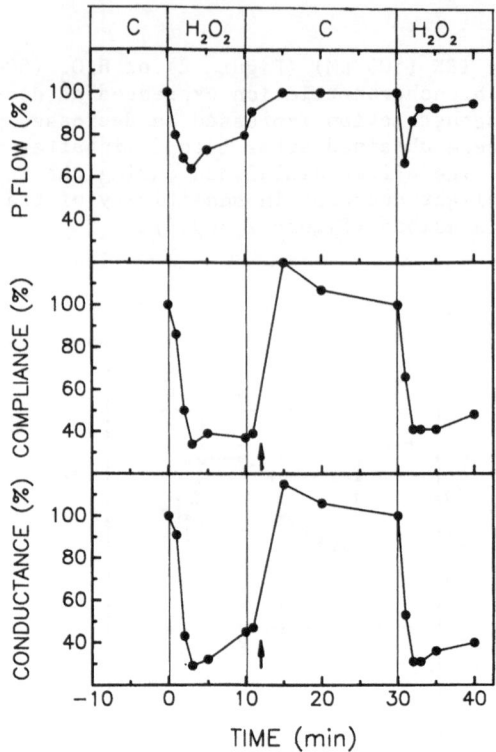

Figure 3.   Effect of H_2O_2 (500 μM) on perfusion flow and lung mechanics in the perfused rat lung. Infusion regiments are indicated on the figure. An augmented breath was induced at the onset of the recovery period between the two H_2O_2-infusions, to recover lung mechanics (↑). Values are means of two experiments.

Table 1.   Changes in lung GSH and conductance following DEM and TBH treatment.

Treatment	GSH nmol/g tissue wet weight	Conductance % change
Control	906 ± 58	0 ± 5
DEM (1 mM infusion)	70 ± 25	0 ± 5
TBH (200 μM infusion)	770 ± 80	30 ± 10
TBH (200 μM) + DEM	N.D.	70 ± 15
TBH (400 μM)	676 ± 97	110 ± 50
TBH (400 μM) + DEM	N.D.	190 ± 120

Values are means ± SEM (n=3–5)
GSH levels were measured after 30 min infusion of DEM or three 10 min infusions of TBH with 10 min intervals. The change in conductance was estimated at the maximum response after the first TBH infusion.

Infusion of TBH also caused pulmonary edema and following two 10 min infusions of TBH (500 μM) the lung weight was almost doubled (Table 2). Infusion of the cyclooxygenase inhibitor indomethacin (28 μM) prior and during the second TBH infusion mostly prevented the edema formation (Table 2) and also attenuated the broncho- and vasoconstriction induced by TBH (Figure 4). The lipoxygenase inhibitor NDGA (25 μM) had no effect on either TBH induced bronchoconstriction or decreased pulmonary circulation.

Table 2. Wet to dry weight ratios at the end of rat lung perfusion.

Treatment	Wet/dry weight
Control	7.36 ± 0.13
TBH	13.6 ± 0.80
TBH + indomethacin	8.74 ± 0.22

Values are means ± SEM (n=3)
TBH (500 μM) was infused twice for 10 min each time with a 10 min interval. Indomethacin (28 μM) was present during the interval and during the latter TBH infusion. The lungs were weighed and dried immediately after the perfusion.

DISCUSSION

When either the organic hydroperoxide TBH or H_2O_2 was infused into the circulation of a perfused rat lung a dose-related broncho- and vasoconstriction was noted. The fact that either hydroperoxides induced these effects indicate the importance of the hydroperoxide moiety for the response.

Stimulated arachidonate metabolism through the cyclooxygenase pathway appears to be of importance for the hydroperoxide induced broncho- and vasoconstriction since indomethacin, a well known blocker of this pathway, almost completely attenuated the effects in our lung model. Gurtner and associates found using an isolated perfused rabbit lung model that indomethacin blocked TBH induced vasoconstriction.[4] They have later attributed the TBH effect to an increased thromboxane production.[5] Whether or not a stimulated thromboxane production is of importance in our system remains to be established.

We did not observe any protective effects of NDGA, an inhibitor of lipoxygenase, on hydroperoxide induced broncho- and vasoconstriction in our rat lung model. This was in contrast to recent findings by Gurtner et al.[10] who using different inhibitors of this pathway, showed that both edema and vasoconstriction were inhibited. The reason for this discrepancy is not know but may depend on species variation.

Recently, it has been shown that TBH can stimulate release of arachidonate from lipid stores of endothelial cells and human embryo lung fibroblasts.[11,12] Also, TBH seems to stimulate prostaglandin synthesis.[11,12] The reason to the decrease in bronchoconstriction noted during the later part of each TBH infusion, may be due to depletion of an arachidonate pool. Since repeated infusion of TBH (500 μM) during 10 min at 10 min intervals caused comparable or only slightly decreased broncho- and vasoconstrictions, such a pool seems to be reestablished during this short time interval.

TBH was found to be a potent agent to reduce the GSH-content of the lungs. However, depletion of GSH does not lead to bronchoconstriction. This was clearly shown by infusion of DEM, which almost completely depleted the lungs of GSH but no bronchoconstriction was observed. However, the increased sensitivity of the lungs, obtained from rats pretreated with DEM, to TBH can be due to the fact that in these GSH depleted lungs TBH will not be inactivated by GSH as can occur in lungs with intact GSH content (t – BuOOH + 2 GSH $\longrightarrow$ t – BuOH + H_2O + GSSG). Thus, in the GSH depleted lungs, more TBH will be

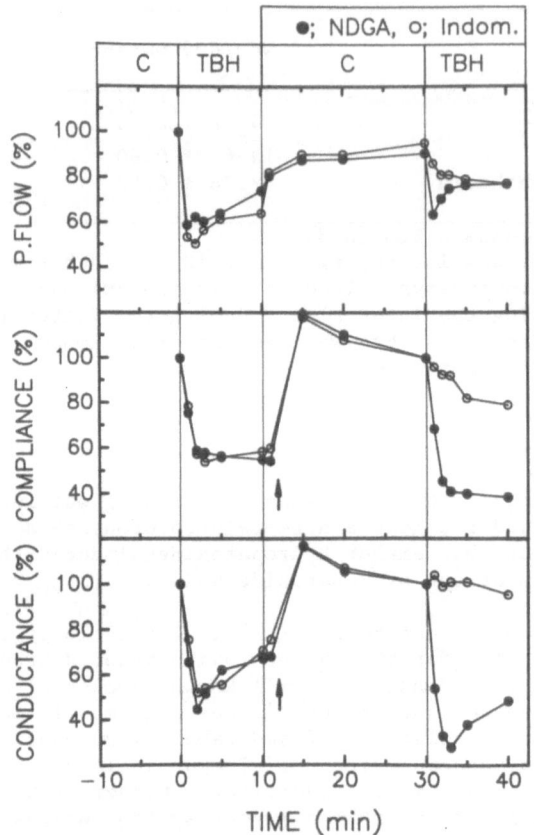

Figure 4. Effect of indomethacin (28 µM) and NDGA (25 µM) on TBH (500 µM) induced changes in perfusion flow and lung mechanics in the perfused rat lung. The agents were administered for 30 min starting with the 20 min recovery period between the two TBH infusions. Values are means of two experiments.

available to stimulate release of arachidonate. This is supported by the fact that a more pronounced difference in sensitivity between DEM-pretreated and not pretreated lungs was found at the TBH concentration 200 µM than at 400 µM. The latter TBH concentration may be high enough that GSH depletion will be of less importance.

More work is needed to elucidate the mechanism by which the hydroperoxides stimulate arachidonate metabolism in the perfused rat lung. We believe it is a selective process possibly involving receptor oxidation rather than stimulated lipid peroxidation. We have for instance not been able to detect any lipid peroxidation products in either perfusate or lung following TBH treatment.

REFERENCES

1. J. C. Fantone, and P. A. Ward, Am. J. Pathol. 107:396 (1982)
2. M. Kuroda, K. Murakami, and Y. Ishikawa, Am. Rev. Resp. Dis. 136:1435 (1987).
3. Å. Ryrfeldt, F. Kröll, M. Berggren, and P. Moldéus, Life Sciences 42:1429 (1988)
4. G. H. Gurtner, A. Knoblauch, P. L. Smith, H. Sies, and N. F. Adkinson, J. Appl. Physiol. 55:949 (1983).
5. I. S. Farrukh, J. R. Michael, W.R. Summer, N. F. Adkinson, and G. H. Gurtner, J. Appl. Physiol. 58:34 (1985).
6. F. Kröll, J. A. Karlsson, E. Nilsson, C. G. A. Persson, and Å. Ryrfeldt, Acta Physiol. Scand. 127:1 (1986).
7. M. O. Amdur, and J. Mead, Am. J. Physiol. 192:364 (1958).
8. J. Waldeck, Thesis, Lunds Technical University (1988).
9. D. J. Reed, J. R. Babson, P. W. Beatty, A.E. Brodie, W.W. Ellis, and D. W. Potter, Anal. Biochem. 106:55 (1980).
10. G. H. Gurtner, I.S. Farrukh, N. F. Adkinson, A. M. Sciuto, J. M. Jacobson, and J. R. Michael, Am. Rev. Resp. Dis. 136:480 (1987).
11. L. Taylor, M. J. Menconi, and P. Polgar, J. Biol. Chem. 258:6855 (1983).
12. R. M. Jackson, D. B. Chandler, and J. D. Fulmer, J. Appl. Physiol. 61:584 (1986).

PATHOPHYSIOLOGICAL RELEVANCE OF FREE RADICALS TO THE ETHANOL-INDUCED DISORDERS IN MEMBRANE LIPIDS

Roger Nordmann, Catherine Ribière and Hélène Rouach

Department of Biomedical Research on Alcoholism
Université René Descartes, 45 rue des Saints-Pères
75270 Paris Cedex 06 , France

INTRODUCTION

Numerous studies have shown that ethanol administration is followed by changes in the membranous lipids, especially at the level of the phospholipid polyunsaturated fatty acid side chains. These changes in the membrane lipid composition have been suggested to modify membrane fluidity and thereby play a role in prominent manifestations of alcohol abuse such as tolerance to or dependence on ethanol (1). They have also been implied as important causative factors in many aspects of ethanol toxicity.

RELEVANCE OF FREE RADICALS TO THE ETHANOL-INDUCED DISORDERS IN HEPATIC MEMBRANE LIPIDS

That an oxidative stress could be, at least partly, involved in the genesis of lipid membrane disorders was first shown at the hepatic level.

After having shown that the development of the fatty liver induced by the administration of a large single dose of ethanol can be attenuated by administering anti-oxidants (2), Di Luzio and Hartman (3) reported that ethanol increases the content of lipid peroxides in the rat liver. It could therefore be suggested that ethanol produces free radical disturbances in the liver that result in an enhanced membrane lipid peroxidation.

These pioneer observations were followed by many controversial results concerning the onset of hepatic lipid peroxidation during acute as well as chronic ethanol intoxication. However it appears now established that lipid peroxidation takes place in the liver following ethanol administration, at least in most experimental conditions currently used. The data allowing such a conclusion have been recently reviewed by Dianzani (4). Some of these data have been obtained by non-invasive techniques such as low-level chemiluminescence assay or determination of alkane generation in perfused rat liver or isolated hepatocytes. Moreover Reinke et al. (5) using electron spin resonance (e.s.r.) spectroscopy succeeded recently in detecting in the liver of ethanol-fed rats receiving a spin trapping agent e.s.r. signals which appeared most likely issued from carbon-centered radical adducts of membranous lipids.

The mechanisms responsible for hepatic lipid peroxidation during ethanol intoxication remain controversial. Due to its high reactivity, the hydroxyl radical ($^{\cdot}$OH) is the most likely candidate having a cytotoxic role (6). It can contribute to induce an oxidative stress, i.e. a disturbance in the pro-oxidant/anti-oxidant systems in favour of the former (7). Membranous lipids represent an essential target of such a free radical attack which results in lipid peroxidation. Superoxide radicals ($O_2^{\cdot-}$), although being themselves rather inactive, are of prominent importance by contributing to the biosynthesis of $^{\cdot}$OH through the iron-catalyzed Haber-Weiss reaction.

These considerations led to consider the possible role of $^{\cdot}$OH as initiator of ethanol-induced lipid peroxidation, especially at the liver <u>microsomal</u> level.

As shown by Cederbaum (8), microsomal $^{\cdot}$OH generation is significantly increased after chronic ethanol consumption. This increase is likely related to an enhancement of $O_2^{\cdot-}$ radical production which can occur at the site of either the flavoenzyme NADPH-cytochrome P450 reductase (8) or cytochrome P450 itself (9). Since microsomal lipid peroxidation can be almost completely abolished by antibodies directed against the ethanol-inducible cytochrome P450 isoforms (10), these isoforms appear of primary importance in the onset of microsomal lipid peroxidation in the liver of ethanol-treated rats. However these ethanol-induced P-450 forms appear specially effective in ethanol oxidation and NADPH oxidation, but not in $^{\cdot}$OH production (11).

Since other experimental results also show that lipid peroxidation in rat liver microsomes isolated after chronic alcohol feeding is unrelated to $^{\cdot}$OH generation (12), it appears that the radical species initiating lipid peroxidation may not be represented by $^{\cdot}$OH itself. Alternate initiators consisting in complexes between oxygen and iron such as perferryl ions or ferrous-dioxygen-ferric complexes have therefore been suggested (13).

Another possibility to be considered is the involvement of free radical species issued from ethanol itself during its microsomal oxidation. As hydrogen-atom abstraction from ethanol leads to α-hydroxy ethyl (CH_3-$C^{\cdot}HOH$), β-hydroxy ethyl ($C^{\cdot}H_2$-CH_2OH) or ethoxyl radicals (CH_3-$CH_2O^{\cdot}$), Slater (14) suggested as early as 1972 that ethanol can be metabolized in the liver to a free radical intermediate that might provoke damaging consequences. The generation of α-hydroxy ethyl radicals in isolated rat liver microsomes incubated with ethanol in the presence of a NADPH generating system and a spin trapping agent has been recently ascertained using e.s.r. spectroscopy by Albano et al. (15) as well as by Reinke et al. (5). Since ethanol is a well-known scavenger of $^{\cdot}$OH (8) it is likely that the α-hydroxy ethyl radical is generated during the oxidation of ethanol initiated by $^{\cdot}$OH. This α-hydroxy ethyl radical could contribute to the initiation of microsomal lipid peroxidation during acute as well as chronic ethanol intoxication either by itself or through the generation of a more reactive ethanol-derived radical. This last one could be represented by the α-hydroxy ethyl peroxyl radical (CH_3-CH-OH) which is easily generated from the α-hydroxy
$\quad\quad\quad\quad\quad\quad\quad\quad\quad\quad$ O-O$^{\cdot}$
ethyl radical in the presence of dioxygen (16).

Lipid peroxidation, which is well ascertained at the microsomal level, appears also to effect the <u>mitochondria</u> of the hepatocytes during acute or chronic ethanol intoxication. We have thus shown (17) that

310

an acute ethanol intoxication results in an increased susceptibility of liver mitochondria to peroxidation associated with a decrease in their content in α-tocopherol, the main endogenous membranous anti-oxidant. Long-term exposure of rats to ethanol vapors elicits an increase in the sensitivity of liver mitochondria to peroxidative attacks, increase which is still apparent 24 hrs after ethanol withdrawal [18]. Taken together these results suggest that lipid peroxidation may be implicated in the production of the functional and structural abnormalities of liver mitochondria resulting from alcohol consumption. This assumption is strengthened by the prevention by desferrioxamine of the hepatic ultrastructural mitochondrial lesions induced in the rat by chronic ethanol administration [19], desferrioxamine acting altogether as an iron-chelator and a scavenger of O_2^- radicals [20].

The mechanisms involved in the ethanol-induced liver mitochondrial lipid peroxidation are still controversial. It is however well known that the mitochondrial respiratory chain represents one of the main sources of superoxide [21]. Furthermore we have reported that an acute ethanol load results in a significant increase in this mitochondrial O_2^- production [22]. This increase which is likely related to a desorganizing effect of ethanol on the mitochondrial membranes may contribute to the ethanol-induced enhancement in liver mitochondrial lipid peroxidation.

Another site of production of active species able to participate to the onset of lipid peroxidation is represented by the cytosolic enzyme xanthine oxidase. Ethanol administration promotes the conversion of xanthine dehydrogenase to the O_2^- generating xanthine oxidase [23, 24]. At the same time it provides substrates for this oxidase, i.e. purines resulting from an increased ATP breakdown and acetaldehyde [23].

It appears thus that an enhanced hepatic O_2^- production may occur altogether at the microsomal, mitochondrial and cytosolic levels following ethanol administration. This increase in O_2^- production is likely able to contribute to the generation through iron-catalyzed reactions of more reactive species such as ˙OH or other aggressive oxidizing radicals.

The implication of iron in this generation led us to study whether ethanol administration affects the hepatic distribution of iron and especially of the low molecular weight chelatable iron (LMWC-iron) which appears to represent the catalytically active decompartmentalized iron [25]. During preliminary studies concerned with acute ethanol administration [26] we observed an increase in the hepatic total non-heme iron content, predominantly located in the microsomes and the cytosol. Interestingly we also found a significant increase in the LMWC-iron fraction in the cytosol, increase which may be linked to excessive mobilization of iron from its ferritin stores. Such a mobilization may be favoured by the reductive environment induced by the oxidation of ethanol through alcohol dehydrogenase (ADH) and/or by the generation of O_2^- during ethanol oxidation by non-ADH pathways.

To further assess the possible links between ethanol toxicity and iron disturbances we studied the effects of chronic iron-overload on the ethanol-induced membranous disorders. Feeding rats a carbonyl iron supplemented diet during 7 weeks and administering subsequently an acute ethanol load, we observed an additive effect of iron overload and ethanol administration on the changes in the liver lipid peroxide content as well as in the mitochondrial susceptibility to peroxidation and vitamin E content [17, 27]. Such an additive effect may be of importance in alcoholic liver diseases since increased iron stores have often been reported in the liver of these patients [28].

Having considered the changes in the production of free radicals and the distribution of iron which may contribute to ethanol-induced lipid peroxidation in the liver, we are to take also into account the disorders in the cellular anti-oxidant defence. As a matter of fact an oxidative stress occurs only when this defence is insufficient to cope with the free radicals that are generated. We have already mentioned the ethanol-induced decrease in the mitochondrial α-tocopherol content, decrease which appears of special importance in the onset of lipid peroxidation due to the prominent role of α-tocopherol in the membranous anti-oxidant defence. Much attention has also been given to changes in the liver reduced glutathione (GSH) content, since GSH is a substrate of glutathione peroxidase, an enzyme metabolizing hydrogen peroxide as well as lipoperoxides. In most conditions ethanol administration is followed by a decrease in liver GSH (29). Several factors are able to contribute to this decrease. It could be mediated by the formation of adducts between GSH and aldehydes, such as acetaldehyde. Higher aldehydes, like 4-hydroxy nonenal, originating from the breakdown of polyunsaturated fatty acids are also specially able to react with GSH, thus contributing to the spreading of lipid peroxidation (30). We have shown that desferrioxamine administration prevents the ethanol-induced decrease in liver GSH (31), a finding that suggests that this decrease is, at least partly, related to the direct scavenging effect of GSH on free radicals.

RELEVANCE OF FREE RADICALS TO THE ETHANOL-INDUCED DISORDERS IN MEMBRANE LIPIDS IN EXTRA-HEPATIC TISSUES

Whereas the first reports concerning ethanol-induced lipid peroxidation suggested that this disorder occurs only in the liver, where ethanol is actively metabolized, more recent data have shown that ethanol administration to rodents elicits also an oxidative stress in several extra-hepatic tissues.

Among these tissues, the central nervous system appears of special importance, since it possesses, at least in vitro, a high ability to undergo lipid peroxidation (32). This ability is linked to its high oxygen consumption and its high content in easily oxidizable substrates (mainly polyunsaturated fatty acids and catecholamines) contrasting with the low activity of some enzymes involved in the cellular anti-oxidant defence (33). We therefore studied whether acute ethanol administration induces lipid peroxidation in the cerebellum. Cerebellum was chosen since the ability to undergo lipid peroxidation in vitro is higher in this brain area than in others. Our data (34) show that lipid peroxidation is significantly enhanced in rat cerebellum following an acute ethanol load. They represent, to our knowledge, the first report of an enhancement in lipid peroxidation in the nervous system induced by the sole administration of ethanol. The acute ethanol administration elicited at the same time a significant increase in the cerebellar non-heme iron content as well as a decrease in the cerebellar level of the main anti-oxidant substrates, i.e. α-tocopherol and ascorbate (35). Whether such an oxidative stress also occurs during chronic alcohol consumption remains to be ascertained. An argument favouring its occurrence and contribution to the disorders in the central nervous system is represented by the reduction in severity of physical dependence on ethanol in mice receiving desferrioxamine during chronic ethanol inhalation (36).

Ethanol-induced lipid peroxidation has also been reported in cardiac tissue (5, 37) and related to free radicals generated either at the level of xanthine oxidase (23) or of the peroxisomal acyl-CoA oxidase (37).

Furthermore an enhancement in lipid peroxidation has been suggested to contribute to the _testicular_ (38) and _renal_ (39) injury following ethanol administration.

CONCLUSION

Taken together the reported data show that an ethanol-induced oxidative stress occurs altogether in the liver and in extra-hepatic tissues, such as cerebellum, heart and testes. It is likely that the disorders in membrane lipids resulting from such an oxidative stress represent an important causative factor in ethanol-induced cellular injury. This view opens attractive pharmacological possibilities for the prevention of alcoholic diseases (35) especially since primates appear to be more susceptible than rodents to lipid peroxidation, at least at the hepatic level (40).

ACKNOWLEDGEMENTS

Work from the authors' Laboratory was supported by Fondation de France, INSERM (CRE 872010) and IREB (Grant 86/10-07).

REFERENCES

1. D. B. Goldstein, Ethanol-induced adaptation in biological membranes, Ann. N.Y. Acad. Sci. 492 : 103-111 (1987).
2. N. R. Di Luzio, Prevention of the acute ethanol-induced fatty liver by the simultaneous administration of anti-oxidants, Life Sci. 3 : 113-118 (1964).
3. N. R. Di Luzio and A. D. Hartman, Role of lipid peroxidation in the pathogenesis of ethanol-induced fatty liver, Federation Proc. 26 : 1436-1442 (1967).
4. M. U. Dianzani, Lipid peroxidation in ethanol poisoning : a critical reconsideration, Alcohol Alcoholism 20 : 161-173 (1985).
5. L. A. Reinke, E. K. Lai, C. M. Du Bose and P. B. Mc Cay, Reactive free radical generation in vivo in heart and liver of ethanol-fed rats : Correlation with radical formation in vitro, Proc. Natl. Acad. Sci. USA, 84 : 9223-9227 (1987).
6. B. Halliwell and J. M. C. Gutteridge, Oxygen free radicals and iron in relation to biology and medicine : some problems and concepts, Arch. Biochem. Biophys. 246 : 501-514 (1986).
7. H. Sies, Oxidative stress : Introductory remarks, in : "Oxidative Stress", H. Sies, ed., Acad. Press, London, pp. 1-8 (1985).
8. A. I. Cederbaum, Microsomal generation of hydroxyl radicals : its role in microsomal ethanol oxidizing system (MEOS) activity and requirement for iron, Ann. N.Y. Acad. Sci. 492 : 35-49 (1987).
9. G. Ekström and M. Ingelman-Sundberg, Mechanisms of lipid peroxidation dependent upon cytochrome P-450 LM(2), Eur. J. Biochem. 158 : 195-201 (1986).
10. M. Ingelman-Sundberg, G. Ekström and N. Tindberg, Lipid peroxidation dependent on ethanol-inducible cytochrome P-450 from rat liver, in : "Advances in the Biosciences, vol. 71 : Alcohol Toxicity and Free Radical Mechanisms", R. Nordmann, C. Ribière and H. Rouach, eds., Pergamon Press, Oxford, pp. 43-48 (1988).
11. G. Ekström, T. Cronholm and M. Ingelman-Sundberg, Hydroxyl-radical production and ethanol oxidation by liver microsomes isolated from ethanol-treated rats, Biochem. J. 233 : 755-761 (1986).
12. S. Shaw, E. Jayatilleke and C. S. Lieber, The effect of chronic alcohol feeding on lipid peroxidation in microsomes : Lack of

relationship to hydroxyl radical generation, Biochem. Biophys. Res. Comm. 118 : 233-238 (1984).

13. G. Minotti and S. D. Aust, The requirement for iron (III) in the initiation of lipid peroxidation by iron (II) and hydrogen peroxide, J. Biol. Chem. 262 : 1098-1104 (1987).

14. T. F. Slater, Hepatotoxic effects of alcohol, in : "Free Radical Mechanisms in Tissue Injury", Pion Ltd., London, pp. 171-197 (1972).

15. E. Albano, A. Tomasi, L. Goria-Gatti, G. Poli, V. Vannini and M. U. Dianzani, Free radical metabolism of alcohols by rat liver microsomes, Free Radical Res. Comm. 3 : 243-249 (1987).

16. T. F. Slater, Free radical mechanisms in tissue injury with special reference to the cytotoxic effects of ethanol and related alcohols, in : "Advances in the Biosciences, vol. 71 : Alcohol Toxicity and Free Radical Mechanisms", R. Nordmann, C. Ribière and H. Rouach, eds., Pergamon Press, Oxford, pp. 1-9 (1988).

17. H. Rouach, M. K. Park, M. T. Orfanelli, B. Janvier, P. Brissot, M. Bourel and R. Nordmann, Effects of ethanol on hepatic and cerebellar lipid peroxidation and endogenous antioxidants in naive and chronic iron overloaded rats, in : "Advances in the Biosciences, vol.71: Alcohol Toxicity and Free Radical Mechanisms", R. Nordmann, C.Ribière and H.Rouach, eds., Pergamon Press, Oxford, pp.49-54 (1988).

18. H. Rouach, M. Clément, M. T. Orfanelli, B. Janvier, J. Nordmann and R. Nordmann, Hepatic lipid peroxidation and mitochondrial susceptibility to peroxidative attacks during ethanol inhalation and withdrawal, Biochim. Biophys. Acta 753 : 439-444 (1983).

19. J. Sinaceur, C. Legendre, J. Montagne, M. Jagueux, L. Orcel and R. Nordmann, Prevention by desferrioxamine (DFO) of the hepatic ultrastructural lesions induced in the rat by chronic ethanol administration, Alcohol Alcoholism 22 : A 10 (1987).

20. C. Ribière, D. Sabourault, J. Sinaceur, R. Nordmann, C. Houée-Levin, and C. Ferradini, Radiolysis study of the reaction of desferrioxamine with $O_2^{\bar{}}$ free radicals, in : "Superoxide and Superoxide Dismutase in Chemistry, Biology and Medicine", G. Rotilio, ed., Elsevier, Amsterdam, pp. 47-49 (1986).

21. H. J. Forman and A. Boveris, Superoxide radical and hydrogen peroxide in mitochondria, in : "Free Radicals in Biology", vol.5, W. A. Pryor, ed., Academic Press, New-York, pp. 65-90 (1982).

22. J. Sinaceur, C. Ribière, D. Sabourault and R. Nordmann, Superoxide formation in liver mitochondria during ethanol intoxication : possible role in alcohol hepatotoxicity, in : "Free Radicals in Liver Injury", G. Poli, K. H. Cheeseman, M. U. Dianzani and T. F. Slater, eds., IRL Press, Oxford, pp. 175-177 (1988).

23. H. H. H. Oei, H. C. Zoganas, J. M. McCord and S. W. Schaffer, Role of acetaldehyde and xanthine oxidase in ethanol-induced oxidative stress, Res. Comm. Chem. Pathol. Pharmacol. 51 : 195-203 (1986).

24. L. G. Sultatos, Effects of acute ethanol administration on the hepatic xanthine dehydrogenase-oxidase system in the rat, J. Pharmacol. Exp. Ther. 246 : 946-949 (1988).

25. A. Jacobs, Low molecular weight intracellular iron transport compounds, Blood 50 : 433-439 (1977).

26. H. Rouach, P. Houzé, M. T. Orfanelli, M. Gentil, R. Bourdon and R. Nordmann, Effect of acute ethanol administration on the subcellular distribution of iron in rat liver and cerebellum, Submitted for publication.

27. R. Nordmann, C. Ribière and H. Rouach, Involvement of iron and iron-catalyzed free radical production in ethanol metabolism and toxicity, Enzyme 37 : 57-69 (1987).

28. M. G. Irving, J. W. Halliday and L. W. Powell, Association between alcoholism and increased hepatic iron stores, Alcoholism Clin. Exp. Res. 12 : 7-13 (1988).

29. L. A. Videla, V. Fernandez, A. de Marinis, N. Fernandez and

A. Valenzuela, Liver lipoperoxidative pressure and glutathione status following acetaldehyde and aliphatic alcohols pretreatments in rat, Biochem. Biophys. Res. Comm. 104 : 965-970 (1982).

30. A. Müller and H. Sies, Alcohol, aldehydes and lipid peroxidation : current notions, in : "Advances in Biomedical Alcohol Research", K. O. Lindros, R. Ylikhari and K. Kiianmaa, eds., Pergamon Press, Oxford, pp. 67-74 (1987).

31. H. Antébi, C. Ribière, J. Sinaceur, C. Abu-Murad and R. Nordmann, Involvement of oxygen radicals in ethanol oxidation and in the ethanol-induced decrease in liver glutathione, in : "Oxygen Radicals in Chemistry and Biology", W. Bors, M. Saran and D. Tait, eds., De Gruyter, Berlin, pp. 757-760 (1984).

32. B. Halliwell, Free radicals and metal ions in health and disease, Proc. Nutrition Soc. 46 : 13-26 (1987).

33. P. M. Sinet, R. E. Heikkila and G. Cohen, Hydrogen peroxide production by rat brain in vivo, J. Neurochem. 34 : 1421-1428 (1980).

34. R. Nordmann, Oxidative stress from alcohol in the brain, in : "Advances in Biomedical Alcohol Research", K. O. Lindros, R. Ylikhri and K. Kiianmaa, eds., Pergamon Press, Oxford, pp. 75-82 (1987).

35. R. Nordmann, C. Ribière and H. Rouach, Free radicals and oxidative stress : their implication in the metabolism and toxicity of ethanol, in : "Biomedical and Social Aspects of Alcohol and Alcoholism", K. Kuriyama, A Takada and H. Ishii, eds., Excerpta Medica, Amsterdam, pp. 17-27 (1988).

36. C. Abu-Murad and R. Nordmann, Reduction in severity of physical dependence on ethanol in mice caused by desferrioxamine administration, Pharmacol. Biochem. Behav. 18 : 515-517 (1983).

37. L. F. Panchenko, S. V. Pirozhkov, S. V. Popova and V. D. Antonenkov, Effect of chronic ethanol treatment on peroxisomal acyl-CoA oxidase activity and lipid peroxidation in rat liver and heart, Experientia 43 : 580-581 (1987).

38. E. Rosemblum, J. S. Gavaler and D. H. Van Thiel, Lipid peroxidation : a mechanism for ethanol-associated testicular injury in rats, Endocrinology 116 : 311-318 (1985).

39. Y. Kera, S. Komura, Y. Ohbora, T. Kiriyama and K. Inone, Ethanol induced changes in lipid peroxidation and non protein sulhydryl content, Res. Comm. Chem. Pathol. Pharmacol. 47 : 203-209 (1985).

40. S. Shaw, K. P. Rubin and C. S. Lieber, Depressed hepatic glutathione and increased diene conjugates in alcoholic liver disease. Evidence of lipid peroxidation, Digest. Diseases Sciences 28 : 585-589 (1983).

LIPID PEROXIDATION IN EXPERIMENTALLY PRODUCED LIVER INJURY,

LIVER TUMORS AND IN LIVER REGENERATION

T.F. Slater

Dept. of Biology & Biochemistry, Brunel University
Uxbridge, Middx, U.K.

INTRODUCTION

Lipid peroxidation is a free radical-mediated process leading to an oxidative degradation of lipid materials including triglycerides, phospholipids, cholesterol and its esters, and unsaturated fatty acids. It may be a non-enzymic process (eg. following the impact of radiation or with a transition metal-catalysed reaction) or it may be catalysed by enzymes, either specifically as with cyclo-oxygenase and lipoxygenases, or non-specifically as with the role of the NADPH-cytochrome P_{450} reductase in maintaining an iron chelate in a reduced state[1]. In this article lipid peroxidation will be considered in relation to the peroxidation of polyunsaturated fatty acids (PUFA's), both free and esterified. More general background reviews on lipid peroxidation that may be consulted for reference are references [2,3].

The initial reaction in the peroxidation of PUFA's involving an oxidising free radical R· is the abstraction of a hydrogen atom:

$$PUFA(H) + R \longrightarrow PUFA\cdot + RH \qquad (1)$$

where PUFA· is a carbon-centred radical. This is seen more clearly in Fig.1 where hydrogen atom abstraction from a $C_{18:2}$ fatty acid is considered. Fig.1 also shows that initiation is followed by bond rearrangement producing a conjugated diene. This diene radical can react very rapidly with molecular oxygen to yield a peroxyl-radical. In conditions where the oxygen tension is very low, and where there is a high concentration of a reducing material, the conjugated diene radical may also be reduced to the parent conjugated diene fatty acid[4]. Fig.1 only shows early steps in the complex sequence of reactions that are associated with lipid peroxidation; subsequent reactions of the lipid peroxyl radicals are considered later in this article.

Since the initiation of lipid peroxidation requires a reactive oxidising free radical (there is also a different route[5] involving singlet oxygen that is not considered here) it is pertinent to ask: how can reactive free radicals be produced in our cells and tissues? Major pathways that can result in the production of free radicals are (i) following exposure to radiation, which can be (a) ionizing (b) ultra-violet (c) visible light (d) thermal and (e) ultra-sonic; (ii) by reactions involving the exchange of a single-electron, so-called redox reactions, which can be catalysed by (a) transition metal ions such as Fe^{2+} or Cu^{2+} and (b) enzymes. Since this topic is dealt with in detail in many other reviews (see[6,7], for example) only brief comment is required here. An example of free radical production by ionising radiation is the radiolysis of aerated water:

$$\mathbf{\wedge\!\!\!\wedge\!\!\blacktriangleright} H_2O + O_2 \longrightarrow H_3O^+ + H^{\boldsymbol{\cdot}} + HO^{\boldsymbol{\cdot}} + e^- + O_2^{\boldsymbol{\cdot}-} \qquad (2)$$

Ultra-violet radiation of wave-length less than about 300 nm has sufficient intrinsic energy to cause homolysis of covalent bonds[6]. A very clear example is the homolysis of trichlorobromomethane by uv-light to produce $\cdot CCl_3$ that can be trapped by phenyl butyl nitrone to produce a radical adduct with a characteristic electron spin resonance (esr) spectrum[8].

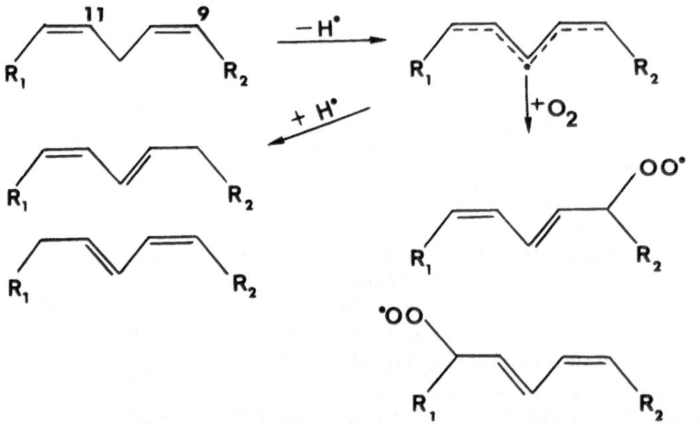

Fig. 1.   The initiation of lipid peroxidation in linoleic acid (R_1 = $CH_3(CH_2)_4$; R_2 = $(CH_2)_4$ COOH) by hydrogen atom abstraction to produce a carbon-centred free radical with electron delocalisation over carbon atoms 9 to 13. This can then add back a hydrogen atom provided by a suitable reducing donor and give dienes in which the double bonds are conjugated. Alternatively in the presence of oxygen, peroxyl free radicals are formed. The Figure does not include illustrations of stereoisomers that involve spatial changes in the positions of R_1 and R_2 relative to a neighbouring double bond.

Even radiation of inherently low intrinsic energy can result in the formation of free radicals if an appropriate photosensitiser is present[9]. Many clinical cases of photosensitivity involve porphyrins as the photosensitising agent[10]; in consequence, the exciting wavelength of light is in the Soret (approx. 410 nm) region. Photosensitisation of skin by porphyrins results in a very rapid damage to epidermal lysosomes[12] and lipid peroxidation. A general reaction mechanism (see [12]) is shown in Fig.2. It is interesting to note that photosensitivity type reactions are now being used quite frequently in treating certain ulcerative and malignant lesions (see [13]).

318

$$^1P \xrightarrow{\text{light}} {}^1P^* \longrightarrow {}^3P^*$$

$$^3P^* + {}^3O_2 \longrightarrow {}^1P + {}^1O_2$$

$$^1O_2 + \text{PUFA} \longrightarrow \text{PUFA-OOH}$$

$$\text{PUFA-OOH} + \text{transition metal ions} \longrightarrow \text{PUFAOO}^\cdot, \text{PUFAO}^\cdot$$

$$\text{PUFAOO}^\cdot, \text{PUFAO}^\cdot \longrightarrow \text{lipid peroxidation}$$

Fig. 2. A scheme (see refs.[11,12]) to illustrate the sequence of events in porphyrin (P) photosesitised lipid peroxidation refs.[9-13]). A superscript* indicates an excited state; PUFA is a polyunsaturated fatty acid.

In Fig. 2, a porphyrin is excited by incident visible radiation to an excited singlet state ($^1P^*$) that can either decay back to the original ground-state singlet or can make a so-called forbidden transition to an excited triplet state ($^3P^*$). A triplet state has two unpaired (but closely coupled) electrons. The excited triplet porphyrin can then interact readily with normal ground-state oxygen, which is unusual in being in a triplet state: the result is the production of a very reactive form of oxygen, singlet oxygen, that can initiate lipid peroxidation.

Transition metal ions, and many metal chelates, can catalyse free radical formation by participating in single electron transfer reaction especially with hydrogen peroxide and lipid hydroperoxides (LOOH):

$$Fe^{2+} + H_2O_2 \longrightarrow Fe^{3+} + HO^- + HO^\cdot \qquad (3)$$

$$Fe^{2+} + \text{LOOH} \longrightarrow LO^\cdot + HO^- + Fe^{3+} \qquad (4)$$

$$Fe^{3+} + \text{LOOH} \longrightarrow LOO^\cdot + Fe^{2+} + H^+ \qquad (5)$$

Free radical species can also be produced during a variety of enzyme catalysed reactions. In some cases, the free radical is an intermediate in a specific metabolic reaction pathway, and does not signficantly 'escape' from the enzyme before being converted to a more stable and non-radical product. In many other cases, however, the initial free radical species is able to diffuse away from its site of formation and may then cause damage to the cell and its surroundings. This latter sequence of events whereby a molecule is enzymically converted to a more reactive and often more damaging form is known as metabolic activation[14]; products of metabolic activation can be N-hydroxy derivatives, epoxides, carbonium ions etc. as well as free radicals. Examples of substances that can be metabolically activated to free radical intermediates include a variety of halogeno-alkanes, quinones, hydrazines, and nitro-compounds. For general reviews on this subject see [15,16]. A specific instance of metabolic activation that has been studied in great depth by many investigators for a long time[14,17-20] concerns carbon tetrachloride; this will be discussed in the following section in some detail as a number of important general concepts have emerged from these studies.

CARBON TETRACHLORIDE

A single dose of CCl_4 administered to a rat, or to individual animals of many other species including man, produces a centrilobular necrosis and fatty degeneration of the liver[21]. Repeated doses can produce cirrhosis and liver cancer; see[6]. Although it is a powerful solvent for lipids, and after administration is widely distributed throughout the body, its main point of attack is the liver; within the liver the first signs of toxic disturbance are seen in the endoplasmic reticulum[22]. This led to the hypothesis[6,14] that CCl_4 is metabolically activated to the trichloromethyl free radical. It took more than ten years, however, to establish unequivocally[23,24] that $\cdot CCl_3$ is indeed formed in liver in vivo, and in liver fractions in vitro, using esr-spin trapping techniques. In connection with this example of metabolic activation it is interesting to note that CCl_3F, which is relatively non-toxic[25], does not yield a significant esr spin-trap adduct either with microsomal suspensions or liver plasma membrane suspensions[26]. Moreover, when isolated hepatocytes are incubated with CCl_3F there is no significant blebbing of the plasma membrane unlike the situation seen[27] following similar incubations with CCl_4. These studies with CCl_3F, which is also a powerful lipid solvent, are not consistent with the view[28] that a significant part of the cytotoxic action of CCl_4 on liver cells is due to a direct lipophilic solvent action

An early finding in studies on the hepatotoxicity of CCl_4 was that CCl_4 stimulates lipid peroxidation both in vivo and in vitro; Table 1 summarises the major contributions to this view. Two points require special emphasis here (a) CCl_4 stimulates lipid peroxidation in vivo at very early times following administration when liver structure is intact; the increased lipid peroxidation is not due,therefore,simply to a disruption of structure as occurs in preparing tissue homogenates; (b) the introduction[32,33] of the washed microsomal model demonstrated the dependence of activation on an NADPH-requiring process, allowing the activation to be identified with the NADPH-cytochrome P_{450} electron transport chain.

Table 1   Significant events in relation to the metabolic activation of CCl_4, and its stimulatory action on lipid peroxidation in rat liver.

Metabolic activation	Reference
Butler (1961)	29
Slater (1966)	14
Lipid peroxidation in vitro	
Comporti, Saccoci and Dianzani (1965)	30
Ghoshal and Recknagel (1965)	31
Slater (1966, 1967)	32, 33
Lipid peroxidation in vivo	
Loss of fatty acids:	
Comporti, Burdino and Ugazio (1969)	34
Increased diene conjugation:	
Klaassen and Plaa (1969)	35
Increased malonaldehyde:	
Jose and Slater (1972)	36
Increased alkane exhalation:	
Riely, Cohen and Lieberman (1974)	37

Following the production of $\cdot CCl_3$ by metabolic activation it was easy to visualise how lipid peroxidation can be initiated:

$$PUFA(H) + \cdot CCl_3 \longrightarrow PUFA\cdot + CHCl_3 \qquad (6)$$

; chloroform is a known[38] product of the metabolism in vivo of CCl_4. Later studies, however, showed that $\cdot CCl_3$ is relatively not very reactive unless traces of oxygen are present when the highly reactive trichloromethylperoxyl radical is formed:

$$\cdot CCl_3 + O_2 \longrightarrow CCl_3OO\cdot \qquad (7)$$

; this reaction has a large 2nd order rate constant[40] of $9 \times 10^9 M^{-1}s^{-1}$. rate constants for the interactions of $CCl_3OO\cdot$ with biomolecules including arachidonic acid are shown in Table 2; it is clear that $CCl_3OO\cdot$, if formed in the neighbourhood of a PUFA, can initiate lipid peroxidation.

The studies illustrated in Table 2 on the high reactivity of $CCl_3OO\cdot$, and the relatively low reactivity of $\cdot CCl_3$ referred to above help to explain a long standing difficulty in relation to the hepatotoxicity of CCl_4. It has been known for a long time from studies by Rees and colleagues (see [6] for review) that Promethazine has a strong protective action in relation to the onset of necrosis, but very little effect on the development of a fatty liver. Arguements (reviewed in [18]) have subsequently occurred as to whether the stimulating action of CCl_4 on lipid peroxidation is or is not more important than covalent binding of CCl_3 to lipids and proteins. Since $CCl_3OO\cdot$ can react with PUFA's and can be scavenged by Promethazine it seems as if the effects produced by $CCl_3OO\cdot$ are relevant to necrosis. Promethazine, however, has very little influence on covalent binding, and reacts only slowly with $\cdot CCl_3$. It has been concluded[18] that the action of CCl_4 on the liver is multicausal involving not only $\cdot CCl_3$ and $CCl_3OO\cdot$, but also related species of free radicals, and results in lipid peroxidation, covalent binding and disturbances of transport processes.

Table 2   Some second-order rate constants ($M^{-1}s^{-1}$) for the interaction of the hydroxyl and trichloromethylperoxyl radicals with various substances. The data are from references[41-45].

Substance	$HO\bullet$	$CCl_3OO\bullet$
Promethazine	10^{10}	4.5×10^8
Glutathione	10^{10}	1.4×10^7
Cysteine	4×10^{10}	5×10^7
Tryptophan	10^{10}	1.2×10^8
Linoleic acid	--	4×10^6
Arachidonic acid	10^{10}	7×10^6
Docosahexaenoic acid		-
α-Tocopherol	10^{10}	5×10^8
β-Carotene	10^{10}	1.5×10^9
Catechin	10^{10}	5×10^7
Desferrioxamine	5×10^9	-
Indomethacin	6.2×10^9	10^8
Flurbiprofen	7.8×10^9	$<10^6$

Very high rate constants as shown in Table 2 ensure that reactive free radicals such as HO· and $CCl_3OO·$ cannot diffuse significantly from their site of formation within a cell or tissue; because of their very high reactivity they are essentially restricted to their micro-environment[46]. This fundamental concept has important consequences for applications of free radical scavengers. Firstly, it is clear from the data of Table 2 that there can be no selective scavengers of HO·; secondly, because of the restricted diffusion of reactive free radicals it is necessary for an effective scavenger to reach the precise site within a cell or tissue where the free radical is formed. Moreover, because of the transient existence of reactive free radicals, and the fact that scavenging in a biological environment is essentially a competition, the scavenger must get to the right site at the right time, and in a sufficiently high concentration[47].

BIOLOGICAL CONSEQUENCES OF LIPID PEROXIDATION

As mentioned above, a number of toxic agents such as CCl_4 can undergo metabolic activation to reactive free radical intermediates; these can initiate lipid peroxidation. It has become apparent over the last ten years or so that some products of lipid peroxidation have very high and potentially very important biological effects (see [48]). Lipid hydroperoxides are known to affect the prostaglandin cascade at sub-micromolar concentrations[49]; epoxy-derivatives of arachidonate have powerful effects on the release of some hormones[50], and 4-hydroxy-alkenals have a multitude of effects[51] including inhibitory actions on DNA-synthesis. The production of such compounds as by-products of lipid peroxidation allows a considerable diffusion of damaging intermediates from a precise site of free radical initiation[52]. Since the 4-hydroxy alkenals can strongly inhibit DNA-synthesis at low concentration (see[51]) it might be expected that dividing cells should have a relatively low rate of lipid peroxidation[53]. This is examined briefly below in relation to liver tumours and regenerating liver.

LIVER TUMOURS

It is well known that liver tumours in general have a low rate of lipid peroxidation (see[54]). Evaluation of the reasons for this have shown[54,55] that there are a number of contributory factors: a decreased level of substrate PUFA's; a very much decreased content of cytochrome P_{450} that can act normally as an initiating site for free radical production; a decreased concentration of NADPH that can act as a reducing source for iron chelates etc.; and an increased content of α-tocopherol. The latter aspect appears to be of special significance[54,55] in the Novikoff and Yoshida rat liver tumours; in these liver tumours, as in normal liver, α-tocopherol is the major lipid-soluble chain-breaking antioxidant present. Measurements of α-tocopherol per cell at different periods of growth of Yoshida tumour cells in the peritoneal cavity of mice show clearly that the tumour cell mass is accumulating substantial amounts of α-tocopherol during growth[56]. These data are consistent with the view that dividing cells 'benefit' from a low rate of lipid peroxidation[53]. Further studies of this view have been with regenerating liver.

LIVER REGENERATION

When part (eg 2/3) of the total liver mass of a rat is surgically removed the remaining liver undergoes a hyperplasia that, rather quickly, restores total liver mass although not original shape[57]; this process is often referred to as liver regeneration. If rats are entrained to a precise light-dark and feeding regimen then the cell division and DNA-synthesis associated with regeneration occur in well separated cycles[58]. Using this approach we have been able to show[59] that peaks of DNA-synthesis (actually expressed as thymidine kinase activity) coincide with minimum activities of

lipid peroxidation provoked by pro-oxidant stress. These changes in
thymidine kinase activity and lipid peroxidation are very closely linked in
time[60].

The down-regulations of lipid peroxidation that occur at approximately
24-hr intervals during liver regeneration are associated with corresponding
fluctuations of antioxidants, including α-tocopherol[61]. The details of the
mechanisms that control this variation in lipid peroxidation in relation to
liver growth have still to be clarified. Nonetheless, the data obtained so
far point to the conclusion that liver cell division in both normal and
tumour cell populations is associated with a down-regulation of lipid peroxidation: this may well have important biological complications.

ACKNOWLEDGEMENTS

I am grateful to and dependent on all my co-workers and colleagues who
have contributed to the studies reported here, especially Dr K H Cheeseman,
Sean Emery and Professor K U Ingold. Our work has been generously supported
by the Cancer Research Campaign, The Association for International Cancer
Research and the Wellcome Trust.

REFERENCES

1. P. Hochstein, K.Nordenbrand and L.Ernster, Evidence for the involve-
    ment of iron in the ADP-activated peroxidation of lipids in micro-
    somes and mitochondria, _Biochem. Biophys. Res. Commun._, 14: 323-
    328 (1964).
2. M. Comporti, Biology of disease. Lipid peroxidation and cellular
    damage in toxic liver injury, _Lab. Invest._ 53:599-623 (1985).
3. D. L Tribble, T. Y. Aw and D. P. Jones, The pathophysiological sig-
    nificance of lipid peroxidation in oxidative cell injury,
    Hepatology, 7:377-386.
4. T. L. Dormandy, Free radical activity and diene conjugation in man,
    _in_ "Free Radicals in Liver Injury", G. Poli, K. H. Cheeseman,
    M. U. Dianzani and T. F Slater, eds. IRL Press, Oxford, 167-173,
    (1985).
5. C. S. Foote, Mechanism of addition of singlet oxygen to olefins and
    other substances, _Pure Appl. Chem._ 27:635, (1971)
6. T. F. Slater, Free radical mechanisms in tissue injury, Pion, London,
    1-283 (1972).
7. W. A. Pryor, Free Radicals, McGraw Hill Book Co., New York, 1-354
    (1966).
8. M. Davies and T. F Slater, Electron spin resonance spin trapping
    studies on the photolytic generation of halocarbon radicals,
    Chem. Biol. Interactions, 58: 137-147 (1986).
9. H. F. Blum, Photodynamic action and diseases caused by light,
    Hafner, New York, 1-309 (1964).
10. I. H. Magnus, Dermatological Photobiology, Blackwells, Oxford, 1-292,
    (1976)
11. T. F. Slater and P. A. Riley, Photosensitisation and Lysosomal
    Damage, _Nature,Lond._, 209:151-154 (1966).
12. R. K. Clayton, "Light and Living Matter: A Guide to the Study of
    Photobiology. Volume 2: The Biological Part", 1-243 (1971).
13. D. Kessel and T J Dougherty (eds), Porphyrin Photosensitization,
    Plenum Press, New York, 1-294 (1983).
14. T. F. Slater, Necrogenic action of carbon tetrachloride in the rat:
    a speculative mechanism based on activation, _Nature,_ Lond., 209:
    36-40 (1966).
15. J. R. Mitchell, S. D. Nelson, S. S. Thorgeirsson, R. J. McMurtry and
    E. Dybing, Metabolic Activation: Biochemical basis for many drug-
    induced liver injuries, H. Popper and F. Schaffner eds, _Progr._

Liver Disease 5:259-279, Grune and Stratton, New York (1976).

16. M. A. Trush, E. G. Mimnaugh and T. E. Gram, Activation of pharmacological agents to radical intermediates. Implications for the role of free radicals in drug action and toxicity, Biochem. Pharmacol. 31:3335-3346 (1982).

17. R. O. Recknagel, Carbon tetrachloride toxicity, Pharmacol. Rev. 19: 145-208 (1967).

18. T. F. Slater, Activation of carbon tetrachloride: chemical principles and biological significance, in "Free radicals, lipid peroxidation and cancer", D. C. H. McBrien and T. F. Slater eds, Academic Press, London, 243-270 (1982).

19. T. F. Slater, K. H. Cheeseman and K. U. Ingold, Carbon tetrachloride toxicity as a model for studying free-radical mediated liver injury, Phil. Trans. R. Soc. Lond, B311:633-645 (1985).

20. M. U. Dianzani and G. Ugazio, Lipid peroxidation in "Biochemical Mechanisms of Liver Injury", T. F. Slater ed., Academic Press, London, 669-707 (1978).

21. G. R. Cameron and W. A. E. Karunaratne, Carbon tetrachloride cirrhosis in relation to liver regeneration, J. Path. Bact., 42:1-21 (1936).

22. E. A. Smuckler, O.A.Iseri and E. P. Benditt, An intracellular defect in protein synthesis induced by carbon tetrachloride, J. Exp. Med. 116:55-72, (1962)

23. J. L. Poyer, R. A. Floyd, P. B. McCay, E. G. Janzen and E. R. Davis, Spin-trapping of the trichloromethyl radical produced during enzymic NADPH oxidation in the presence of carbon tetrachloride or bromotrichloromethane, Biochim. Biophys. Acta, 539:402-409 (1978).

24. E. Albano, K. A. K. Lott, T. F. Slater, A. Stier, M. C.R. Symons and A. Tomasi, Spin-trapping studies on the free-radical products formed by metabolic activation of carbon tetrachloride in rat liver microsomal fractions, isolated hepatocytes and in vivo in the rat, Biochem, J. 204:593-603 (1982).

25. T. F. Slater, A note on the relative toxic activities of tetrachloromethane and trichlorofluoromethane on the rat, Biochem. Pharmacol. 14: 178-181 (1965).

26. R. N. Le Page, K. H. Cheeseman, N. Osman and T. F. Slater, Lipid peroxidation in purified plasma membrane fractions of rat liver in relation to the hepatotoxicity of carbon tetrachloride, Cell Biochemistry and Function 6:87-99 (1988).

27. R. Carini, K. H. Cheeseman and T. F. Slater, Morphological and biochemical studies on the effects of carbon tetrachloride and trichlorofluoromethane on isolated rat hepatocytes, (unpublished data, 1989).

28. M. L. Berger, H. Bhatt, B. Combes and R W Estabrook, CCl_4-induced toxicity in isolated hepatocytes: the importance of direct solvent injury, Hepatology, 6:36-45 (1986).

29. T. C. Butler, Reduction of carbon tetrachloride in vivo and reduction of carbon tetrachloride and chloroform in vitro by tissues and tissue constituents, J. Pharmacol. Exp. Therap., 134: 311-319 (1961).

30. M. Comporti, C. Saccocci and M. U. Dianzani, Effect of CCl_4 in vitro and in vivo on lipid peroxidation of rat liver homogenates and subcellular fractions, Experientra, 29:185-204, (1965).

31. A. K. Ghoshal and R. O. Recknagel, Positive evidence of acceleration of lipoperoxidation in rat liver by carbon tetrachloride: in vitro experiments, Life Sciences, 4:1521-1530 (1965).

32. T. F. Slater, In vitro effects of carbon tetrachloride on rat liver microsomes, Biochem. J., 101:16-17P (1966).

33. T. F. Slater, Stimulatory effects of CCl_4 in vitro on lipid peroxi-

dation in rat liver microsomes, Proc. 4th FEBS Symposium, Oslo; Abstract 214 (1967)

34. M. Comporti, E. Burdino and G. Ugazio, Alterazioni della composizione in acidi grassi dei lipidi del fegato e dei microsomi epatici nel ratto intossicato contetracloruro di carbonio, Boll. Soc. Ital. Biol. Sper., 45:700-703 (1969).

35. C. D. Klaassen and G. L. Plaa, Comparison of the biochemical alterations elicited in livers from rats treated with carbon tetrachloride, chloroform, 1,1,2-trichloroethane and 1,1,1-trichloroethane, Biochem. Pharmacol., 18:2019-2027 (1969).

36. P. J. Jose and T. F. Slater, Increased concentrations of malonaldehyde in the livers of rats treated with carbon tetrachloride, Biochem. J., 128: 141P (1972).

37. C. A. Riely, G. Cohen and M. Lieberman, Ethane Evolution: A new index of lipid peroxidation, Science, 183:208-210 (1974).

38. J. S. L. Fowler, Carbon tetrachloride metabolism in the rabbit, Brit. J. Pharmacol., 37:733-737 (1969).

39. J. E. Packer, T. F. Slater and R. L. Willson, Reactions of the carbon tetrachloride-related peroxy free radical ($CCl_3O_2\cdot$) with amino acids: pulse radiolysis evidence, Life Sciences, 23:2617-2620, (1978).

40. J. Monig, D. Barnemann and K D Asmus, One electron reduction of CCl_4 in oxygenated aqueous solutions: a $CCl_3O_2\cdot$-free radical mediated formation of Cl- and CO_2, Chem. Biol. Interactions, 45:15-27 (1983).

41. R. L. Willson, Electrophilic free radicals and nucleic acid damage: pulse radiolysis studies, Panminerva Medica 18:391-402 (1976).

42. J. E. Packer, R. L. Willson, D. Bahnemann and K. D. Asmus, Electron transfer reactions of halogenated aliphatic peroxyl radicals: measurement of absolute rate constants by pulse radiolysis, J. Chem. Soc. Perkin II, 296-299 (1980).

43. K. O. Hiller, P. L. Hodd and R. L. Willson, Anti-inflammatory drugs: protection of a bacterial virus as an in vitro biological measure of free radical activity, Chem. Biol. Int., 47:293-305 (1983).

44. L. G. Forni, J. E. Packer, T. F. Slater and R. L. Willson, Reactions of the trichloromethyl and halogen-derived peroxy radicals with unsaturated fatty acids: a pulse radiolysis study, Chem. Biol. Interactions, 45:171-177 (1983).

45. T. F. Slater, K. H. Cheeseman, M. J. Davies and J. S. Hurst, Free radical mechanisms in relation to cell injury and cell division, in "Drug Metabolism: from Molecules to Man", D. J. Benford, J. W. Bridges and G. G. Gibson eds., Taylor and Francis, London, 679-689, (1987).

46. T. F. Slater and K. H. Cheeseman, Free radical mechanisms of tissue injury and mechanisms of protection in "Reactive Oxygen Species in Chemistry, Biology and Medicine", Q. Quintanilha ed., 1-14 (1988).

47. T. F. Slater, Free radical scavengers in "International workshop on (+)- cyanidanol-3 in diseases of the liver", H. O. Conn ed, Royal Society of Medicine Int. Congress and Symposium, Series No.47, Royal Society of Medicine, London, 11-15 (1981).

48. T. F. Slater, Free Radical Mechanisms in Tissue Injury, Biochem. J. 222:1-15 (1984).

49. M. E. Hemler, H. W. Cook and W. E. M. Lands, Prostaglandin synthesis can be triggered by lipid peroxides, Archs. Biochem. Biophys.193: 340-345 (1979).

50. G. D. Snyder, J. Capdevila, N. Chacos, S. Manna and J. R. Falck, Action of luteinizing hormone - releasing hormone: involvement of novel arachidonic acid metabolites, Proc. Natn. Acad. Sci., U.S.A., 80: 3504-3507 (1983).

51. H. Esterbauer, H. Zollner and R. J. Schaur, Hydroxyalkenals: cytotoxic products of lipid peroxidation, ISI Atlas of Science: Biochemistry 311-317 (1988).

52. T. F. Slater, Biochemical Pathology in Microtime, Panminerva Medica, 18:381-390 (1976).

53. T. F. Slater, C. Benedetto, G. W. Burton, K. H. Cheeseman, K. U. Ingold and J. T. Nodes, Lipid peroxidation in animal tumours: a disturbance in the control of cell division? in "Icosanoids and Cancer, H. Thaler-Dao, A. Crastes de Paulet and R. Paoletti eds., Raven Press, New York, 21-29 (1984).

54. K. H. Cheeseman, M. Collins, K. Proudfoot, T. F. Slater and G. W. Burton, A. C. Webb and K. U. Ingold, Studies on lipid peroxidation in normal and tumour tissues. The Novikoff rat liver tumour, Biochem. J., 235:507-514 (1986).

55. K. H. Cheeseman, S. Emery, S. P. Maddix, T. F. Slater, G. W. Burton and K. U. Ingold, Studies on lipid peroxidation in normal and tumour tissues. The Yoshida rat liver tumour, Biochem. J., 250: 247-252 (1988)

56. S. Emery, K. H. Cheeseman and T. F. Slater, Effects of vitamin E deficiency on the growth in vivo of the Yoshida rat liver tumour in the rat (unpublished data, 1989).

57. M. R. Allison, Regulation of hepatic growth, Physiol. Revs., 66: 499-541 (1986).

58. H. A. Hopkins, H. A. Campbell, B. Barbiroli and R. Van Potter, Thymidine kinase and deoxyribonucleic acid metabolism in growing and regenerating livers from rats on controlled feeding schedules. Biochem. J. 136:955-966 (1973).

59. K. H. Cheeseman, M. Collins, S. Maddix, A. Milia, K. Proudfoot, T. F. Slater, G. W. Burton, A. Webb and K U. Ingold, Lipid peroxidation in regenerating rat liver, FEBS Letters, 209:191-196 (1986).

60. T. F. Slater, K. H. Cheeseman, S. Emery, S. Maddix, M. Collins and A. Milia, Lipid peroxidation and thymidine kinase activity in regenerating rat liver: correlation of changes in time (unpublished data, 1989).

61. T. F. Slater, K. H. Cheeseman, C. Benedetto, M. Collins, S. Emery, S. P. Maddix, J. T. Nodes, K. Proudfoot, G. W. Burton, A. Webb and K. U. Ingold, Studies on the hyperplasia (regeneration) of the rat liver following partial hepatectomy: changes in lipid peroxidation and general biochemical aspects (submitted for publication, 1989).

LIPID PEROXIDATION AND CELLULAR FUNCTIONS: IN VITRO MODELS AND RELATION TO IN VIVO OBSERVATIONS

[1] J.C. Mazière, [1] S. Salmon, [2] R. Santus, [1] C. Candide
[2] J.P. Reyftmann, [3] P. Morlière, [1] C.Mazière, and
[3] L. Dubertret

[1] Laboratoire de Biochimie, Faculté de Médecine
Saint-Antoine, 27 rue Chaligny, Paris; [2] Laboratoire de Physico-Chimie de l'Adaptation Biologique,
Muséum d'Histoire Naturelle de Paris, INSERM U 312,
45 rue Cuvier, 75005 Paris; [3] Laboratoire de Recherche Dermatologique, INSERM U 312, Hôpital Henri
Mondor, 94010 Créteil (France)

SUMMARY

The consequences of lipid peroxidation on various cell metabolisms are reviewed with special emphasis on low density lipoprotein catabolism and its relation to atherosclerosis. We also present results concerning an original model developed in our laboratories for the study of the effects of singlet oxygen on lipid peroxidation. In this experimental model, lipoproteins are used as a lipidic environment for porphyrins generating singlet oxygen during their photoactivation. We demonstrate that singlet oxygen attack results in the appearance of fatty acid and cholesterol peroxidation products and in alterations of apolipoproteins, but that apolipoprotein alterations markedly differ between low density and high density lipoproteins. Besides its theoretical interest for the study of lipid oxidation in lipid-protein complexes, this model brings new data concerning the consequences of the photoactivation of anticancer porphyrins which are carried by plasma lipoproteins, mainly LDL and HDL.

I. THE ACTIVE SPECIES OF OXYGEN AND THEIR PRODUCTION BY VARIOUS TYPES OF CELLS

Active oxygen species responsible for lipid peroxidation are a/ these which are involved in nucleophilic or free radical attack, and b/ singlet oxygen, which implicates allylic transposition of lipid double bonds.

In the first group, superoxide anion or its protoned form and hydrogen peroxide have to be mainly considered as precursors of hydroxyl radical, which is much more active towards lipids, although direct nucleophilic attack by superoxide anion appears to be possible. Superoxide anion can be generated by mitochondrial and microsomal electron transport chains, by amino acid oxidases, and by the NADPH oxidase of phagocytes during their activation. It is also produced by oxidation of Fe^{2+} in aqueous solution, a mechanism which is probably implicated in the so-called "autooxidation" of lipids and lipoproteins. Hydrogen peroxide is the product of the ubiquitous enzyme superoxide dismutase, and it can also be produced by the xanthine oxidase system and other flavinic dehydrogenases. The hydroxyl radical (OH·) can be generated from superoxide anion and hydrogen peroxide by the Haber-Weiss reaction, in the presence of transition metal ions such as Fe^{2+} or Cu^{2+}. These mechanisms are reviewed in [1, 2].

Singlet oxygen is enzymatically produced in activated neutrophils by myeloperoxidase and lactoperoxidase, involving halogens such as Cl^- or Br^- (3). It is also generated during photoactivation of porphyrins, by energy transfer from the activated triplet state of the porphyrin to ground state triplet oxygen (4).

The hydroxyl radical has a very short lifetime, due to its high reactivity, and thus cannot be secreted by cells. H_2O_2, which is a stable molecule, readily diffuses across the membrane and can be secreted by cells in relatively high amounts. Superoxide anion has a short lifetime, but sufficient to be secreted, although it slowly crosses the membrane through anionic channels.

Secretion of superoxide anion has been described in various types of cultured cells: endothelial cells (5), smooth muscular cells (6), and phagocytes (7). Important differences may exist, depending upon cell types and animal species (8). For example, bovine endothelial cells have a very low level of superoxide production as compared to rabbit endothelial cells, and rabbit smooth muscle cells are more active than rabbit endothelial cells (8). The appearance of superoxide anion in the extracellular space is closely related to the composition of the culture medium, and especially to the presence of transition metal cations (Fe^{2+}, Cu^{2+}). In particular models such as arterial smooth muscle cells, it also depends upon the concentration of cystine in the culture medium (10). Indeed, Heinecke et al. demonstrated that in this experimental model an extracellular production of superoxide anion can occur, depending upon the uptake of cystine and subsequent secretion of thiols such as cysteine or glutathion in the culture medium: these thiols, in the presence of traces of transition metals, can reduce molecular oxygen to superoxide anion in the extracellular compartment. This secretion of thiols in the culture medium appears to be dependent upon divalent cations such as Ca^{2+} and Mg^{2+}. Thus, in addition to cellular production and secretion, extracellular formation of superoxide anion can occur in some experimental models.

II. EFFECTS OF LIPID PEROXIDATION PRODUCTS ON CELL METABOLISMS

Peroxidation of polyunsaturated fatty acids (PUFA) gives complex mixtures containing alkanes, alcenes, aldehydes, hydroperoxides and hydroxyacids. The effects of these peroxidation products of PUFA, especially aldehydes and hydroperoxides, have been extensively studied during the last years. Strong reduction of the energetic metabolism by aldehydes (malondialdehyde, hydroxypentenal, hydroxynonenal), and by various hydroperoxides has been described: inhibition of glycolytic enzymes, especially glyceraldehyde dehydrogenase (10), and reduction of the mitochondrial respiration (10). Other membrane-bound enzymes such as glucose 6-phosphatase are also affected (11). Extracellular enzymes such as lecithine:cholesterol acyl-transferase (12) and lipoprotein lipase (13) have been described to be inhibited by linoleic and arachidonic hydroperoxides, respectively. Protein and RNA synthesis are reduced by PUFA peroxidation products in various types of cells (14, 15)[for a review see (16)]. Aldehydes also react with DNA (17,18) and their mutagenic effects are well documented (19,20), suggesting a potential role of these lipid peroxidation products in carcinogenesis [for a review, see (21)].

Oxysterols are also toxic towards cells, resulting in inhibition of membrane systems such as Na^+, K^+ - dependent ATPase, 5' nucleotidase, deoxyglucose transport (22) and alterations in DNA synthesis (23). The effects on calcium flux depends upon the nature of the oxysterols: for example, 7-beta hydroxycholesterol and 25-hydroxy-cholesterol are potent inhibitors of calcium transport, whereas 22-hydroxy cholesterol or cholestane triol increases it (24). The cytotoxicity of oxysterols has been described in various types of cells (25-28), and especially in endothelial cells, which could be relevant to the atherogenic process, as suggested by Imai et al. (29).

Thus, lipid peroxidation products potentially affect all aspects of the cell machinery, and their metabolic effects could be involved in numerous pathological processes such as drug-induced toxicity, carcinogenesis and atherosclerosis. However, it must be stressed that most of these results have been obtained on _in vitro_ models, using relatively high concentrations (often in the range of 0.1 to 1mM), especially regarding the effects of aldehydes. Even if it can be conceived that local concentration of lipophilic aldehydes in cell membranes could be higher than expected, this has to be taken into account before drawing any conclusion.

III. LOW DENSITY LIPOPROTEIN MODIFICATION AND ITS RELEVANCE IN ATHEROGENESIS

1. In vitro modification of LDL by cultured cells

The first observation by Henriksen et al. (30) that LDL incubation with endothelial cells resulted in alterations of the particle give rise to numerous works in the last years. The main characteristics of endothelial cell-modified LDL are: an increase in the negative net charge resulting in increased electrophoretic mobility, a decrease in the recognition by the apo B/E receptor of fibroblasts, and, by contrast, an increase in the uptake and degradation by the "scavenger" receptor of macrophages. As a result, due to the fact that cholesteryl ester formation is not down-regulated in macrophages, these cells turn into foam cells, a process which is well admitted to be involved in the appearance of atherosclerotic plaques (31, 32).

Such modification of LDL has also been reported in other experimental models: macrophages (33), stimulated neutrophils (34), stimulated monocytes (35), and arterial smooth muscle cells (6). It is inhibited by antioxidants such as butylhydroxytoluene, suggesting an oxidative mechanism. It is also inhibited by superoxide dismutase and well correlated to the ability of cells to produce superoxide anion (8), which suggests the implication of superoxide anion in LDL modification. Moreover, factors which increase superoxide production such as transition metals (Fe^{2+}, Cu^{2+}) and cystine, are necessary to promote LDL modification (9).

Modification of LDL by endothelial cells, arterial smooth muscle cells or macrophages is also accompanied by the appearance of PUFA peroxidation products (thiobarbituric acid reactive substances: "TBARS", [36,37]). Such modified LDL has been described to be cytotoxic towards cultured fibroblasts or endothelial cells (38, 39). This cytotoxic effect of modified LDL could be related to the presence of toxic PUFA peroxidation products such as aldehydes or hydroperoxides (see chapter II), or to the presence (not demonstrated up to now) of cholesterol oxidation products. Thus, another factor involved in the atherogenic process could be the cytotoxicity of modified LDL towards endothelial cells. However, it must be stressed that the actual importance of the cytotoxic effect of oxidatively modified LDL towards endothelial cells remains disputed (40): high density lipoprotein (HDL), which are present _in vivo_ have been demonstrated to protect endothelial cells against cytotoxicity of modified LDL (41), and sera from hypercholesterolemic patients had no cytotoxic effects towards endothelial cells (for a review, see [40]).

LDL modification by various types of cells has been also demonstrated to result in phospholipid, especially phosphatidylcholine (PC) hydrolysis (42, 43). Up to 50% of the PC of the LDL particle can be degraded, resulting in the appearance of large amounts of lyso PC. LDL modification by macrophages does not occur when phospholipase A2 (PLA2) inhibitors are added to the incubation medium (43), suggesting the involvement of such phospholipasic activity in the modification process. This PLA2 or PLA2-like activity appears to be intrinsic to the LDL, as PC hydrolysis also takes place in LDL oxidatively modified in the absence of cells and is also prevented by PLA2 inhibitors [43] ("autooxidation" of the LDL in the presence of Cu^{2+}). Whether or not this PLA2-like activity is primarily or secondarily involved in LDL modification is not known. It must be also noted that in vitro treatment of LDL with lipooxigenase and PLA2 results in LDL modification which mimics that obtained in the presence of cells: alteration of the electrophoretic mobility, and subsequent recognition and degradation by macrophages (44).

The appearance of lysoderivatives as a consequence of PC hydrolysis could be of importance in view of the fact that recent works demonstrated that such lysoderivatives have a chemo-attractive action towards monocytes and, conversely, immobilize macrophages (45, 46), which could be involved in the development of atherosclerotic plaques. It must also be remembered that some PUFA peroxidation products such as aldehydes (4 hydroxy nonenal, 4 hydroxy octenal) have been described to have a chemotactic activity towards neutrophils (47).

2. Do other types of LDL modification exist ?

Besides the well documented oxidation which results from incubation of LDL with various types of cultured cells, other types of modification have been suggested. One of them is the possibility of LDL modification by malondialdehyde (MDA) which is produced during platelet activation. This has been suggested by Fogelman et al. (48) based mainly on the facts that a/ platelet activation results in the production of large amounts of MDA, and b/ MDA can react in vitro with LDL, leading to increased electrophoretic mobility of the particle and to its recognition by the scavenger receptor of the monocyte/macrophage (48). If at present time there is no direct evidence for LDL modification by MDA originating from platelets, Haberland et al. recently demonstrated the existence of MDA-modified apolipoprotein B100 in atherosclerotic lesions of Watanabe hyperlipidemic (WHHL) rabbits (49). Another interesting hypothesis has been proposed by YAGI et al., which consists in the possibility of LDL modification by hydroperoxides. Preincubation of LDL with linoleic acid hydroperoxides results in subsequent uptake by macrophages and their conversion into foam cells (50). If it can be conceivable that hydroperoxides could appear in the extracellular medium and react with LDL, the biochemical mechanism of this interaction of hydroperoxides with the LDL particle has not been precised.

3. Can LDL modification occur in vivo ?

Indirect evidence has been reported concerning the possibility of _in vivo_ LDL modification, such as the existence of lipofuscin-like pigments in atherosclerotic plaques. Hoff et al. has isolated a LDL presenting increased electrophoretic mobility from atherosclerotic lesions (51). Clevidence et al. also showed that LDL-like particles isolated from atherosclerotic plaques can be recognized and metabolized by macrophages (52). More recently, Avogaro et al. found in sera from normolipidemic patients a LDL sub-fraction presenting some degree of modification as compared to native LDL: increased electrophoretic mobility, decreased interaction with the apo B/E receptor, and uptake by the scavenger pathway of macrophages leading to cholesteryl ester accumulation (53). Although no data concerning hypercholesterolemic patients are available at the present time , these works strongly suggest the possibility of _in vivo_ LDL modification.

This is also supported by recent works concerning the mechanism of the anti-atherogenic effect of the hypocholesterolemic drug probucol. In hypercholesterolemic WHHL rabbits, probucol treatment results only in moderate decrease (10 to 15%) of cholesterolemia, but in strong reduction (50 to 70%) of the extent of atherosclerotic lesions in the whole aorta. Concerning the anti-atherogenic potency, probucol even appears to be more effective than hydroxymethyl glutaryl Coenzyme A inhibitors such as lovastatin, despite the fact that lovastatin has a stronger effect on cholesterolemia (54). This is a clear demonstration that in some cases, no direct relationship exists between cholesterolemia and the appearance of atherosclerotic lesions. In order to explain this quite surprising observation, recent works from the team of Steinberg demonstrated that when added to the incubation medium, probucol is able to inhibit _in vitro_ LDL modification by endothelial cells or Cu^{2+}-induced LDL oxidation (55). In the same work, these authors showed that LDL re-isolated from sera of patients treated with probucol are much more resistant to subsequent oxidation than LDL from untreated subjects (55). The apparent relationship between the protective effect of probucol against LDL oxidation and its anti-atherogenic effect indirectly but strongly suggests the importance of LDL oxidative modification in the appearance of atherosclerotic lesions.

IV. NEW IN VITRO MODELS FOR LIPOPROTEIN OXIDATION PORPHYRIN-INDUCED PHOTOPEROXIDATION OF LIPOPROTEINS

As described in section I, porphyrin photoexcitation can result in singlet oxygen production. Singlet oxygen is very active towards lipids, leading to the appearance of PUFA peroxidation products and to specific cholesterol peroxidation products such as 5 alpha-hydroperoxides. In the last years some of us reported that intraveinously injected anticancer porphyrins are carried in plasma by lipoproteins, mainly LDL and HDL (56). We also demonstrated that anticancer porphyrins are delivered to cells with great efficiency by the LDL (apo B/E) receptor pathway (57, 58). In the present paper we will not consider the intracellular effects of porphyrins, but only the experimental model constituted by LDL loaded with porphyrins and subsequently illuminated. This relatively simple model allows the study of the effect of singlet oxygen on lipid-protein complexes. Moreover, in view of the important role of LDL and HDL as carriers of anticancer porphyrins in plasma, it was of interest to investigate the consequences of the presence of these drugs inside the lipoprotein particles, especially after irradiation. We thus studied the influence of irradiation on the appearance of lipid peroxidation products and on apolipoprotein modification in LDL or HDL preloaded with anticancer porphyrins.

A. Porphyrin-induced photoperoxidation of LDL

Irradiation at 405nm of LDL loaded (57,58) with the anticancer porphyrin derivative Photofrin II (P2) or with protoporphyrin (PP) leads to the appearance of lipid peroxidation products, depending upon the time of light exposure. This is presented in Figure 1, which shows the increase in the amount of thiobarbituric reactive substances (TBARS, [Fig.1a]) and lipofuscin-like pigments (Fig.1b) as a function of the irradiation time. Figure 2 presents the HPLC analysis of lipid extracts from irradiated LDL unloaded (a) or loaded (b) with porphyrins. 4 hydroxy 2,3 transnonenal, an aldehyde resulting from arachidonic acid peroxidation, can be detected (peak 2), as well as sterol hydroperoxides, especially the 5 alpha derivative (peak 3), which is typically produced by the singlet oxygen attack on cholesterol, and epimeric 6 alpha, beta hydroperoxides (peaks 4 and 5, [59]).

334

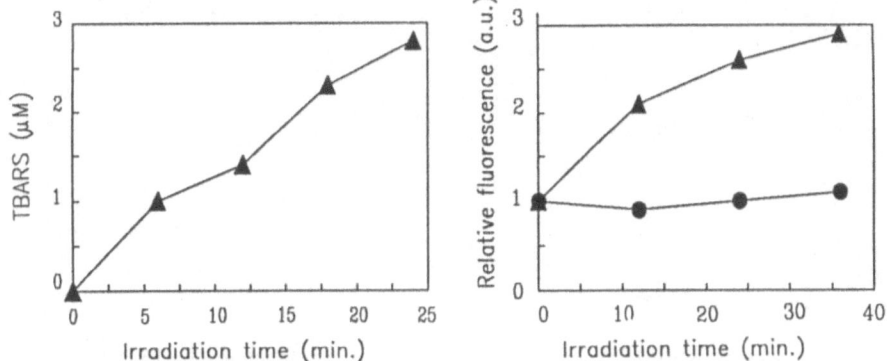

Figure 1. appearance of thiobarbituric reactive substances (TBARS) [1A] and lipofuscin—like pigments [1B] as a function of the irradiation time in LDL loaded with Photofrin II. 1B: ▲ : whole LDL; ● : delipidated LDL.

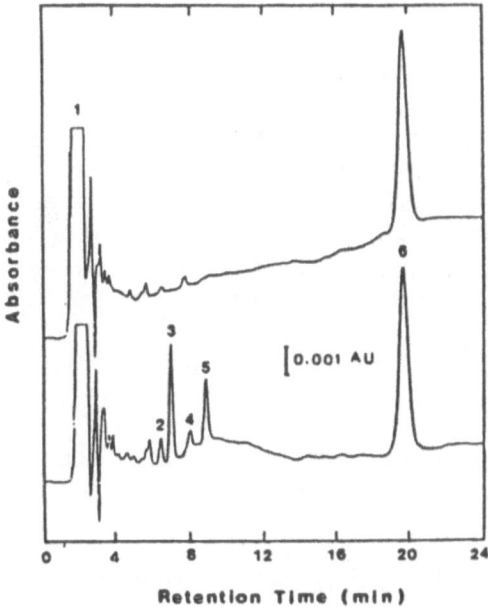

Figure 2. HPLC analysis of a lipid extract of protoporphyrin—loaded LDL (absorbance at 210nm). Top: before irradiation; bottom: after 1h irradiation at 405nm. 1: solvent; 2: 4 hydroxy 2,3 transnonenal; 3: 5 alpha—hydroperoxicholesterol; 4,5: putative epimeric 6 alpha, 6 beta hydroperoxicholesterol; 6: cholesterol.

Figure 3 shows that the lysine and tryptophan contents of the apolipoprotein B100 (apo B100) are decreased as a function of irradiation time in P2 and PP-loaded LDL. However, the decrease in lysine content of apo B100 is dependent upon lipid peroxidation, as no effect was observed in delipidated LDL. By contrast, the tryptophan content of apo B100 was decreased even in delipidated LDL (data not shown). These results can be easily explained since Trp residues are photodynamic substrates whereas Lys are insensitive to the photodynamic reaction.

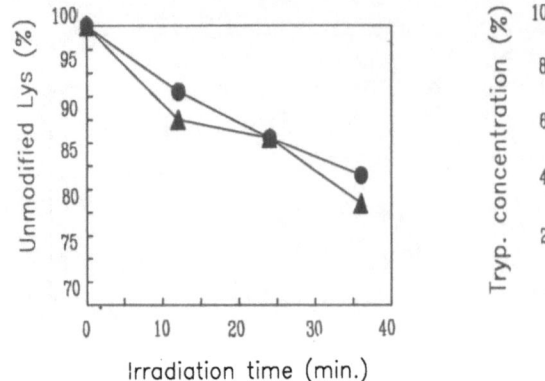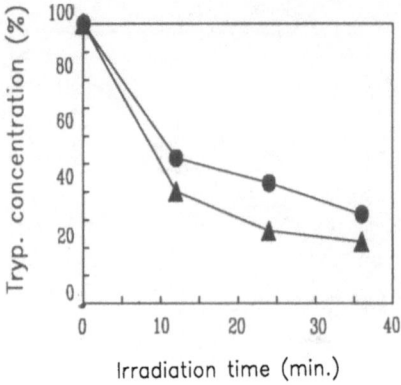

Figure 3. lysine (3A: Lys) and tryptophan (3B: Tryp.) content of porphyrin — loaded LDL as a function of irradiation time at 405nm. ▲ : protoporphyrin—loaded LDL; ● : Photofrin II— loaded LDL.

As cited in section III, Fe^{2+} and Cu^{2+} are known to enhance oxidative modification of LDL. We thus investigated the role of these cations in singlet oxygen-mediated modification of porphyrin-loaded LDL. Table I shows that Fe^{2+} or Cu^{2+} did not significantly modify the lysine content of LDL during photosensitization.

Table I. Lysine content (percent of control) after photosensitization of LDL in presence of metal ions.

Addition	–	$5\mu M\ Cu^{2+}$	$10\mu M\ Cu^{2+}$	$10\mu M\ Fe^{2+}$
Protoporphyrin	77	–	85	77
Photofrin II	81	81	84	71

Duration of irradiation at 405nm: 36 min. Protoporphyrin concentration: $10\mu M$; Photofrin II concentration: $18\mu g/ml$.

Thus singlet oxygen-induced modification of LDL markedly differs from cell-induced LDL modification probably mediated by superoxide anion. In particular, singlet oxygen attack does not involve metal ions, and is achieved on a very short time scale: the same extent of lysine derivatization which results from a 24h autoxidation of LDL in the presence of Cu^{2+} (60) is obtained in our model after only 30 min. photoperoxidation of porphyrin-loaded LDL. Moreover, under our experimental conditions, substantial amounts of sterol hydroperoxides are produced during LDL peroxidation. It must be emphasized that this is the first direct demonstration of the appearance of oxysterols during LDL oxidation.

B. Porphyrin-induced photoperoxidation of HDL:

Fig.4A shows that substantial TBARS are produced in photosensitized HDL, as observed in photosensitized LDL (see Fig.1a). It can be seen in Figure 4B that lipofuscin-like pigments are also formed in photosensitized HDL. These pigments found in all degenerative processes accompanying aging, result from Schiff base formation between free amino groups of lipoproteins and aldehydic groups of decomposition products of lipid peroxides (60). But in sharp contrast with LDL, lysine residues are *not modified* during photosensitization of HDL, although tryptophan residues, which respond to photosensitization, are still rapidly destroyed (Fig.5).

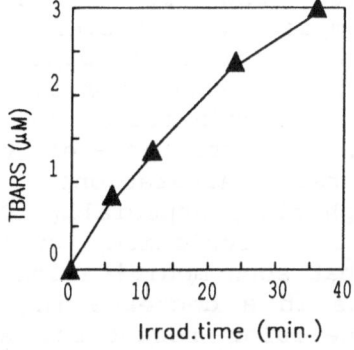 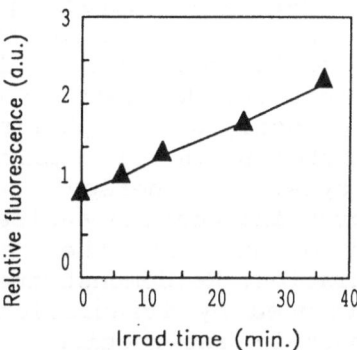

Figure 4. appearance of TBARS (4A) and lipofuscin—like substances (4B) as a function of irradiation time in porphyrin — loaded HDL 3; 4A: HDL loaded with Photofrin II; 4B: HDL loaded with protoporphyrin.

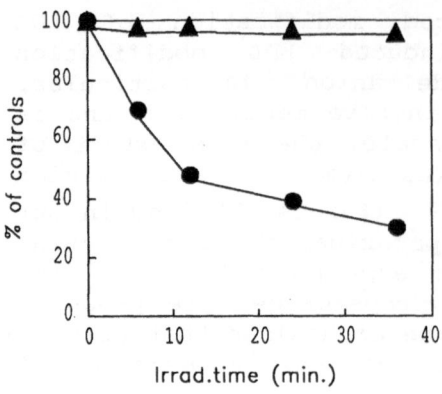

Figure 5: lysine (▲) and tryptophan (●) content of Photofrin II — loaded HDL as a
function of irradiation time.

To our knowledge, this is the first observation
concerning HDL photoperoxidation. The fact that, in
contrast with LDL, lysine residues of HDL apolipoproteins
are not affected despite the appearance of lipofuscin-like
pigments and TBARS, is quite surprising. One can suppose
that lipid - protein interactions in HDL differs from lipid
- protein interaction in LDL, and that, as a consequence,
lysine residues of HDL apolipoproteins could be
unaccessible to lipid degradation products.

In conclusion, porphyrin-induced photosensitization of
lipoproteins appears to be a rapid and convenient model for
the study of lipid peroxidation and lipid-protein
interactions in lipid-protein complexes. Moreover, as
lipoproteins are closely involved in anticancer porphyrin
transport and delivery to cells, a better knowledge of the
consequences of porphyrin-induced photosensitization of
lipoproteins is of importance in the strategy of the
photodynamic therapy of tumors. Alterations of
lipoproteins bearing anticancer porphyrins, especially LDL,
could occur in the peritumoral circulation during
irradiation. We demonstrated (59) that such modification of
LDL induced by irradiation results in a decrease in the
lipoprotein interaction with its cellular receptors, and
this fact has to be taken into account in the pharmacology
of anticancer porphyrin derivatives.

ACKNOWLEDGMENTS: J.C. and C.Mazière thank La Ligue
Nationale Contre Le Cancer, Comité de Paris, for financial
support.

REFERENCES

1. Kappus, H. Lipid peroxidation: Mechanisms, Analysis, Enzymology, and Biological Relevance. In "Oxidative Stress", H.Sies ed., Academic Press Inc., London; 273 (1985).
2. Halliwell, B., and Gutteridge, J.M.C. Oxygen toxicity, oxygen radicals, transition metals and disease. Biochem.J., 219:1 (1984).
3. Kanofsky, J.R. Singlet oxygen production by lactoperoxidase: halide dependence and quantitation of yield. J.Photochemistry 25: 105 (1984).
4. Foote, C.S. Photooxidation of Biological Model Compounds. In "Oxygen and Oxy-radicals in Chemistry and Biology", Rodgers, M.A. and Powers, E.L. eds., Academic Press, N.Y.; 425 (1981).
5. Rosen, G.M., and Freeman, B.A. Detection of superoxide generated by endothelial cells. Proc.Natl.Acad.Sci.USA, 81: 7269 (1984).
6. Heinecke, J.W., Baker, L., Rosen, H., and Chait, A. Superoxide-mediated modification of low density lipoprotein by arterial smooth muscle cells. J.Clin.Invest., 77: 757 (1986).
7. Scully, S.P., Segel, G.B., and Lichtman, M.A. Relationship of superoxide production to cytoplasmic free calcium in human monocytes. J.Clin.Invest., 77, 1349-1356 (1986).
8. Steinbrecher, U.P. Role of superoxide in endothelial-cell modification of low-density lipoproteins. Biochim. Biophys. Acta, 959, 20-30 (1988).
9. Heinecke, J.W., Rosen, H., Suzuki, L.A., and Chait, A. The role of sulfur-containing amino acids in superoxide production and modification of Low Density Lipoprotein by arterial smooth muscle cells. J.Biol.Chem. 262: 10098 (1987).
10. Schauenstein, E., Esterbauer, H., and Zollner, H. In "Aldehydes in Biological Systems: Their Natural Occurrence and Biological Activities", Pion Ldt. London (1977).
11. Benedetti, A., Comporti, M., and Esterbauer, H. Identification of 4-hydroxynonenal as a cytotoxic product originating from the peroxidation of liver microsomal lipids. Biochim.Biophys.Acta 620, 281-296 (1980).
12. Takatori, T., and Prevett, O.S. Studies on serum lecithin-cholesterol acyl transferase activity in rat: effect of vitamin E deficiency, oxidized dietary fat or intravenous administration of ozonides or hydroperoxides. Lipids, 9, 1018- 1023 (1974).
13. Wada, K., Miki, H., Etoh, M., Okuda, F., Kumada, T., and Kusukawa, R. The inhibitory effect of lipid peroxide on the activity of the membrane-bound and the solubilized lipoprotein lipase. Japan.Circ.J., 47, 837-842 (1983).
14. Dianzani, M.U., Bertone, G.F., Bonelli, G., Canuto, R.A., Feo, F., Gabriel, L., Gravela, E., and Pernigotti, L. Interaction of aldehydes and other derivatives of lipid peroxidation with cell structures. Med.Biol.Environ., 4, 345-365 (1976).
15. Dianzani, M.U. Biological activity of methyl-glyoxal and related activities. In "Submolecular Biology and Cancer" (Ciba Symposium 67), Amsterdam, Excerpta Medica, 245-270 (1979).
16. Esterbauer, H. In "Free Radicals and Liver Injury", Poli, G., Cheeseman, K.H., Dianzani, M.U., and Slater, T.F. eds., IRL Press Limited, Oxford, England, 29-47 (1985).
17. Reiss, U., Tappel, A.L., and Chio, K.S. DNA-malondialdehyde reaction: formation of fluorescent products. Biochem.Biophys.Res.Commun., 921-926 (1972).
18. Nair, V., Cooper, C.S., Vietti, D.E., and Turner, G.A. The chemistry of lipid peroxidation metabolites: crosslinking reactions of malondialdehyde. Lipids, 21, 6-10 (1986).

19. Mukai, F.H., and Goldstein, B.D. Mutagenicity of malondialdehyde, a decomposition product of peroxidized polyunsaturated fatty acids. Science, 191, 868-869.

20. Cajelli, E., Ferraris, A., and Brambilla, G. Mutagenicity of 4-hydroxynonenal in V79 Chinese hamster cells. Mut.Res., 190, 169-171 (1987).

21. Vaca, C.E., Wilhelm, J., and Harms-Ringdahl, M. Interaction of lipid peroxidation products with DNA. A review. Mut.Res., 195, 137-149 (1988).

22. Peng, S.K., and Morin, R.J. Effects on membrane function by cholesterol oxidation derivatives in cultured aortic smooth muscle cells. Artery 14: 85 (1987).

23. Astruc, M., Rousillon, S., Defay, R., Descomp,B., and Crastes de Paulet, A. DNA and cholesterol biosynthesis in synchronized embryonic rat fibroblasts. Biochim.Biophys.Acta 763: 11 (1983).

24. Neyses, L., Lochern, R., Stimpel, M., Streuli, R., and Vetter, W. Stereospecific modulation of calcium channel in human erythrocytes by cholesterol and its oxidized derivatives. Biochem.J. 227: 105 (1985).

25. Cheng, K.P., Nagano, H., Luu, B., Ourisson, G., and Beck, J.P. Chemistry and Biochemistry of Chinese Drugs, Part I. Sterol derivatives cytotoxic to hepatoma cells, isolated from the drug Bombyx cum botryte. J.Chem.Res (S) 217: 2501 (1977).

26. Maltese, W.A., Reitz, B.A., and Volpe, J.L. Selective decrease in the viability and the sterol content on proliferating versus quiescent glioma cells exposed to 25-hydroxycholesterol. Cancer Res. 41: 3448 (1981).

27. Peng, S.K., Tham, P., Taylor, C.B., and Mikkelson, B. Cytotoxicity of oxidation derivatives of cholesterol on cultured aortic smooth muscle cells. Am.J.Nutr. 32: 1033 (1979).

28. Hietter, H., Trifilieff, E., Richert, L., Beck, J.P., Luu, B., and Ourisson, G. Antagonistic action of cholesterol towards the toxicity of hydroxysterols on cultured hepatoma cells. Biochem.Biophys.Res. Commun. 120: 657 (1984).

29. Imai, H., Werthessen, N.T., Subramanyam, V.S., LeQuesne, P.W., Soloway, A.H., and Kanisawa, M. Angiotoxicity of oxygenated sterols and possible precursors. Science, 207, 651-652 (1980).

30. Henriksen, T., Mahoney, E.M., and Steinberg, D. Enhanced macrophage degradation of low density lipoprotein previously incubated with cultured endothelial cells: recognition by receptors for acetylated low density lipoproteins. Proc.Natl.Acad.Sci.U.S.A., 78, 6499-6503 (1981).

31. Goldstein, J.L., Ho, Y.K., Basu, S.K., and Brown, M.S. Binding sites on macrophages that mediates uptake and degradation of acetylated LDL producing massive cholesterol deposition. Proc.Natl.Acad.Sci.U.S.A., 76, 333-337 (1979).

32. Gerrity, R.G. The role of the monocyte in atherogenesis. I. Transition of blood borne monocytes into foam cells in fatty lesions. Am.J.Pathol., 103, 181-190 (1981).

33. Parthasarathy, S., Printz, D.J., Boyd, D., Joy, L., and Steinberg, D. Macrophage oxidation of low density lipoprotein generates a modified form recognized by the scavenger receptor. Arteriosclerosis, 6, 505-510 (1986).

34. Cathcart, M.K., Morel, D.W., and Chisolm, G. Monocytes and neutrophils oxidize low density lipoprotein making it cytotoxic. J.Leuk.Biol., 38, 341-350 (1985).

35. Hiramatsu, K., Rosen, H., Heinecke, J., Wolfbauer, G., and Chait, A. Superoxide initiates oxidation of low density lipoprotein, by human monocytes. Arteriosclerosis, 7, 55-60 (1987).

36. Morel, D.W., Hessler, J.R., and Chisolm, G.M. Low density lipoprotein cytotoxicity induced by free radical peroxidation of lipid. J.Lip.Res., 24, 1070-1076 (1983).

37. Morel, D.W., DiCorleto, P.E., and Chisolm, G. Endothelial and smooth muscle cells alter low density lipoprotein in vitro by free radical oxidation. Arteriosclerosis, 4, 357-364 (1984).

38. Henriksen, T., Evensen, S.A., and Carlander, B. Injury to human endothelial cells in culture induced by low density lipoproteins. Scand.J.Clin.Lab.Invest., 39, 361-368 (1979).

39. Evensen, S.A., Nilsen, E., and Galdal, K.S. Injury to cultured human fibroblasts induced by low density lipoproteins: potentiating and protective factors. Scand.J.Clin.Lab.Invest., 42, 285-290 (1982).

40. V.W.M. Van Hinsbergh. LDL cytotoxicity. The state of the art. Atherosclerosis, 53, 113-118 (1984).

41. Hessler, J.R., Robertson, A.L.Jr., and Chisolm, G.M. LDL-induced cytotoxicity and its inhibition by HDL in human vascular smooth muscle and endothelial cells in culture. Atherosclerosis, 32, 213-229 (1979).

42. Steinbrecher, U.P., Parthasarathy, S., Leake, D.S., Witzum, J.L., and Steinberg, D. Modification of low density lipoprotein by endothelial cells involves lipid peroxidation and degradation of low density lipoprotein phospholipids. Proc.Natl.Acad.Sci.U.S.A., 81, 3883-3887 (1984).

43. Parthasarathy, S., Steinbrecher, U.P., Barnett, J., Witzum, J.L., and Steinberg, D. Essential role of phospholipase A_2 activity in endothelial cell-induced modification of low density lipoprotein. Proc.Natl.Acad. Sci.U.S.A., 82, 3000-3004 (1985).

44. Sparrow, C.P., Parthasarathy, S., and Steinberg, D. Enzymatic modification of low density lipoprotein by purified lipoxygenase plus phospholipase A_2 mimics cell-mediated oxidative modification. J.Lip.Res., 29, 745-753.

45. Quinn, M.T., Parthasarathy, S., Fong, L.G., and Steinberg, D. Oxidatively modified low density lipoproteins: a potential role in recruitment and retention of monocyte/macrophages during atherogenesis. Proc.Natl.Acad. Sci.U.S.A., 84, 2995-2998.

46. Quinn, M.T., Parthasarathy, S., and Steinberg, D. Lysophosphatidylcholine: a chemotactic factor for human monocytes and its potential role in atherogenesis. Proc.Natl.Acad.Sci.U.S.A., 85, 2805-2809 (1988).

47. Curzio, M., Esterbauer, H., Di Mauro, C., Cecchini, G., and Dianzani, M. Chemotactic activity of the lipid peroxidation product 4-hydroxynonenal and homologous hydroxyalkenals. Biol.Chem.Hoppe-Seyler, 367, 321-329 (1986).

48. Fogelman, A.L., Shechter, I., Seager, J., Hokom, M., Child, J.S., and Edwards, P.A. Malondialdehyde alteration of low density lipoproteins leads to cholesteryl ester accumulation in human monocyte-macrophages. Proc.Natl. Acad. Sci.U.S.A., 77, 2214-2218 (1980).

49. Haberland, M., Fong, D., and Cheng, L. Malondialdehyde-altered protein occurs in atheroma of Watanabe heritable hyperlipidemic rabbits. Science, 241, 215-218 (1988).

50. Yagi, K., Inagaki, T., Sasaguri, Y., Nakano, R., and Nakashima, T. Formation of lipid-laden cells from cultured aortic smooth muscle cells and macrophages by linoleic acid hydroperoxide and low density lipoprotein. J.Clin.Biochem. Nutr., 3, 87-94 (1987).

51. Hoff, H.F., Bradley, W.A., Heideman, C.L., Gabatz, J.W., Karagas, M.D., and Gotto, A.M. Characterization of low density lipoprotein-like particle in the human aorta from grossly normal and atherosclerotic regions. Biochim. Biophys.Acta., 573, 361-374 (1979).

52. Clevidence, B.A., Morton, R.E., West, G., Dusek, D.M., and Hoff, H.F. Cholesterol esterification in macrophages. Stimulation by lipoproteins containing apo-B isolated from human aortas. Arteriosclerosis, 4, 196-207 (1984).

53. Avogaro, P., Bittolo Bon, G., and Cazzolano, G. Presence of modified low density lipoprotein in humans. Arteriosclerosis, 8, 79-87 (1988).

54. Kita, T., Nagano, Y., Yokode, M., Ishii, K., Kume, N., Ooshima, A., Yoshida, H., and Kawai, C. Probucol prevents the progression of atherosclerosis in Watanabe heritable hyperlipidemic rabbit, an animal model for familial hypercholesterolemia. Proc.Natl. Acad.Sci.U.S.A., 84, 5928-5931 (1987).

55. Parthasarathy, S., Young, S.G., Witzum, J.L., Pittman, R.C., and Steinberg, D. Probucol inhibits oxidative modification of low density lipoprotein. J.Clin.Invest., 77, 641-644 (1986).

56. Reyftmann, J.P., Morlière, P., Goldstein, S., Santus, R., Dubertret, L., and Lagrange, D. Interaction of human serum low density lipoproteins with porphyrins: a spectroscopic and photochemical study. Photochem.Photobiol., 40, 721-729 (1984).

57. Candide, C., Morlière, P., Mazière, J.C., Goldstein, S., Santus, R., Dubertret, L., Reyftmann, J.P., and Polonovski, J. In vitro interaction of the photoactive anticancer porphyrin derivative Photofrin II with low density lipoprotein and its delivery to cultured human fibroblasts. FEBS Lett., 207, 133-138 (1986).

58. Morlière, P., Kohen, E., Reyftmann, J.P., Santus, R., Kohen, C., Mazière, J.C., Goldstein, S., Mangel, W.F., and Dubertret, L. Photosensitization by porphyrin delivered to L cells by human serum low density lipoproteins. A microspectrofluorometric study. Photochem.Photobiol., 46, 183-191 (1987).

59. Candide, C., Reyftmann, J.P., Santus, R., Mazière, J.C., Morlière, P., and Goldstein, S. Modification of Σ-amino group of lysines, cholesterol oxidation and oxidized lipid-apoprotein cross-link formation by porphyrin-photosensitized oxidation of human low density lipoproteins. Photochem.Photobiol., 48, 137-146 (1988).

60. Steinbrecher, U.P. Oxidation of human low density lipoprotein results in derivatization of lysine residues of Apolipoprotein B by lipid peroxide decomposition products. J.Biol.Chem., 262: 3603 (1987).

BIOCHEMICAL MECHANISMS OF OXIDANT-INDUCED CELL INJURY

Charles G. Cochrane, Paul A. Hyslop, Janis
H. Jackson, Ingrid U. Schraufstatter

Scripps Clinic and Research Foundation
Department of Immunology
10666 North Torrey Pines Road
La Jolla, CA 92037

INTRODUCTION

With the current knowledge that oxidants are generated in inflammatory responses of several kinds and participate in the development of tissue injury, it is clearly important to gain greater insight into the mechanisms by which oxidants damage cells and extracellular tissues. For the past few years we have studied the effects on target cells of oxidants that are generated by stimulated leukocytes, in order to gain insight into the mechanisms by which externally generated oxidants cause functional and structural damage to these target cells.

Initially we observed, as had many others before, that the generation of O_2^-, H_2O_2 and the spectrum of oxidants released by stimulated leukocytes would, after a latent period of 2-3 hours, cause death of target cells. However, by adding catalase at various times following exposure of the target cells to the oxidants, it became clear that the effective damage was initiated fully in the first 15-30 minutes. This indicated that oxidants render major biochemical mischief within minutes in cells contacting stimulated leukocytes.

Free Radicals, Lipoproteins, and Membrane Lipids
Edited by A. Crastes de Paulet *et al.*
Plenum Press, New York, 1990

It was then clear that the examination of biochemical systems in cells within minutes of exposure to external oxidants would produce information essential to understanding the mechanisms of oxidant-injury. Such became the goal of these investigations. In these studies we have used as target cells primarily the murine macrophage cell line, P388D1, (which has no measurable PMA-stimulated oxidase system), and in addition, bovine aortic endothelial cells, GM1380 human fibroblast cell line, human peripheral lymphocytes, monocytes and neutrophils, and rabbit alveolar macrophages. Oxidants utilized include the spectrum of oxidants generated by stimulated neutrophils, chemically generated oxidants from xanthine oxidase + purine, reagent H_2O_2 and reagent HOCl.

Effect of external oxidants on the glutathione-redox system

A major cellular defense against oxidants lies in the capacity of glutathione (GSH) to reduce peroxides and free radicals rapidly through the catalytic activity of the silenium enzyme, glutathione peroxidase (1-7). GSH is maintained in the reduced form by electrons obtained from NADPH and the hexose monophosphate shunt. Each member of this GSH-redox system was examined kinetically in cells exposed to oxidants (8).

When P388D1 cells were exposed to increasing doses of reagent H_2O_2 (0.01 to 5 mM), little loss of reduced glutathione was observed in the presence of glucose. However removal of glucose resulted in the loss of GSH and the instantaneous appearance of GSSG and glutathione bound to protein (8). H_2O_2 exposure within seconds in the absence of glucose caused a drop in NADPH levels from 55 to 22 pmol/10^6 cells. This was prevented by inhibiting turnover of GSH with buthionine sulfoxamine or (1,3-bis) 2 chloroethyl-1-nitrosourea. The hexose monophosphate shunt (HMPS) was simultaneously stimulated 5-10 fold (measured by the conversion of C_1 vs. C_6 labelled glucose to CO_2) within a few seconds of exposure of the cells to H_2O_2. The data indicated that cells such as P388D1 cells are able to activate their HMPS immediately upon exposure to externally generated oxidant and, in the presence of glucose, maintain activity of the pentose pathway and sufficient reducing equivalents in the glutathione cycle to keep glutathione in the reduced form.

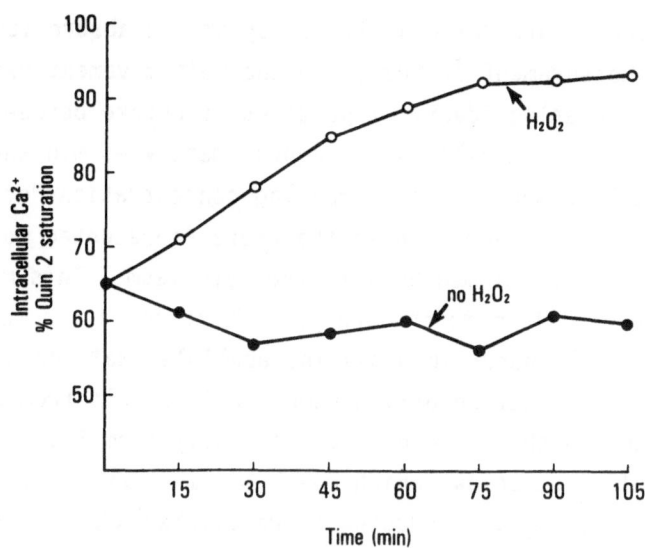

Fig 1   Intracellular concentration of Ca^{++} in P388D$_1$ cells
treated with 1 mM H$_2$O$_2$ for time intervals shown in the
figure.

Effects of external oxidants on cellular calcium homeostasis

Extensive investigations of Orrenius and co-workers (9-11) have
revealed important relationships between the derangement of
intracellular calcium homeostasis and exposure of cells to agents that
increase intracellular Ca^{++} levels over a period of hours (Orrenius,
this volume). It was not known whether calcium homeostasis was
similarly affected by external oxidants and hence we examined changes
in intracellular Ca^{++} over a time course in cells exposed to varying
concentrations of H$_2$O$_2$ (12).

Two phases of change in Ca^{++} homeostasis were observed when
P388D1 cells were exposed to H$_2$O$_2$. The first phase occurred in the
first 30 min and involved a translocation of intracellular Ca^{++} from
bound to free cytosolic form. This was measured as increased
fluorescence of cells preloaded with Quin-2 and as a fall in total
cellular Ca^{++} by atomic absorption analysis (1). Presumably the
Ca^{++} released from the intracellular stores was pumped out of the
cells. A net loss of Ca^{++} during oxidative stress was also observed
in whole liver perfusion studies by Sies et al. (7). In our studies
(12) the second phase of the changes in Ca^{++} homeostasis began after

30 min when intracellular calcium levels by atomic absorption analysis began rising. Measurement of transmembrane Ca^{++} movement using external $^{45}Ca^{++}$ revealed identical patterns of uptake between control and H_2O_2-exposed cells until approximately 45 min when the H_2O_2-exposed cells demonstrated increasing concentrations of $^{45}Ca^{++}$ (Fig. 1). This second phase therefore represented movement of Ca^{++} from the external medium into the cytoplasm. This may well reflect a diminution in membrane ion pump function, a conclusion in keeping with a simultaneous increase intracellular Na^+ and a loss of K^+. Such a loss in function of ion pumps could result from a direct action of oxidant on the proteins and indirectly from loss of intracellular ATP (see below) which reach minimal levels after 15-30 minutes' exposure to H_2O_2. Reducing intracellular ATP levels by the metabolic inhibitors 2-deoxy-D-glucose and carbonyl cyanide m-chlorophenylhydrazone led to an increase in intracellular Ca^{++} after ATP levels reached <0.1 m mol/10^6 (12).

Changes in cytoskeleton in cells exposed to oxidant

Using the fluorescent derivative NBD-phallicidin which binds f-actin with significantly higher affinity than G-actin, an increase in cytoplasmic fluorescence was observed in H_2O_2-exposed cells after 30 minutes (13). These studies involved both direct visual observation of cells and flow cytometric analysis. Transmission electron microscopy revealed clusters or bundles of microfilaments linked in side-to-side fashion. These changes were noted using single doses of H_2O_2 above 500 uM. When the cells were solubilized in Triton X-100 and the centrifuged sediments were examined by acrylamide electrophoresis, increasing amounts of f-actin were observed on reduced gels as early as 15-30 minutes, and in non-reduced gels, increasing amounts of aggregated proteins appeared that failed to enter the gel. A link between microfilament aggregation and fall in ATP levels was also observed in cells depleted of glucose (14). When ATP levels reached <10% of control values, changes in the microfilaments developed similar to those in H_2O_2-exposed cells.

Associated with the changes in cytoskeletal elements was a marked blebbing of the cytoplasm. This blebbing which precedes lysis, was similar to that in cells undergoing injury induced by internally-generated oxidants as studied by Orrenius and co-workers (this volume).

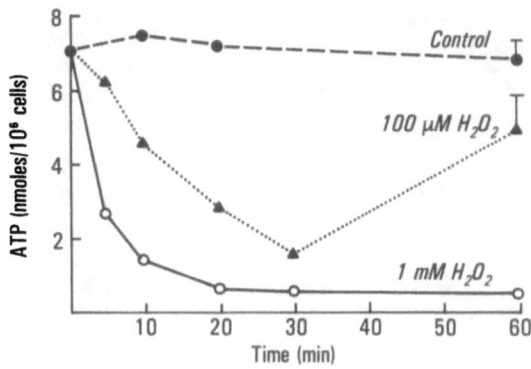

Fig 2   Diminution of ATP in P388D₁ cells exposed to 100 uM and 1 mM H₂O₂ for the time intervals shown.

The data suggest that oxidants whether generated externally or internally induce cross-linking of actin-rich microfilaments of the cytoskeleton. These are dissociated in the presence of reducing agent. Whether the increase in NBD-phallicidin binding represents an increase in quantity of f-actin is uncertain. And while a single exposure of H_2O_2 of 500 uM was required for clear demonstration of these changes in the microfilaments, it is likely that at lower concentrations of H_2O_2, fine changes in cytoskeletal elements may occur. The importance of such changes in a number of cellular functions could be significant.

<u>Oxidant effects on cellular energy systems</u>

Early observations from our laboratory indicated that ATP levels fall in P388D1 and cultured endothelial cells exposed to oxidants. Initial reduction occurred at 2-3 min, followed by a dose-dependent decrease reaching levels below 10% of control values (Fig. 2) (15).

The loss of ATP could result from either accelerated consumption or diminished formation. To examine the rate of consumption in P388D1 cells exposed to oxidants, the rate of loss of ATP was followed in cells exposed to H_2O_2 in which both glycolytic and mitochondrial synthesis were blocked (2 deoxyglucose + atractyloside). A calculated loss of 2.3 nmoles ATP/min/10^6 cells compared favorably with the rate in cells not exposed to H_2O_2 and to the rate of synthesis in control cells of 2.1 nmol/min/10^6 cells (16). Thus increase in catabolism was an unlikely explanation of the drop in ATP levels.

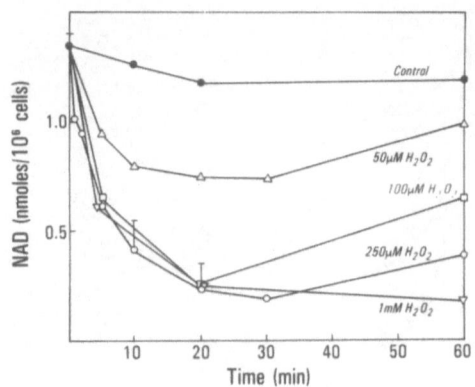

Fig 3   Diminution in levels of mistinamide adenine dinucleotide (NAD)
         in P388D$_1$ cells exposed to various initial concentrations of
         H$_2$O$_2$.

By contrast, both glycolytic and mitochondrial synthesis were
profoundly affected by exposure to oxidants (H$_2$O$_2$, O$_2^-$, and
HOCl) (16). With H$_2$O$_2$ in P388Dl cells, the glycolytic pathway was
inhibited principally at the level of glyceraldehyde-3-phosphate
dehydrogenase (GAPDH) (4). Of the eleven enzymes of the glycolytic
pathway analyzed, only GAPDH showed a significant decrease in Vmax. Km
of each enzyme was affected by less than 10%. Vmax of GAPDH in cells
exposed to 5 mM H$_2$O fell from 78.1 $\pm$ 8.7 to 0.8 $\pm$ 0.1
nmoles/min/10^6 cells and nearly total inhibition of activity at 1 mM
H$_2$O$_2$. This was accompanied by an elevation of the components of
the glycolytic pathway immediately preceding the GAPDH step,
glyceraldehyde-3-phosphate and dihydroxyacetone phosphate. The
combined levels were increased approximately 6 fold in the 10 minutes
after exposure to H$_2$O$_2$. Lactate production decreased in proportion
(16).

The inhibitory effect on the GAPDH step resulted from three
independent factors: a direct attack of H$_2$O$_2$ on GAPDH, a decrease
in its essential cofactor, nicotinamide adenine dinucleotide (NAD) and
a small shift in pH in the cell responding to H$_2$O$_2$. As to the
first, not only a decrease in Vmax of GAPDH was measured in
oxidant-exposed cells with an IC$_{50}$ of approximately 100 uM H$_2$O$_2$,
but isolated rabbit muscle GAPDH also was inactivated by H$_2$O$_2$ with
an IC$_{50}$ of 10 uM H$_2$O$_2$. Further studies have indicated that loss
of two SH groups is associated with loss of activity of GAPDH and that

S-S bonds are formed. The SH groups involved were found to be cysteines 149 and 153 which form crystallographic studies are known to be spatially opposed in one turn of an alpha helix. Disulfide formation between these cysteines inhibits activity (Hyslop et al., unpublished observations), possibly by preventing binding of NAD.

The second major influence on GAPDH activity was found to be a fall in levels of the essential cofactor NAD (Fig. 3) (16,17). NAD levels fall relative to NADH in cells that contain poly ADP ribose polymerase (polymerase). This latter enzyme becomes activated during H_2O_2 exposure and consumes NAD, its substrate. The two products of NAD during hydrolysis by the polymerase are nicotinamide (NA) and ADP-ribose which becomes polymerized and binds to proteins in the vicinity of its formation. NA formation in P388D1 cells paralleled stoichiometrically the loss of NAD. Ribosylation of proteins simultaneously occurred. These proteins were indistinguishable from nucleohistones by HPLC analysis.

The effects of the direct inactivation of GAPDH by H_2O_2, diminishing levels of cofactor NAD relative to NADH and slight, but significant lowering of pH would appear to combine to produce the inhibition of the glycolytic pathway at this step.

A relationship between the fall in NAD levels and activation of poly ADP ribose polymerase was observed employing inhibitors of the polymerase. 3-aminobenzamide, nicotinamide and theophylline each inhibited the polymerase as measured by NAD formation and protein ribosylation and prevented the loss of NAD (18), with P388D1 cells, inactivation of the polymerase with 3 aminobenzamide was associated with not only maintenance of levels of NAD, but ATP as well. These cells also did not develop a flux of extracellular Ca^{++} and viability was maintained (Fig. 4).

It should be noted that treatment of cells with 3 aminobenzamide did not affect the half-life of H_2O_2 and did not affect certain polymerase-independent effects of H_2O_2 such as translocation of intracellular Ca^{++}. Thus in P388D1 cells, NAD cofactor of the GAPDH steps of the glycolytic pathway appeared under the conditions of these

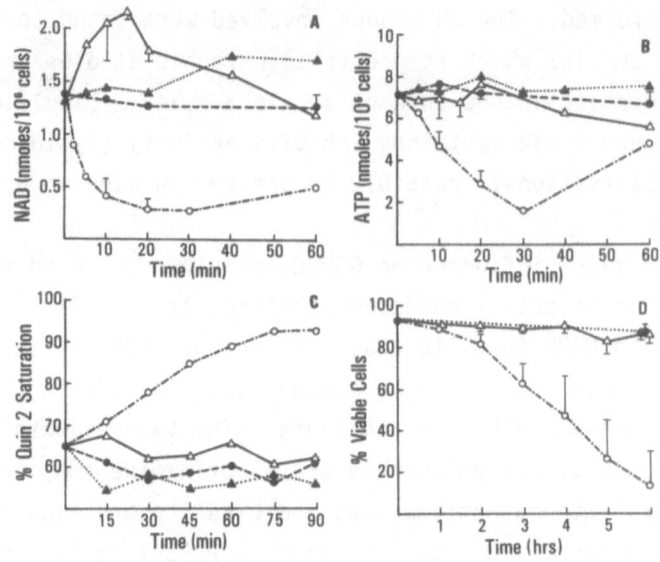

Fig 4   The capacity of 2.5 mM3-aminobenzamide (3 ABA) to inhibit
        biochemical alterations in $2x10^6$ target P388D$_1$ cells
        induced by H_2O_2.

   A.   NAD levels ●---● untreated; o-.-o exposed to 250 uM H_2O_2;
        ▲....▲ treated with 2.5 mM 3 ABA; △——△ treated with 2.5
        mM 3 ABA followed by 250 uM H_2O_2.

   B.   ATP levels. Symbols same as in A.

   C.   Intracellular Ca^{2+} levels as indicated by quin-2
        fluorescence. Symbols same as in A.

   D.   Viability. Symbols same as in A.

experiments to be a critical modulating factor in oxidant damage to the
energy producing system and viability of the cells.

     The second ATP generating system, i.e., oxidative phosphorylation
by the mitochondria, was also affected by exposure to oxidants, and
specifically, H_2O_2. Of interest, at H_2O_2 concentrations of
100-300 uM, an increase in net mitochondrial synthesis of ATP
occurred. Nevertheless, over the same concentration range of H_2O_2
and above, ATP synthetase (ADP phosphorylation) activity consistently
declines (16).

     Thus both glycolytic and oxidative phosphorylation pathways of ATP
formation are affected by exposure of cells to oxidants. The balance
of effect on one vs. the other pathway will likely vary with cell

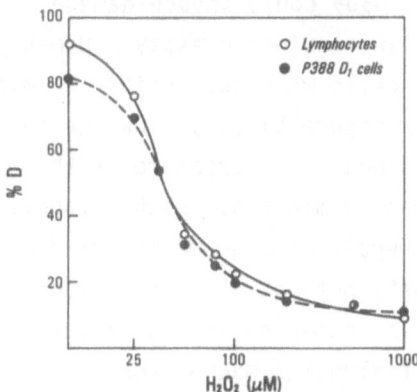

DNA STRAND BREAKS INDUCED
BY EXPOSURE TO H_2O_2 FOR 5 MIN

Fig 5   DNA single-strand breaks as determined by alkaline unwinding
         technique. $P388D_1$ cells (●) or human peripheral lymphocytes
         (o) (2×10^6/ml MGB) were incubated with different doses
         H_2O_2 for 5 min at 37°C, H_2O_2 was removed by
         centrifugation, the cell pellet was resuspended in 450 ul
         ice-cold meso-inositol buffer, and 200 ul samples were lysed
         in alkaline medium. Ethidium bromide fluorescence was
         determined (excitation, 520 nm; emission, 590 nm) after a 45
         min incubation at 15°C. D represents percent
         double-stranded DNA at the end of this incubationperiod as
         described. Mean of three experiments determined in duplicate.

type. Those cells heavily dependent upon mitochondrial oxidative
phosphorylation will not be as subject to ATP loss caused by inhibition
of GAPDH as those more dependent upon glycolysis. The reverse would be
true for cells more dependent upon glycolysis.

Oxidant damage of cellular DNA

     Since poly(ADP-ribose) polymerase is activated under conditions of
DNA strand breakage, we reasoned that DNA damage might have occurred in
cells exposed to oxidants. This has proved to be the case in all cell
types assayed to date, and the DNA damage occurs within seconds of
exposing the cells to oxidants (17).

     A major unanswered question was that of the mechanisms by which
oxidants damage DNA. Information available from the literature
suggested that O_2^- (19), as well as H_2O_2 and hydroxyl radical
(·OH), could be responsible, although little information about HOCl,

singlet O_2 ($^1O_2$), or N-chloramines was available. We first determined that reagent H_2O_2 could induce damage of cellular DNA, as measured by an alkaline unwinding assay. Human peripheral lymphocytes and P388D1 cells were susceptible to micromolar concentrations of H_2O_2 (Figure 5) [17]. Damage to cellular DNA developed within 20 seconds after exposure of the cells to H_2O_2 and reached a maximal effect in minutes. With catabolic removal of the H_2O_2 from the medium, repair of the DNA occurred over the succeeding hours in most, but not all, cell types. Exposure of several cell types (P388D1 cells, human peripheral lymphocytes, GM1380 fibroblasts, and rabbit alveolar macrophages) to H_2O_2 revealed varied susceptibility of the DNA to damage by the H_2O_2. If H_2O_2 were the major oxidant species penetrating the cells to damage the DNA, then specific inhibition of catalase in the cytoplasm of the various cell types should produce cells of equal susceptibility. This proved to be the case. When these various cell types were first exposed to 3-amino-1,2,4-triazole to inhibit catalase (as well as myeloperoxidase in the neutrophils) and were then treated with various concentrations of H_2O_2, the cells showed greater susceptibility to DNA damage, and the degree of susceptibility was the same for all the cell types. In addition, in the absence of aminotriazole, the susceptibility of the cellular DNA to various concentrations of H_2O_2 was inversely proportional to the activity of catalase in the cells [20].

Several oxidants generated by human monocytes were then assessed for their capacity to damage DNA in peripheral human lymphocytes. The ratio of monocytes to lymphocytes was 1:3. A strong correlation existed between the concentration of H_2O_2 in the extracellular medium and the degree of DNA strand breaks. Addition of catalase to the medium completely prevented the strand breaks in each system, but addition of superoxide dismutase (SOD) did not. These data suggest that H_2O_2 in the medium to which the target cells were exposed was responsible for initiating the DNA damage [20].

HOCl, N-chloramine, H_2O_2, and $^1O_2$ were then compared for their capacity to damage isolated PM2 DNA. Isolated DNA was chosen to avoid intracellular variation in oxidant activity, such as from inactivation by cellular inhibitors and natural scavengers. While 100 uM H_2O_2 plus 30 uM iron induced an average of 55% strand breaks, 100 uM HOCl induced 5%, N-chloramine (100 uM HOCl plus 500 uM taurine) induced 6%, and a combination of 100 uM H_2O_2 and 100 uM HOCl (to generate $^1O_2$) induced 20%. Similarly, addition of 100 uM HOCl to P388D1 cells failed to induce DNA damage, as measured by the alkaline unwinding assay.

These results support the conclusion that H_2O_2 is the dominant oxidant leading to DNA strand breaks. The data on cells in which intracellular catalase was inhibited by aminotriazole suggest that H_2O_2 penetrates the cell, thereby gaining access to DNA. The H_2O_2 could conceivably induce damage directly or indirectly. Therefore, experiments were undertaken to examine this aspect.

Participation of the hydroxyl radical in DNA damage

Our attention turned to ·OH as the free radical most likely generated by the presence of H_2O_2 in the vicinity of nuclear DNA. H_2O_2 alone, at concentrations as high as 100 uM, failed to induce significant numbers of strand breaks in isolated PM2 DNA. However, abundant strand breaks were observed in the presence of Fe^{2+} (10 to 30 uM), especially when either ascorbate or O_2^- was added to reduce Fe^{3+} to Fe^{2+}. These results, in combination with previous studies (e.g. references 21-23), strongly implicated the generation of ·OH from H_2O_2 as a mechanism. In whole P388D1 cells, DNA damage, measured as strand breaks, was prevented by the chelation of iron using phenanthroline or desferrithiocin.

The apparent association of ·OH formation by the Fenton reaction and DNA damage was to this point based on indirect evidence. To determine more accurately if this association exists, we conducted experiments to measure ·OH levels at a time when DNA strand breaks occurred. We used electron paramagnetic resonance (EPR) spectroscopy with the spin trap dimethylpyroline-N-oxide (DMPO) to detect the presence of ·OH. With H_2O_2, Fe^{2+}, and PM2 DNA, a 4-line spectrum appeared that was characteristic of the DMPO adduct with ·OH (24). A similar spectrum was observed utilizing cigarette smoke (a rich source of H_2O_2) and asbestos (a rich source of iron). A correlation was observed between the amount of DMPO-OH observed by EPR and the percentage of DNA with strand breaks (Figure 6). Since the

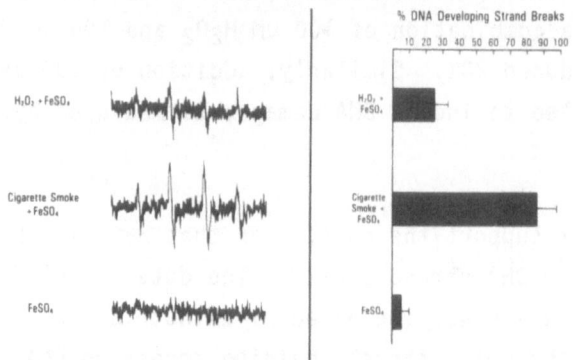

EPR PATTERNS AND STRAND BREAKS
OF TREATED PM-2 DNA

Fig 6   Electron paramagnetic resonance spectral patterns developing
         from mixtures of 0.8 ug PM2 DNA, 0.05 mg crocidolite asbestos,
         10 ul cigarette smoke in PBS, 65 uM H_2O_2 and/or Fe SO_4
         in a final volume of 100 ul of PBS, pH 7.4. 100 mM DMPO was
         added to form adducts with HO·. Astrisks in the upper and
         lower panels indicate the DMPO-OH signal, characteristic of
         the adduct formed between OH· and DMPO.

DMPO-OH spectrum can be produced by mechanisms not involving ·OH, the
secondary ·OH traps ethanol and dimethylsulphoxide (DMSO) were added
to the mixtures described above. The resulting EPR spectra revealed
the 6-line DMPO-alpha-hydroxyethyl signal and the 6-line DMPO-methyl
signal, respectively. These patterns develop only with DMPO-OH is
generated from ·OH, confirming that ·OH had been formed. Further
confirmation was obtained by the observation that catalase (but not
SOD), ·OH scavengers, and iron chelators inhibited both ·OH
detection and DNA strand breaks. Complete inhibition of DNA strand
breaks was observed with the addition of 3.5 ug catalase, 100 mM DMSO,
100 mM mannitol, 100 mM sodium benzoate, 4 mM phenanthroline, and 4 mM
desferrithiocin (24).

DNA damage by products of lipid peroxidation

     The question was then raised whether damage by oxygen free radicals
was limited to single strand breaks in DNA, or if modifications in
bases took place as well. To examine this, calf thymus DNA was exposed
to neutrophils stimulated with phorbol myristate acetate in the
presence or absence of Fe. The DNA was then analyzed by gas
chromatography-mass spectrometry following hydrolysis and

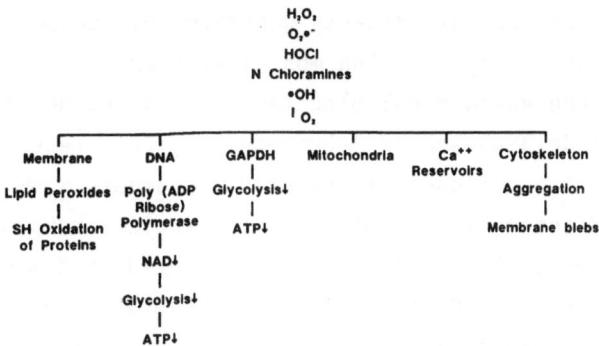

Fig 7   Functionally important primary targets of oxidants in target
         cells and biochemical consequences of the oxidant interaction
         with these targets.   GAPDH = glyceraldehyde-3-phosphate
         dehydrogenase.

derivatization. Hydroxylation was observed in all four bases and the
addition of Fe increased the yeild such that 7 out of every 1000 bases
were modified (25).

DNA base hydroxylation by leukocytic oxidants

Damage to DNA by the H_2O_2/·OH reaction may indeed play a
major role in cellular injury and in altered function in the
inflammatory response. In addition, products of lipid peroxidation
induced by the interaction of ·OH with polyunsaturated fatty acids in
the membranes of target cells may also contribute to the DNA damage.
Ochi and Cerutti (25) emphasized the relationship of lipid
hydroperoxides to the clastogenic agents released by stimulated
monocytes. These investigators have described the capacity of such
factors, at concentrations of 10 to 20 uM, to induce DNA strand breaks
in target cells. Our preliminary data support their conclusions.
Unsaturated fatty acids (oleic, linoleic, linolenic, and arachidonic
acids, 164 uM) induced DNA damage in GM1380 fibroblasts, and the amount
of damage (leading to DNA strand breaks) corresponded to the degree of
fatty acid unsaturation. The mechanism of DNA damage by this means
remains uncertain, but the effect of the unsaturated fatty acids is
clear. The importance of this mechanism in the inflammatory response
remains to be determined.

Conclusions

The studies reviewed and the data described here support the hypothesis that, of various oxidants generated by leukocytes during the inflammatory response, H_2O_2 is the most likely oxidant to penetrate cells and reach the various cellular targets, including cell membranes, DNA, glyceraldehyde-3-phosphate dehydrogenase, Ca^{2+} reservoirs, and mitochondria. A tabulation of intracellular targets shown in those studies to be affected by oxidants, and in particular H_2O_2 or H_2O_2--$\cdot$OH, is shown in Figure 7. Considering oxidant damage to DNA, concentrations of H_2O_2 as low as 25 uM cause damage to DNA in target cells. Such concentrations are physiologically relevant since millimolar concentrations of H_2O_2 can be generated in the immediate vicinity of the target cells by stimulated neutrophils. Studies suggest that H_2O_2 acts indirectly through $\cdot$OH to damage DNA.

Acknowledgments

This work was supported by United States Public Health Service Grants HL23584 and GM37696 and Office of Naval Research 105-837.

REFERENCES

1. Nathan, CF, Silverstein SC, Brukner LH, Cohn ZA. Extracellular cytolysis by activated macrophages and granulocytes. II. Hydrogen peroxide as a mediator of cytotoxicity. J. Exp. Med. 1979:149:100-113.
2. Simon RH, Scoggin CH, Patterson D. Hydrogen peroxide causes the fatal injury to human fibroblasts exposed to oxygen radicals. J. Biol. Chem. 1981: 256:7181-7186.
3. Weiss SJ, Young J., LoBuglio AF, Slivka A. Role of hydrogen peroxide in neutrophil-mediated destruction of cultured endothelial cells. J. Clin. Invest. 1981:68:714-724.
4. Jarrick BA, Nathan CF, Griffith OW, Cohn ZA. Glutathione depletion sensitizes tumor cells to oxidative cytolysis. J. Biol. Chem. 1982:257(3):1231-1237.
5. Harlan JM, Levine JD, Callahan KS, Schwartz BR, Harker LA. Glutathione redox cycle protects cultured endothelial cells against lysis by extracellularly generated hydrogen peroxide. J. Clin. Invest. 1984:73:706-713.
6. Sies H, Gerstenecker C, Menzel H, Flohe L. Oxidation in the NADP-system and release of GSSG from hemoglobin-free perfused rat liver during peroxidative oxidation of glutathione by hydroperoxidase. FEBS Lett. 1972:27:171-175.
7. Sies H, Grafp, Estrela, JM. Hepatic calcium efflux during cytochrome P-450-dependent drug oxidations at the endoplasmic reticulum in intact liver. Proc. Natl. Acad. Sci. 981:78:3358.
8. Schraufstatter IU, Hyslop PA, Spragg RG, Cochrane CG. Glutathione cycle activity and pyridine nucleotide levels in oxidant-induced injury of cells. J. Clin. Invest. 1985:76:1131-1139.
9. Orrenius S, Jewell SA, Bellomo G, Thor H, Jones DP, Smith MT. Regulation of calcium regulation in the hepatocyte-a critical role of glutathione. In: Functions of Glutathione: Biochemical, Physiological, Toxicological and Clinical Aspects. A. Larsson, S.

Orrenius, A. Holmgren, and B. Mannervik, editors. Raven Press, NY, 261-273, 1983.

10. Jewell SA, Bellomo G, Thor H, Orrenius, S, Smith, MT. Bleb formation in hepatocytes during drug metabolism is caused by distrubances in thiol and calcium ion homeostasis. Science. 1982:217:1257.

11. Bellomo G, Jewell SA, Thor H, Orrenius S. Regulation of intracellular calcium compartmentation: Studies with isolated hepatocytes and t-butyl hydroperoxide. Proc. Natl. Acad. Sci. 1982:79:6842.

12. Hyslop PA, Hinshaw DB, Schraufstatter IU, Sklar, LA, Spragg RG, Cochrane CG. Intracellular calcium homeostasis during hydrogen peroxide injury to cultured P388D11 cells. J. Cell Physiol. 1986:129:356.

13. Hinshaw DB, Sklar LA, Bohl BP, Schraufstatter IU, Hyslop PA, Rossi MW, Spragg RG, Cochrane CG. Cytoskeletal and morphologic impact of cellular oxidant injury. Am. J. Path. 1986:123:454-464.

14. Hinshaw DB, Armstrong BC, Burger JM, Beals TF, Hyslop PA. ATP and microfilaments in cellular oxidant injury. Am. J. Path. 1988:132:479-488.

15. Spragg RG, Hinshaw DB, Hyslop PA, Schraufstatter IU, Cochrane CG. Alterations in adenosine triphosphate and energy charge in cultured endothelial and P388D1 cells following oxidant injury. J. Clin. Invest. 1985:76:1471-1476.

16. Hyslop PA, Hinshaw DB, Halsey WA Jr. et al. Mechanisms of oxidant mediated cell injury: The glycolytic and mitochondrial pathways of ADP phosphorylation are major intracellular targets inactivated by hydrogen peroxide. J. Biol. Chem. 1988:263:1665.

17. Schraufstatter IU, Hinshaw DB, Hyslop PA, Spragg RG, Cochrane CG. Oxidant injury of cells: DNA strand-breaks activate polyadenosine diphosphate-ribose polymerase and lead to depletion of nicotinamide adenine dinucleotide. J. Clin. Invest. 1986:77:1312.

18. Schraufstatter IU, Hyslop PA, Hinshaw DB, Spragg RG, Sklar LA, Cochrane CG. Hydrogen peroxide-induced injury of cells and its prevention by inhibitors of poly(ADP-ribose) polymerase. Proc. Natl. Acad. Sci. 1986:83:4908.

19. Birnboim HC, Kanabus-Kominska M. The production of DNA strand breaks in human leukocytes by superoxide anion may involve a metabolic process. Proc Natl Acad Sci 1987:82:6820-4.

20. Schraufstatter IU, Hyslop PA, Jackson JH, Cochrane CG. Oxidant-induced DNA damage of target cells. J. Clin. Invest. 1988:1040-1050.

21. Floyd RA. DNA-ferrous iron catalyzed hydroxyl free radical formation from hydrogen peroxide. Biochem Biophys Res Comm 1981:1209-15.

22. Frenkel K, Chrzan K, Troll W, Teebor GW, Steinberg JJ. Radiation-like modification of bases in DNA exposed to tumor promoter-activated polymorphonuclear leukocytes. Cancer Res 1986:46:5533-40.

23. de Mello Filho AC, Meneghini R. Protection of mammalian cells by o-phenanthroline from lethal and DNA-damaging effects produced by active oxygen species. Biochim Biophys Acta 1985:847:82-9.

24. Jackson J, Schraufstatter IU, Hyslop PA, Vosbeck K, Sauerheber R, Weitzman SA, Cochrane CG. Role of oxidants in DNA damage: Hydroxyl radical mediates the synergistic DNA damaging effects of asbestos and cigarette smoke. J Clin Invest 1987:80:1090-1095.

25. Jackson JH, Gajewski E, Fuciarelli AE, Schraufstatter, IU, Hyslop, PA, Cochrane, CG, Dizdarogler M. Damage to the bases in DNA induced by stimulated neutrophils. J. Clin. Invest. 1988:84.

26. Ochi T, Cerutti PA. Clastogenic action of hydroperoxy-5, 8,11,13-icosatetranoic acids on the mouse embryo fibroblasts C3H/10 Tl/2. Proc Natl Acad Sci 1987:84:990-4.

DEFENSE MECHANISMS

CYTOCHROME P-450 AND VITAMIN E FREE RADICAL REDUCTASE:
FORMATION OF AND PROTECTION AGAINST FREE RADICALS

Aalt Bast and Guido R.M.M. Haenen

Department of Pharmacochemistry, Faculty of Chemistry, Vrije Universiteit

De Boelelaan 1083, 1081 HV Amsterdam, the Netherlands

I INTRODUCTION

In studies on lipid peroxidation, liver microsomes are frequently employed. Already in 1963, Hochstein and Ernster[1] reported that in their experiments lipid peroxidationin rat liver microsomes required molecular oxygen and NADPH and that lipid peroxidation intensified by the availability of ADP and Fe^{2+}. This process is referred to as enzymic lipid peroxidation since it appears to depend on the presence of the haemoprotein cytochrome P-450 and the flavoprotein NADPH cytochrome P-450 reductase, both residing in the microsomal membrane[2]. The precise contribution of both proteins to NADPH dependent microsomal lipid peroxidation is still a matter of debate[2]. We described that cytochrome P-450 might be involved in both lipid hydroperoxide (LOOH) independent and LOOH dependent lipid peroxidation[2].

In this paper the remarkably varied utilization of oxygen by cytochrome P-450[3] is summarized and the possible ways of free radical formation by the haemoprotein will be indicated.

The assumed extensive free radical generation in liver microsomes helps to explain the existence of an apparently unique effective glutathione (GSH) dependent protective factor against lipid peroxidation, with which the microsomal membrane seems to be equipped. It is exemplified that GSH inhibits microsomal lipid peroxidation via a vitamin E free radical reductase[4]. Other explanations which have been provided for the protective action of GSH do not seem valid.

II ROLE OF CYTOCHROME P-450 IN FREE RADICAL FORMATION

The haemoprotein cytochrome P-450 is widely distributed throughout the phylogenetic scale. Distinct cytochrome P-450 isozymes have been described and are distributed from

$$RH + NADPH + H^+ + O_2 \rightarrow ROH + NADP^+ + H_2O \quad \text{monooxygenase}$$
$$NADPH + H^+ + O_2 \rightarrow NADP^+ + H_2O_2 \quad \text{oxidase}$$
$$RH + XOOH \rightarrow ROH + XOH \quad \text{peroxidase}$$

Fig. 1   Stoichiometry of various functions of P-450. RH, ROH and XOOH represent the substrate, the product and a peroxy compound respectively.

bacteria (often in a soluble form) to mammals (particularly localized in the membrane of the endoplasmatic reticulum). Discrete genes for cytochrome P-450 have been characterized as belonging to different families, which form a cytochrome P-450 multigene superfamily[5]. The vast number of different P-450 enzymes enables the P-450 to catalyze a wide variety of reactions and to cope with many xenobiotics[3]. P-450 may function as a monooxygenase, a dioxygenase (not studied extensively), an oxidase and as a peroxidase (Fig. 1).

IIa The Monooxygenase Activity of P-450

The monooxygenase function of P-450 depends on electron transfer. Electrons, both the first and second electron, are delivered via NADPH cytochrome P-450 reductase (Fig. 2). Alternatively the second electron can be provided via cytochrome b_5, which can be reduced by NADH cytochrome b_5 reductase of NADPH cytochrome P-450 reductase. The monooxygenase activity of P-450 may result in free radical metabolites of xenobiotics.

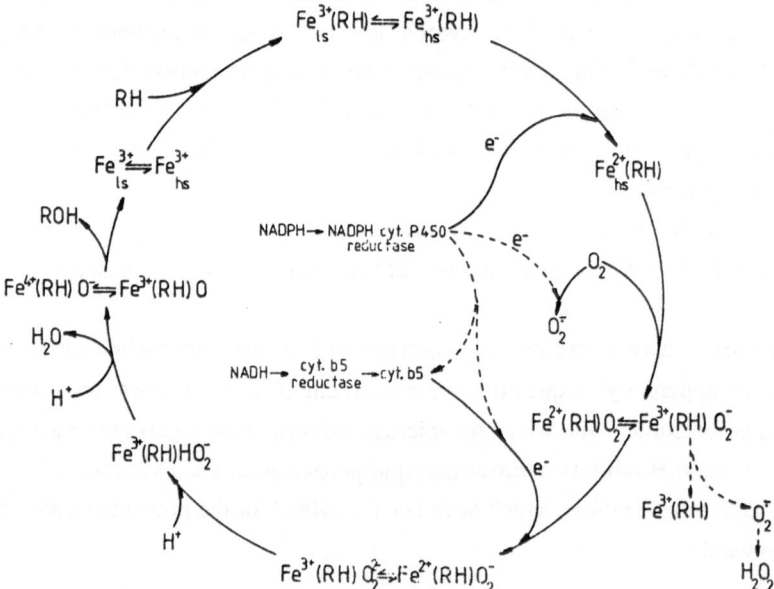

Fig. 2   Reaction cycle of P-450. Notice the spin state changes (high spin (hs) and low spin (ls)) that can occur during interaction of P-450 (indicated as Fe) with a substrate RH.

Moreover, if the coupling efficiency for the electron flow to P-450 is low, the NADPH cytochrome P-450 reductase may give electrons to dioxygen which result in superoxide anion radicals ($O_2^-\cdot$). This apparently occurs in reconstituted micellar systems or in the presence of high amounts of chelated iron[6].

IIb The Oxidase Activity of P-450

P-450 can also donate electrons to dioxygen. This so-called oxidase activity of P-450 (Fig. 1) may include both one, two and four electron reduction pathways.

The one electron reduced oxygen in the P-450 reaction cycle (Fig. 2) can dissociate from the haemoprotein and upon subsequent dismutation $O_2^-\cdot$ can result in H_2O_2.

$$Fe^{3+} - O_2^- \rightarrow Fe^{3+} + O_2^-\cdot$$

In the two electron pathway, direct protonation of released two electron reduced oxygen will give H_2O_2.

$$Fe^{3+} - O_2^{2-} + 2H^+ \rightarrow H_2O_2 + Fe^{3+}$$

This pathway is considered to be less likely than the one electron reduction mechanism[3].

Rather than the introduction of a second electron, it has been suggested[7] that disproportionation between two molecules ferrocytochrome P-450 - O_2 ($Fe^{2+} - O_2$) may yield $Fe^{2+} - O_2^-$

$$Fe^{2+} - O_2 + Fe^{2+} - O_2 \rightarrow Fe^{2+} - O_2^- + Fe^{3+} + O_2$$

A four electron oxidase pathway has also been reported, in which two molecules H_2O are formed[8].

$$O_2 + 2NADPH + 2H^+ \xrightarrow{\text{4e oxidase of P-450}} 2H_2O + 2NADPH^+$$

Some xenobiotics inhibit the microsomal H_2O_2 production. With a series of histamine H_2-antagonists we showed that attenuation of endogenous microsomal H_2O_2 production was in proportion to the extent of ligand binding to the haem iron of P-450[9]. We concluded that interaction with haem iron prevents oxygen binding[9]. Under well defined circumstances the stimulation by some substrates of the microsomal H_2O_2 production (probably formed via the one electron reduction pathway with subsequent dismutation of $O_2^-\cdot$) can be explained as well[3]. With a homogenous series of 5-ethyl-5-alkyl substituted barbiturates, the enhancement of H_2O_2 formation could be bilinearly correlated to log P' (logarithm of apparent partition coefficient, determined as octanol-buffer partition) (Fig. 3). A similar relationship between affinity of substrate binding and log P' for this series of compounds

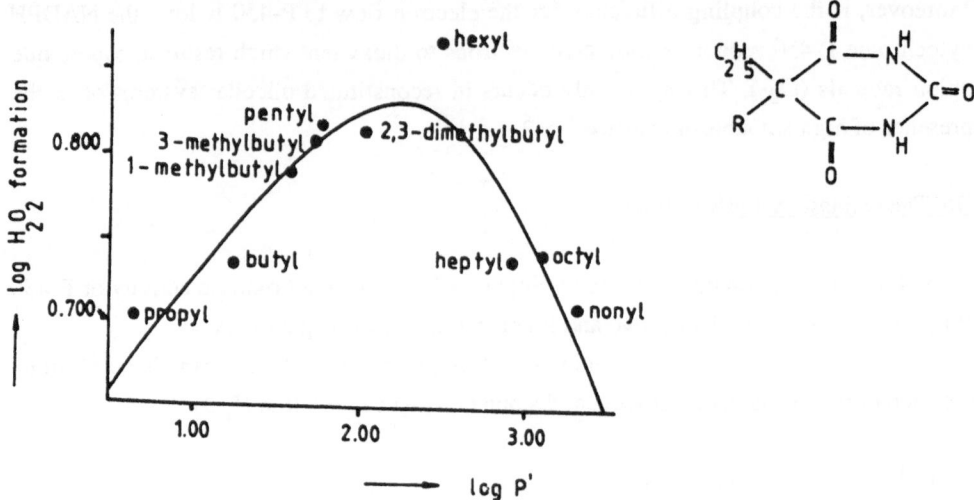

Fig. 3   Bilear relationship between log P' (octanol-buffer) and log microsomal H_2O_2 formation in hepatic mircrosomes of phenobarbital pretreated rats of a series of 5-ethyl-5-alkyl substituted barbiturates.

has been substantiated too. We concluded[3] that substrate interaction characteristics anticipate the stability of the ferric cytochrome P-450 - superoxide anion complex (Fe^{3+} - $O_2^-\cdot$).

In our opinion the oxygen species $O_2^-\cdot$ and H_2O_2 and derived hydroxyl radical may eventually become recognized as playing a crucial role in the general activity of P-450, enhancing its ability to metabolize a vast majority of xenobiotics. Hydroxyl radicals are highly reactive and secure the possibility for P-450 to react with all xenobiotics that are encountered.

It is common knowlegde that the Fenton reaction (Fe^{2+} + H_2O_2) may give rise to the reactive hydroxyl radical. The finding that modulation of microsomal hydroxyl radical formation (via enhancement and inhibition of microsomal H_2O_2 generation) did not correlate with the extent of lipid peroxidation[10,11] and the inability of catalase to inhibit microsomal lipid peroxidation[12] are arguments against the involvement of hydroxyl radicals in NADPH dependent microsomal lipid peroxidation.

IIc The Peroxidase Activity of P-450

For the peroxidase function of P-450 both heterolytic and homolytic cleavage mechanisms of the peroxide have been suggested[2]. If we assume homolytic cleavage of the organic peroxide (Fig. 4) we see that suitable hydrogen donors, viz. NADPH or oxidizable substrates (RH) lead to innoxious decomposition of the peroxide. If a suitable hydrogen donor is absent P-450 may initiate a LOOH dependent lipid peroxidation as depicted in Fig. 4.

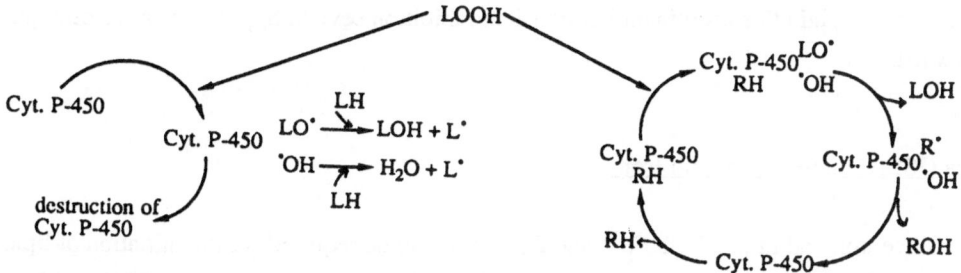

Fig. 4 Cytochrome P-450 as peroxidase. In the absence of a suitable hydrogen donor the
        peroxidase activity of P-450 may give lipid peroxidation. In the presence a hydrogen
        donor (RH) the peroxide becomes decomposed.

Taken together, it shows that P-450 can exert different modes of action. Consequently,
enzymatic activity displayed within the endoplasmatic reticulum entails an abundance of free
radicals in this organelle.

III ROLE OF GLUTATHIONE IN PROTECTION AGAINST MICROSOMAL LIPID PEROXIDATION

Free radicals may initiate lipid peroxidation. Glutathione in its reduced form (GSH)
protects against NADPH / Fe^{2+} dependent lipid peroxidation (Fig. 5).

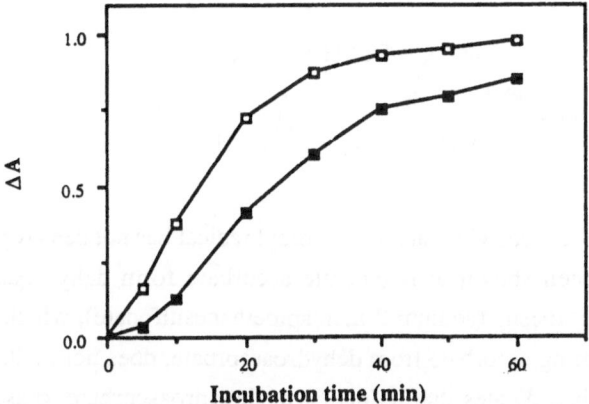

Fig. 5 Protection of 0.1 mM GSH (■) in NADPH (regenerating system) / Fe^{2+} (10 μM)
        induced time dependent liver microsomal lipid peroxidation. Control time course of
        lipid peroxidation is indicated as (□). Lipid peroxidation is measured as formation of
        thiobarbituric acid reactive material (ΔA 535-600 nm). Microsomes derived from
        1/8 g liver in 1 ml were incubated in a 50 mM Tris-HCl, 150 mM NaCl buffer
        (pH 7.4).

In order to explain this prominent but transient inhibition several hypotheses have been put forward.

IIIa GSH Complexation with Iron.

Since iron and in fact both Fe^{2+} and Fe^{3+} seems to be required for the initiation of lipid peroxidation [13], it has been suggested that "complex formation between GSH and iron cannot be ruled out, thereby lowering induction reactions[14]". It is known that heating of the microsomes abolished the GSH dependent protection[4]. Heating as such will not prevent complex formation between GSH and iron, which makes this suggestion, even as an additional explanation for the GSH dependent inhibition of lipid peroxidation[14], very improbable.

IIIb Interaction between GSH and Ascorbate

During Fe^{2+} / ascorabate (AH$^-$) induced lipid peroxidation the ascorbyl radical (A$\cdot$) and dehydroascorbate are formed[15]. Theoretically it can be assumed that GSH regenerates ascorbate from the ascorbyl radical or from dehydroascorbate[15], thus modulating the pro- and antioxidant capacity of ascorbate in favour of the antioxidant activity. The redox couples of ascorbate (AH$^-$/A$\cdot$) and GSH (GSH/GS$\cdot$) are linked as established in test tube conditions[16]. However the linkage which presumably couples hydrogen transfer to electron transfer describes the reaction of ascorbate with the thiyl radical,

$$
\begin{array}{ccc}
\text{R.} & \text{GSH} & \text{A}^- \\
\text{hydrogen} & & \text{electron} \\
\text{transfer} & & \text{transfer} \\
\text{RH} & \text{GS.} & \text{AH}^-
\end{array}
$$

whereas the interaction between GSH and the ascorbyl radical has not been reported.

GSH has indeed been shown to regenerate ascorbate form dehydroascorbate very effectively[15]. However, mesna (sodium 2-mercaptoethanesulfonate), which also displays this feature of regenerating ascorbate from dehydroascorbate, does not inhibit microsomal lipid peroxidation. This indicates that reduction of dehydroascorbate to ascorbate is not involved in the protective effect of GSH.

In conclusion, the GSH dependent transient inhibition of microsomal lipid peroxidation is not explained by regeneration of ascorbate from the ascorbyl radical or from dehydroascorbate. Finally, the finding that GSH protects in ascorbate independent systems as well (Fig. 5) strongly indicates that interaction between GSH and ascorbate does not play a role.

It has been suggested that the GSH effect proceeds via the microsomal GSH S-transferase. It has been stated that a case where the GSH S-transferase and its activation might be operative "is during lipid peroxidation, where a GSH-dependent protective factor has been observed in rat liver microsomes[17]". A series of inhibitors of GSH S-transferase like Rose Bengal, tributyltin acetate, S-hexylglutathione, indomethacin, Cibacron Blue and bromosulphophtalein showed inhibition towards the GSH dependent protection of lipid peroxidation as well[18]. In the view of Mousialou et al.[18] this stengthened the supposition that the GSH dependent inhibition of microsomal lipid peroxidation is catalyzed via the microsomal enzyme GSH S-transferase, which possesses GSH peroxidase enzymatic activity as well. We think that this hypothesis is unlikely since the rank order of IC_{50} values for both enzymatic activites differed. Moreover N-ethyl maleimide known to activate microsomal GSH S-transferase[19], thereby enhancing the GSH peroxidase activity too[18], completely blocks the GSH dependent inhibition of lipid peroxidation (Fig. 6). Also other manipulations such as CCl_4 treatment and limited proteolysis known to activate the microsomal GSH S-transferase[20], inhibit the GSH dependent protection against lipid peroxidation[4,21]. Our conclusion therefore is that the microsomal GSH S-transferase (also indicated as a Se-independent GSH peroxidase) is not responsible for the GSH effect on microsomal lipid peroxidation.

Since the GSH S-transferase is not involved, the conclusion reached by Yonaha and Tampo[22] stating: "It seems likely that the vitamin E radical that is produced by scavenging lipid radicals during peroxidation is regenerated to vitamin E by the microsomal GSH S-transferase in the presence of GSH", surely is invalid.

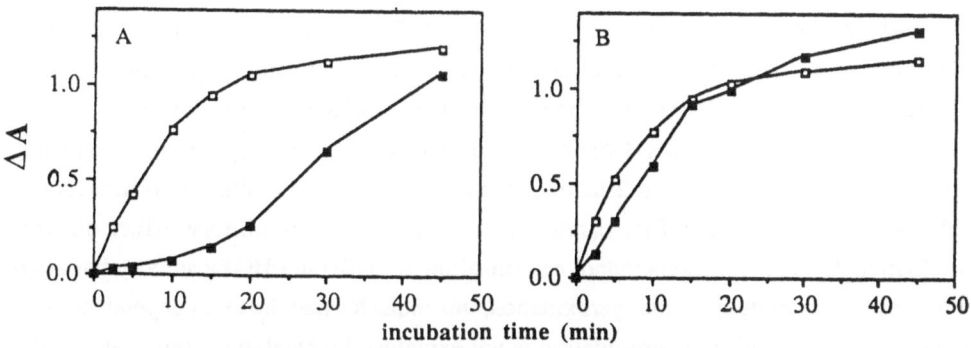

Fig. 6 Time course of lipid peroxidation (see legend fig. 5) in control (A) and 0.1 mM N-ethyl maleimide pretreated microsomes (B). Microsomes were preincubated for 30 min. at 37°C in 50 mM sodium phosphate, 0.1 mM EDTA buffer (pH 7.4) with vehicle (A) or 0.1 mM N-ethyl maleimide (B). Incubation as in Fig. 5 with (■) or without (□) 1 mM GSH using 10 μM $FeSO_4$ and 0.2 mM ascorbate to stimulate lipid peroxidation.

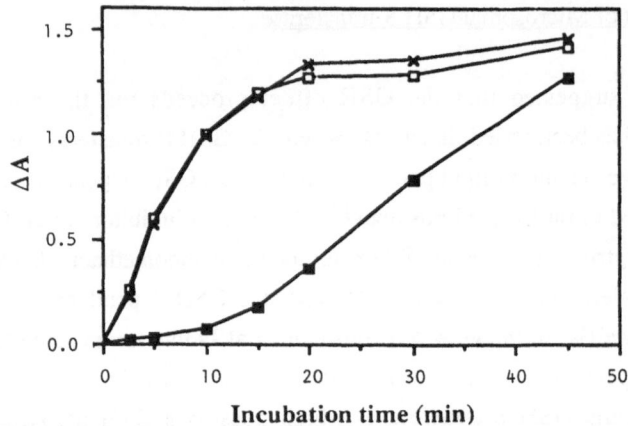

Fig. 7   Time course of lipid peroxidation. Incubation as in Fig. 5 but lipid peroxidation was stimulated with 10 μM FeSO$_4$ and 0.2 mM ascorbate. Symbols represent lipid peroxidation in control microsomes (□), in the presence of 1 mM GSH (■) and in the presence of 1 mM 2-mercaptoethanol (x).

IIId Involvement of a Phospholipid Hydroperoxide Glutathione Peroxidase (PHGPx)

In 1982 Ursini and coworkers[23] reported the isolation of a cytosolic protein that was involved in the inhibition of lipid peroxidation. At that time the protein was denominated as PIP, abbreviation for Peroxidation Inhibiting Protein. After establishing the catalytic activity of the enzyme, it was characterized as a phospholipid hydroperoxide glutathione peroxidase (PHGPx). The enzyme has been identified in for example liver, brain, heart, kidney and lung. Ursini et al.[24] were of the opinion "that the peroxidase activity of PHGPx accounts for the almost complete inhibition of lipid peroxidation, as a component of an antiperoxidant system where a key role is played also by vitamin E". This implies that the GSH dependent protection is in fact due to the activity of PHGPx. Initial experiments using Fe^{3+}-triethylenetetramine promoted peroxidation of phosphatidylcholine liposomes showed that in the protection by PHGPx, GSH as cofactor could be replaced by 2-mercaptoethanol. However, we found (Fig. 7) that 2-mercaptoethanol did not give inhibition of microsomal lipid peroxidation and therefore cannot replace GSH in its inhibitory effect on lipid peroxidation. Since 2-mercaptoethanol can substitute for GSH in PHGPx activity but not in the microsomal inhibition of lipid peroxidation, this indicates that the GSH dependent effect on hepatic microsomal lipid peroxidation is not explained by catalytic activity of PHGPx. Moreover, PHGPx is Se-dependent[24], whereas a Se deficient diet does not affect the GSH dependent protection[25]. Recent data (Chapter by Ursini et al. in this volume) indicate that PHGPx only slowly and partly looses its Se, which weakens this reasoning. It might implicate that although a Se deficient diet has been given, PHGPx might still retain its Se content.

A further argument against a role of PHGPx in the GSH effect on microsomal lipid peroxidation is its seemingly ubiquitous presence, whereas the GSH protection seems to be limited to liver microsomes. We were unable to detect any effect of GSH in brain, heart and lung (data not shown).

IIIe Characteristics of the GSH dependent Protective Effect in Liver Microsomes

The foregoing paragraphs disprove the assumed involvement of iron complexation by GSH, the interaction with available ascorbate, the microsomal GSH S-transferase or the phospholipid hydroperoxide GSH peroxidase in the GSH dependent protective activity in microsomal lipid peroxidation. In our opinion some indisputable characteristics of the GSH dependent inhibition of microsomal lipid peroxidation can be provided as well. The effect is heat and proteolytic labile[4,21] and is vitamin E dependent[25]. This made us hypothesize the existence of a heat labile GSH dependent vitamin E reductase[4].

One particular aspect of the GSH effect puzzled us for some time. GSH only prolonged the lag phase of lipid peroxidation and did not prevent the oxidative deterioration of polyunsaturated fatty acids completely. Remarkably, the GSH consumption was minimal[4]. However, although the GSH level was still high enough to maintain inhibition, lipid peroxidation commences. During lipid peroxidation, membranous vitamin E levels decrease to almost zero[26]. If we assume that the microsomal factor were performing a GSH dependent recycling of vitamin E (Fig. 8) then the presence of GSH in the microsomal system would prevent vitamin E consumption. Apparently this is not the case. The vitamin E concentration declines whereas the GSH level is still high enough to prevent vitamin E consumption. We therefore assumed that the factor is vulnerable to oxidative stress[4]. Indeed, applying a limited level of lipid peroxidation already leads to destruction of the GSH dependent effect[4]. Recently[19], we corroborated this finding by showing that 4-hydroxy-2,3-trans- nonenal, an aldehyde formed in large quantities during microsomal lipid peroxidation inhibits the GSH effect. We suggest that the GSH-vitamin E radical oxidoreductase contains an essential sulfhydryl moiety which is sensitive to oxidative stress. 4-Hydroxy-2,3-trans- nonenal might alkylate this SH-moiety which leads to a blockade of its catalytic activity (Fig. 8).

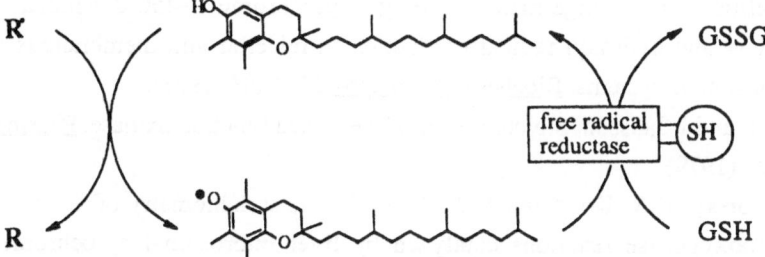

Fig. 8 Proposed mechanism for the GSH dependent protection against microsomal lipid peroxidation. Free radicals (R•) in the membrane may arise from the cytochrome P-450 system and are scavenged by vitamin E, which is regenerated by a vitamin E free radical reductase. This GSH dependent enzyme might contain a functionally important sulflhydryl moiety.

IV CONCLUSIONS

Liver microsomes seem to be furnished with a very effective system to protect the membrane from oxidative damage. The plethora of reactive oxygen species emanating from the P-450 system teleologicly necessitates the presence of such an antioxidant system. The membrane bound vitamin E free radical reductase offers the possibility for an interplay between the cytosolicly located GSH and the membrane bound P-450. The cooperation between GSH and P-450 which reside in different cell compartments, seems to imply a living-apart-together (LAT) relationship between GSH and P-450. This might be an unique feature of the hepatic endoplasmatic reticulum. The vitamin E free radical reductase might even function as a more general free radical reductase, signifying the hypothesized intimate association of this factor to the presence of P-450. The unstable character of this presumed vitamin E free radical reductase obviously will hamper its purification.

REFERENCES

1. P. Hochstein and L. Ernster, ADP-activated lipid peroxidation coupled to the TPNH oxidase system of microsomes. Biochem. Biophys. Res. Commun. 12:388 (1963).

2. A. Bast and G.R.M.M. Haenen, Cytochrome P-450 and glutathione: What is the significance of their interrelationship in lipid peroxidation? Trends Biochem. Sci. 9:510 (1984).

3. A. Bast, Is formation of reactive oxygen by cytochrome P-450 perilous and predictable? Trends Pharmacol. Sci. 7:266 (1986).

4. G. R.M.M. Haenen and A. Bast, Protection against lipid peroxidation by a microsomal glutathione-dependent labile factor. FEBS Lett. 159:24 (1983).

5. D. W. Nebert, M. Adesnik, M.J. Coon, R.W. Estabrook, F.J. Gonzalez, F.P. Guengerich, I.C. Gunsalus, E.F. Johnson, B. Kemper, W. Levin, I.R. Phillips, R. Sato and M.R. Waterman, The P-450 gene superfamily: recommended nomenclature. DNA 6:1 (1987).

6. Y. Terelius and M. Ingelman-Sundberg, Cytochrome P-450-dependent oxidase activity and hydroxyl radical production in micellar and membranous types of reconstituted systems. Biochem. Pharmacol. 37: 1383 (1988).

7. G. Powis and I. Jansson, Stoichiometry of the mixed function oxidase. Pharmac. Ther. 7:297 (1979).

8. L. D. Gorsky, D.R. Koop and M.J. Coon, On the stoichiometry of the oxidase and monooxygenase reactions catalyzed by liver microsomal cytochrome P-450. Products of oxygen production. J. Biol. Chem. 259:6812 (1984).

9. A. Bast, E.M. Savenije-Chapel and B.H. Kroes, Inhibition of mono-oxygenase and oxidase activity of reat-hepatic cytochrome P-450 by H_2-receptor blockers. Xenobiotica 14:399 (1984).

10. A. Bast, J.W. Brenninkmeijer, E.M. Savenije-Chapel and J. Noordhoek, Cytochrome P-450 oxidase activity and its role in NADPH dependent lipid peroxidation. FEBS Lett. 151:185 (1983).

11. A. Bast and M.H.M. Steeghs, Hydroxyl radicals are not involved in NADPH dependent microsomal lipid peroxidation. Experientia 42:555 (1986).

12. L. A. Morehouse, M. Tien, J.R. Bucher and S.D. Aust, Effect of hydrogen peroxide on the initiation of microsomal lipid peroxidation. Biochem. Pharmacol. 32:123 (1983).

13. G. Minotti and S.D. Aust, The role of iron in the initiation of lipid peroxidation. Chem. Physics of Lipids 44:191 (1987).

14. H. Wefers and H. Sies, The protection by ascorbate and glutathione against microsomal lipid peroxidation is dependent on vitamin E. Eur. J. Biochem. 174:353 (1988).

15. A. Bast, G.R.M.M. Haenen and E.M. Savenije-Chapel, Inhibition of rat hepatic microsomal lipid peroxidation by mesna via glutathione. Arzneim.-Forsch./Drug Res. 37:1043 (1987).

16. L. G. Forni and R.L. Willson, Vitamin C and consecutive hydrogen atom and electron transfer reactions in free radical protection: A novel catalytic role for glutathione, in: "Protective agents in cancer", D.C. McBrien and T.F. Slater eds., Academic Press, New York (1983).

17. R. Morgenstern, G. Lundqvist, V. Hancock and J.W. DePierre, Studies on the activity and activation of rat liver microsomal glutathione transferase, in particular with a substrate analogue series. J. Biol. Chem. 263:6671 (1988).

18. E. Mousialou and R. Morgenstern, Studies on the glutathione dependent inhibition of lipid peroxidation. Abstract 3.15 in Eur. Workshop on Drug Metab., Univ. of Konstanz, F.R.G. (1988).

19. G. R.M.M. Haenen, J.N.L. Tai Tin Tsoi, N.P.E. Vermeulen, H. Timmerman and A. Bast, 4-Hydroxy-2,3-trans-nonenal stimulates microsomal lipid peroxidation by reducing the gluthathione-dependent protection. Arch. Biochem. Biophys. 259:449 (1987).

20. R. Morgenstern, H. Wallin and J.W. DePierre, Mechanisms of activation of the microsomal glutathione transferase, in "Glutathione S-transferases and carcinogenesis", T.J. Mantle, C.B. Pickett and J.D. Hayes eds., Taylor and Francis, London, New York and Philadelphia (1987).

21. R. F. Burk, Glutathione-dependent protection by rat liver microsomal protein against lipid peroxidation. Biochim. Biophys. Acta 757:21 (1983).

22. M. Yonaha and Y. Tampo, Studies on protection by glutahione against lipid peroxidation in rat liver microsomes. Effect of bromosulfophtalein. Chem. Pharm. Bull. 34:4195 (1986).

23. F. Ursini, M. Maiorino, M. Valente, L. Ferri and C. Gregolin, Purification from pig liver of a protein which protects liposomes and biomembranes from peroxidative

degradation and exhibits glutathione peroxidase activity on phosphatidylcholine hydroperoxides. Biochim. Biophys. Acta 710:197 (1982).

24. F. Ursini and A. Bindoli, The role of selenium peroxidases in the protection against oxidative damage of membranes. Chem. Physics of Lipids. 44:255 (1987).

25. C. C. Reddy, R.W. Scholz, C.E. Thomas and E.J. Massaro, Vitamin E dependent reduced glutathione inhibition of rat liver microsomal lipid peroxidation. Life Sci. 31:571 (1982).

26. M. E. Murphy and J.P. Kehrer, Simultaneous measurement of tocopherols and tocopheryl quinones in tissue fractions using high-performance liquid chromatography with redoxcycling electrochemical detection. J. Chromatogr. (Biomed. Applicat.) 421:71 (1987).

FREE RADICAL FORMATION AND ANTIOXIDANT PROTECTION IN EXTRACELLULAR FLUIDS

John M. C. Gutteridge*

Molecular Toxicology Research Group
Oklahoma Medical Research Foundation
825 N. E. 13th Street, Oklahoma City, Oklahoma 73104, USA

OXYGEN TOXICITY

The present oxygen concentration of dry air is 21% (v/v) and aerobic life processes utilize molecular oxygen for the controlled oxidation of carbon-containing molecules in order to release energy and heat. Molecular oxygen is itself a radical with two unpaired electrons with parallel spins. Most incoming electrons would have opposite spins and therefore be restricted in their reactions with oxygen. This restriction considerably slows down the reaction of oxygen with non-radicals. Unfortunately, it also results in the formation of univalent reduction intermediates with unpaired electrons such as the hydroperoxyl radical (HO_2^-), superoxide anion (O_2^-) and the hydroxyl radical ($\cdot OH$). At physiological pH values HO_2^- will yield the superoxide anion which is also formed in numerous metabolic processes (for reviews see Halliwell and Gutteridge 1985)[1,2]. Superoxide is able to do some direct damage in biological systems[3] but is not considered a particularly aggressive oxidant. Nevertheless generation of O_2^- has consistently been shown to accompany significant damage to biological molecules.[4,5] The currently preferred explanation is that damage is usually mediated by a highly aggressive oxidant such as the hydroxyl radical. The chemical reaction sequence leading to $\cdot OH$ formation has been shown to involve hydrogen peroxide (H_2O_2) and a transition metal complex, with a variable oxidation number, and is known as the 'Superoxide-driven Fenton' reaction.

Lipid Peroxidation

Most complex organic molecules are susceptible to damage by oxidant species such as $\cdot OH$ and, one of the most studied of these reactions is the auto-oxidation of polyunsaturated fatty acid (PUFA) components of cell and organelle membranes. Abstraction of a hydrogen atom from an unsaturated lipid initiates peroxidation. Oxygen uptake by the resulting carbon-centred radical ($R\cdot$) rapidly follows to form an organic peroxy radical (RO_2^-) and this radical can itself abstract a hydrogen atom from another

---

*Permanent Address:
Division of Chemistry, National Institute for Biological
Standards and Control, Blanche Lane, South Mimms,
Potters Bar, Hertfordshire, EN6 3QG, UK.

PUFA forming a stable lipid peroxide (ROOH). By this process a propagation
of radical damage takes place as the carbon-centred radical (R·) formed in
the latter sequence continues the process described above (for a review see
Gutteridge 1988,)[6]. Most of the lipid peroxidation detected in vitro and,
possibly in vivo, probably results from metal stimulated decomposition of
existing lipid peroxides (Eq.1.).

$$2ROOH \xrightarrow{\text{Transition metal complexes}} RO· + RO_2^· + H_2O \quad [1]$$

Transition Metals Ions In Biological Fluids

As previously mentioned, molecular oxygen reacts only slowly with non-
radicals because of the spin restriction. However, there is no such restriction when molecules, atoms or ions can change their oxidation number by a
single electron transfer. Transition metal ions, particularly iron and
copper are for this reason intimately involved in all aspects of oxygen
metabolism. Uncertainty about the availability of suitable metal ions to
catalyse ·OH radical formation in biological systems has largely been
resolved by the observation that superoxide, hydrogen peroxide and organic
peroxides can 'release' iron from iron-containing proteins, such as
ferritin, haemoglobin and myoglobin (for a review see Ref. 6), that can be
complexed by desferrioxamine, apotransferrin, bleomycin and other
chelators. This released iron can stimulate radical formation [7,8].

BIOLOGICAL ANTIOXIDANT PROTECTION

The term anti-oxidant can be used to describe any substance that
inhibits or delays an oxidative sequence when present at concentrations
considerably lower than the oxidizable substrate. Several ways in which
anti-oxidants can inhibit the oxidation of lipid are shown in the Table
with examples of biological mechanisms operating in vivo.

Table

(a) Some important ways in which antioxidants can act.

1. Removal of oxygen

2. Removal of metal catalysts

3. Removal of key intermediates ie. H_2O_2, O_2^-

4. Breaking the chain reaction of an initiated sequence

(b) Levels of protection in living systems

1. Prevention of radical formation (2-3 above)

2. Interception of formed radicals (4,5 above)

3. Repair of oxidative damage

4. Elimination and non-repair of excessively damaged
    molecules

Protection Inside Cells

Oxygen is metabolised inside cells and it is here that protein anti-oxidants have evolved to deal specifically and speedily (catalytically) with products of reduced oxygen. Enzymes such as the superoxide dismutases (SOD) catalase and glutathione peroxidase (selenium enzyme) function in a co-ordinated way to eliminate reduction intermediates of oxygen (Eq 2-4)

$$2\ O_2^- + 2H^+ \quad \xrightarrow{\hspace{1em} SOD \hspace{1em}} \quad H_2O_2 + O_2 \quad [2]$$

$$2\ H_2O_2 \quad \xrightarrow{\hspace{1em} CATALASE \hspace{1em}} \quad 2H_2O + O_2 \quad [3]$$

$$2\ GSH + H_2O_2 \quad \xrightarrow{\hspace{1em} GSHPase \hspace{1em}} \quad 2H_2O + GSSG \quad [4]$$

The safe removal of oxygen intermediates should allow a low molecular mass pool of iron, required for the synthesis of iron-containing proteins, to safely exist inside the cell.

Membrane Protection

Different types of radicals will be formed in the lipid interior of membranes and special anti-oxidants are required to deal with them. Lipid soluble molecules such as α-tocopherol (vitamin E), β-carotene and possibly cholesterol fulfill some of these functions in animal and plant membranes. An extremely important feature of membrane protection appears to be the "structural integrity" of the membrane, requiring that the correct ratios of phospholipid and cholesterol are present as well as the correct phospholipids and their fatty acid side chains[9].

Extracellular Anti-oxidant Protection

We do not find enzymes like the intracellular SODs, catalase and glutathione peroxidase in extracellular fluids and the concentrations of GSH are very low. Nevertheless, extracellular fluids are often subjected to fluxes of superoxide and hydrogen peroxide by 'activated' phagocytic cells and by substrate oxidations. As a result of studies in our own laboratories over the last 15 years we have proposed that extracellular anti-oxidant defences substantially depend on mechanisms that remove or inactive reactive transition metal complexes, particularly iron, before they can form more aggressive and damaging oxidants[9,10]. In this way it may be possible to carefully control the levels of O_2 and H_2O_2 in extracellular fluids so that they can act as signal or trigger molecules between cells. Numerous low molecular mass chemicals exist in extracellular fluids such as ascorbate, urate, glucose, and bilirubin which have the ability to act as scavengers and chain-breaking anti-oxidants.

MAJOR PROTEIN ANTI-OXIDANTS IN EXTRACELLULAR FLUIDS

(A) Proteins that bind metal complexes

Transferrin. Apotransferrin binds two moles of ferric ion per mole of protein with high affinity producing a coloured complex absorbing at 460 nm. Under normal physiological conditions transferrin is only partly (one third) loaded with iron retaining a substantial iron-binding capacity. It is this iron-binding potential which accounts almost entirely for its anti-oxidant properties in iron-stimulated free radical reactions[9,10]. The protein continues to function effectively as an anti-oxidant even when subjected to considerable oxidant damage[11].

Several reports suggest that iron-loaded transferrin can promote Fenton chemistry or lipid peroxidation (for a review see Ref. 12). However, a recent report has shown that iron-loaded transferrin does not accelerate radical reactions unless an iron chelator is present and the protein has been incorrectly loaded with iron[12].

Patients with active arthritic diseases, when compared with normal patients, often show decreased plasma levels of transferrin which has a lower percentage iron saturation. The increased iron-binding capacity together with raised levels of the protein caeruloplasmin make the arthritic group appear to have higher levels of anti-oxidant activity. However, when knee-joint synovial fluids are taken and examined from the same group of patients, some 40% show the presence of iron complexable to bleomycin under the assay conditions used[13]. Patients whose fluids release iron to bleomycin form a distinct group with generally lower levels of transferrin, lactoferrin and caeruloplasmin. These proteins are important extracellular anti-oxidants and measured anti-oxidant activities are extremely low in the iron-releasing fluids as shown in Figure 1.

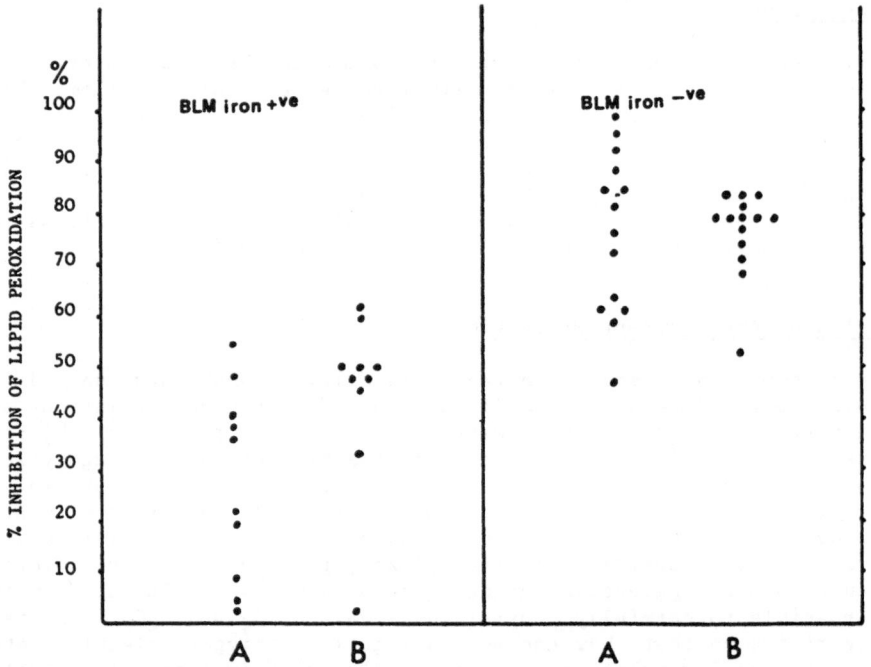

Figure 1   The ferroxidase 'A' and iron-binding 'B' antioxidant activities of synovial fluids from arthritic patients. Results are shown as the % inhibition of peroxidising phospholipid stimulated by iron salts.

BLM iron = bleomycin-detectable iron

(+) (−) = the presence (+) or absence (−) of detectable iron
(Data from Ref. 13).

In the iron overload condition idopathic haemochromatosis the serum transferrin levels are low and the protein is often at or close to saturation with iron. Hence, iron-binding anti-oxidant activities are extremely low and complexable iron is often present[9,14] in plasma. The plasma levels of caeruloplasmin are, however, elevated producing increased ferroxidase I activity and giving a substantial anti-oxidant activity[9,10]. Patients with thalassaemia undergoing treatment by blood transfusions develop a transfusional iron overload which saturates their transferrin with iron greatly reducing the iron-binding anti-oxidant activity of their plasma[9,10].

Lactoferrin. Lactoferrin is actively secreted by neutrophils into surrounding fluids and, like transferrin can bind two moles of iron per mole of protein with high affinity but unlike transferrin can hold on to this iron down to pH values of 4.0[15,16]. Lactoferrin is normally only partly loaded with iron and like transferrin it displays potent anti-oxidant activities when radical formation is stimulated by iron[9,10].

Haptoglobins. Haptoglobins are glycoproteins found in the α-1-globulin fraction of serum that respond as 'acute-phase' proteins. They bind haemoglobin (both oxy and met) in a 1:1 ratio to form a stable complex which has one of the strongest non-covalent protein bonds known. The normal level of circulating haptoglobin is sufficient to bind some 3g of haemoglobin making sure that no free haemoglobin is normally present in plama[17]. Free haemoglobin has the potential to stimulate lipid peroxidation as well as to be degraded and release low molecular mass iron able to participate in Fenton Chemistry[6]. The ability of haptoglobins to inhibit haemoglobin stimulated lipid peroxidation is shown in Figure 2.

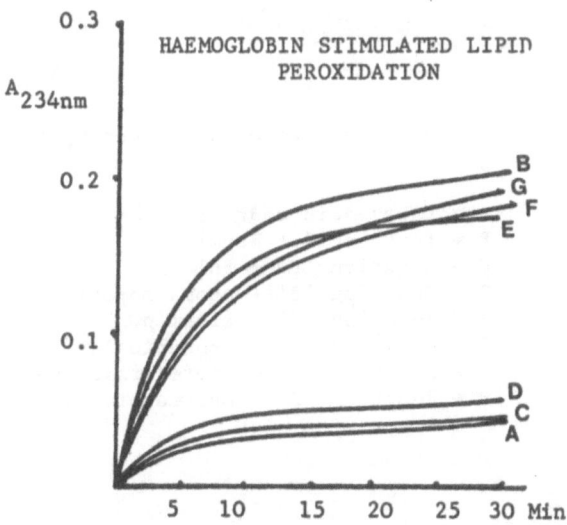

Figure 2   Lipid peroxidation (measured as diene conjugation) stimulated by haemoglobin.

        A = Fatty acid control
        B = Fatty acid + methaemoglobin + desferrioxamine
        C = Reaction 'B' + haptoglobins
        D = Reaction 'B' + BHT
        E = Reaction 'B' + apohaemopexin
        F = Reaction 'B' + apotransferrin
        G = Reaction 'B' + albumin

(Data abstracted from Gutteridge, J.M.C. (1987) Biochim. Biophys. Acta. 917, 219-223).

Haemopexin. Haemopexin is a plasma β-glycoprotein that binds haem tightly in a 1 : 1 ratio to form a pink-coloured complex. When delivering haem to cells the haemopexin molecule is not degraded and returns to the circulation as an intant protein[18]. Haem iron is active in several radical reactions (for references see 19) including lipid peroxidation; a process which is strongly inhibited by haemopexin (Figure 3)[19].

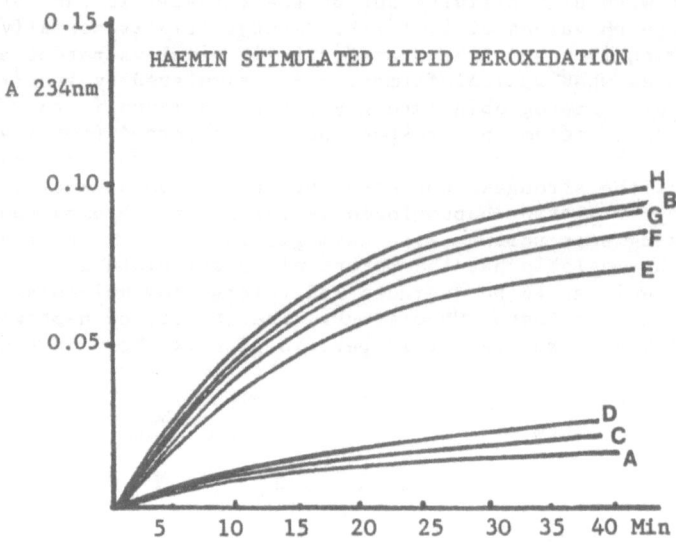

Figure 3   Lipid peroxidation (measured as diene conjugation) stimulated by haemin.

```
A = fatty acid control
B = fatty acid + haemin
C = Reaction 'B' + BHT
D = Reaction 'B' + apohaemopexin
E = Reaction 'B' + albumin
F = Reaction 'B' + apohaptoglobin
G = Reaction 'B' + desferrioxamine
H = Reaction 'B' + apotransferrin
```

(Data abstracted from Gutteridge, J.M.C. and Smith, A. Biochem. J. (1988) 256, 861-865).

Albumin.   Albumin is a highly water-soluble protein and a major constituent of human plasma, synovial fluid and other extracellular fluids. It has important binding, transporting, 'solubilizing' and osmotic properties in the body. Human albumin has one high affinity copper-binding site but, like most other proteins readily and non-specifically binds copper ions at many other sites. Albumin like most other proteins will therefore effectively inhibit copper stimulated radical damage to any non-protein detector molecule. However, under such circumstances radical damage usually takes place at or close to the copper binding site on the protein. Copper binding by proteins provides the clearest evidence of site-specific metal-dependant radical damage[20] although, the same principles apply to iron-binding[9]. At present it is not clear whether damage is done by the ·OH radical or a higher oxidation state of iron or copper such as ferryl (iron IV) or cupryl (copper III). Albumin can also scavenge hypochlorous acid (HOCl)[21], peroxy radicals[22] and decrease damage done by an iron salt[23]. The large amount of albumin present in fluids, its high turnover and the resiliance of the molecule to impaired biological functions make it an important sacrificial anti-oxidant for protecting critical biological targets from oxidative damage[9].

(B) Proteins that inactivate metal complexes or oxygen intermediates

Caeruloplasmin. Caeruloplasmin is found in the α-2-globulin fraction of mammalian plasma and is a glycoprotein containing 6-7 copper ions per molecule. Caeruloplasmin can catalyse the oxidation of many polyamines and polyphenols although these oxidations have no known biological significance. The catalytic activity of caeruloplasmin to oxidise ferrous ion (ferroxidase activity) was first described in 1966[24] and suggested to be an important physiological mechanism for incorporating iron into transferrin[25]. However, the true physiological role for caeruloplasmin is still the subject of debate. The oxidation of ferrous ions catalysed by caeruloplasmin results in the complete co-reduction of oxygen to water (addition of four electrons) ensuring that no reactive oxygen intermediates are released into free solution. This has been considered an important anti-oxidant role for caeruloplasmin during normal iron metabolism[26]. (Eq. 5-6).

$$Fe^{2+} + Cu^{2\pm}Cp \text{ (Caeruloplasmin)} \text{ --------} > + Cu^{\pm}Cp + Fe^{3+} \quad [5]$$

$$4 Cu^{\pm}Cp + O_2 + 4 H^+ \text{ ---------------} > \quad 4 Cu^{2\pm}Cp + 2H_2O \quad [6]$$

Fractionation studies on human serum have shown that caeruloplasmin and transferrin, representing only 4% of the total proteins present in serum, account for most of the anti-oxidant activity of human serum (when tested for its ability to inhibit the peroxidation of a bovine brain homegenate)[27].

Feroxidase Activity. The ferroxidase activity of freshly taken human plasma can be almost entirely attributed to caeruloplasmin (ferroxidase I). A second ferroxidase activity 'ferroxidase II'[28] appears and increases in human plasma as an artefact of storage[29]. Ferroxidase II is a copper-lipid-protein complex and is probably formed when copper is proteolytically released from caeruloplasmin[29].

A major anti-oxidant activity of caeruloplasmin can be attributed to its ferroxidase activity removing ferrous ions before they can participate in lipid peroxidation or Fenton chemistry (section I). Whatever tissues are damaged it is highly likely that metal ions will be released from safely sequestered sites (mal-placed iron) and radical reactions become more likely. By responding to tissue damage as an acute phase protein caeruloplasmin plays an important role as a protective anti-oxidant. Such a protective role has been proposed for caeruloplasmin, in rheumatioid arthritis[26]. Low levels of caeruloplasmin, hence low levels of anti-oxidant activity are seen in newborn infants and patients with Wilson's disease[26].

Scavenging of Radicals. A catalytic role for caeruloplasmin as a scavenger of superoxide radicals has been proposed[30]. However, others have found no such role[31,32] and concluded that the reaction is stoichiometric in nature[33]. It is likely that batches of caeruloplasmin will be contaminated with traces of extracellular SOD and this could account for some scavenging detected in certain studies[34].

Copper Inactivation. Caeruloplasmin has multiple binding sites for divalent transition metal ions[35] in addition to the specific site for its ferroxidase activity i.e., ferrous ions. Caeruloplasmin appears to be able to effectively inhibit copper stimulated lipid peroxidation several hundred times more effectively than an equimolar amount of albumin[36] although the mechanisms involved have not yet been identified.

Hydrogen Peroxide Removal. Recent work has shown that one of the 'blue-coppers' in caeruloplasmin can not be re-oxidized by oxygen but can be by hydrogen peroxide whereas, the other blue-copper is re-oxidized by both oxygen and hydrogen peroxide[37]. When copper ions are redox cycling in caeruloplasmin, hydrogen peroxide can be consumed. This would explain why caeruloplasmin is such a good inhibitor of the Fenton reaction since both ferrous ions and hydrogen peroxide are removed.

Extracellular (EC) Superoxide Dismutase (EC-SOD) and Glutathione Peroxide. A superoxide dismutase distinct from all other known SODs was recently described[38]. It has a molecular mass similar to that of caeruloplasmin and is also a glycoprotein. However, there are no antigenic or structural similarities. The concentration of EC-SOD in plasma is extremely low and the true site of location may be on the endothelial cell surface[34].

Similarly an extracellular glycosylated form of glutathione peroxidase has also recently been isolated which is immunologically distinct from the intracellular enzyme[39,40] but, also contains selenium at its active centre. The enzyme is present in low concentrations in human plasma and its substrate GSH is not normally available in plasma making its physiological role at present unclear.

SUMMARY

It appears that a main function of extracellular anti-oxidant activity is the efficient inactivation or removal of metal catlaysts likely to stimulate the formation of damaging and reactive forms of oxygen. This may allow superoxide and hydrogen peroxide to act as trigger or signal molecules between cells.

REFERENCES

1. B. Halliwell and J. M. C. Gutteridge, Free Radicals in Biology and Medicine, Oxford University Press, Oxford (1985).
2. B. Halliwell and J. M. C. Gutteridge, Mol. Aspects Med, 8; 89-193.
3. I. Fridovich, Arch. Biochem. Biophys. 247; 1-11 (1988).
4. I. Fridovich, Adv. Enzymol. 41; 35-48 (1974).
5. I. Fridovich, Science 209, 875-877 (1978).
6. J. M. C. Gutteridge, In Oxygen Radicals and Tissue Injury, Proc. UpJohn Symposium, Michigan, pp.9-19 (1987).
7. J. M. C. Gutteridge, FEBS Lett. 201, 291-295 (1986)
8. A. Puppo and B. Halliwell, Free Rad. Res. Commun. 4, 415-422 (1988)
9. J. M. C. Gutteridge, T. Westermarck and B. Halliwell, In Free Radicals, Aging and Degenerative Diseases, J. E. Johnson, Jr., R. Walford, D. Harmar and J, Miquel, eds., Alan R. Liss Inc., New York, pp. 99-139. (1985).
10. J. M. C. Gutteridge and B. Halliwell, In Cellular Antioxidant Defense Mechanisms, H. Chow, ed., Vol. II, in press (1988).
11. J. M. C. Gutteridge, Biochim. Biophys. Acta., 869, 119-127, (1986).
12. O. I. Aruoma and B. Halliwell, Biochem. J. 241, 273-278 (1987).
13. J. M. C. Gutteridge Biochem. J. 245, 415-421 (1987).
14. R. G. Batey, P. L. C. Fong, S. Shamir and S. Sherlock, Dig. Dis. Sci. 25, 40-60 (1980).
15. M. L Groves, J. Amer. Chem. Soc. 82, 3345-3350 (1960).
16. B. C. Johansson, Acta Chem. Scanda. 14, 341-350 (1960).
17. F. W. Putnam, In the Plasma Protiens, F. W. Putnam, ed., Academic Press, New York, pp. 1-50 (1975).
18. A. Smith and W. T. Morgan, Biochem. J. 182, 47-54 (1979).
19. J. M. C. Gutteridge and A. Smith, Biochem. J., In Press, (1988).

20. A. Samuni, M. Chevion and G. Czapski, J. Biol. Chem. 256, 12632-12635 (1981).
21. M. Wasil, B. Halliwell, D. C. Hutchison and H. Baum, Biochem. J. 243, 219-227 (1987).
22. D. D. M. Wayner, G. W. Burton, K. U. Ingold and S. Locke, FEBS Lett. 187, 33-37 (1985).
23. F. J. Carver, D. Farb and E. Frieden, Biol. Tr. Elem. Res. 4, 1-19 (1981).
24. S. Osaki, D. A. Johnson and E. Frieden, J. Biol. Chem. 241, 2746-2751 (1966).
25. E. Frieden and S. Osaki, In Effects of metals subcellular elements and macromolecules, J. Mariloff, J. R. Coleman and M. W. Miller, eds., (C. C. Thomas, Springfield, III.) Chapter 3 (1970).
26. J. M. C. Gutteridge and J. Stocks, CRC Crit. Rev. Clin. Lab. Sci. 14, 257-329 (1981).
27. J. Stocks, J. M. C. Gutteridge, R. J. Sharp and T. L. Dormandy, Clin. Sci. Mol. Med. 47, 223-233 (1974).
28. R. W. Topham and E. Frieden, J. Biol. Chem. 245, 6698-6705 (1970).
29. J. M. C. Gutteridge, P. G. Winyard, D. R. Blake, J. Lunec, S. Brailsford and B. Halliwell, Biochem. J. 230, 517-523 (1985).
30. I. M. Goldstein, H. B. Kaplan, H. S. Edelson and G. Weissmann, J. Biol. Chem. 254, 4040-4045 (1979).
31. D. R. Blake, N. D. Hall, D. A. Treby, B. Halliwell and J. M. C. Gutteridge, Clin. Sci. 61, 483-486 (1981).
32. J. V. Bannister, W. H. Hill, H. A. O. Hill, J. F. Mahood, R. L. Wilson and B. S. Wolfenden, FEBS Lett. 118, 127-129 (1980).
33. B. Halliwell and J. M. C. Gutteridge, Lancet ii, 556. (1982).
34. S. L. Markland, J. Free Rad. Biol. Med. 2, 255-260 (1986).
35. D. J. McKee and E. Frieden, Biochemistry 10, 3880-3883.
36. J. M. C. Gutteridge, C. Hill and D. R. Blake, Clin. Chim. Acta. 139, 85-90 (1984).
37. L. Calabrese and M. Carboraro, Biochim. J. 238, 291-295 (1986).
38. S. L. Markland, E. Holme and L. Hellner, Clin. Chim. Acta. 126, 41-51 (1982).
39. K. R. Maddipati, C. Gasparski and L. J. Marnett, Arch. Biochem. Biophys. 254, 9-17 (1987).
40. K. Takahashi, N. Avissar, J. Whitin and H. Cohen, Arch. Biochem. Biophys. 256, 677-686 (1987).

QUINONE REDOX CYCLING AND THE PROTECTIVE EFFECT OF DT DIAPHORASE

Helmut Sies

Institut für Physiologische Chemie I, Universität Düsseldorf
Moorenstrasse 5, D-4000 Düsseldorf, West-Germany

The one-electron oxidation-reduction cycling of quinones, semiquinones and hydroquinones contributes to the generation of reactive oxygen species and, consequently, toxicity, carcinogenicity and cell damage, e.g. to enzymes, DNA and membranes. In this process of redox cycling, interest has focused on the biochemistry, molecular biology, toxicology and physiology of the enzyme, NADPH:quinone oxidoreductase, also known as DT diaphorase, catalyzing the two-electron reduction of quinones.

A variety of endogenous and exogenous compounds exert cytotoxic effects via oxygen reduction. In general, these compounds are reduced by intracellular reductases in one-electron transfer reactions before they in turn reduce O_2 to $O_2^{\cdot}$, the superoxide anion radical. Thus, a cycle is formed of O_2 uptake at the expense of cellular reducing equivalents, notably NADPH, generating reactive oxygen species. Structures capable of redox cycling also include catechols, iron chelates and aromatic nitro compounds. Several anticancer agents and some mutagens operate on this principle, and their toxic effects may be explained in part by redox cycling. The particular importance of hypoxic conditions for deleterious O_2 effects is given by the concomitant flux through reductive as well as oxidative pathways (1). There are multiple lines of defense against toxic oxygen intermediates, and one of them is the NADPH:quinone oxido-reductase, also known as DT diaphorase. This enzyme activity circumvents the semiquinone reduction step by reducing the quinone directly to the hydroquinone in one two-electron reduction (2). The hydroquinone may then be eliminated either directly or after processing by conjugation to glucuronic acid or sulfate. The quinone reductase has been studied from many points of view (3), and a brief overview of this field has been given (4).

Redox cycling quinones occur in biological systems mainly as derivatives of benzoquinone, naphthoquinone, and anthraquinone. These include anticancer drugs such as mitomycin C, quinoneimines, and the antimicrobial drug beta-lapachone, different anthraquinone compounds such as adriamycin, the herbicide paraquat, and a number of aromatic nitro compounds such as nitrofurantoin, nifurtimox, benznidazole, misonidazole, metronidazole, nitrazepam. A common denominator of these compounds seems to reside in their capability of redox cycling via enzymatic reduction and superoxide anion radical formation. There has been a correlation between the rate of redox cycling and toxicity. However, toxicity is dependent on other parameters such as the hydrophobicity ratio (5) and also on the specific targeting of the reactive species. For example, the bleomycin-iron-complex is localized specifically in DNA, so that strand-breaks can form at position C4' of the deoxyribose.

There is a sequence of reactive oxygen species which lead to DNA damage, lipid peroxidation, enzyme inactivation and also carbohydrate damage. The processes resulting are cytotoxicity (6,7), mutagenicity (8,9), carcinogenicity and cell necrosis (For review, see 10).

The effects may be mediated by the production of reactive oxygen species, but as studied with menadione, other potential deleterious routes may include the electron bypass in the mitochondrial respiratory chain. The oxygen dependence of this route was studied by using an oxystat system. The menadione-mediated extra oxygen uptake in the intact hepatocyte predominantly (90 %) results from electron transfer in the mitochondrial respiratory chain (11).

Further pathways of toxicity include the secondary effects dependent on the generation of glutathione disulfide, GSSG. This follows from the generation of H_2O_2 from the superoxide anion radical via superoxide dismutase, and the subsequent reduction of H_2O_2 as catalyzed by glutathione peroxidase. The transport of a bile acid, taurocholate, across the canalicular membrane of the hepatocyte into the biliary canaliculi is inhibited during menadione metabolism and is attributable to the formation of GSSG (12).

This flavoprotein, also known as DT diaphorase, menadione reductase and vitamin K reductase reduces a variety of structures including quinones, tetrazolium compounds, nitroso compounds etc.. The enzyme occurs widely in most tissues and is present at high activity in tumors. Recently, it was shown that the m-RNA is increased up to 7-fold in hepatocyte nodules induced by carcinogens (13). The enzyme has been purified and characterized in two isofunctional forms from mouse liver (14). The protective effect of this enzyme against menadione-induced generation of reactive oxygen species has been studied by measuring low-level chemiluminescence (15, 16). The inhibition of the enzyme by dicoumarol increased the level of photoemissive species.

The cloning of the gene was performed (13,17). The cDNA clone of 1501 bp has an open reading frame (bases 75-899) and a nucelotide sequence consistent with a new gene family (17). On Northern blot analysis, a single 3-methylcholanthrene-inducible rat liver m-RNA (about 1.6 kb) hybridised to the cDNA probe. On Southern blot analysis a total of 14-16 kb of rat genomic DNA fragments hybridized to the cDNA probe, indicating one or a small number of genes in this family. The aminoacid sequence (274 aminoacid residues) and a molecular weight of 30,9 kDa compare to the size estimated from sodium dodecylsulfate-polyacrylamide gel electrophoresis. The aminoacid sequence has been determined.

It is assumed that an autoregulatory repressor model can explain the responses of the enzyme activity under the control of inducers like 2,3,7,8-tetrachlorodibenzo-p-dioxin (TCDD), which also uses specific forms of cytochrome P-450 and the UDP glucuronosyltransferase activity (18). An effect of DNA hypomethylation was postulated from an increase in activity and an increase in mRNA in intact animals after treatment with 5-azacytidine or feeding a methionine-choline deficient diet (19).

Following a first report that cytosolic quinone reductase of rat mammary gland and fat rose to more than twice the control values after feeding of a single oral dose of 7,12-dimethyl-benz(a)anthracene (20), a wide variety of compounds including polycyclic aromatic hydrocarbons such as 3-methyl-cholanthrene or aromatic amines such as 3- and 6-amino-chrysenes has been observed. It was particularly interesting that some other compounds such as dioxins (TCDD) were extremely active. TCDD was calculated to be about 17,000 times more potent than 3-methylcholantrene in elevating quinone reductase. Antioxidants such as BHA (2(3)-tert-butyl-4-hydroxyanisole) more selectively induced the enzyme and afforded protection against quinone toxicity. A group of compounds of chemotherapeutic and chemoprotective properties, 1,2-dithiol-3-thiones, induced the enzyme and elevated glutathione levels in hepatoma cells in culture and thereby enhanced the detoxification potential (21).

Redox cycling has been studied in the past (1) and more recently in toxicological experiments, majmly using menadione. There is a sequence of events, including GSH depletion as a result of extensive GSH oxidation to GSSG, and the formation of glutathione-protein mixed disulfides and glutathione thioethers (22,23). Menadione was found to lead to surface blebbing. Recently, the cytoskeletal changes in cellular oxidant injury was identified as an increase in F-actin (24), and menadione-induced bleb formation was found to be associated with the oxidation of thiol groups in actin (25). Studies on the effect of menadione on mitochondrial calcium sequestration and oxidation of nicotinamide adenine nucleotides and the inhibitory effect on the endoplasmic reticular calcium sequestration have been described (26). Protection against any single mechanism of the damage did not result in complete prevention of toxicity indicating that, as the oxidative damage proceeds, a recruitment of additional toxic mechanisms occurs which finally leads to irreversible cell damage and death. However, when the cytotoxicity of benzoquinones was compared, it was found that they did not mediate oxidative stress but arylated and still were highly toxic (27). Thus, the relationship of DT diaphorase to quinone toxicity holds for menadione but not for many other quinones. It was concluded that unsubstituted quinones that readily arylate GSH and protein are highly toxic, whereas the more highly substituted quinones that participate in redox cycling and oxygen activation are particularly toxic if the cells are compromised in their antioxidant defense by the inactivation by catalase and glutathione reductase.

Many enzymes get inactivated by oxidative stress (28,29), and one example of a functional inactivation is the impairment of the function of the endoplasmic reticulum by oxidation of sulfhydryl groups in the calcium ATPase (30). The elevation of cytosolic calcium and the stimulation of a non-lysosomal proteolytic system relate to this area (31). Naphthoquinone toxicity (32) and the possible role of 4,4-diethyl-stilbestrol quinone in carcinogenesis (33) provide further examples.

There has been an interest in the use of quinones as anti-cancer agents (34,35) especially in brain tumor therapy. In the studies on mechanisms of action of antitumor agents (36,37), redox cycling has received attention. In particular, this applies to hypoxia-mediated nitro-heterocyclic drugs in the radio- and chemotherapy of cancer (38,39).

Experimental exposure of cells and organs to oxidative stress with quinones of different kinds will shed further light on mechanisms of cell death and, in particular, on target organ toxicity.

Thus, a number of endogenous quinones in specialized cells, for example in dopaminergic neurons, will lead to a better understanding of toxicity of endogenous and exogenous compounds.

The advances in molecular biology of antioxidant enzymes like quinone reductases and, importantly, other enzymes like glutathione S-transferases will continue to form an area of active research. Here, the different layers of organisation of antioxidant defense comprise a multitude of open questions. One model system is the one in micro-organisms associated with the oxyR regulon, and the effects in spontaneous mutagenesis observed in deletions of this regulon (40). The inducibility of quinone reductase by carcinogens as well as by anti-oxidants, and the different types of regulatory response such as the Ah locus or the effects of DNA (5-cytosine)methylation currently form a bewildering array of potential sites of control.

A new area of research emerges in the analysis of DNA damage by redox cycling compounds. There are two types of mechanism, one being the hydroxyl radical induced single-strand breaks, exemplified by bleomycin, and the other being the calcium-dependent activation of an endonuclease (41).

ACKNOWLEDGEMENT

Supported by Deutsche Forschungsgemeinschaft, National Foundation for Cancer Research, Jung-Stiftung für Wissenschaft und Forschung.

REFERENCES

1. H. Kappus and H. Sies, Toxic drug effects associated with oxygen metabolism: redox cycling and lipid peroxidation, Experientia 37:1233 (1981).

2. C. Lind, P. Hochstein and L. Ernster, DT diaphorase as a quinone reductase: a cellular control device against semiquinone and superoxide radical formation, Arch. Biochem.Biophys. 216:178 (1982).

3. L. Ernster, R.W. Estabrook, P. Hochstein and S. Orrenius (eds.), DT Diaphorase. A Quinone reductase with special functions in cell metabolism and detoxication, Chemica Scripta 27A:1 (1987).

4. H. Sies, Oxidative Stress: Quinone Redox Cycling, ISI Atlas of Science: Biochemistry 1:109 (1988).

5. G. Powis, E.M. Hodnett, K.S. Santone, K.S. See and D.C. Melder, Role of metabolism and oxidation-reduction cycling in the cytotoxicity of antitumor quinoneimines and quinonediimines, Canc. Res. 47:2363 (1987).

6. S. Orrenius, Oxidative stress studied in intact mammalian cells, Phil. Trans. R. Soc. Lond. B311:673 (1985).

7. A.S. Atallah, J.R. Landolph, L. Ernster and P. Hochstein, DT-diaphorase in C3H/10T1/2 mouse embryo cells, Biochem.Pharmacol. 37:2451 (1988).

8. P.I. Chesis, D.E. Levin, M.T. Smith, L. Ernster and B.N. Ames, Mutagenicity of quinones: pathways of metabolic activation and detoxication. Proc. Natl. Acad. Sci. USA, 81:1696 (1984).

9. S. De Flora, C. Bennicelli, A. Camoirano, D. Serra and P.Hochstein, Influence of DT diaphorase on the mutagenicity of organic and inorganic compounds, Carcinogenesis 9:611 (1988).

10. H. Sies. Biochemistry of oxidative stress, Angew. Chem. Int. Ed. Engl. 25:1058 (1986).

11. H. De Groot, T. Noll and H. Sies, Oxygen dependence and subcellular partitioning of hepatic menadione-mediated oxygen uptake, Arch. Biochem. Biophys. 243:556 (1985).

12. T.P.M. Akerboom, T. Bultmann and H. Sies, Inhibition of biliary taurocholate excretion during menadione metabolism in perfused rat liver, Arch. Biochem. Biophys. 263:in press (1988).

13. J.B. Williams, A.Y.H. Lu, R.G. Cameron and C.B. Pickett, Rat-liver NADPH-quinone reductase - construction of a quinone reductase cDNA clone and regulation of quinone reductase messenger-RNA by 3-methyl-cholanthrene and in persistent hepatocyte nodules induced by chemical carcinogens, J. Biol. Chem. 261:5524 (1986).

14. H.J. Prochaska and P. Talalay, Purification and characterization of two isofunctional forms of NAD(P)H-Quinone reductase from mouse liver, J. Biol. Chem. 261:1372 (1986).

15. H. Wefers, T. Komai, P. Talalay and H. Sies, Protection against reactive oxygen species by NAD(P)H: quinone reductase induced by the dietary antioxidant butylated hydroxyanisole (BHA). Decreased hepatic low-level chemiluminescence during quinone redox cycling, FEBS Lett. 169:63 (1984).

16. H.J. Prochaska, P. Talalay and H. Sies, Direct protective effect of NAD(P)H: Quinone reductase against menadione-induced chemilumi-nescence of postmitochondrial fractions of mouse liver, J. Biol. Chem. 262:1931 (1987).

17. J.A. Robertson, H.C. Chen and D.W. Nebert, NADPH-menadione oxido-reductase - novel purification of enzyme, cDNA and complete amino-acid sequence and gene regulation, J. Biol. Chem. 261:15794 (1986).

18. J.A. Robertson and D.W. Nebert, Autoregulation plus positive and negative elements controlling transcription of genes in the (Ah) battery, Chem. Scripta 27A:83 (1987).

19. G. Wagner, U. Pott, M. Bruckschen and H. Sies, Effects of 5-azacyti-dine and methyl-group definiciency on NAD(P)H: quinone oxido-reductase and GSH S-transferase in liver, Biochem. J. 251:825 (1988).

20. H.G. Williams-Ashman and C. Huggins, Oxydation of reduced pyridine nucleotides in mammary gland and adipose tissue following treatment with polynuclear hydrocarbons, Med.Exper. 4:223 (1961).

21. M.J. De Long, P. Dolan, A.B. Santamaria and E. Bueding, 1,2-Dithiol-3-thione analogs: effects on NAD(P)H: quinone reductase and glutathione levels in murine hepatoma cells, Carcinogenesis 7:977 (1986).

22. D. Di Monte, D. Ross, G. Bellomo, L. Eklöw and S. Orrenius,
Alterations in intracellular thiol homeostasis during the metabolism
of menadione by isolated rat hepatocytes,
Arch. Biochem. Biophys. 235:334 (1984).

23. G. Bellomo, F. Mirabelli, D. Di Monte, P. Richelmi, H. Thor,
C. Orrenius and S. Orrenius, Formation and reduction of
glutathione-protein mixed disulfides during oxidative stress,
Biochem. Pharmacol. 36:1313 (1987).

24. D.B. Hinshaw, L.A. Sklar, B. Bohl, I.U. Schraufstatter, P.A. Hyslop,
M.W. Rossi, R.G. Spragg and C.G. Cochrane,
Cytoskeletal and morphologic impact of cellular oxidant injury,
Am. J. Path. 123:454 (1986).

25. F. Mirabelli, A. Salis, V. Marinoni, G. Finardi, G. Bellomo,
H. Thor and S. Orrenius, Menadione-induced bleb formation in
hepatocytes is associated with the oxidation of thiol groups in
actin, Arch. Biochem. Biophys. 264:261 (1988).

26. B. Frei, K.H. Winterhalter and C. Richter, Menadione-(2-Methyl-1,4-
naphtoquinone-) Dependent Enzymatic Redox Cycling and Calcium
Release by Mitochondria, Biochemistry 25:4438 (1985).

27. L. Rossi, G.A. Moore, S. Orrenius and P.J. O'Brien,
Quinone toxicity in hepatocytes without oxidative stress,
Arch. Biochem. Biophys. 251:25 (1986).

28. C.N. Oliver, B. Ahn, E.J. Moerman, S. Goldstein and E.R. Stadtman,
Age-related changes in oxidized proteins, J. Biol. Chem. 262:5488
(1987).

29. K.J.A. Davies, Protein damage and degradation by oxygen radicals,
J. Biol. Chem. 262:9895 (1987).

30. N.M. Scherer and D.W, Deamer, Oxidative stress impairs the function
of sarcoplasmic-reticulum by oxidation of sulfhydryl-groups in the
Ca^{2+}ATPase, Arch. Biochem. Biophys. 246:589 (1986).

31. P. Nicotera, P. Hartzell, C. Baldi, S.A. Svensson, G. Bellomo and
S. Orrenius S, Cystamine induces toxicity in hepatocytes through the
elevation of cytosolic Ca^{2+} and the stimulation of a nonlysosomal
proteolytic system, J.Biol. Chem. 261:14628 (1986).

32. M.G. Miller, A. Rodgers and G.M. Cohen, Mechanism of toxicity of
naphthoquinones to isolated hepatocytes, Biochem. Pharm. 35:1177
(1986).

33. J.G. Liehr, Possible role of 4,4-Diethylstilbestrol quinone in
diethylstilbestrol carcinogenesis. J. Tox. Env. H. 16:693 (1985).

34. M.T. Smith, Quinones as mutagens, carcinogens, and anticancer agents
- introduction and overview, J. Tox. Env. H. 16:665 (1985).

35. M.S. Berger, Use of quinones in brain-tumor therapy - preliminary
results of preclinical laboratory investigations,
J. Tox. Env. H. 16:713 (1985).

36. A. Begleiter, Studies on the mechanism of action of quinone antitumor agents, Biochem. Pharmacol. 34:2629 (1985).

37. S.R. Keyes, S. Rockwell and C. Sartorelli, Enhancement of mitomycin C cytotoxicity to hypoxic tumor cells by dicoumarol in vivo and in vitro, Cancer Res. 45:213 (1985).

38. G.E. Adams and I.J. Stratford, Hypoxia-mediated nitro-heterocyclic drugs in the radio- and chemotherapy of cancer, Biochem. Pharmacol. 35:71 (1986).

39. J.E. Biaglow, M.E. Varnes, L. Roizen-Towle, E.P. Clark, E.R. Epp, M.B. Astor and E.J. Hall, Biochemistry of reduction of nitro heterocycles, Biochem. Pharmacol. 35:77 (1986).

40. G. Storz, M.F. Christman, H. Sies and B.N. Ames, Spontaneous mutagenesis and oxidative damage to DNA in Salmonella typhimurium, Proc. Natl. Acad. Sci. USA 84:8917 (1987).

41. D.J. McConkey, P. Hartzell, P. Nicotera, A.H. Wyllie and S. Orrenius, Stimulation of endogenous endonuclease activity in hepatocytes exposed to oxidative stress, Toxicol. Lett. 42:123 (1988).

VITAMIN E - VITAMIN C SYNERGESTIC EFFECT

TOWARDS PEROXYL RADICALS: INFLUENCE

OF DEHYDRO-ASCORBIC ACID

D. Jore, M.N. Kaouadji
and C. Ferradini

Laboratoire de Chimie Physique
Université René Descartes
45, rue des Saints Pères
75270 Paris Cedex 06, France

SUMMARY

The Vitamin E - Vitamin C synergestic effect towards peroxyl radicals
has been characterized by γ radiolysis in ethanolic medium. Vit E - Vit C
aerated solutions have been irradiated in the presence or in the absence
of dehydro-ascorbic acid for ratios (Vit E)/(Vit C) between 1 and 27. The
experimental results point out that the regeneration of Vit E from its oxidised radical by Vitamin C ($k = 6 \times 10^5 mol^{-1}.dm^3.s^{-1}$) may be in competition
with its consumption through the reaction of dehydro-ascorbic acid with the
tocopheroxyl radical ($k = 2 \times 10^7 mol^{-1}.dm^3.s^{-1}$).

INTRODUCTION

It is now well established that Vitamin E represents the last line
of defence against lipid peroxidation by scavenging the $LO_2^{\cdot}$ peroxyl radicals[1,2]. The use of several model systems[3-6] allowed the characterization
of the monoelectronic exchanges involved in this antioxidant function.
Furthermore the capability of Vitamin C (AH^-) in acting synergestically
with Vitamin E (TH) (fig.1) as been often invoked[7,8]. Some studies examined the radical mechanisms responsible of this synergestic effects either
by ESR or pulse radiolysis[9,3]. We have ever studied the oxidation - reduction of Vit E and Vit C by γ radiolysis in ethanolic medium[5,10], we intent
now to characterize quantitatively by the same method the Vit E - Vit C
interactions, focusing on the influence of the oxidized form of Vit C the
dehydro-ascorbic acid (A) (fig.1).

Free Radicals, Lipoproteins, and Membrane Lipids
Edited by A. Crastes de Paulet *et al.*
Plenum Press, New York, 1990

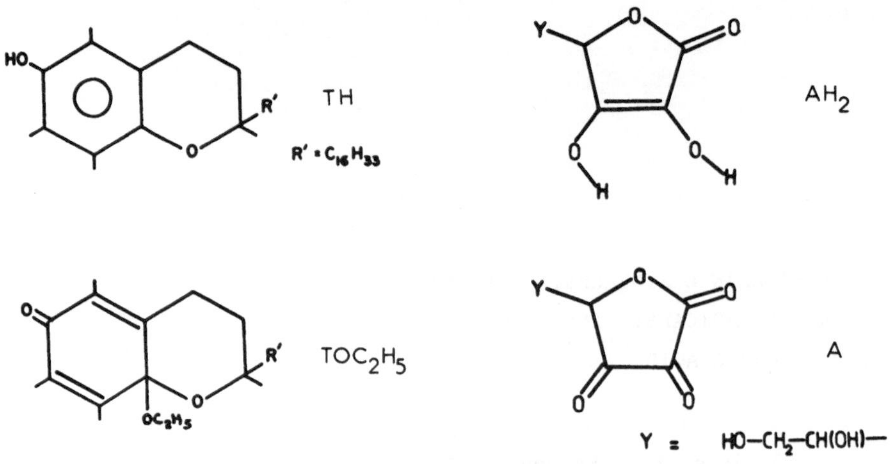

Fig. 1. Molecular structures : TH, AH$^-$, A, TOC$_2$H$_5$.

MATERIALS AND METHODS

α tocopherol (Vit E), ascorbic acid (Vit C) and dehydro-ascorbic acid are Merck "for analysis" reagents. The ethanol used is Absolute Ethanol Normapur from Prolabo.

The γ irradiations have been performed with a ^{60}Co irradiator. The dosimetry nas been determined by the Fricke's method : radio-oxidation of aerated ferrous sulfate solutions, 10^{-3} mol.dm^{-3} in the presence of H_2SO_4, 0.4 mol. dm^{-3} taking G(Fe^{3+}) = 15.6 molec/100eV, λ max (Fe^{3+}) = 304 nm, ε_{304}(Fe^{3+}) = 2204 mol^{-1}dm^3.cm^{-1} at 25°c. The dose rate was equal to 1 x 10^{18}eV.cm^{-3}.h^{-1}. The doses thus measured have been used for G values calculation without correction. After cleaning, the irradiation vessels have been heated at 400°C for 4 hours. The UV absorption spectra have been recorded with a Beckman 35 spectrophotometer : λ max (TH) = 292nm, ε_{292} = 3,1 x 10^3mol^{-1}dm^3.cm^{-1} ; λ max (AH$^-$) = 245nm, ε_{245} = 8.5 x 10^3mol^{-1}dm^3.cm^{-1} ; A does not exhibit a significant absorption in the 240-300nm region.

ETHANOL RADIOLYSIS

Radiolysis of ethanol[11] provides in the nanosecond time scale homogeneous solutions of solvated electrons (e$^-$solv) and α hydroxyethyl radicals (R$^\cdot$). After a few microseconds, a steady state is established with known yields :

$$G_{e^-solv} = 1.7 \text{ molec/100eV}$$
$$G_{R^\cdot} = 4.8 \text{ molec/100eV}$$

When γ irradiations are performed under aerobic conditions, the following reactions occur :

(1) e$^-$solv + O_2 $\longrightarrow$ $O_2^{\cdot -}$ k_1 = 1.9 x 10^{10}mol^{-1}.dm^3.s^{-1}

(2) R$^\cdot$ + O_2 $\longrightarrow$ RO$_2^\cdot$ k_2 = 4.6 x 10^9 mol^{-1}.dm^3.s^{-1}

This $RO_2^{\cdot}$ radical species is a good model of the peroxyl radicals involved in peroxidation chains.

RADIOLYSIS OF VIT E AERATED ETHANOLIC SOLUTIONS

Pulse and γ radiolysis experiments performed in aerated ethanol have allowed the characterization of the monoelectronic exchanges involved in Vit E oxidation by $RO_2^{\cdot}$ peroxyl radicals[5,6]. TH reacts with $RO_2^{\cdot}$ according to (3) and leads to the formation of the tocopheroxyl radical $T^{\cdot}$. This radical is oxidized by $RO_2^{\cdot}$ (4)

(3) $\quad TH + RO_2^{\cdot} \longrightarrow T^{\cdot} + RO_2H \qquad\qquad k_3 = 9.4 \times 10^4 mol^{-1}.dm^3.s^{-1}$

(4) $\quad T^{\cdot} + RO_2^{\cdot} \xrightarrow{H^+} T^+ + RO_2H \qquad\quad k_4 = 2.5 \times 10^6 mol^{-1}.dm^3.s^{-1}$

or disproportionates according to (5) :

(5) $\quad T^{\cdot} + T^{\cdot} \xrightarrow{H^+} T^+ + TH \qquad\qquad k = 1 \times 10^4 mol^{-1}.dm^3.s^{-1}$

(The acido basic equilibria are not taken in account in the formulation of the above reactions). T^+ reacts with ethanol (6) and provides

(6) $\quad T^+ + C_2H_5OH \longrightarrow TOC_2H_5 + H^+$

TOC_2H_5 as final oxidation product (λ max = 242nm, ε_{242} = 8.6 $\times 10^3 mol^{-1}.dm^3.cm^{-1}$, MW = 474). The above kinetic scheme explains the TH radio-oxidation with a constant yield $G(-TH) = G(TOC_2H_5) = 2.4$ molec/100 eV for (TH) concentrations above $2 \times 10^{-4} mol\ dm^{-3}$. This yield value, equal to $G(RO_2^{\cdot})/2$, corresponds to the scavenging of all $RO_2^{\cdot}$ radicals present.

RADIOLYSIS OF VIT C AERATED ETHANOLIC SOLUTIONS

The experimental data already obtained[10] when aerated ethanolic solutions of ascorbic acid have been irradiated, $G(-AH^-) = 3.2$ molec/100eV, may be explained by the fact that AH^- is oxidised by $RO_2^{\cdot}$ and $O_2^{\cdot-}$ according to reactions (7) and (8).

(7) $\quad AH^- + RO_2^{\cdot} \longrightarrow A^{\cdot-} + RO_2H$

(8) $\quad AH^- + O_2 \xrightarrow{H^+} A^{\cdot-} + H_2O_2$

The obtained ascorbyl radical may then disproportionate (9) and lead to the formation of dehydro-ascorbic acid (A)

(9) $\quad A^{\cdot-} + A^{\cdot-} \xrightarrow{H^+} AH^- + A$

RADIOLYSIS OF VIT E - VIT C MIXTURES

γ irradiations of Vit E - Vit C mixtures in aerated ethanolic solutions have been performed for 2 different values of the ratio $(TH)/(AH^-)$: 1 and 27, in the presence or in the absence of dehydro-ascorbic acid.

Ratio $(TH)/(AH^-)$ = 27 without A

Our previous experiments performed for various concentrations (TH) and (AH^-) with the same ratio $(TH)/(AH^-) = 27$[12] indicated that, after a lag phase, the disappearance of TH was a linear function of the absorbed dose $(G(-TH) = 2.4 \text{ molec/100eV})$. As can be seen on figure 2 the lag phase is proportional to (AH^-) concentration.

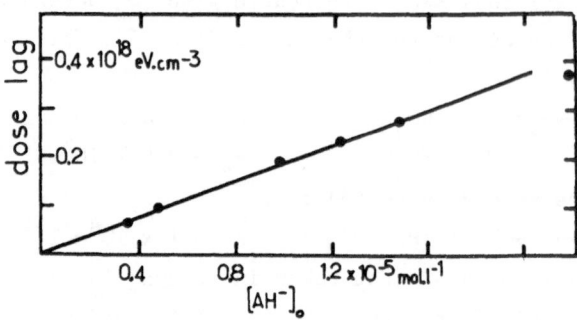

Fig. 2 . γ Irradiations of TH – AH^- ethanolic solutions with a ratio $(TH)/(AH^-) = 27$: lag phase plotted versus (AH^-)

Ratio $(TH)/(AH^-) = 1$ without A

As an example, fig. 3 shows the evolution of the absorption spectra obtained when a Vit E – Vit C mixture $((TH) = 2.4 \times 10^{-4} \text{mol.dm}^{-3}$, $(AH^-) = 2.4 \times 10^{-4} \text{mol.dm}^{-3})$ is irradiated with doses between 0.3 and 2.4×10^{18} eV cm^{-3}. In this case, no lag phase is obtained. The differential optical density $\Delta OD'_{292}$, the evolution of which represents TH oxidation, increases linearly with the dose (fig. 3 inset) . The slope of this curve indicates that the yield $G(-TH)$ is equal to 1 molec/100 eV. Such experiments were performed for other values $(TH) = (AH^-)$ between 2.4×10^{-4} and 6.5×10^{-4} mol.l^{-1}.

Fig. 4 shows in those cases the evolution of ΔOD_{292} plotted versus the dose. The yield $G(-TH)$ varies from 1 to 1.3 molec/100 eV which is considerably lower than the value obtained when TH is irradiated alone in the same conditions $(G(-TH) = 2.4 \text{ molec/100 eV})$. Furthermore it can be noticed that no lag phase is observed in any case.

Ratio $(TH)/(AH^-) = 27$ with A

In order to determine the possible influence of A present in the medium through reaction (9), γ irradiations of Vit E – Vit C mixtures $((TH = 2.7 \times 10^{-4} \text{mol.dm}^3$, $(AH^-) = 10^{-5} \text{mol.dm}^{-3})$ were performed in the presence of various amounts of A.

392

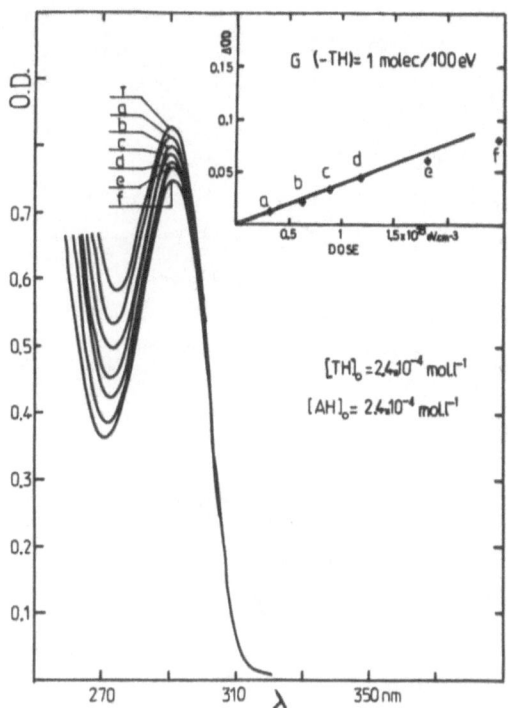

Fig. 3. Absorption spectra of a mixture (TH) = 2.4 x 10^{-4}mol.dm^{-3}
(AH$^-$) = 2.4 x 10^{-4}mol.l^{-1} irradiated in aerated ethanol;doses
(10^{18}eV.cm^{-3}):(T)initial solution, (a) 0.33, (b) 0.62, (c) 0.89,
(d) 1.20, (e) 1.81, (f) 2.48

Inset : differential optical density ΔOD_{292} plotted versus dose.

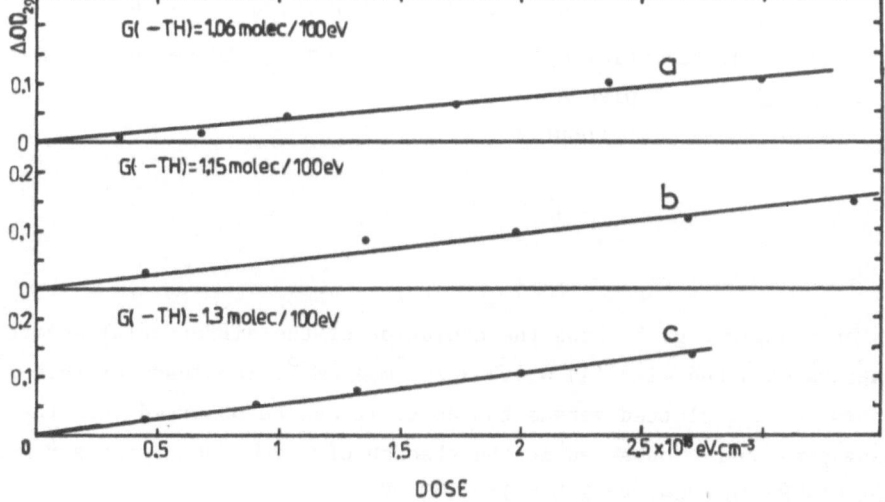

Fig. 4. irradiations of TH–AH$^-$ ethanolic solutions with a ratio (TH)/(AH$^-$)
= 1 : ΔOD_{292} plotted versus dose for : a : (TH)=3.4 x 10^{-4}mol.dm^{-3}
b : (TH) = 4.4 x 10^{-4}mol.dm^{-3}, c : (TH) = 6.5 x 10^{-4}mol.dm^{-3}

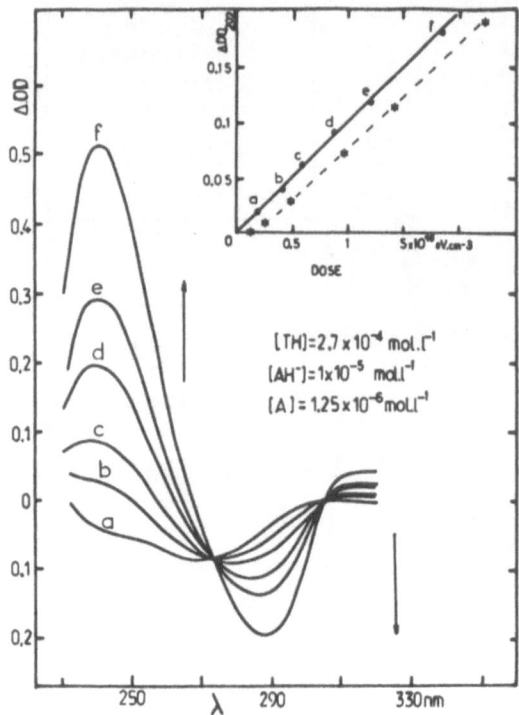

Fig. 5. Differential absorption spectra of a mixture (TH) = 2.7 x 10^{-1}mol. dm^{-3}, (AH^{-}) = 1 x 10^{-5}mol.dm^{-3} irradiated in aerated ethanol in presence of A. (A) = 1.2 x 10^{-6}mol.dm^{-3}

reference : initial solution

doses a : 0.2 x 10^{18} eV.cm^{-3} d : 0.9 x 10^{18} eV.cm^{-3}

     b : 0.4   "   e : 1.2   "

     c : 0.6   "   f : 1.8   "

inset : differential optical density ΔOD_{292} plotted versus dose

————.———.———.   with A

x-----x-----x   without A

As an example fig. 5 shows the evolution of the differential absorption spectra obtained with (A) = 1.2 x 10^{-6}mol.dm^{-3}. The inset of this figure shows ΔOD_{292} plotted versus the dose. It can be observed that the lag phase previously observed in the absence of A does not exist any more, and that G(-TH) is equal to 2.5 molec/100 eV.

Other analogous irradiations were done with (A) values between 6 x 10^{-8} and 10^{-5}mol.1^{-1}. The results indicate that the lag phase disappears for (A) $\geqslant$ 1.2 x 10^{-6}mol.1^{-1} (fig.6).

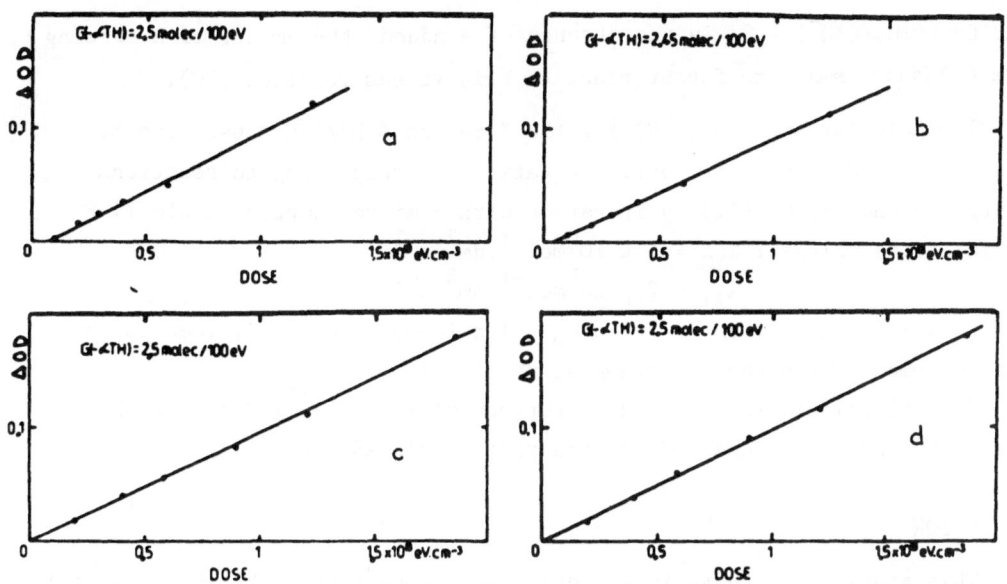

Fig. 6.   irradiations of TH – AH⁻ ethanolic solutions with a ratio
(TH)/(AH⁻) = 27 in presence of A :

(TH) = 2.7 x 10^{-4}mol.dm^{-3} (AH⁻) = 1 x 10^{-5}mol.dm^{-3}

a (A) = 6.5 x 10^{-8}mol.dm^{-3} c (A) = 5 x 10^{-6}mol.dm^{-3}

b (A) = 1.3 x 10^{-7}mol.dm^{-3} d (A) = 1 x 10^{-5}mol.dm^{-3}

DISCUSSION

When A is not added before irradiation, the disappearance of the lag
phase when decreasing the ratio (TH)/(AH⁻) from 27 to 1 indicates that no
competition exists between reactions (3) and (7) for (TH)/(AH⁻) = 27. The
lag phase can therefore only be explained by a regeneration reaction of TH
such as 10 :   (10) T˙ + AH⁻ ⟶ TH + A⁻

which consumes ascorbic acid. During the lag phase, the competition between
the reactions (10) and (5) is developed and leads finally to the observed
yield G(-TH) equal to 2.4 molec/100eV. Addition of A leading, for (TH)/(AH⁻)
= 27, to the decrease of this lag phase indicates that A is involved in a
reaction which competes with (10) (We have checked that A does not react with
RO_2˙; furthermore the low (A) concentration does not favor its reaction with
R˙ in the presence of O_2,(O_2)∿ 1.3 x 10^{-3}mol.dm^{-3}).Reaction (11) in fact
          (11) T˙ + A ⟶ T⁺ + A⁻

would explain the experimental results showing the TH consumption by A.

Such a competition between reactions (10) and (11) also explains the data
obtained for (TH)/(AH⁻) = 1. In this case A arising from reaction (9) oxidizes T˙, which results in a G(-TH) value which is lower than 2.4 molec/100eV
but can no more be equal to 0 as in the lag phases. It is important to notice

that for $(TH)/(AH^-) = 27$ in the absence of A added, the amount of A arising from (9) is too small to favour reaction (11) versus reaction (10).

The calculation of G (–TH) for the 2 ratios $(TH)/(AH^-)$ was done by solving the system of differential equations corresponding to reactions (1) to (5) and (7) to (11) by iteration with a micro computer Apple II–E. The following values : $k_{10} = 5 \times 10^6 mol^{-1}.dm^3.s^{-1}$
$$k_{11} = 2 \times 10^7 mol^{-1}.dm^3.s^{-1}$$
with $k_7 = 1.2 \times 10^4 mol.dm^3.s^{-1}$ and $k_9 = 1 \times 10^2 mol^{-1}.dm^3.s^{-1}$, lead to the best agreement with the experimental results :

G (–TH) calculated = G (–TH) experimental = 1 molec/100 eV for the ratio $(TH)/(AH^-) = 1$, and = 0 for the ratio $(TH)/(AH^-) = 27$

CONCLUSION

This study shows that, in a model homogeneous system, Vit E is in fact regenerated from its oxidized radical by Vit C. Furthermore dehydro–ascorbic acid is capable of oxidizing the tocopheroxyl radical leading to a consumption of Vit E. Therefore the presence of dehydroascorbic acid may widely decrease the synergestic effect of Vit E – Vit C towards peroxyl radicals.

REFERENCES

1. L.A.Wittig, Vitamin E and lipid antioxidants in free radical initiated reactions, in Free radicals in Biology, Pryor, W.A., ed, Academic Press, New York, 4 : 295 (1980).

2. G.W.Burton, A.Joyce and K.U.Ingold, Is Vitamin E the only lipid soluble chain breaking antioxidant in Human Blood plasma and erythrocyte membranes ?, Arch. Biochem. Biophys., 221 : 281 (1983).

3. J.E.Packer, T.F.Slater and R.L.Willson, Direct observation of a free radical interaction between vitamin E and vitamin C, Nature, 278 : 737 (1979).

4. M.G.Simic, Vitamin E radicals, in Oxygen and Oxy–Radicals in chemistry and biology, Rodgers, M.A.J. and Powers E.L.,eds, Academic Press, New York, 109 (1981).

5. D.Jore and C.Ferradini, Radiolytic study of α tocopherol in ethanolic solution, Febs. Lett., 183 : 299 (1985).

6. D.Jore, L.K.Patterson and C.Ferradini, Pulse radiolytic study of α tocopherol radical mechanisms in ethanolic solution, J. Free Rad. Biol. Med. 2 : 405 (1986).

7. A.L.Tappel, Will antioxidant nutrients slow aging processes ?, Geriatrics, 23 : 97 (1968).

8. P.B.McCay, Vitamin E : interactions with free radicals and ascorbate, Ann. Rev. Nutr., 5 : 323 (1985).

9. M.Scarpa, A.Rigo, M.Maiorino, F.Ursini and C.Gregolin, Formation of α tocopherol radical and recycling of α tocopherol by ascorbate during peroxidation of phosphatidylcholine liposomes : an electron paramagnetic resonance study, Biochim. Biophys. Acta, 801 : 215 (1984).

10. M.N.Kaouadji, D.Jore, L.K.Patterson and C.Ferradini, Radiolytic scanning of vitamin E -, vitamin C oxidation - reduction mechanisms, Bioelectrochem. Bioenerg., 18 : 59 (1987).

11. G.R.Freeman, Radiation Chemistry of Ethanol, Nat. Stand. Ref. Data Serv., Nat. Bur. Stand. (U.S.) 48 (NSRDS-NBS 48), (1974).

12. D.Jore, N.Kaouadji and C.Ferradini, The Vitamin E - Vitamin C synergestic effect studied by γ radiolysis in ethanol, presented at International Conference on medical, biochemical and chemical aspects of free radicals, IVth biennal general meeting of the Society for free radical research, Kyoto, April 9 to 13,(1988), (proceeding in press).

CONTRIBUTORS

ALBANO E.
Dept. Exper. Medicine & Oncology,
Univ. Turin, Corso Raffaello 30
10125 TURIN, Italy

ASTRUC M.E.
INSERM U 58
60 rue de Navacelles
34090 MONTPELLIER, France

BABIOR B.M.
Dept. Molecular & Experimental Medicine
Research Institute of Scripps Clinic
LA JOLLA, CA 92037, USA

BARRERA G.
Dept. Exper. Medicine & Oncology,
Univ. Turin, Corso Raffaello 30
10125 TURIN, Italy

BASCOUL J.
INSERM U 58
60 rue de Navacelles
34090 MONTPELLIER, France

BAST A.
Dept. Pharmacology, Faculty Chemistry
Vrije Universiteit de Boelelaan 1083
1081 HV AMSTERDAM,
The Nertherlands

BAZZONI F.
Instituto di Patologia Generale
Universita di Verona, Strada Le Grazie
37134 VERONA, Italy

BEDWELL S.
Cell Biol. Res. Group, Dept. Biol.
and Biochem., Brunel University
UXBRIDGE, Middlesex, UK

BELLAVITE P.
Instituto di Patologia Generale
Universita di Verona, Strada le Grazie
37134 VERONA, Italy

BERGGREN M.
Department of Toxicology
Karolinska Institutet
S-104 01 STOCKHOLM, Sweden

BERTRAND Y.
Clinique St-Jean
Rue des Marais 104,
1000 BRUXELLES, Belgium

BIASI F.
Dept. Exper. Medicine & Oncology
Univ. Turin, Corso Raffaello 30
10125 TURIN, Italy

BOES M.
Lab. Bioch. & Radiobiol. Univ. Liège
Institut de Chimie, B6, Sart Tilman
4000 LIEGE I, Belgium

BOLLI R.
Baylor College of Medicine,
HOUSTON,TX, USA

BORS W.
Institut für Strahlenbiologie
GSF Forschungszentrum
D-8042 NEUHERBERG, FRG

BOURDON-NEURAY J.
Lab. Bioch & Radiobiol. Univ. Liège
Institut de Chimie, B6, Sart Tilman
4000 LIEGE I, Belgium

BRAQUET P.
I.H.B. Research Labs.
17 avenue Descartes
92350 LE PLESSIS ROBINSON, France

CAMUS G.
Institut Supérieur d'Education Physique
B 21, Sart Tilman,
4000 LIEGE I, Belgium

CANDIDE C.
Laboratoire de Biochimie
Faculté de Médecine Saint-Antoine
27 rue Chaligny,
75012 PARIS, France

CHEESEMAN K.H.
Department of Biology and Biochemistry
Brunel University
UXBRIDGE, England UB8 3PH

CHIARPOTTO E.
Dept.Exper. Medicine & Oncology
Univ. Turin, Corso Raffaello 30
10125 TURIN, Italy

COCHRANE C.G.
Scripps Clinic and Research Foundation
Department of Immunology
10666 North Torrey Pines Road
LA JOLLA, CA 92037, USA

CORNWELL D.G.
Dept. of Physiological Chemistry and
Internal Medicine ,The Ohio State Univ.
COLOMBUS, Ohio, USA

COTTALASSO D.
Institute of General Pathology,
University of Genoa, via Alberti 2
16132 GENOA, Italy

CRASTES DE PAULET A.
INSERM U 58
60 rue de Navacelles
34090 MONTPELLIER, France

DAVIS W.B.
Dept. of Physiological Chemistry and
Internal Medicine, The Ohio State Univ.
COLOMBUS, Ohio, USA

DEAN R.T.
Heart Research Institute,
SYDNEY, Australia

DEBY C.
Lab. Bioch. & Radiobiol. Univ. Liège
Institut de Chimie, B6, Sart Tilman,
4000 LIEGE I, Belgium

DEBY-DUPONT G.
Lab. Bioch. & Radiobiol. Univ. Liège
Institut de Chimie, B6, Sart Tilman
4000 LIEGE I, Belgium

DEFAY R.
INSERM U 58
60 rue de Navacelles
34090 MONTPELLIER, France

DIANZANI M.U.
Dept. Exper. Medicine & Oncology,
Univ. TURIN, Corso Raffaello 30,
10125 TURIN, Italy

DOUSSET N.
INSERM U.101, Biochimie des Lipides
C.H.U. Purpan
37 allée J. Guesde
31059 TOULOUSE Cedex, France

DOUSTE-BLAZY L.
INSERM U.101, Biochimie des Lipides
C.H.U. Purpan
37 allée J. Guesde
31059 TOULOUSE Cedex, France

DUBERTRET L.
Laboratoire de Recherche Dermatologique
INSERM U 312, Hôpital Henri Mondor,
94010 CRETEIL, France

DUSI S.
Instituo di Patologia Generale
Universita di Verona, Strada Le Grazie
37134 VERONA, Italy

EATON J.W.
University of Minnesota Medical School,
and Dight Laboratories,
MINNEAPOLIS, MN 55455, USA

EMERIT I.
Institut Biomédical des Cordeliers
Université Paris VI
15 rue de l'Ecole de Médecine
75006 PARIS, France

ERBEN-RUSS M.
Sektion Physik , LMU Müchen,
Coulombwall,
8046 GARCHING, FRG

ESTERBAUER H.
Institute of Biochemistry,
University of Graz Schubertstrasse 1
A-8010 GRAZ, Austria

FERRADINI C.
Lab. Chimie Physique, UA 400
Univ. R. Descartes,
45 rue des St- Pères
75270 PARIS Cedex 06, France

GARDES-ALBERT M.
Lab. Chimie Physique, UA 400
Univ. R. Descartes,
45 rue des St-Pères
75270 PARIS Cedex 06, France

GRANGER D.N.
Dept. Physiol.& Biophysics
LSU Medical Center PO Box 33932
SHREVEPORT, Louisiana 71130-3932, USA

GRISHAM M.D.
Dept. Physiol.& Biophysics
LSU Medical Center PO Box 33932
SHREVEPORT, Louisiana 71130-3932, USA

GUILBAUD J.
Burn Centre, H.I.A. Percy
92141, CLAMART, France

GUTTERIDGE J.M.C.
Molecular Toxicology Research Group
OKLAHOMA Medical Research Foundation
825 N.E. 13th street, OKLAHOMA City,
OKLAHOMA 73104, USA

HAENEN G.R.M.M.
Dept. of Pharmacochemistry, Faculty of
Chemistry, Vrije Universiteit ,
De Boelelaan 1083,
1081 AMSTERDAM, The Netherlands

HOSFORD D.
I.H.B. Research Labs.
17 avenue Descartes
92350 LE PLESSIS ROBINSON, France

HOUEE-LEVIN C.
Lab. Chimie Physique, UA 400
Univ. R. Descartes,45 rue des St-Pères,
75270 PARIS Cedex 06, France

HYSLOP P.A.
Scripps Clinic and Research Foundation
Department of Immunology
10666 North Torrey Pines Road
LA JOLLA, CA 92037, USA

JACKSON J.H.
Scripps Clinic and Reseach Foundation
Department of Immunology
10666 North Torrey Pines Road
LA JOLLA, CA 92037, USA

JEROUDI M.O.
Baylor College of Medicine
HOUSTON, TX, USA

JESSUP W.
Cell Biol. Res. Group, Dept. Biol.
and Biochem.,Brunel University
UXBRIDGE, Middlesex , UK

JORE D.
Lab. Chimie Phys., Univ. Descartes
45 rue des St-Pères
75270 PARIS Cedex 06, France

KAOUDJI M.N.
Lab. Chimie Phys., Univ. Descartes
45 rue des St-Pères
75270 PARIS Cedex 06, France

KENSESE S.M.
Dept. Hum. Biol. Chem. & Gen.
University of Texas Med. Branch
GALVESTON, TX 77550, USA

KHAN S.
Institut Biomédical des Cordeliers
Université Paris VI
15 rue de L'Ecole de Médecine
75006 PARIS, France

KOLTZ P.
Burn Centre, H.I.A. Percy,
92141 CLAMART, France

KONDO M.
First Department of Medicine
Kyoto Prefectural University of Medicine
Kamigyo-ku,
KYOTO 602, Japan

LAI E.K.
Oklahoma Med. Res. Foundation
825 N. E. 13th Street
OKLAHOMA City, OK 73104,USA

LAMY M.
Service d'Anesthésiologie, CHU, B33
Sart Tilman,
4000 LIEGE I, Belgium

LEVADE T.
INSERM U.101, Biochimie des Lipides
C.H.U. Purpan
37 allée J. Guesde
31059 TOULOUSE Cedex, France

LEVY A.
Institut Biomédical des Cordeliers
Université Paris VI
15 rue de l'Ecole de Médecine
75006 PARIS, France

LISMONDE M.
Service D'Anesthésiologie, CHU, B 33,
Sart Tilman,
4000 LIEGE I, Belgium

LOPEZ M.
INSERM U.101, Biochimie des Lipides
C.H.U. Purpan
37 allée J. Guesde
31059 TOULOUSE Cedex, France

MARET A.
Lab. Biochimie Méd.
Fac. Méd. Toulouse Purpan
37 allées J. Guesde
31073 TOULOUSE Cedex, France

MARINARI U.M
Institute of General Pathology,
University of Genoa, via Alberti 2,
16132 GENOA, Italy

MAZIERE C.
Laboratoire de Biochimie,
Faculté de Médecine Saint-Antoine
27 rue Chaligny
75012 PARIS, France

MAZIERE J.C
Laboratoire de Biochimie
Faculté de Médecine Saint-Antoine
27 rue Chaligny
75012 PARIS, France

MAZZONE A.
Dept. of Internal Medicine & Therapeutics
Section of Medical Pathology,
University of Pavia, IRCCS S. Matteo Hospital,
27100 PAVIA, Italy

McCAY P.B.
OKLAHOMA Med. Res. Foundation
825 N.E. 13 th street
OKLAHOMA City, OK 73104, USA

McCORD J.M.
Dept. Bioch. College of Medicine
University of South Alabama
MOBILE, AL 36688, USA

MICHEL C.
Institut Für Strahlenbiologie
GSF Forschungszentrum
D-8042 NEUHERBERG, FRG

MIRON S
Instituto di Patologia 'Generale
Universita di Verona, Strada le Grazie
37134 VERONA, Italy

MOLDEUS P.
Department of Toxicology
Karolinska Institutet
S-104 01 STOCKHOLM, Sweden

MORLIERE P.
Laboratoire de Recherche Dermatologique
INSERM U 312. Hôpital Henri Mondor
94010 CRETEIL, France

NAITO Y.
First Department of Medicine
Kyoto Prefectural University of Medicine
Kamigyo-ku,
KYOTO 602, Japan

NANNI G.
Institute of General Pathology,
University of Genoa, via Alberti 2,
16132 GENOA, Italy

NEGRE A.
INSERM U.101, Biochimie des Lipides
C.H.U. Purpan
37 allée J. Guesde
31059 TOULOUSE Cedex, France

NORDMANN R.
Dept. of Biomedical Research on Alcoholism,
Univ. R. Descartes, 45 rue des St-Pères,
75270 PARIS Cedex 06, France

NOTARIO A.
Dept. of Internal Med. and Therapeutics.
University of Pavia,
IRCCS S. Matteo Hospital
27100 PAVIA, Italy

OLAFSDOTTIR K.
Department of Toxicology
Karolinska Institutet
S-104 01 STOCKHOLM, Sweden

ORRENIUS S.
Department of Toxicology,
Karolinska Institutet,
S-104 01 STOCKHOLM, Sweden

PANGANAMALA R.V.
Department of Physiological Chemistry
and Internal Medicine,
The Ohio State University,
COLUMBUS, Ohio, USA

PARADISI L.
Dept. Exper. Medicine & Oncology,
Univ. Turin, Corso Raffaello 30
10125 TURIN, Italy

PAROLA M.
Dept. Exper. Medicine & Oncology,
Univ. Turin, Corso Raffaello 30
10125 TURIN, Italy

PATEL B.S.
Baylor College of Medicine,
HOUSTON, TX, USA

PAUBERT-BRAQUET M.
Burn Centre, H.I.A. Percy,
92141 CLAMART, France

PETERSON D.A.
University of Minnesota Medical School,
and Dight Laboratories,
MINNEAPOLIS, MN 55455, USA

PIERAGGI M.T.
Anatomie Pathologique, CHU Rangueil,
31073 TOULOUSE, France

PINCEMAIL J.
Lab. Bioch. & Radiobiol., Univ. Liège
Institut de Chimie, B6, Sart Tilman
4000 LIEGE I, Belgium

POLI G.
Dept. Exper. Medicine & Oncology
Univ. Turin, Corso Raffaello 30
10125 TURIN, Italy

PRONZATO M.A.
Institute of General Pathology,
University of Genoa, via Alberti 2
16132 GENOA, Italy

REYFTMANN J.P.
Lab. Physico-Chimie de l'Adaptation Biol.
Muséum d'Histoire Naturelle de Paris,
INSERM U 312, 45 rue Cuvier
75005 PARIS, France

RIBIERE C.
Dept. of Biomedical Research on Alcoholism,
Univ. R. Descartes, 45 rue des St-Pères
75270 PARIS, France

RICE-EVANS C.
Dept.of Biochemistry and Chemistry
Royal Free Hospital School of Medicine
LONDON, NW3 2PF, England

RICEVUTI G.
Dept. of Internal Medicine and Therapeutics.
Section of Medical Pathology,
University of Pavia, IRCCS S. Matteo Hospital
27100 PAVIA, Italy

ROSSI A.
Dept. Exper. Medicine & Oncology
Univ. Turin, Corso Raffaello 30
10125 TURIN, Italy

ROTHENEDER M.
Institut of Biochemistry,
University of Graz, Schubertstrasse 1
A-8010 GRAZ, Austria

ROUACH H.
Dept. of Biomedical Research on Alcoholism,
Univ. R. Descartes, 45 rue des St-Pères
75270 PARIS, France

RYRFELDT Å.
Department of Toxicology,
Karolinska Institutet,
S-104 01 STOCKHOLM, Sweden

SADRZADEH S.M.H.
University of Minnesota Medicál School
and Dight Laboratories,
MINNEAPOLIS, MN 55455, USA

SALMON S.
Laboratoire de Biochimie,
Faculté de Médecine Saint-Antoine
27 rue Chaligny
75012 PARIS, France

SALVAYRE R.
INSERM U.101, Biochimie des Lipides
C.H.U. Purpan
37 allée J. Guesde
31059 TOULOUSE Cedex, France

SANTUS R.
Lab. Physico-Chimie de l'Adaptation Biol.
Muséum d'Histoire Naturelle de Paris,
INSERM U 312, 45 rue Cuvier
75005 PARIS, France

SARAN M.
Institut für Strahlenbiologie
GSF Forschungszentrum
D-8042 NEUHERBERG, FRG

SCHRAUFSTATTER I.U.
Scripps Clinic and Research Foundation
Department of Immunology
10666 North Torrey Pines Road
LA JOLLA, CA 92037, USA

SEKAKI A.
Lab. Chimie Physique, UA 400
Univ. Paris V, 45 rue des St-Pères
75270 PARIS Cedex 06, France

SERRA M.C.
Instituto di Patologia Generale
Universita di Verona,
37134 VERONA, Italy

SIES H.
Institut für Physiologische Chemie I
Universität Düsseldorf, Moorenstrasse 5
D-4000 DUSSELDORF, FRG

SLATER T.F
Dept. Biology & Biochemistry,
Brunel University,
UXBRIDGE, Middlesex, UK

SMITH L.L.
Dept. Human Biol. Chem. and Genetics,
University of Texas Medical Branch
GALVESTON, TX 77550, USA

STEINBRECHER U.P.
Department of Medicine
University of British Columbia
2211 Wesbrook Mall
VANCOUVER, British Columbia
V6T 1W5 Canada

STRIEGL G.
Institute of Biochemistry,
University of Graz, Schubertstrasse I
A-8010 GRAZ, Austria

SUGINO S.
First Department of Medicine
Kyoto Prefectural University of Medicine
Kamigyo-ku,
KYOTO 602, Japan

TANIGAWA T.
First Department of Medicine
Kyoto Prefectural University of Medicine
Kamigyo-ku,
KYOTO 602, Japan

TURRENS J.F.
Dept. Bioch. College of Medicine
University of South Alabama
MOBILE, AL 36688, USA

WHISLER R.L.
Dept. Physiological Chemistry and
Internal Medicine, The Ohio State University
COLUMBUS, Ohio, USA

YOSHIKAWA T
First Department of Medicine
Kyoto Prefectural University of Medicine
Kamigyo-ku,
KYOTO 602, Japan

ZHANG H.
Dept. Physiological Chemistry and
Internal Medicine, The Ohio State University
COLUMBUS, Ohio, USA

ZIMMERMAN B.J.
Dept. Physiol. & Biophysics
LSU Medical Center PO Box 33932
SHREVEPORT, Louisiana 71130-3932, USA

AUTHOR INDEX

SUBJECT INDEX

Endonucleases 257, 384
Eosinophils 17, 23, 45, 128
Erythrocytes 73, 106, 269
Ethanol 309, 390
Exocytosis 55
Extracellular fluids 85, 371

Fatty liver 189,309, 321
Fenton reaction 8, 105, 128, 204, 353, 362,
    374
Ferryl species 269
Flavonoids 10, 244
FMLP 46, 55
Foam cells 193, 331
Free radicals
    ethanol induced disorders 309
    metabolism activation 133
    production 23, 45, 65, 84, 89, 359
    protection 359
    scavenger 20, 85, 322

Galactosyl transferase 184
Glutathione , 258, 303, 312, 321, 344, 359, 382
Glutathione peroxidase 9, 66, 200, 373
Glutathione S-transferase 365
Golgi apparatus 183
Granulocytes 55,81, 178, 301

Haber Weiss reaction 17, 84, 105, 249, 270,
    310, 328
Heart 68, 89
Hemoglobin 73, 269
Hepatocytes 172, 183, 257, 292, 309,320, 382
Hydrogen donors 3, 66, 362
Hydrogen peroxide 23, 55, 65, 83, 259, 276,
    282, 301, 328, 371
Hydroperoxides 6, 76, 84, 119, 241, 259
Hydroperoxyeicosatetraenoic acid (HPETE)
    18, 221
Hydroxyeicosatetraenoic acid (HETE) 18
Hydroxyl radical 74, 84, 128, 240, 270, 301,
    310, 328, 362, 371, 390
Hydroxynonenal (4-) , 103, 148, 171, 196, 312,
    329, 367
Hyperoxia 68

Inflammation 343
Injury 68
Integrins 57
Intestine 81, 99
Iron , 5, 19, 73, 84, 249, 269, 310, 364,372

Irradiation UV 249
Ischemia 68
Ischemia reperfusion 81, 89, 100, 105

Lactoferrin 375
Leukotrienes 86
Linoleic acid 282
Linolenic acid 282
Linoleic acid hydroperoxide 282
Linolenic acid hydroperoxide 282
Lipid hydroperoxide 359
Lipid peroxidation 1, 17, 75, 125, 171, 279,
    317,327,359
Lipoperoxidation in vivo 163
Lipoprotein
    UV treated LDL 249
    low density (LDL) 203, 327
    oxidation 193, 203, 215
    peroxidized 249
    secretion 183
    structure 239
Lipoxygenases 9, 125, 317
Liver
    Golgi apparatus 183
    injury 317
    membrane lipids 309
    microsomal system 125, 261, 359
    plasma membranes 144, 171
    regeneration 317
    tumors 317
Luciferin analog (Cyprinida) 153
Lung 301
Lymphoid cells 250

Malonldialdehyde , 18, 146, 163, 193, 320,
    329, 332
    adducts 200
Mediators (endogenous) 301
Membrane
    glycoproteins 55
    lipids 309
    oxidation 269
    phospholipids 105, 309
Menadione 259, 382
Metal chelation. 239 see Desferoxamine
Microfilament 260 ,346
Mitochondrial transport system 89
Mitochondrion 45, 347
Monocytes 23, 45, 56, 101, 331, 352
Monoxygenase 360